CISM COURSES AND LECTURES

Series Editors:

The Rectors of CISM
Sandor Kaliszky - Budapest
Mahir Sayir - Zurich
Wilhelm Schneider - Wien

The Secretary General of CISM
Giovanni Bianchi - Milan

Executive Editor
Carlo Tasso - Udine

The series presents lecture notes, monographs, edited works and proceedings in the field of Mechanics, Engineering, Computer Science and Applied Mathematics.
Purpose of the series in to make known in the international scientific and technical community results obtained in some of the activities organized by CISM, the International Centre for Mechanical Sciences.

INTERNATIONAL CENTRE FOR MECHANICAL SCIENCES

COURSES AND LECTURES - No. 361

THEORY AND PRACTICE OF ROBOTS AND MANIPULATORS

Proceedings of RoManSy 10: The Tenth CISM-IFToMM Symposium

EDITED BY

A. MORECKI
WARSAW UNIVERSITY OF TECHNOLOGY

AND

G. BIANCHI
POLYTECHNICAL UNIVERSITY OF MILAN

AND

K. JAWOREK
WARSAW UNIVERSITY OF TECHNOLOGY

SPRINGER-VERLAG WIEN GMBH

Le spese di stampa di questo volume sono in parte coperte da
contributi del Consiglio Nazionale delle Ricerche.

This volume contains 237 illustrations

Originally published by Springer-Verlag Wien New York in 1995

In order to make this volume available as economically and as rapidly as possible the authors' typescripts have been reproduced in their original forms. This method unfortunately has its typographical limitations but it is hoped that they in no way distract the reader.

ISBN 978-3-211-82697-3 **ISBN 978-3-7091-2698-1 (eBook)**
DOI 10.1007/978-3-7091-2698-1

PREFACE

The CISM-IFToMM RoManSy Symposia have played a dynamic role in the development of the theory and practice of robotics. The proceedings of the ten symposia to date present a world view of the state of the art, including a unique record of the results achieved in central and eastern Europe.

RoManSy 10, held September 12-15, 1994 in Gdansk, Poland was attended by 64 participants from seventeen countries.

The proceedings of this tenth edition of RoManSy focus mainly on problems of mechanical engineering and control.

In his opening lecture, A. Morecki reviews the history of the symposia and examines the main trends, past, present and future, of research in robotics.

The fifty-one papers illustrate significant contributions in mechanics (ten papers), control of motion (fourteen), synthesis and design (fourteen), applications and performance evaluation (four), sensing and machine intelligence (three), and the biomechanical aspects of robots and manipulators (six). They appear here in the order and form in which they were presented in the various working sessions.

RoManSy 11 will be held in Udine, Italy in September 1996.

The Editors express their thanks to A. Ferenc of the Warsaw University of Technology for his help in the preparation of this manuscript.

A. Morecki
G. Bianchi
K. Jaworek

CONTENTS

In Memory of Ichiro Kato

Ichiro Kato

was born in Chiba Prefecture (Tateyama City)
on May 2nd 1925

ICHIRO KATO

Ichiro Kato began his studies of robotics over thirty years ago.
His work on artificial arms and hands dates from 1967, and incorporates the technological assets gained from developing active prostheses, a program initiated three years earlier. At first his aim was to develop machines that could replace persons in the performance of manual labor, and concentrated on the development of artificial hand mechanisms. More recently, this objective had been extended to develop robots which can perform intelligent work as well as manual labor.

The emphasis in the laboratory has always been on developing robotics with a target of twenty to thirty years and themes themes on the basis of the belief that the twenty-first century will be an era of personal robots and cyborgs. Consequently Kato became very interested not only in the function of robots, but in their physical form as well. His endeavors have not been limited to the manufacturing sector, but have included tertiary industries and the medical field as well.

WAM-1 was first developed in 1967; it used WASEDA-type artificial muscles of rubber and featured seven degrees of freedom (DOF), four in the hand and three in the arm. In 1969 the computer-controlled WAM-2, with five electrically activated DOF in the arm was assembled. A further refinement, WAM-3, was developed in 1970. Both WAM-2 and WAM-3 had position sensors and pressure sensors on their fingers, so that they could automatically grasp and transport objects.

If, as forecasts say, robots will penetrate society in the twenty-first century, substituting human labor in industrial and service sectors, they should be able to deal with information as well as a person does. The anthropomorphic intelligent robot WABOT (WAseda roBOT) was undertaken to create a "personal robot" which resembled a person as much as possible.

Four laboratories in the Waseda University School of Science and Engineering joined to set up "The Bioengineering Group", which began the WABOT project in 1970. Kato's laboratory was in charge of the system of artificial limbs and their control.

Many different models of biped walking robots were built and tested (WL-1, WL-3, WAP-1, WAP-2, WAP-3, WL-5, WL-9DR, WL-10DR, WL-10RD). The development of a software system for the anlysis and design of universal multi-link systems was undertaken in 1979. CAD systems for artifical limbs (1980-1981), CAD systems for robotics (1982-1983), a CAD composition system for walking pattern - Walk Mater-2 (1984): these are a few examples of his investigations of biped locomotion.

Kato also made numerous contributions to biomedical engineering. There are the many models and prototypes of his EMG-controlled upper limb prosthesis, the Multifunctional A/K Prosthesis, the FES Automatic Palpation System for the diagnosis of breast cancer...

At the International Science and Technology Exposition in 1985 on the theme of Dwelling Surroundings: Science and Technology for Man at Home, three robots were demonstrated at the "Man and Scientific Technology " area in the pavillion of the Japanese government. These were a quadruped walking robot, a biped walking robot, and a robot that played a keyboard instrument, in an implicit evocation of human evolution.

Both the biped walking robot WHL-11 and the robot musician WASUBOT were developed from studies conducted in Kato's laboratory. They quickly became a center of attention as sumbols of the most advanced technology in robotics.

The WASUBOT performed with the Symphony Orchestra of the Japan Broadcasting Corporation NHK, playing the Bach Aria on the G-string.

Ichiro Kato's principal studies after 1985 focused on the creation of piano-playing mobile manipulators and the Multifunctional Active A/K Prosthesis.

His contributions to the science of robotics, in the areas of medical manipulators, FES, and bioelectrical control are acknowledged throughout the world.

He published many books and papers in these fields, and was the supervisor of many M.Sc. and Ph.d. dissertations.

He was the founder of the Japanese Society for Biomechanisms, a Member and Chairman of many ISIR and ICAR Symposia.

He was also an active member of the Japanese IFToMM Committee, and one of the founders of the RoManSy Symposia.

Today, together with our sorrow and sense of loss, remains the memory of his outstanding personality, of his excellence in engineering, science, and design, and may I add, for all who had the honor and pleasure to work with him, of a warm and valued friendship.

A. Morecki

Opening lecture

Past, Present and Future of RoManSy Symposia
Where are we going ?

A. Morecki

Warsaw University of Technology
Warsaw, Poland

Abstract

This paper presents the history of the RoManSy Symposia, describing the main scientific results obtained over the last twenty years and published in the RoManSy Proceedings. Some proposals for topics of research for the coming years are given in conclusion.

Introduction

During the Second World Congress of IFToMM, held in September 1971 in Kupari (Yugoslavia), the organizers, Artobolevski, Kobrinski, Konstantinov, Roth, Bianchi and Sobrero, decided to hold a symposium on the theory and practice of robots and manipulators. The symposium would be called RoManSy for Robot and Manipulator Systems.

A. Morecki was asked to prepare the first meeting of the program, and the Organizing Committee met in Nieborow, Poland in May 1972.
Kobrinski (Chairman), Artobolevski, Kostantinov, Roth, Thring, Morecki, Vukobratovic, Kato, and Bianchi, members of IFToMM, took part, together with Sobrero and Bertozzi of CISM. At the next meeting, in Split, Yugoslavia, in April 1973, Bazjanac and Jelovac of Yugoslavia and Romiti from Italy also attended. All the papers submitted were carefully examined; 45 were selected for presentation and later published by Springer Verlag.

Salient Periods in RoManSy Activty

1. The First Period: 1973-1983

The first CISM-IFToMM Symposium was held in Udine in September 1973. The main topics were: walking machines, kinematics and dynamics, the biomechanics of motion, man-machine systems, artificial intelligence, applications, and control of motion.

Three subjects were discussed in the field of kinematics and dynamics: the design of computer-controlled manipulators proposed at Stanford University, algorithms for solving some dynamical problems in manipulators, and algorithms for the kinematical investigation of manipulators.

In the field of walking machines some important topics were presented: the analysis of the gait of six-legged vehicles performed at the Institute for the Study of Machines in Moscow, the construction of the Biped Walking Machine WL-5 developed by the group of Bio-engineering at Waseda University (one of the first presentations concerning walking machines).

During its first meeting the Program and Organizing Committee had decided to include biomechanical aspects of manipulation and locomotion in the list of topics of RoManSy Symposia.

Interesting results obtained at Rome and Warsaw Universities were presented during the first Symposium: the biomechanics of climbing stairs and of constructing artificial walking systems, as well as an analysis of the distribution of drives in "live" manipulators and pedipulators, and in medical manipulators. The Tokyo Institute of Technology presented the results of the first biomechanical study of serpentine locomotion.

Interesting results obtained in the field of man-machine systems were: the design of master-slave manipulators, man-machine interaction in intelligent robotic systems, and the application of sensory information and multifunction learning.

Artificial intelligence was another important subject included in the program. Among the related problems discussed during the debates: the structure of the highest level of control in robot manipulators, adaptive robots, visual feedback, finite state machines, and optical sensors.

A number of papers were dedicated to applications.

Some topics were associated with space systems, such as the remote control of manipulators in space, and remotely controlled systems for operation and exploration in space. Other impressive achievements dealt with remote surgery, remote mining, manipulator systems developed by General Electric (Quadruped, Handyman, Hardiman, and others), the synthesis and scaling of advanced manipulator systems.

The final topic was the the control of motion. Some of the results were: a master-slave remote-control manipulator, synergic rate control, rate control of the Rancho arm, control of anthropomorphic, active mechanisms (Mihajlo-Pupin Institute, Belgrade), vision as a factor in the control of motion, the integration of control software.

The success of the Symposium was extremely encouraging. The participation exceeded all expectations, gathering scientists from all over the world. It was decided to organize the next Symposium in Poland in 1976. The chairmanship was handed over to Roth (USA); Kedzior (Poland) was appointed Scientific Secretary, and Bertozzi (Italy) Secretary of the Program and Organzing Committee.

The next three Symposia were held in 1976, 1978 and 1981. The main topics of the first Symposium - mechanics, biomechanical aspects of manipulation and locomotion, control of motion, synthesis and design, sensors and artificial intelligence - were also the main topics of these Symposia.

New participants, from the USA, the former USSR, Japan, France, the UK, Italy and Germany in particular, presented interesting new subjects.

Mathematical models, the ranges of motion of manipulators, identification problems, stiffness constraints, the choice of geometric parameters for walking machines are some of the solutions of mechanical problems presented in the years from 1976 through 1981. Myoelectric control, microcomputer-aided prothesis control, a head manipulator for bilaterally disarticulated children, medical rehabilitation, computer control of a multitask exoskeleton for paraplegic, manipulators for tetraplegics are major examples of medical manipulators presented during the three Symposia.

Particularly interesting results were obtained in the area of computer-controlled support and substitution of lost functions of human upper and lower limbs.

New solutions were presented concerning artificial skeletal activity, walker's motion control, and a four-legged machine. New results were obtained in the design, or computer-aided design of manipulators and robots: the Olivetti-Sigma industrial robot, to give just one example.

In the field of control we observed new trends concerning force-reflecting manipulator systems, sensing robots, problem solving in cognitive robotic systems, electronic logic control, and supervisory control. Important problems of social impact, education, training, terminology, and symbols were also discussed.

Academician Artobolevski, one of the founders of the Symposia, died in September 1977. It was a great loss for all of us.

RoManSy '78 presented new studies addressing the grasping process and grippers with elastic fingers, optimization of manipulator performance, the concept of anthropomorphism, dimensional synthesis, the application of bond graphs, decoupled feedback control, dynamic control, real-time computer control, imaging sensors, a microprocessor-based telemanipulator system, a remote inspection vehicle for pipe, underwater manipulators, computer graphics design of industrial robot applications.

L. Sobrero, founder and host of the first RoManSy, died in 1979. This loss was followed in 1980 by that of W. Olszak, Rector of CISM and invaluable supporter of the Symposia.

New problems were considered during RoManSy '81: the theory of parts-mating for assembly automation. the dynamics of computer controlled robotic manipulators, the modeling and control of flexible manipulators, high accuracy position control, optimal control. servo-micro-manipulator design (Tiny-Micro-Mark 1), a robot for ultrasonic inspection. proximity sensors in computer enhanced telemanipulation, a remote manipulator for severely handicapped operators, computer simulation of the dynamics of robotic mechanisms. And a one-legged hopping machine was shown for the first time.

2. Second period of activity: 1984-1993

After four very successful symposia it was decided to institutionalize this series of scientific meetings.

The next three Symposia, RoManSy 4, RoManSy 5 and RoManSy 6, were held in 1984, 1986 and 1988.

Most of the papers presented at RoManSy 4 addressed particular problems of mechanics, control of motion, synthesis and design, locomotion, and man-intelligent machines systems.

Some new subjects also appeared on the scene: the optimal selection and placement of manipulators, trajectory planning of redundant manipulators, computer-aided generation of multibody system equations, application of the NEWEUL formalism, dynamics of robots with closed loops. In the area of control of motion, aspects of optimal dynamic trajectories, robot control and computer languages, and robust control were treated. The mechanical and geometric design of the ASV, and an actuator based on SME were new and interesting subjects in the field of synthesis and design.
Original ideas concerning a hierarchical system for the computer control of a hexapod walking machine, the plane walking of a biped robot, trolling and bounding in a planar two-legged model were new developments in the field of locomotion.
Manual control, communication in space teleoperation, sensory-based control for robots and teleoperators, tele-existence, representing three dimensional shapes, sensor-aided and/or computer-aided bilateral teleoperator systems were issues in the field of man-intelligent machine systems. Studies on the design parameters of I.R. and automatic assembly illustrated the progress made in applications and performance evaluation.

A panel discussion on "The Role of RoManSy Symposia" was held on 29 June 1984 during the final plenary session in Udine at CISM. Reflecting on the experience of the previous ten years, the question now was what to do in the future: in what direction should further Symposia be organized ?

Konstantinov, Morecki, Roth, Vukobratovic and Vertut of the Organzing Committee and other participants were asked to give their opinions on the following important matters:
* should we organize symposia in the future ?
* if we do, what form should these symposia have: small (60-70 participants, with 35-40 invited papers), or large (100-150 participants, with 60-80 papers) ?
* what kind of topics should be included: more theoretically oriented, more application-oriented, or both (and in what proportion) ?
* how frequently should RoManSy Symposia be held: every other year, every third year?
* what is working well, and should be maintained ?
* what is not working well and should be changed ?

Most of the participants agreed that the Symposia should continue to be held every other year, but with restricted attendance and invited papers on an advanced theoretical level in the areas of mechanics , control of motion, synthesis and design, manipulation and locomotion, applications and performance evaluation of manipulators and robots. Only papers which represented good and original theoretical contributions to these topics would be selected for presentation. Survey papers, panel discussion, and discussion following presenations would also play an important role in future symposia.

J. Vertut died before RoManSy 6. A special commemorative session devoted to advanced teleoperations was held during the Symposium. Followers of Jean Vertut from the French school of teleoperation, which had begun its work in 1980 with the ARA program, presented new and important results.

Mechanics was once again a major subject. Papers dealt with redundant manipulators, position and orientation accuracy problems, the modeling of flexible robot arms, gripping. Special attention was given to synthesis and design problems: smart hand systems, micro-mechanical grippers, the analytical design of two-revolute open chain, robotic aids for the severely disabled, collision avoidance.

General languages for mobile robots, mobile robots for use in unstructured terrain, and a wall-climbing vehicle were the main topics in the area of locomotion.

In the field of the control of motion the problems of expert systems for strategic level control, time-optimal robotic manipulator task planning, robust control, self-learning algorithms were discussed.

Sensing and machine intelligence were at the focus of attention too, and especially: force feedback in telemanipulators, optical fiber sensors, bilateral remote control, active force control, and a three-dimensional optical syntaxer.

Some new results obtained in the field of applications were: the automation of mine support erection technology and the minimization of vibrations.

RoManSy 7 was held in 1988.

In the field of mechanics, the studies presented concerned special open kinematic chains, including sigularities; inverse kinematic problems; computational algorithms for the identification of inertia; dynamically decoupled motion; and flexible multibody systems.

Multi-mode manual control in telerobotics, the active control of robot arms, active end-effector control, mobile robot computer-aided teleoperation, the compensation of Coulomb friction in position control, trajectory control for elastic manipulators were the main subjects.

We noted further development in biped locomotion, elastic manipulators combined with a quadruped machine, and articulated body mobile robots.

The design of a piano playing manipulator, parallel manipulators, a compliant direct drive, and knowledge-based sequence planning systems dominated in the field of synthesis and design. Other papers concerned sensor-guided robot control using fuzzy logic, imaging sensors, and dynamic tactile sensing.

Robot controllers, software tools, shape memory alloy for clean gripper actuators, assembly with single and double arm robots, an instruction system for robotics, the planning of squeeze-grasp are examples of the new applications presented.

RoManSy 8 was held in Cracow in 1990.

An opening lecture in the field of mechanics reviewed studies in multibody systems and robot dynamics.

Many new and valuable results were reported at this Symposium. We mention only a few of them: a general solution of the inverse kinematics of all series chains, a general model of special stiffness, the stability analysis of a force control robot, pose and twist estimation, the kinematic behavior of mapped complex joint angles, the mechanics of a multi-limk elastic robot system, the kinematic solution of a special parallel robot.

In the area of the control of motion, new results were obtained in force feedback control, the control of space manipulators, bilateral control in master-slave manipulators, direct compliance control, tele-robotics, a neuromorphic suboptimal controller.

Examples selected from the area of mobile robotics are: the dynamic behavior of a simulated 3D biped; a combined wheeled and legged vehicle, the quadru-truck crawler vehicle Helios II; walking without impact; the walking of a stick insect; a walking machine that jumps over obstacles.

Grasping with mechanical intelligence programming languages, and sensor-based control were described in the field of sensing and machine intelligence. Selecting examples in the field of synthesis and design selected examples we have: a master-slave gripper controller and a cable-circuit drive system.

Other interesting applications were the decoration of scale models and a 3D scanner dedicated to robotic safety.

RoManSy 9, held in Udine in 1992, closed the second period in the life of the RoManSy Symposia.

M. Kostantinov, one of the founders of IFToMM, died in 1991. This loss was followed by the death of A. Kobrinskji, first chairman of the Organizing Committee of the CISM IFToMM RoManSy Symposia, in 1992.

RoManSy 9 opened with a special lecture on Micromechanisms and Microbiotics. The rather new and very exciting subject is expected to play an important role in the coming decade.

New and interesting results arrived in the field of mechanics: inverse kinematic solutions, symbolic kinematic redundancy, stiffness matrices, geometrical decomposition of robot elasticity.

Hybrid force-position control, the parallel computing of symbolic robot models, decoupling and optimal control were topics in the field of the control of motion.

Topics in sensing and machine intelligence addressed the problems of multi-functional tactile sensing, new languages, and a flexible controller equipped with sensors.

The design and testing of new manipulators, grippers, and motors were described: a parallel actuated manipulator, dextrous manipulators, a spherical ultrasonic motor, power grasping, a carbon fiber robot, and six DOF telerobotic joysticks.

In the area of the biomechanical aspects of robotics, particularly interesting results were biped walking without impact, the modeling of upper limb muscles, a coupled tendon drive manipulator, and the concept of super rabbit.

Robotic training stations, interactive planning, the robotic milking of cows were some of the new applications.

A round table session was dedicated to the rule of dynamic control in robotics, present and future.

Tables One, Two and Three present the principal data illustrating all these Symposia.

Table One shows the number of participants, countries, and authors
in general, together with the number of papers presented in each field. Over the last 20 years, some 700 participants from 20 countries have attended the RoManSy Symposia; 430 papers , published in nine volumes by Springer-Verlag, PWN-Elsevier, MIT Press, and Hermes, have been presented. Two special issues of the MMT Journal were dedicated to selected papers which could not be presented at the Sympsia. The number of authors per Symposium varies between 80 and 150; their papers reflect the complexity and interdisciplinary character of the problems treated in the Symposia.

Table Two shows the number of papers by countries.

Here it can be seen that four hundred eight papers come from the ten countries which have contributed from the beginning to our Symposia: the USA, Japan, Russia and the former USSR, France, Germany, Poland, Italy, Yugoslavia, the UK, and Bulgaria. The remaining twenty-one papers have arrived from ten countries which are expected to become increasingly active in the near future.

Table Three lists the members of the CISM-IFToMM Program and Organizing Committee from 1971 to 1994. In this period of time, thirty-two distinguished scientists from prominent universities and institutes in twelve countries have contributed to the work of this committee. It has been a great honour for CISM and IFtoMM to have had their assistance in organizing and selecting the papers presented and published over the last twenty years.

3. The third period of activity: 1994 -

RoManSy 10 opens the third period of activity of the RoManSy Symposia. Of course, it is not easy to predict the future, but in my opinion the theory and practice of robots and manipulators will continue to be of great importance for future developments in science and technology.

Judging from the number of papers collected for this tenth Symposium, we assume that there is still a great interest in the fields of mechanics, control, sensing, and the design of robots and manipulators.

In the field of mechanics the following subjects are investigated: the generation of symbolic equations for multiloop robot kinematics, singularity analysis, redundancy problems, the dynamics of elastic robot arms, anthropomorphic telemanipulation, force control in telerobotics, modeling and control synthesis of flexible multibody systems, constrained robots, robot control based on qualitative models, vibration control of flexible arms. These topics have played an important role in the history of the RoManSy Symposia.

The optimization of inertial properties in the design of manipulators, multiobjective optimization, optimal synthesis, the design of underwater robots, the design of micro-robots with parallel links, new parallel robots, the design of novel robot structures are the main topics in the area of synthesis and design.

In the field of the biomechanical aspects of robots and manipulators, the propblems addressed are: fuzzy logic control, distal learning, the modeling of walking machines, robot systems for surgery, and rehabilitation manipulators.

Sensing position in underwater vehicles, neural- based sensing, and LED Array-Based Imaging Sensors are some examples of the problems treated in the area of associated sensors and artificial intelligence. The sector of application and performance evaluation investigates problems such as: a polymer gel photo micro-actuator, dependent assembly planning, and robust multiaxes controllers.

4. RoManSy Symposia in the next decade

In my opinion mechanics and biomechanics, the control of motionm design, sensing and some application in the field of robots and manipulators will be the main subjects of research in the coming decade.

Some possible topics could be:

* microrobotics and micromobile robots
* mechatronics and biomechatronics in robotics
* zoomorphic robotics
* flexible systems
* unconventional architecture of robots and manipulators
* underwater robotics and robotics in space
* rehabilitation robotics, robots for surgery and therapy
* biorobotics
* mobile vehicles
* service robotics in unstructured terrain in the presence of movable objects (dynamic scene)

Mechanics (structure, kinematics, dynamics, and especially multibody mechanical systems), will continue to be the main topic of research, using modeling, identification, computer-aided analysis, and the synthesis and design of of manipulators and robots. Flexibility, accuracy, sensitivity, vibrations, friction, redundancy will be the main factors determining the success of robots and manipulators.

New methods of control, new types of actuators, grippers, multisensorial systems, and intelligent systems are other examples of possible research in the near future.

Anthropomorphic, bionic and biomechanical robot systems will, I believe, be an interesting subject of research. Crawling, walking, running, hopping mobile robots with legs and wheels, and their control and sensing will be investigated. A new architecture of robots for space and underwater applications, and their kinematics and dynamic properties appear to be promising subjects of research. Medical manipulators for surgery, therapy, and physical exercises will surely be of interest.

Applications are likely to concern the areas of home service, construction, agriculture, fishing, forestry, ocean exploration, and inspection.

As I have said, it is not easy to predict trends in research and applications, but judging from the experience of the past, let us hope that at least some of these problems find solutions in the near future.

Table 1
NUMBER OF PAPERS BY SECTION
(the main data)

RoMאnSy Symposium by years	Number of countries	Number of participants	Total number of papers published	Number of authors	Opening lectures	Mechanics	Control of motion	Mobile robots and walking machines	Sensing and artificial intelligence	Synthesis and design	Application and performance evaluation	Biomechanics of motion (Manip. + Locomotion)	Man-machine system	Computer enhanced teleoperation	Panels (papers)	Movie and video sesions	Publisher of Proceeding
1973	11	60	47	88	2	3	9	4	8	–	7	6	8	–	–	–	Springer–Verlag
1976	15	117	48	116	1	4	6	7	10	6	7	3	4	–	2(3)	2	Elsevier PWN
1978	12	67	34 11*)	78	–	5	8	–	4	4	7	4	2	–	2(3)	2	Elsevier PWN
1981	12	80	38 9*)	90	–	4	10	8	–	6	6	–	–	4	1	1	PWN
1984	14	65	45	123	1	10	13	–	–	6	2	6	7	–	1	1	Kogan–Page Hermes Publishing
1986	16	90	66	151	5	13	11	7	10	15	5	–	–	–	–	3	MIT Press
1988	14	71	49	127	1	7	10	7	5	10	7	2	–	–	1	1	Hermes
1990	16	93	58	132	1	18	14	8	4	5	4	4	–	–	–	–	Warsaw Univ. of Technology Public.
1992	15	51	44	100	1	12	6	–	3	10	6	6	–	–	1	1	Springer–Verlag
Total		694	429+ 20*)		12	76	87	41	44	62	51	31	21	4	8 (10)	12	

*) Not published in the Proceedings, but in special issues of the MMT Journal, 16(1) and 18(4).

Table 2

NUMBER OF PAPERS BY COUNTRIES

Countries	Years 1973	1976	1978	1981	1984	1986	1988	1990	1992	Total
USA	13	11	9	7	14	11	10	12	8	95
Japan	4	5	5	6	6	6	8	6	4	50
USSR, Russia	15	8	3	2	5	7	4	3	2	49
France	1	2	2	7	8	14	3	6	5	48
Germany	2	4	4	4	2	3	9	9	2	39
Poland	2	6	2	3	2	7	3	6	6	37
Italy	2	5	2	2	1	2	3	4	5	26
Yugoslavia	4	2	1	3	4	4	2	2	2	24
UK	3	3	6	1	–	3	1	2	4	23
Bulgaria	1	1	–	1	3	3	4	2	2	17
Czech Republic	–	1	–	2	–	1	–	–	–	4
Australia	–	–	–	–	–	1	1	1	–	3
Romania	–	–	–	–	–	1	1	1	–	3
Canada	–	–	–	–	–	–	–	2	1	3
China	–	–	–	–	–	2	–	–	–	2
Hungary	–	–	–	–	–	1	–	1	–	2
Switzerland	–	–	–	–	–	–	–	1	–	1
Belgium	–	–	–	–	–	–	–	–	1	1
Kazakhstan	–	–	–	–	–	–	–	–	1	1
Spain	–	–	–	–	–	–	–	–	1	1
Total	47	48	34	38	45	66	49	58	44	429
Host country I - Italy, PL - Poland	I	PL	I	PL	I	PL	I	PL	I	

Table 3

MEMBERS OF THE PROGRAMME AND ORGANISING
COMMITTEE OF THE CISM–IFToMM RoManSy SYMPOSIA

Names	Countries	Years									
		1973	1976	1978	1981	1984	1985	1988	1990	1992	1994
1. Artobolevski	Former USSR	M	M	–	+	–	–	–	–	–	–
2. Bertozzi	I	S	S	S	S	S	S	S	S	S	S
3. Bessonov	Former USSR	–	–	M	M	M	M	M	M	M	M
4 Bianchi	I	M	M	M	Vch	Ch	Vch	Ch	Vch	Ch	Vch
5. Guinot	F	–	–	–	–	–	–	M	M	M	M
6. Heimann	GDR (GFR)	–	–	–	–	–	–	M	M	M	M
7. Hewit	UK	–	–	–	–	–	–	M	M	M	M
8 Hirose	Jap	–	–	–	–	–	–	M	M	–	–
9 Jaworek	PL	–	–	–	–	–	–	SS	SS	SS	SS
10 Kędzior	PL	–	SS	SS	SS	SS	SS	–	–	–	–
11. Kato	Jap.	M	M	M	M	M	–	–	–	–	–
12 Khatib	USA	–	–	–	–	–	–	–	–	M	M
13 Kobrynski	Former USSR	Ch	M	M	M	M	M	–	+	–	–
14 Konstantinov	Bul.	M	M	M	M	M	M	M	M	–	–
15. Korendyaschev	Former USSR	–	–	–	–	–	–	M	M	M	M
16. Matsushima	Jap.	–	–	–	–	–	–	M	M	–	–
17. Mc Ghee	USA	–	M	M	M	–	–	–	–	–	–
18 Morecki	PL	M	Vch	Vch	Ch	Vch	Ch	Vch	Ch	Vch	Ch
19 Rankers	Nh	–	–	–	–	M	M	+	–	–	–
20 Romiti	I	M									
21 Roth	USA	M	Ch	Ch	M	M	M	–	–	–	–
22 Schiehlen	GFR	–	–	–	–	–	–	M	M	M	M
23 Schraft	GRF	–	–	–	M	M	M	–	–	–	–
24 Sobrero	I	Vch	Vch	Vch	–	–	–	–	–	–	–
25 Tanie	Jap	–	–	–	–	–	–	–	–	M	M
26 Thring	UK	M	M	M	–	–	–	–	–	–	–
27. Vertut	F	–	M	M	M	M	–	–	–	–	–
28. Volmer	GDR	–	–	–	M	M	–	–	–	–	–
29. Vukobratovič	Yu.	M	M	M	M	M	M	M	M	M	M
30. Waldron	USA	–	–	–	–	–	–	M	M	M	M
31 Warnecke	GFR	M	M	M	–	–	–	–	–	–	–
32 Whitney	USA	M	–	–	–	–	–	–	–	–	–

Ch – Chairman
Vch – Vice-chairman
M – Member
SS – Scientific Secretary
S – Secretary

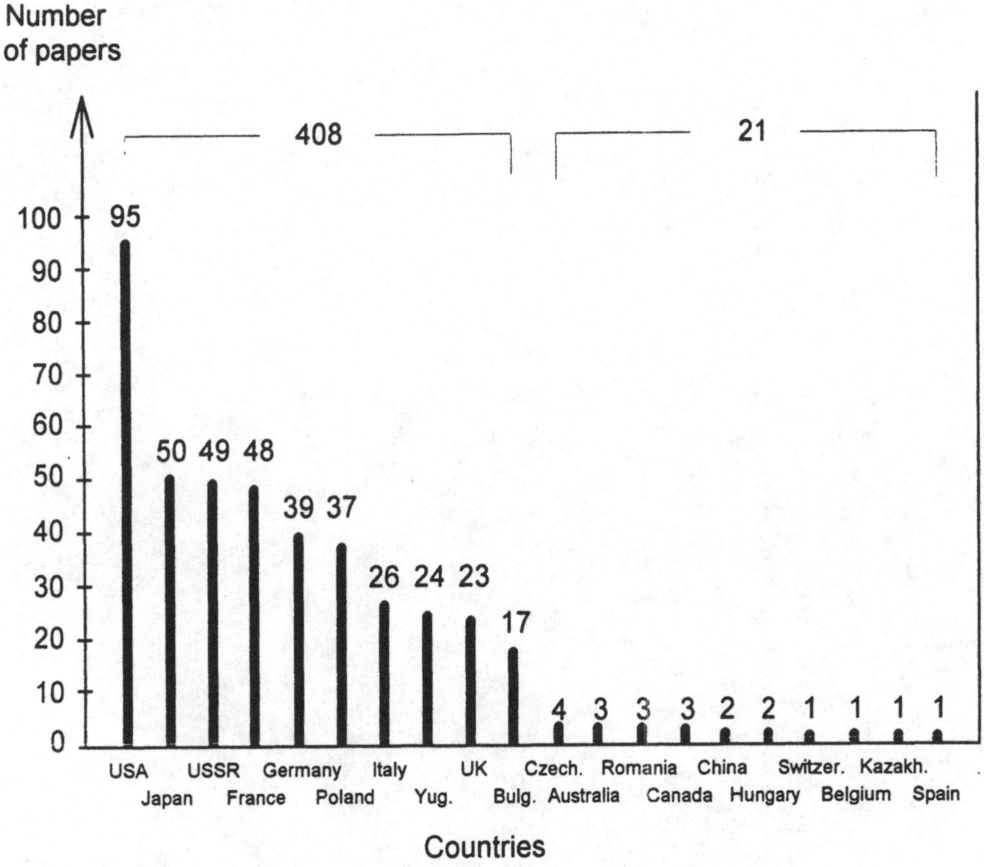

The number of papers by countries presented during RoManSy Symposia in the years 1973-1992.

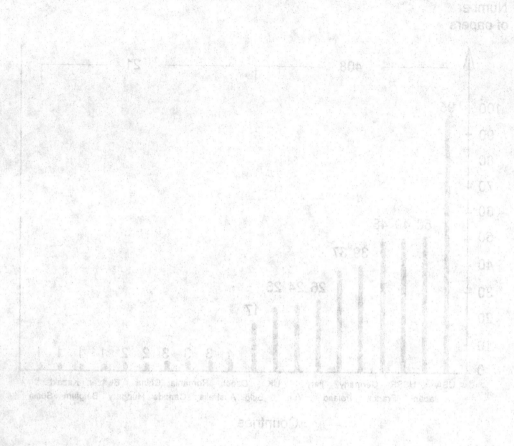

Part I
Mechanics I

POSITION ANALYSIS OF THE RHOMBIC ASSUR GROUP

K. Wohlhart
Graz University of Technology, Graz, Austria

Summary. The paper presents a position analysis of a special higher Assur group , the homogeneous rhombic Assur group of class 4 and order 4. For an Assur group, i.e., a kinematic chain with a certain number of inputs (the class of the group) which becomes rigid if the input parameters are held constant, the position analysis means the determination of all possible positions of this chain which correspond to a given set of input parameters. By solving an algebraic equation of order 20 and two equations of order 18 it is found that the rhombic Assur group 4,4 can theoretically occupy 56 different (real or complex) positions. To know these positions becomes vital if an Assur group for exemple serves as basic mechanism for a redundant planar manipulator whose singular positions must be carefully avoided. Though we were not able to establish an equation which would determine the positions of the general Assur group 4,4, (it would be an algebraic equation of order 56), we nevertheless can conclude that the number of its possible positions must also be 56.

Keywords: Planar linkages, position analysis. Assur groups.

INTRODUCTION

According to the definition given in [1], an Assur group is " the smallest kinematic chain, which when added to, or substracted from, a mechanism results in mechanism that has the same mobility as the original ". L.W Assur in 1914 and 1915 published two great articles [2] - they now belong to the classic literature in the theory of mechanisms - in which he proposed a structural classification of planar linkages with lower kinematic pairs.

Assur distinguishes mainly two types of what now we call (according to N.E. Shukowski) Assur groups: the open and the closed normal chains. The open normal chain consists of N ternary and $N+2$ binary links held together by $3(N+1)$ rotary (or prismatic) joints (Fig 1). Via the $(N+2)$ outer joints or guiding joints, which we shall simply call the "inputs", the group is in contact with the links of any other mechanism. The positions of the outer joints are thought to be given as a function of time. The closed normal chain consists of N ternary and N binary links which are connected by $3N$ rotary joints, N of which, the guiding joints, combine the group with an other mechanism.

If all the guiding joints are held in a fixed position, both Assur groups become rigid structures. With n, the number of links, g, the number of joints and $\sum f_i$, the sum of relative degrees of freedom in a joint, the overall degree of freedom F of a planar mechanism can be found by the topological structure formula of Kutzbach which reads: $F=\sum f_i - 3(g-n+1)$. Substitution of $n=2(N+1)+1, \sum f_i = g = 3(N+1)$ for the open normal chain, or $n=2N, \sum f_i = g = 3N$ for the closed normal chain yields $F=0$. This means that

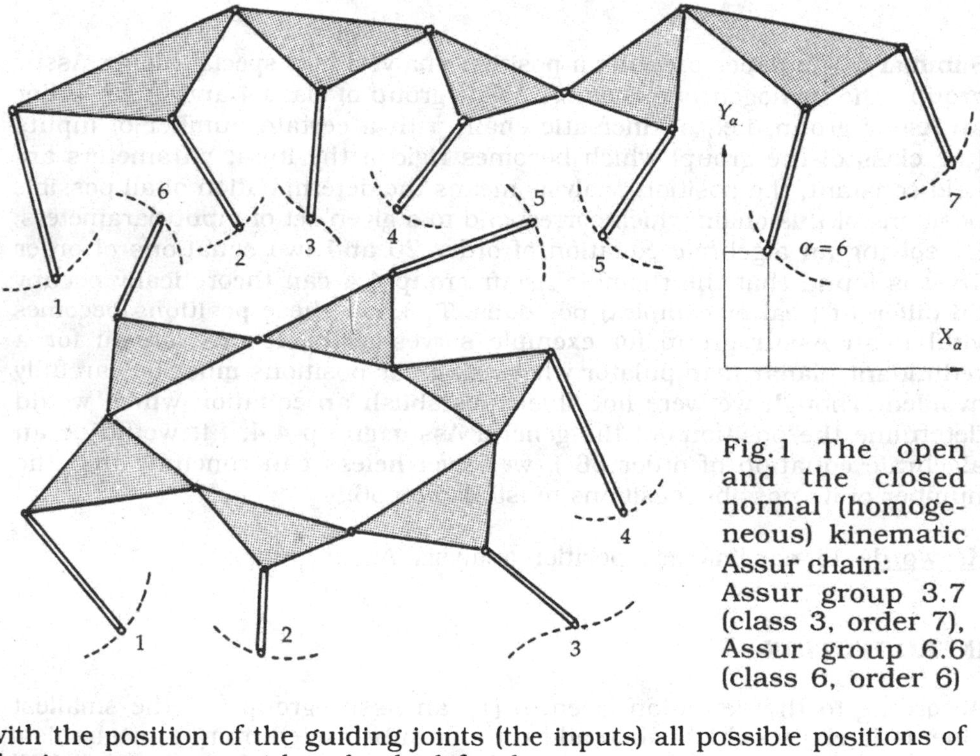

Fig.1 The open and the closed normal (homogeneous) kinematic Assur chain: Assur group 3.7 (class 3, order 7), Assur group 6.6 (class 6, order 6)

with the position of the guiding joints (the inputs) all possible positions of the Assur group must be clearly defined.

I.I. Artobolewski has given the following classification of the Assur groups [3]: The class of the Assur group is equal to the number of kinematic pairs (joints) which forms the greatest simply closed contour in the group, and the order of the group is equal to the number of the guiding joints. An open normal chain then belongs to the class 3 and has the order $N+2$, while the closed normal chain pertain to class N and is of order N.

FORMULATION OF THE PROBLEM

According to the mathematical facilities of their time, Assur and the other scientists were mainly limited to explore the kinematics of a mechanism in a given position, i.e. to the (graphical) determination of the velocities and the accelerations. Assur developed his very efficient own method which now is known as the method of the Assur points [4]. The position analysis of the Assur groups was then, however, feasible only in the more simpler cases. E.E.Paissach gave in [4] the position analysis for the Assur group 3,3. In the context of a planar manipulator a similar analysis has been carried out in [5]. Only with the advent of computer algebra systems as Reduce, Maple, Mathematica, and others, it became, at least partially possible to attack the position analysis of higher Assur groups. In the present paper we analyse a special, closed normal Assur group of class 4 and order 4 : the rhombic Assur group. Starting with the general Assur group 4.4 (Fig. 2) we soon encounter insurmountable mathematical difficulties and are forced to specify some of the system parameters.

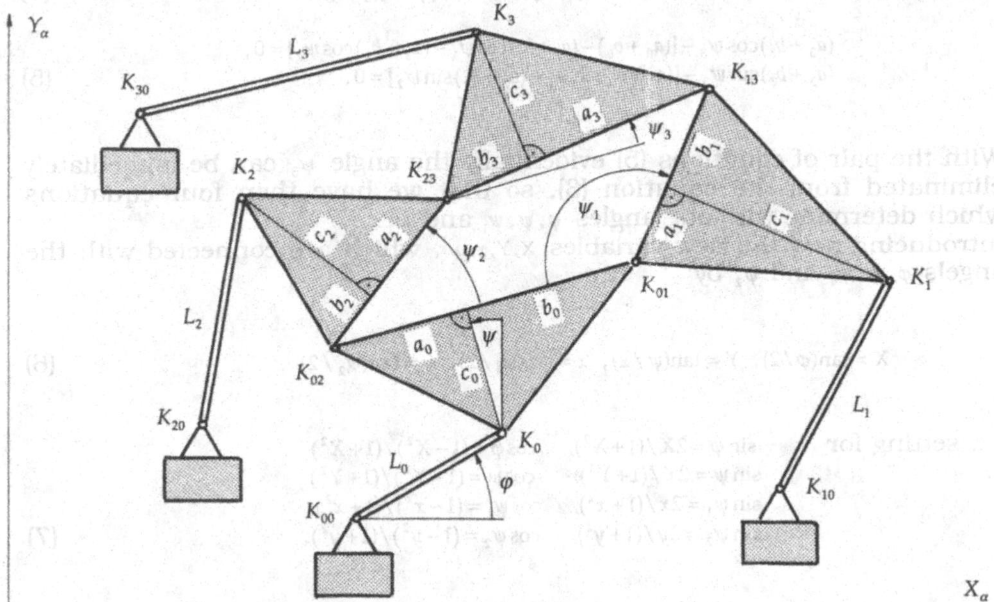

Fig.2 The general Assur group 4.4 with its system-parameters respectively the coordinates of its inputs: $a_i, b_i, c_i, L_i; X_i, Y_i. (i = 0,1,2,3)$. The rhombic Assur group 4.4 is characterized by: $a_0 + b_0 = a_1 + b_1 = a_2 + b_2 = a_3 + b_3$.

THE SET OF EQUATIONS TO SOLVE

The five position parameters $\varphi, \psi, \psi_1, \psi_2$ and ψ_3 are interconnected by five geometrical conditions. The positions of the points K_α $(\alpha = 0 \div 3)$ (Fig.2) can be easily expressed as functions of these angels. The first three conditions the angels φ, ψ, ψ_1, ψ_2 and ψ_3 have to comply with are the distances between the points K_α and the input points $K_{0\alpha}$ $(\alpha = 1 \div 3)$ which are held konstant by the binary links of the lengthes L_α. Two other conditions, which connect only the three angles ψ_1, ψ_2 and ψ_3, results from the closure conditions for the inner polygon. These five conditions read:

$$[X_0 + L_0 \cos\varphi - c_0 \sin\psi + b_0 \cos\psi - a_1 \cos(\psi_1 - \psi) + c_1 \sin(\psi_1 - \psi) - X_1]^2 +$$
$$[Y_0 + L_0 \sin\varphi + c_0 \cos\psi + b_0 \sin\psi + a_1 \sin(\psi_1 - \psi) + c_1 \cos(\psi_1 - \psi) - Y_1]^2 - L_1^2 = 0 \tag{1}$$

$$[X_0 + L_0 \cos\varphi - c_0 \sin\psi - a_0 \cos\psi + b_2 \cos(\psi_2 + \psi) - c_2 \sin(\psi_2 + \psi) - X_2]^2 +$$
$$[Y_0 + L_0 \sin\varphi + c_0 \cos\psi - a_0 \sin\psi + b_2 \sin(\psi_2 + \psi) + c_2 \cos(\psi_2 + \psi) - Y_2]^2 - L_2^2 = 0 \tag{2}$$

$$[X_0 + L_0 \cos\varphi - c_0 \sin\psi - a_0 \cos\psi + (a_2 + b_2)\cos(\psi_2 + \psi)$$
$$+ b_3 \cos(\psi_3 + \psi) - c_3 \sin(\psi_3 + \psi) - X_3]^2 +$$
$$[Y_0 + L_0 \sin\varphi + c_0 \cos\psi - a_0 \sin\psi + (a_2 + b_2)\sin(\psi_2 + \psi)$$
$$+ b_3 \sin(\psi_3 + \psi) + c_3 \sin(\psi_3 + \psi) - X_3]^2 - L_3^2 = 0 \tag{3}$$

$$[(a_0 + b_0) - (a_1 + b_1)\cos\psi_1 - (a_2 + b_2)\cos\psi_2]^2 +$$
$$+ [(a_1 + b_1)\sin\psi_1 - (a_2 + b_2)\sin\psi_2]^2 - (a_3 + b_3)^2 = 0 \tag{4}$$

$$(a_3 + b_3)\cos\psi_3 - [(a_0 + b_0) - (a_1 + b_1)\cos\psi_1 - (a_2 + b_2)\cos\psi_2] = 0,$$
$$(a_3 + b_3)\sin\psi_3 - [(a_1 + b_1)\sin\psi_1 - (a_2 + b_2)\sin\psi_2] = 0. \tag{5}$$

With the pair of equations (5) evidentially the angle ψ_3 can be immediately eliminated from the equation (3), so that we have then four equations which determine the four angles φ, ψ, ψ_1 and ψ_2.
Introducing now the new variables X, Y, x, y, which are connected with the angels φ, ψ, ψ_1 and ψ_2 by

$$X = \tan(\varphi/2), \quad Y = \tan(\psi/2), \quad x = \tan(\psi_1/2), \quad y = \tan(\psi_2/2) \,, \tag{6}$$

i.e. setting for

$$\begin{aligned} \sin\varphi &= 2X/(1+X^2), & \cos\varphi &= (1-X^2)/(1+X^2) \\ \sin\psi &= 2Y/(1+Y^2), & \cos\psi &= (1-Y^2)/(1+Y^2) \\ \sin\psi_1 &= 2x/(1+x^2), & \cos\psi_1 &= (1-x^2)/(1+x^2) \\ \sin\psi_2 &= 2y/(1+y^2), & \cos\psi_2 &= (1-y^2)/(1+y^2), \end{aligned} \tag{7}$$

results in a set of four algebraic equations, quadratic in all of the new variables. Their structure is specific:

$$\begin{aligned} G_1 &(X, Y, x \quad\ ) = 0 \,, \\ G_2 &(X, Y, \quad\ y) = 0 \,, \\ G_3 &(X, Y, x, y) = 0 \,, \\ G_4 &(\qquad\quad x, y) = 0 \,. \end{aligned} \tag{8}$$

This specific structure let expect that a solution could be found by eliminating first the two unknowns x and y. It is possible to derive from the equations

(8) two polynomes of degree 12 and one of degree 14 in both of the variables X and Y. Elimination of one of these two variables by Sylvester's Method [7] leads to a 19×19 matrix, whose determinant equated to zero would yield an algebraic equation of order 238. With the present facitilies it is impossible to estabish this equation which, besides, would also contain a great number of unwanted roots, which would not solve all equations of (8). So we are forced to make some simplifications. If we limit ourselves to the rhombic Assur group 4,4, i.e. to the Assur group 4,4 with an aequilateral rectangle (rhombus) as the inner polygon $(K_{01},K_{02},K_{23},K_{13})$, the reduction of (8) to one single equation containing only one variable becomes practicable due to the fact that the last equation of (8) in this case decays into three factors.

With

$$(a_0+b_0)=(a_1+b_1)=(a_2+b_2)=(a_3+b_3) \tag{9}$$

the last equation of (8) takes the form

$$G_4=(\ 1-x\ y\ )\ x\ y\ =0, \tag{10}$$

and we can distinguish three cases:

$$x\ y\ =1,\quad x=0,\quad y=0. \tag{11}$$

In the following elimination process we shall find in the first case a final polynom of degree 20, and in both other cases a polynom of degree 18.

POSITION ANALYSIS

First case. The equation $x\ y\ =1$ is equivalent to: $\psi_2=\pi-\psi_1$, it is the case in which the rhombic polygon is open. With $y=1/x$ we obtain from (8) three equations in the in the variables X,Y and x for which we can write:

$$\begin{aligned} G_1&=G_{11}x^2+G_{12}x+G_{13}=0,\\ G_1&=G_{11}x^2+G_{12}x+G_{13}=0\\ G_3&=G_{31}x^2+G_{32}x+G_{33}=0, \end{aligned} \tag{12}$$

where the coefficients G_{ij} are quadratic polynoms in X as well in Y. As the simultaneous existence of the three homogeneous equations (12) demands the vanishing of coefficient determinant we get a first equation in X and Y from

$$\det\begin{bmatrix} G_{11} & G_{11} & G_{13}\\ G_{11} & G_{11} & G_{23}\\ G_{31} & G_{32} & G_{33} \end{bmatrix}=0\ . \tag{13}$$

By developing (13) and splitting up a factor $(1+X^2)(1+Y^2)$ we obtain an algebraic equation of order four in X and in Y for which we write

$$A(X,Y)=A_1X^4+A_2X^3+A_3X^2+A_4X+A_5=0. \tag{14}$$

Two other polynoms in X and Y can be found with any two pairs of equations from (12). With $G_1=0$ and $G_2=0$, or $G_2=0$ and $G_3=0$ we get, by removing in both cases a factor $(1+X^2)(1+Y^2)$, from

$$\det\begin{bmatrix} 0 & G_{11} & G_{12} & G_{13} \\ G_{11} & G_{12} & G_{13} & 0 \\ 0 & G_{21} & G_{22} & G_{23} \\ G_{21} & G_{22} & G_{23} & 0 \end{bmatrix}=0\ ,\quad \det\begin{bmatrix} 0 & G_{21} & G_{22} & G_{23} \\ G_{21} & G_{22} & G_{23} & 0 \\ 0 & G_{31} & G_{32} & G_{33} \\ G_{31} & G_{32} & G_{33} & 0 \end{bmatrix}=0: \tag{15}$$

the following algebraic equations of order six which can be written in the form

$$B(X,Y)=B_1X^6+B_2X^5+B_3X^4+B_4X^3+B_5X^2+B_6X+B_7=0, \tag{16}$$

$$C(X,Y)=C_1X^6+C_2X^5+C_3X^4+C_4X^3+C_5X^2+C_6X+C_7=0. \tag{17}$$

The polynom-coefficients A_i, B_i and C_i in (14),(16) and (17) are functions of Y only.
The three equations (14), (16) and (17) have a common solution only if the determinant of their Sylvester matrix M vanishes:

$$\underline{M}=\begin{bmatrix} 0 & 0 & 0 & A_1 & A_2 & A_3 & A_4 & A_5 \\ 0 & 0 & A_1 & A_2 & A_3 & A_4 & A_5 & 0 \\ 0 & A_1 & A_2 & A_3 & A_4 & A_5 & 0 & 0 \\ A_1 & A_2 & A_3 & A_4 & A_5 & 0 & 0 & 0 \\ 0 & B_1 & B_2 & B_3 & B_4 & B_5 & B_6 & B_7 \\ B_1 & B_2 & B_3 & B_4 & B_5 & B_6 & B_7 & 0 \\ 0 & C_1 & C_2 & C_3 & C_4 & C_5 & C_6 & C_7 \\ C_1 & C_2 & C_3 & C_4 & C_5 & C_6 & C_7 & 0 \end{bmatrix},\quad \det\underline{M}=0 \tag{18}$$

By devoloping this determinant, we are led to an algebraic equation of order 40 ; but this equation can be factorized: the first of the three factors is $(1+Y^2)^5$, the second , which summerizes the unwanted roots, is a polynom of order 10, and finally the third factor is a polynom of order 20, and this factor contains the solutions. Factorisation of the final polynom is only possible if the system parameters have been entered as integers. Otherwise one has to solve numerically the final equation of order 30 (the factor $(1+Y^2)^5$ can always be extracted) and then, by resubstitution, find out which roots solve the entire set of equations (8). For the final equation (without unwanted roots) we write:

$$f=f_1Y^{20}+f_2Y^{19}+f_3Y^{18}+f_4Y^{17}+\dots\dots\dots\dots f_{20}Y+f_{21}\ , \tag{19}$$

The coefficients in the equation (19) are pure functions of the system parameters and the inputs coordinates.
To find the values of X which corresponds to the roots of (19) we can proceed as follows. With the column matrix $\underline{X}$

$$\underline{X}^T = (X^7, X^6, X^5, X^4, X^3, X^2, X, 1)$$

and the matrix $\underline{M}$ the three equations (14), (16) and (17) can be written in the form of a homogeneous matrix equation:

$$\underline{M}.\underline{X} = 0. \tag{20}$$

Dropping the first row from the matrix $\underline{M}$ we obtain a 7×8 matrix $\underline{N}$ and with its 8 column matrices $\underline{n}_\alpha$ $(\alpha = 1 \div 8)$ we can write instead of (20):

$$\underline{n}_1 X^7 + \underline{n}_2 X^6 + \underline{n}_3 X^5 + \underline{n}_4 X^4 + \underline{n}_5 X^3 + \underline{n}_6 X^2 + \underline{n}_7 X = -\underline{n}_8,$$

and find $$X(Y) = -\frac{\det[\underline{n}_1 \underline{n}_2 \underline{n}_3 \underline{n}_4 \underline{n}_5 \underline{n}_6 \underline{n}_8]}{\det[\underline{n}_1 \underline{n}_2 \underline{n}_3 \underline{n}_4 \underline{n}_5 \underline{n}_6 \underline{n}_7]}. \tag{21}$$

Finally we have to determine the values of x and $y = 1/x$ which correspond to the pairs of X and Y. We rewrite for that purpose the equations (12) in the form

$$\begin{aligned} G_1 &= x\ (G_{11}x + G_{12}) + G_{13} = 0, \\ G_2 &= x\ (G_{21}x + G_{22}) + G_{23} = 0, \\ G_3 &= x\ (G_{31}x + G_{32}) + G_{33} = 0, \end{aligned} \tag{22}$$

where the coefficient G_{ij} are now given functions of X and Y. The solvability condition for any two of these equations, e.g. the first and the second, leads to

$$x(X,Y) = -\det\begin{bmatrix} G_{12} & G_{13} \\ G_{22} & G_{13} \end{bmatrix} / \det\begin{bmatrix} G_{11} & G_{13} \\ G_{21} & G_{13} \end{bmatrix} = 1/y. \tag{23}$$

Second and third case. These cases correspond to the closed inner polygon of the Rhombic Assur group 4,4 as with $x = 0$ or $y = 0$ we have $\psi_1 = 0$ or $\psi_2 = 0$. For $\psi_1 = 0$ ($\psi_2 = 0$) the joints at K_{01} and K_{23} (K_{02} and K_{13}) becomes passive and the points K_{02} and K_{13} (K_{01} and K_{23}) coincide. These configurations actually are identical with the configuration of an Assur group of class 4 and order 3. The elimination process of the variables down to a univariate polynom is in both cases $x = 0$ and $y = 0$ very similar so that it suffices to sketch it only for $x = 0$. With $x = 0$ the first equation of (12) contains only X and Y , and can be written in the form:

$$G_1(X,Y) = E_1X^2 + E_2X + E_3 = 0, \qquad (24)$$

and if we eliminate from the second and the third equation of (12) the variable y by the Sylvester method we obtain a polynom of degree 6 in X and Y for which we write:

$$F(X,Y) = F_1X^6 + F_2X^5 + F_3X^4 + F_4X^3 + F_5X^2 + F_6X + F_7 = 0. \qquad (25)$$

The coefficients E_i and F_i are functions of Y only. The elimination of X from the equations (24) and (25) by equating the determinant of their 8×8 Sylvester matrix to zero yields, apart from a factor $(1+Y^{2)})^3$, an algebraic equation of order 18 in Y which can be solved numerically. To find the corresponding values for X and then for y one can proceed in just the same way as shown explicitely above.

CONCLUSIONS

The position analysis shows that the rhombic Assur group can take alltogether $20+18+18=56$ different positions for a given set of input point positions. This result is valid also for the general Assur group 4.4 because a slight change of one of the polygon system parameters, e.g., $a_1 \Rightarrow a_1 + \varepsilon$ will displace only slightly the roots: i.e. the $20+18+18$ roots of the rhombic Assur group differ from the 56 roots of the general Assur group only by quantities of order ε. How many of the 56 positions are real positions, however, remains an open question. By removing one of the binary links from the general Assur group 4.4 one obtains, if the remaining three input points are fixed, a movable kinematic chain with one degree of freedom (Fig.3). The point K_3 in Fig.3 and all points of the link to which K_3 belongs too, describe a "higher coupler curve". Only if we would be able to find a special system parameter set of the Assur group 4.4 for which all the 56 positions would turn out to be real, we could conclude that these coupler curves are fullcircular algebraic curves of order 56. Unless such a system parameter set is found it can only be stated that the order of these higher coupler curves is above 23 and less than 56.

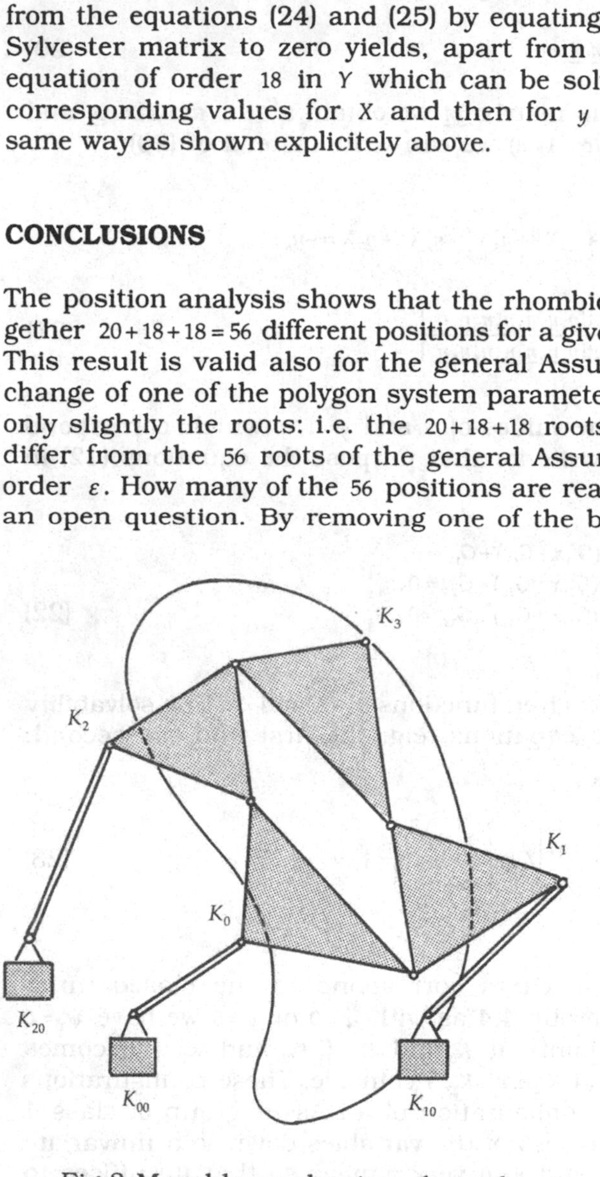

Fig.3 Movable mechanism derived from the general Assur group 4.4

NUMERICAL EXEMPLE

Following up the procedure described above we obtain for the system parameters,respective the input point coordinates:

$$a_0=4,\ b_0=2,\ c_0=3,\ L_0=4,\ X_0=10,\ Y_0=1,$$
$$a_1=4,\ b_1=2,\ c_1=2,\ L_1=4,\ X_1=17,\ Y_1=8,$$
$$a_2=3,\ b_2=3,\ c_2=3,\ L_2=5,\ X_2=3\ ,\ Y_2=2,$$
$$a_3=4,\ b_3=2,\ c_3=5,\ L_3=6,\ X_3=6\ ,\ Y_3=9,$$

the 56 values of φ,ψ,ψ_1 and ψ_2 which are collected in Tab.1. Ten of the positions of the rhombic Assur group are real. Fig.4 shows these positions.

φ^0	ψ^0	ψ_1^0	ψ_2^0
52.16 + 51.94 i	-142.2 - 63.5 i	-96.71 + 29.98 i	-83.29 - 29.98 i
52.16 - 51.94 i	-142.2 + 63.5 i	-96.71 - 29.98 i	-83.29 + 29.98 i
-27.05	-34.99	3.081	176.9
-133.9 - 1.757 i	-17.41 - 18.93 i	90.03 - 58.09 i	89.97 + 58.09 i
-133.9 + 1.757 i	-17.41 + 18.93 i	90.03 + 58.09 i	89.97 - 58.09 i
36.98	-12.05	18.91	161.1
-116.7 - 17.18 i	-1.793 - 63.72 i	72.98 - 96.12 i	107. + 96.12 i
-116.7 + 17.18 i	-1.793 + 63.72 i	72.98 + 96.12 i	107. - 96.12 i
-39.54	20.21	145.2	34.85
40.97	35.41	123.1	56.93
-49.01 + 82.72 i	36.49 - 17.86 i	79.67 - 96.74 i	100.3 + 96.74 i
-49.01 - 82.72 i	36.49 + 17.86 i	79.67 + 96.74 i	100.3 - 96.74 i
138.3 + 67. i	64.93 - 68.32 i	-125.2 - 4.395 i	-54.76 + 4.395 i
138.3 - 67. i	64.93 + 68.32 i	-125.2 + 4.395 i	-54.76 - 4.395 i
-20.25 - 57.24 i	63.02 - 49.79 i	109.7 - 67.09 i	70.26 + 67.09 i
-20.25 + 57.24 i	63.02 + 49.79 i	109.7 + 67.09 i	70.26 - 67.09 i
126.3 + 11.52 i	105.8 - 70.77 i	-101.2 + 12.79 i	-78.82 - 12.79 i
126.3 - 11.52 i	105.8 + 70.77 i	-101.2 - 12.79 i	-78.82 + 12.79 i
-53.45 - 4.281 i	96.96 - 26.65 i	-146.6 + 23.77 i	-33.41 - 23.77 i
-53.45 + 4.281 i	96.96 + 26.65 i	-146.6 - 23.77 i	-33.41 + 23.77 i
-161.9 + 40.75 i	-43.55 - 5.569 i	52.64 + 21.26 i	0
-161.9 - 40.75 i	-43.55 + 5.569 i	52.64 - 21.26 i	0
177.6 + 28.46 i	-8.014 - 58.68 i	-175.5 - 88.7 i	0
177.6 - 28.46 i	-8.014 + 58.68 i	-175.5 + 88.7 i	0
-57.12	33.84	122.2	0
168.6 + 30.84 i	53.83 - 47.41 i	-171.7 - 110.9 i	0
168.6 - 30.84 i	53.83 + 47.41 i	-171.7 + 110.9 i	0
-33.38	52.11	129.3	0
111.9 + 37.79 i	61.98 - 7.708 i	-158.6 + 28.72 i	0
111.9 - 37.79 i	61.98 + 7.708 i	-158.6 - 28.72 i	0
77.75	64.57	-123.9	0
10.05	70.7	-129.5	0
-22.16 + 22.44 i	73.55 - 1.455 i	150.4 - 17.77 i	0
-22.16 - 22.44 i	73.55 + 1.455 i	150.4 + 17.77 i	0
4.893 - 41.8 i	83.12 - 33.4 i	-116.2 - 49.53 i	0
4.893 + 41.8 i	83.12 + 33.4 i	-116.2 + 49.53 i	0
101. - 60.2 i	163.3 - 37.78 i	-86.02 - 85.02 i	0
101. + 60.2 i	163.3 + 37.78 i	-86.02 + 85.02 i	0
143.4 - 8.532 i	-91.94 - 32.78 i	0	-120.7 + 38.12 i
143.4 + 8.532 i	-91.94 + 32.78 i	0	-120.7 - 38.12 i
126.6 + 25.42 i	-74.39 - 24.06 i	0	-115.9 - 12.48 i
126.6 - 25.42 i	-74.39 + 24.06 i	0	-115.9 + 12.48 i
168.1 + 36.74 i	-73.5 - 35.51 i	0	-30.04 + 15.34 i
168.1 - 36.74 i	-73.5 + 35.51 i	0	-30.04 - 15.34 i
-23.14	-32.99	0	175.7
-93.59 - 58.5 i	-43.02 - 69.4 i	0	137.3 + 101.8 i
-93.59 + 58.5 i	-43.02 + 69.4 i	0	137.3 - 101.8 i
41.35 + 28.66 i	-17.4 - 22.12 i	0	169.4 + 64.15 i
41.35 - 28.66 i	-17.4 + 22.12 i	0	169.4 - 64.15 i
-67.68 - 78.37 i	-21.26 - 76.1 i	0	-132.8 + 92.69 i
-67.68 + 78.37 i	-21.26 + 76.1 i	0	-132.8 - 92.69 i
15.01	-13.13	0	151.
28.23 + 47.42 i	22.09 - 22.6 i	0	-147.1 + 59.02 i
28.23 - 47.42 i	22.09 + 22.6 i	0	-147.1 - 59.02 i
-9.401 - 98.28 i	37.52 - 71.67 i	0	105.3 + 69.66 i
-9.401 + 98.28 i	37.52 + 71.67 i	0	105.3 - 69.66 i

Tab.1 Numerical results: Real and the complex position angels

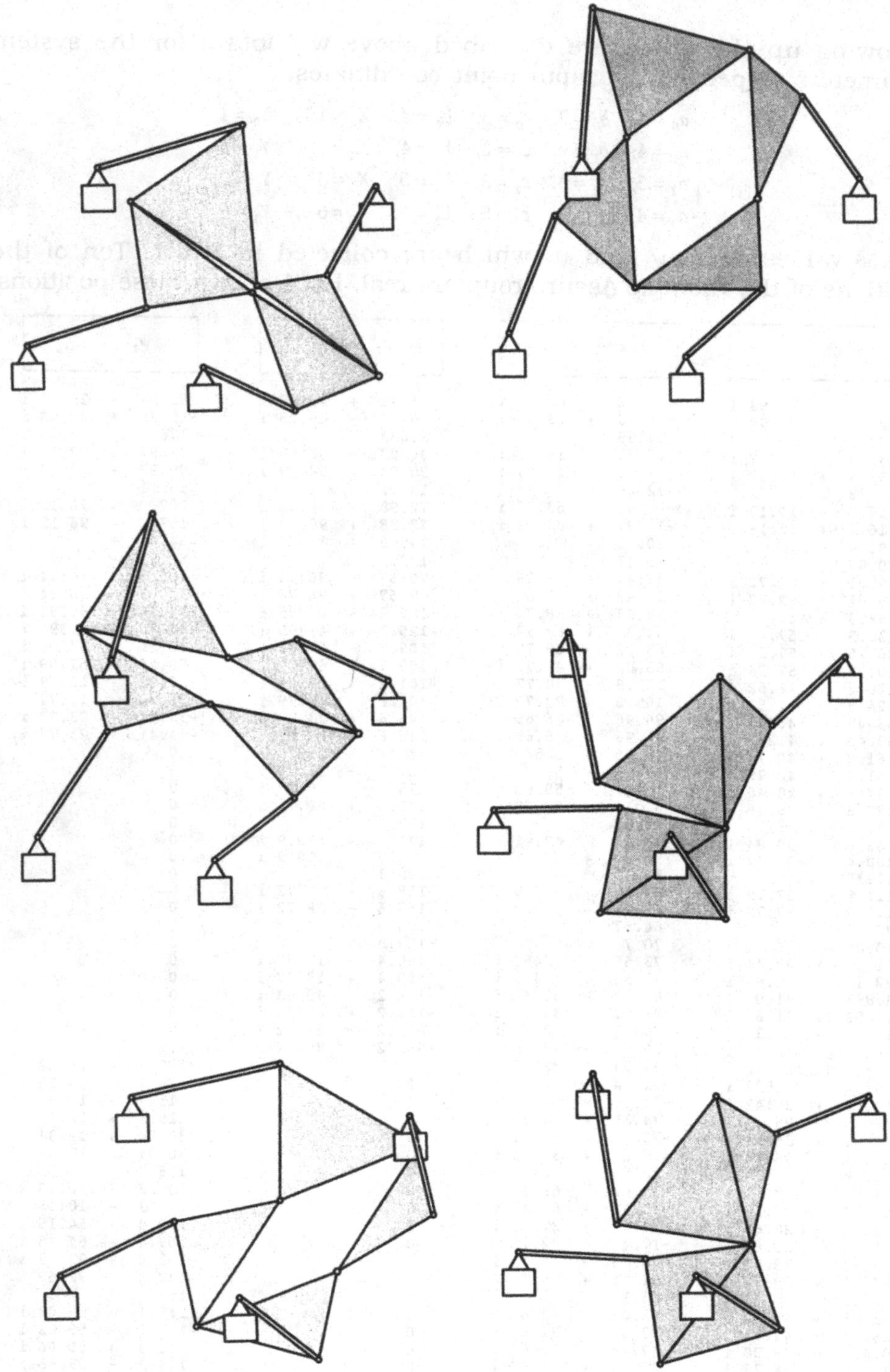

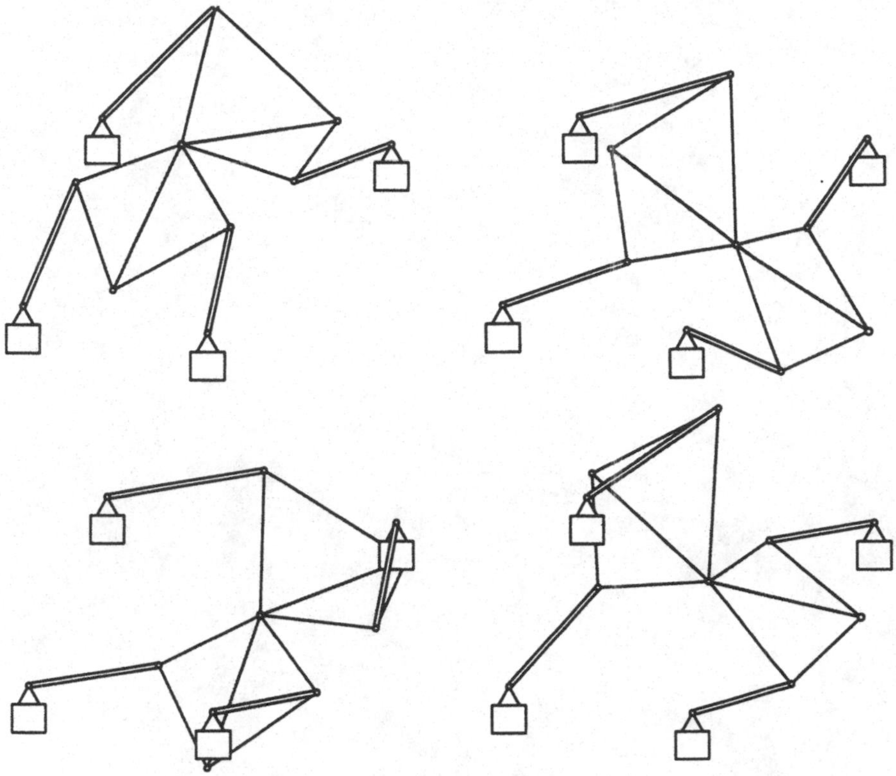

Fig.4 The 10 real positions of the rhombic Assur group 4.4.

REFERENCES

[1] Leinonen ,T. ed.: Terminology for the theory of machines and mechanisms, Mechanism amd Machine Theory, Vol. 26, No.5, p.450, 1991.

[2] Assur, L.W.: Isledowanie ploskich sterschnich mechanismow s nischnimi parami..............Tschast I ,II, Iswestia Petrogradskowo Politechn.1914,1915.

[3] Artobolevski, I.I.: Théorie des Méchanismes et de Machines, Edition Mir, p.50, 1977.

[4] Assur, L.W.: Kratnie skorostei i uskorenii totschek ploskich mechanismow, S. Petersburg, 1911.

[5] Paissach, E.E.: Opredelenie poloschenie swenev trechpowodkowoi i dwuchpowodkowoi tschetirechswennich grupp Assura.....Maschinowedenie, No.5, pp. 55-61, 1985.

[6] Wohlhart, K.: Direct kinematic solution of the general planar Stewart platform, Proc. Intern. Conf. on Computer Integrated Manufacturing, Zakopane, pp. 403-411, 1992.

[7] van der Waerden, B.L.: Algebra I, Springer 1971.

Fig. 4. The closed solutions of the rhombic Assur Group 4.

REFERENCES

[1] Lampner, Fred: Terminology of the theory of machines and mechanisms, *Mechanism and Machine Theory*, Vol. 28 No. 2, pp. 119, 1991.

[2] Assur, L: Wandelbarem...

[3] Artobelevsky...

[4] ...

[5] Peisach...

[6] Wohlhart, K: ...

A NEW METHOD FOR THE INVERSE KINEMATICS

S. Regnier, C. Vougny, F. Ben Ouezdou and D. Duhaut
University of Paris 6 - University of Versailles St. Quentin en Yvelines, Velizy, France

Abstract:

Thanks to various studies on the subject, it is now obvious that multi-agent systems have significant advantages to solve complex problems. Experiments on multi-agent systems has now started to develop in the field of robotics. Research on the CEBOT [1] system has demonstrated the self-organisation of a group of heterogeneous robotics agents. Experiments carried out by Beni and Hackwood [2] about swarms of robots have shown a high level of cooperation under simulated conditions. Brook's researches [3] have consisted in studying multi-agent teams of more than twenty little robots. Arkin [4] has shown the impact on performances when a series of multi-agent robots is used in retrieval tasks.

As far as we are concerned, our essay is about the inverse kinematics [9] To do so, we have not resorted to classical mathematical algorithms, but to a method of distributed resolution. This method requires no previous symbolic computations and is directly applicable to all kinds of serial manipulators by simply changing the initial parameters. This paper deals with this new method to solve inverse kinematics.

INTRODUCTION

It seems that robotics can be defined as all the theories and means aiming at making use of an automation of tasks, so as to replace man in some functions. A class among industrial robots is represented by manipulators made up of several poly-articulated bodies. An operator wishing to control the terminal organ of a serial manipulator must calculate the joint variable of each body that compose the robot This particular problem is called inverse kinematics. The resolution of the inverse kinematics allows the determination of joint parameters which are necessary to place the end effector in both the desired position and orientation. This problem is complex since it involves the calculation of ordered products of matrix (4x4) linked to each body of the manipulator This has been the object of numerous studies [6] [7], which up to now have provided no general solutions to be applied to all kinds of manipulators. In this particular instance of a general 6R manipulator, the technique used implies the resolution of a system of non linear equations by a mathematical analysis called "dyalitic elimination". The use of a symbolic calculator followed by numerical programming on a computer [8] is a possible method to determine a set of joint variables of a configuration. At any rate, a long mathematical process, quite costly because of the time spent on calculation, is requested.

Our purpose is to solve the problem of the inverse kinematics by a multi-agent method which could be applied to all serial manipulators. Our idea was to consider each body of the manipulator as an agent usefully cooperating towards the same collective purpose, that is to say the position and the orientation of the end effector Thanks to this dialogue between agents, the placing of the end effector is performed by successive stages.

The first part of this essay is devoted to an introduction to the multi-agent systems and to the mechanical aspect of the problem of the inverse kinematics. We have focused on the different communication architecture and on the complexity of the kinematics inverse. Our second part concerns our approach to the problem. Then, the results obtained by this new system have been studied. In our conclusion, we have emphasised both the prospects and the limits of our method.

MULTI-AGENT SYSTEMS

In the last few years a new research paradigm has come into the international area. The distributed artificial intelligence (DAI) and multi-agent system (MAS) has gained major importance. What makes the area so fascinating and attractive ? The primary reason for its interest is that multi-agent system concentrate on a them of central importance to human beings, namely, their communicate, cooperative and social nature. A second reason is that the world is very complicated. We need new concepts and practical methods to coordinate groups of men or computer machines.

The word agent is used to designate an intelligent entity, acting rationally and intentionally with respect to its own goals and the current state of its knowledge.

Bond and Gasser [5] speak about intelligent system. Multi-agent system are also concerned with coordinating intelligent behaviour among autonomous intelligent agent. It's the intentional agent.

In an other part, recently, there has been a debate with the reactive system. In this system, knowledge about other agents and the environments is explicitly represented such that the agent is able to reason about how some goal can be achieved. It's the reactive agent.

Each agent owns two types of knowledge : its internal knowledge (its goals, its engagements, its model) and its external knowledge (its environment and the others agents).

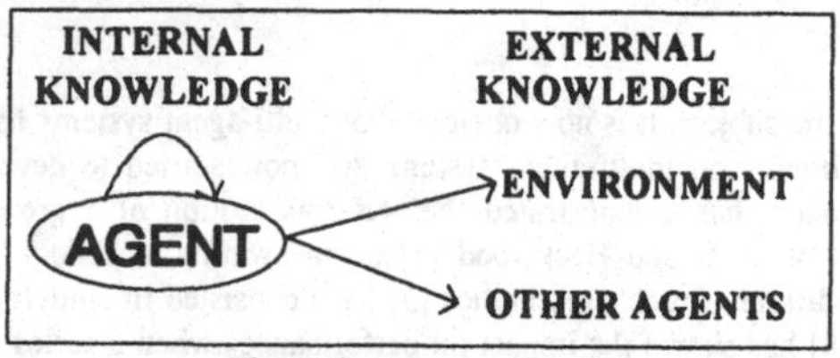

Fig. 1 Knowledge of an agent

We consider several autonomous intelligent agents which coexist and may collaborate with others agents in a common world. Each agent may accomplish its own task and cooperates with other agent to perform a personal or a global task.

In order that the agents have to solve complex problems, there exists a structure which manages the participation of the agents at the life of the group: the multi-agent system. In a multi-agent system, agents must communicate and synchronize their actions to solve conflicts.

Two paradigms of communications emerge:

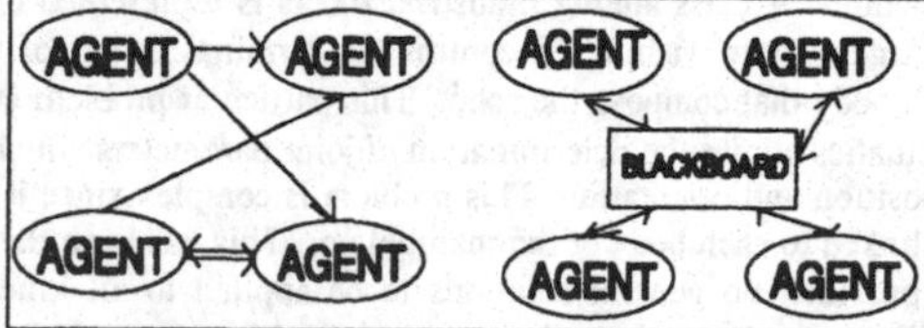

Fig. 2 Types of communications for the agents

The actor paradigm is based on an object-oriented language where each object has an independent life and communicates with others by sending messages. Various kind of message can be imagined as direct addressing, local or global broadcasting.

The blackboard paradigm makes agent communicate by writing on a shared structure called a blackboard. The blackboard can be structured for organizing communication at various levels of abstraction. An agent communicates with another one by writing on the blackboard.

THE INVERSE KINEMATICS

We use Denavit-Hartenberg [10] formalism to describe the kinematics of the manipulator. A coordinate system is attached to each link for describing the relative arrangement among the various link.

The 4x4 transformation matrix relating i^{th} coordinate system to the $i\text{-}1^{th}$ coordinate system is:

$$T^{i}_{i-1} = \begin{bmatrix} C\theta_i & -C\alpha_i S\theta_i & -S\alpha_i S\theta_i & a_i C\theta_i \\ S\theta_i & C\alpha_i C\theta_i & -S\alpha_i C\theta_i & a_i S\theta_i \\ 0 & S\alpha_i & C\alpha_i & d_i \\ 0 & 0 & 0 & 1 \end{bmatrix}$$

where $si = \sin(\theta_i)$ and $ci = \cos(\theta_i)$.

θ_i is the ith joint angle (joint variable for a revolute joint).

a_i is the length between the two coordinate systems.
α_i is the ith twist angle between the axis i-1 and i.
d_i is the offset distance at joint i (joint variable for a prismatic joint).

An operator desiring to command the end effector of a serial manipulator defines a task, an orientation and a position. The end effector position and orientation determines a task matrix T^h expressed in the frame R_0.
We present this matrix as:

$$T^h = \begin{bmatrix} A & B & C & Dx \\ E & F & G & Dy \\ H & I & J & Dz \\ 0 & 0 & 0 & 1 \end{bmatrix}$$

The problem of the inverse kinematics correspond to compute the joints θ_i or d_i such that:

$$T_0^1 T_1^2 \dots T_{n-2}^{n-1} T_{n-1}^n = T^h$$

where n is the number of solid.

Thus, the problem corresponds to solve six equations in six unknowns. The complexity of this problem is a function of the geometry of the manipulator. For a serial manipulator with six links, the solution can be expressed in closed form. But, no such formulation is known for a redundant manipulator.

We present now our multi-agent methods which allows the determination of a set of joint parameters for a general serial manipulator.

INVERSE KINEMATICS BY A MULTI-AGENT SYSTEM

We consider each body of the manipulator as an agent. The multi-agent system represents the manipulator. Each agent i possesses the following knowledge:

The knowledge of the agent i:

Internal knowledge

• a task matrix T_i^h which represents its goal personal matrix expressed in its frame.

T_i^h is computed step by step with the final task matrix into two stages:

- the local goal matrix of the agent A_i is computed by the agent A_{i+1}. This local goal matrix expresses "the missing matrix to solve the collective goal if the others agents don't move". This local goal matrix is computed as following: $T_0^{'h} = T_n^i \cdot T^h$ (' expresses the local goal).

- the agent A_i is excited when it receives its local goal matrix. It computes its personal goal matrix as following $T_i^h = T_0^{'h} \cdot T_{i-1}^0 = T_n^i \cdot T^h \cdot T_{i-1}^0$

• its 4x4 transformation matrix relating the ith joint and the $(i-1)^{th}$ joint T_{i-1}^i which expresses its potential action about the world.

External knowledge

• the 4x4 transformation matrix relating the $(i-1)^{th}$ joint and the 1th joint which represents its position in the absolute frame.

Architecture of communications

In the multi-agent approach, we considered that the agents could only communicate with the previous or next agent.

Fig. 3 Type of communications

The last agent A_n which simulate the end effector of the manipulator receives a task matrix from the user. It converts this matrix in its local frame and leads to its goal matrix with a minimization criterion (cf the minimization criterion).

So, it computes its new joint variable θ_i or d_i (according the kind of joint: prismatic or revolute) and moves. Then, it computes the local goal matrix T_{n-1}^h of the previous agent A_{n-1} in the kinematics chain.

Each agent has this behavior and works in turn according a "wave" mechanism of the last agent to the first agent and inversely. The last agent decides only to stop the method if its transformation matrix is equal to the task matrix (with a precision ε determined by the user).

The minimization criterion

Each agent A_i leads to approach towards its local goal matrix T_i^h with its 4x4 transformation matrix .

$$T_i^h = \begin{bmatrix} A & B & C & Dx \\ E & F & G & Dy \\ H & I & J & Dz \\ 0 & 0 & 0 & 1 \end{bmatrix}$$

We have constructed a matrix/matrix norm which represents the difference between terms of the matrices. We want to minimize this norm. So, we square it and after we derive it.

- for a revolute joint, we derive this norm by θ.

$$\frac{\partial N^2}{\partial^2 \theta} = 2\,[\, s\theta\,(A + Fc\alpha - Gs\alpha + a\,Dx) + c\theta\,(-E + Bc\alpha - Cs\alpha - aDy)\,]$$

We find with the equation $\frac{\partial N^2}{\partial^2 \theta} = 0$ the parameter θ:

$$\theta = \arctan\left(\frac{E - Bc\alpha + Cs\alpha + aDy}{A + Fc\alpha - Gs\alpha + aDx}\right)$$

- for a prismatic joint, we derive this norm by d. We find $\frac{\partial N^2}{\partial^2 d} = 2\,[\, Dz - d\,]$ as $\mathbf{d} = \mathbf{Dz}$

This minimization criterion has been chosen because it is very simple. In fact, we have preferred a resolution of the problem with a minimal utilization of mathematical algorithm. We wanted to wander from the classical methods. But, others criterion could have been chosen as mechanical criterion (position of the joint in the mechanical chain, list of the joint axis...), mathematical criterion (resolution more important of the equations in order to have a precise criterion...) or strategic criterion (to impose on manipulator to place it in position and after in orientation...). But this criterion would have generate knowledge of the agents more significant. And our criterion have permitted to solve inverse kinematics (cf RESULTS) so it corresponds to our choices. We are now going to explain the given results.

RESULTS

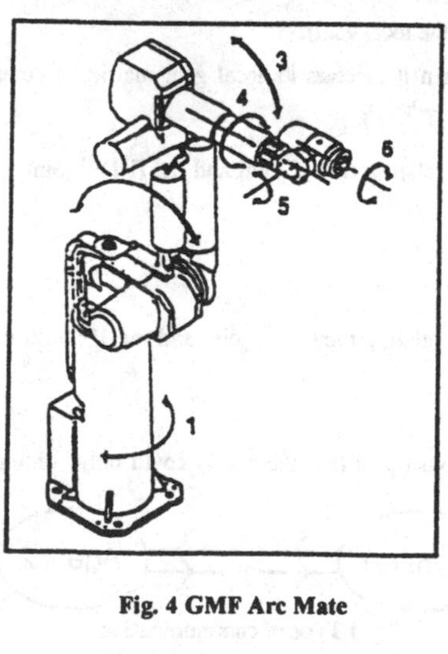

Fig. 4 GMF Arc Mate

We have tested our systems above many classics manipulators (SCARA, PUMA, 6P, 3R , 6R, 5R1P, 4R2P..), redundant (8R, 8P, 6R2P...). The main interest has been in a 6R manipulator because we disposed of results [8] about this type of robot to compare classics approaches and multi-agents approaches.. We have chosen GMF Arc Mate (fig. 3) manipulator which is an industrial 6R manipulator. Its mechanic architecture is similar to others manipulators as PUMA but its inverse kinematics is different and more complex because the last three axis aren't concurrent.

We have carried out 628 various executions. Each execution was various from the others by an increment of 0.01 radians in the parameter θ_1. This type of approach represents for the first joint a circle which inducts many configurations of the manipulators.

In the following figures, we could study the relationship between iteration and error (cf fig.5) and the end effector evolution towards wanted position and orientation (cf fig.6).

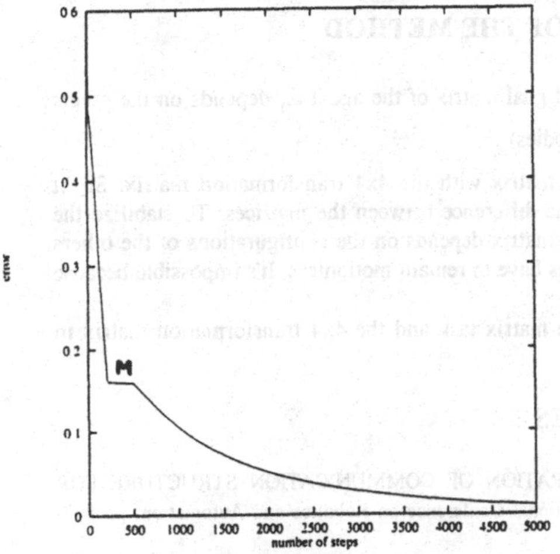

Figure 5
Relationship between iteration and error.

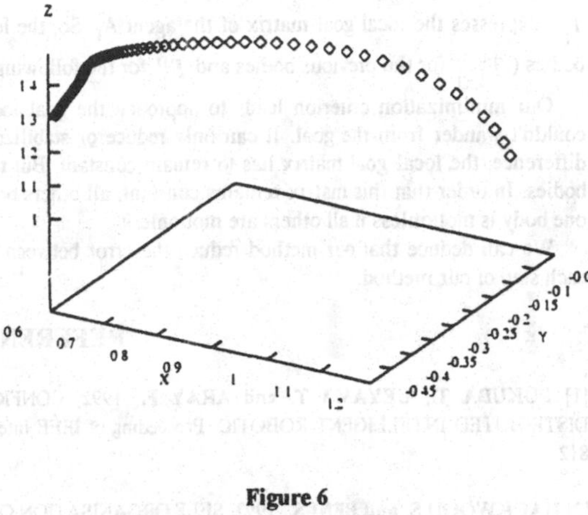

Figure 6
Movements of the end effector.
(scale in metre)

The figures 5 show the error between the task matrix and the transformation matrix relating the last joint. This error represents the average error between each term of the matrixes.

Slopes of the graphs are more important at the beginning because the minimization criterion privileges the wanted position. After, the orientation takes a more important part in the computation and the slope is shorter.

The figure 6 shows a regular convergence. When the end effector is at the near of the wanted position and orientation, the approach is smaller. But the end effector has not direct way because the method solves the inverse kinematics by a "wave" mechanism without searching the shortest way. The point M represents a change of direction towards the wanted position.

We are going now to study the temporal aspect of our method. The precision reprresents the average error between the matrix task T^h and the matrix T_0^n.

Precision	0.1%	0.01%	0.05%
Time	46 seconds	201 seconds	302 secondes

Fig. 7 The relationship between time and precision

Our method solves the inverse kinematics for all configurations of the GMF Arc Mate because the manipulator is never in a singular configuration..

CONCLUSION

The aim of our previous paper was to give a multi-agents method to solve the kinematics inverse. These method allow the determination of a set of joint parameters, so as to place the terminal organ in both a given position and orientation for <u>all</u> serial manipulators. Besides, thanks to the possibility to fasten stops to joints, the inverse kinematics that is found corresponds to a real configuration of the manipulator.

The main flaw of this method lies in the possibility to predict the number of steps to determine the parameters. Yet, this flaw could be corrected by using an abacus that could count the number of necessary steps, depending on the desired position, which seems to exclude the time being its use in a real time.

Resolutions can be perfected as the implemented programs take no account of the particularities of the manipulator (our method is general) and the chosen criterion is very simple.

This method allowed to solve inverse kinematics, without taking advantages of the properties of the multi-agents systems. Thus, we have started to develop a new approach to the inverse kinematics with a parallel method within less limited time conditions.

We are also working to broaden these reasoning methods on inverse kinematics for a closed loop manipulator, on path planning or trajectory generations.

ANNEX 1 CONVERGENCE OF THE METHOD

T_i^h expresses the local goal matrix of the agent A_i. So, the local goal matrix of the agent A_i depends on the others bodies (T_{i-1}^0 for the previous bodies and T_i^n for the following bodies).

Our minimization criterion leads to approach the goal local matrix with the 4x4 transformation matrix. So, it couldn't wander from the goal. It can only reduce or stabilize the difference between the matrices. To stabilize the difference, the local goal matrix has to remain constant. But this matrix depends on the configurations of the others bodies. In order that this matrix remains constant, all others bodies have to remain motionless. It's impossible because one body is motionless if all others are motionless.

We can deduce that our method reduce the error between the matrix task and the 4x4 transformation matrix in each step of our method.

REFERENCES

[1] **FUKUDA T., UEYAMA T. and ARAY F.**, 1992, CONFIGURATION OF COMMUNICATION STRUCTURE FOR DISTRIBUTED INTELLIGENT ROBOTIC Proceeding of IEEE International Conference on Robotics and Automation, pp 807-812

[2] **HACKWOOD S. and BENI S.**, 1992, SELF ORGANISATION OF SENSORS FOR SWARM INTELLIGENCE Proceeding of IEEE International Conference on Robotics and Automation, pp 819-829

[3] **BROOKS R., MAES P., MATARIC M. and MORE G.**, 1990, LUNAR BASE CONSTRUCTION ROBOTS Proceeding of IEEE International Workshop on Intelligent Robot and Systems (IROS) pp 389-392, Tschudiara, Japan, 1990

[4] **ARKIN C., BALCH T. and NITZ E**, 1993, COMMUNICATION OF BEHAVORIAL STATE IN MULTI-AGENTS RETRIAVAL TASKS Proceeding of IEEE International Conference on Robotics and Automation, pp 588-593

[5] **BOND A.H. and GASSER L.**, 1988, READINGS IN DISTRIBUTED ARTIFICIAL INTELLIGENCE, Bond and Gasser editors, Morgan Kaufman

[6] **RAGHAVAN M. ET ROTH B.**, 1990, A GENERAL SOLUTION FOR THE INVERSE KINEMATICS OF ALL SERIES CHAINS Proceeding of the 8th CISM-IFToMM Symposium on robots manipulators, (Romansy 1990), Cracow, Poland

[7] **ANGELES J. and ZANGANESH K.E.**, 1992, A SEMIGRAPHICAL DETERMINATION OF ALL REAL INVERSE KINEMATICS SOLUTIONS OF A GENERAL SIX REVOLUTE MANIPULATORS Proceeding of the 9th CISM-IFToMM Symposium on robots manipulators, (Romansy 1992), Udine, Italy

[8] **MAVROIDIS C., BEN OUEZDOU F. and BIDAUD P.**, 1993, INVERSE KINEMATICS OF SIX DEGREE OF FREEDOM "GENERAL" AND "SPECIAL" MANIPULATORS USING SYMBOLIC COMPUTATION, Accepted to be published in ROBOTICA

[9] **DUHAUT D.**, 1993, USING A MULTI-AGENT APPROACH TO SOLVE THE INVERSE KINEMATICS Proceeding of IROS Intelligent robot and system, pp 2002-2007

[10] **DENAVIT J. ET HARTENBERG R.S.**, 1955, A KINEMATIC NOTATION FOR LOWER PAIR MECHANISMS BASED ON MATRIX. Transactions of the ASME, Journal of Applied Mechanics, Vol. 22, pp 215-221

AN ALGEBRAIC FORMULATION FOR THE WORKSPACE DETERMINATION OF GENERAL 4R MANIPULATORS

M. Ceccarelli
University of Cassino, Cassino, Italy

ABSTRACT - An algebraic approach is proposed to deduce an algorithm for determining the workspace boundary of 4R manipulators. The boundary contour of a workspace is described as a function of the link parameters and one revolute joint angle only. Particularly, the proposed algorithm is useful to trace the cross-section so that the workspace characteristics, as shape, hole and voids, are immediately shown and analysed. Some examples prove the usefulness of the algorithm and illustrate some peculiarities of the workspace topology of general 4R manipulators.

INTRODUCTION

The 4R manipulators, with four revolute joints connecting rigid links in open kinematic chain, Fig.1, are widely used robotic mechanisms for gross motion since they have one redundant degree of freedom useful in avoiding obstacles applications.

The workspace, which is the set of points reachable by the manipulator's hand reference point, is of great importance for manipulation device. Particular interest is addressed to its determination and evaluation with the aim to deduce computational algorithms both for analysis and synthesis purposes. Several techniques have been proposed for workspace analysis of manipulators making use of different physical and computational approaches: by using the mechanics of stable configurations, [1]; by scanning the joint variables to determinate reachable points by means of kinematical matrix methods, [2,3], or through dynamical evaluation of feasible robot configurations, [4]; by using Jacobian analysis of the boundary existence domain via matrix formulation, [5]; by expressing algebraically the characteristic geometry of ring volumes, [7-9]; by using the probabilistic analysis technique of the Monte Carlo method, [10]. In particular, the algebraic approach seems to be the most convenient in order to deduce criteria for workspace analysis, which may be meaningful and useful in robot design procedures, [8].

This investigation, dealing with the workspace of a general 4R manipulator, is an attempt to generalise an algebraic algorithm that has been proposed in [7] for the workspace analysis of 3R manipulators. The 3R ring volume, which is traced by the reference point on the robot hand when the last three revolute joints are rotated, generates the 4R manipulator workspace when the mobility of the first revolute pair is further taken into account in revolving the 3R ring workspace about the first revolute joint axis. Thus, the geometrical properties of the revolution volume can be used in the workspace representation and an algebraic algorithm is deduced, in this paper, to describe and to check the main workspace characteristics.

THE 4R MANIPULATOR WORKSPACE

A general 4R manipulator can be sketched as in Fig.1, where the fixed base frame Z axis is assumed to be coincident to the first link joint axis Z_1, and the chain parameters are represented as

a_i (i=1,..,3): the distance between two consecutive joint axes, measured along their common normal;

a_4 : the distance of the hand reference point H from the last joint, measured along the normal to the last revolute axis;

d_i (i=1,...,4): the distance between two consecutive joint axes, measured along the latter of the joint axes;

θ_i (i=1,...,3): the twist angle between two consecutive joint axes, measured positive clockwise about their common normal;

α_i (i=1,...,4): the joint angle between two consecutive common normals, measured positive counter clockwise about the latter joint axis.

The parametric equation of a 4R manipulator workspace $W_{4R}(H)$ can be expressed as the position vector $\underline{x}$ of H with respect to the base frame in the form

$$\underline{x} = T\,\underline{x}_4 \tag{1}$$

where $\underline{x}_4 = [a_4,0,0,1]^t$ represents the position vector of H with respect to the link 4 frame and t is the transpose operator. The matrix T is given by

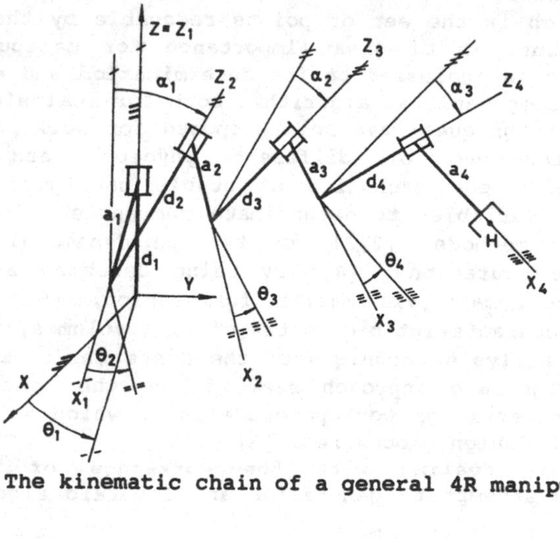

Fig.1. The kinematic chain of a general 4R manipulator.

$$T = T_1 T_2 T_3 T_4 \qquad (2)$$

in which T_i (i= 1,..., 4) is the transformation matrix from the frame attached to the i-th link to the frame on the (i-1)-th link (the frame link is considered 0). The expression of T_i can be given as

$$T_i = \begin{bmatrix} c\theta_i & -s\theta_i & 0 & a_{i-1} \\ s\theta_i c\alpha_{i-1} & c\theta_i c\alpha_{i-1} & s\alpha_{i-1} & d_i\, s\alpha_{i-1} \\ -s\theta_i s\alpha_{i-1} & c\theta_i s\alpha_{i-1} & c\alpha_{i-1} & d_i\, c\alpha_{i-1} \\ 0 & 0 & 0 & 1 \end{bmatrix} \qquad (3)$$

where $\sin\theta_i$ is represented with $s\theta_i$ as well as $\cos\theta_i$ with $c\theta_i$.
A 4R manipulator workspace $W_{4R}(H)$ can be evaluated when $\underline{x}$ is computed assuming all the joint angles as variables.
Referring to Fig.1 and assuming the base frame to be located with respect to the frame on link 1 so that it is α_0=0, d_1, a_0=0, thus the explicit formulation of the coordinates of points belonging to the workspace $W_{4R}(H)$ can be expressed from Eqs.(1)-(3) as

$$\begin{aligned} x = {} & a_4(\Phi_1\, c\theta_2 - \Phi_2\, s\theta_2)c\theta_1 - a_4(\Phi_1\, c\alpha_1\, s\theta_2 + \Phi_2\, c\alpha_1\, c\theta_2 \\ & -\Phi_3\, s\alpha_1)s\theta_1 + (a_1 + \delta_1\, c\theta_2 - \delta_2\, s\theta_2)c\theta_1 \\ & -(\delta_1\, c\alpha_1\, s\theta_2 + \delta_2\, c\alpha_1\, c\theta_2 + d_3\, s\alpha_1 + d_2\, s\alpha_1)s\theta_1 \end{aligned}$$

$$\begin{aligned} y = {} & a_4(\Phi_1\, c\theta_2 - \Phi_2\, s\theta_2)s\theta_1 + a_4(\Phi_1\, c\alpha_1\, s\theta_2 + \Phi_2\, c\alpha_1\, c\theta_2 \\ & -\Phi_3\, s\alpha_1)c\theta_1 + (a_1 + \delta_1\, c\theta_2 - \delta_2\, s\theta_2)s\theta_1 \\ & -(\delta_1\, c\alpha_1\, s\theta_2 + \delta_2\, c\alpha_1\, c\theta_2 + \delta_3\, s\alpha_1 + d_2\, s\alpha_1)c\theta_1 \end{aligned} \qquad (4)$$

$$\begin{aligned} z = {} & a_4(-\Phi_1\, s\alpha_1\, s\theta_2 - \Phi_2\, s\alpha_1\, c\theta_2 - \Phi_3\, c\alpha_1) - d_1\, s\alpha_1\, s\theta_2 \\ & -d_2\, s\alpha_1\, c\theta_2 + d_3\, c\alpha_1 + d_2\, c\alpha_1 + d_1 \end{aligned}$$

in which

$$\begin{aligned} \Phi_1 &= c\theta_3\, c\theta_4 - c\alpha_3\, s\theta_3\, s\theta_4 \\ \Phi_2 &= c\alpha_2\, s\theta_3\, c\theta_4 + c\alpha_2\, c\alpha_3\, c\theta_3\, s\theta_4 - s\alpha_2\, s\alpha_3\, s\theta_4 \\ \Phi_3 &= s\alpha_2\, s\theta_3\, c\theta_4 + s\alpha_2\, c\alpha_3\, c\theta_3\, s\theta_4 + c\alpha_2\, s\alpha_3\, s\theta_4 \\ \delta_1 &= a_2 + a_3\, c\theta_3 - d_4\, s\alpha_3\, s\theta_3 \\ \delta_2 &= a_3\, c\alpha_2\, s\theta_3 + d_3\, s\alpha_2 + d_4\, c\alpha_2\, s\alpha_3\, c\theta_3 + d_4\, s\alpha_2\, c\alpha_3 \\ \delta_3 &= -a_3\, s\alpha_2\, s\theta_3 + d_3\, c\alpha_2 - d_4\, s\alpha_2\, s\alpha_3\, c\theta_3 + d_4\, c\alpha_2\, c\alpha_3 \end{aligned}$$

The analytical complexity of Eqs.(4) justifies the utilization of a laborious algorithm expressed by the matrix formulation of Eqs.(1)-(3), and by adopting a suitable computer graphics technique for visualisation and volume calculation purposes. Moreover, the complexity of the workspace evaluation has given rise to the development of the aforementioned different approaches and computer aided techniques, mostly for synthesis purposes.

AN ALGEBRAIC ALGORITHM

The boundary surface of a $W_{4R}(H)$ solid hollow ring, [9], representing the volume of a 4R manipulator workspace, can be expressed as the envelope surface of the $W_{3R}(H)$ ring, when $W_{3R}(H)$ is rotated about the first joint axis of the 4R manipulator chain. Alternatively, $W_{4R}(H)$ can be considered as the union of all the $W_{3R}(H)$ which are traced because of the mobility of the first three revolute joins when the last one is considered frozen at different values of θ_4. Therefore, the surface contour can be evaluated as a geometrical envelope of the boundary surface of the toroidal surfaces which are traced by the contour points of $W_{3R}(H)$ and which are due to mobility of the first two joints through the angles θ_1 and θ_2. An algebraic formulation can be deduced by using the same algebraic approach that has been proposed in [7] to describe the envelope contour of a 3R manipulator workspace.

Thus, the $W_{4R}(H)$ can be described as a set of points whose radial r_1 and axial z_1 coordinates with respect to the link 1 frame are solved, with the hypotheses that $s\alpha_1 \neq 0$ and $C_1 \neq 0$, as

$$(r_1^2 + z_1^2 - A_1)^2 + (C_1 z_1 + D_1)^2 + B_1 = 0 \tag{5}$$

where

$$\begin{aligned} A_1 &= a_1^2 + r_2^2 + (z_2 + d_2)^2 \\ B_1 &= -4\, a_1^2\, r_2^2 \\ C_1 &= 2\, a_1 / s\alpha_1 \\ D_1 &= -2\, a_1 (z_2 + d_2)\, c\alpha_1 / s\alpha_1 \end{aligned} \tag{6}$$

The quantities r_2 and z_2 are the radial and axial distances of H from the Z_2 revolute axis of 4R manipulator. Their values can be calculated from the 3R manipulator workspace envelope contour and they can be expressed, with the hypotheses $s\alpha_2 \neq 0$ and $C_2 \neq 0$, as, [7]:

$$r_2 = \left[A_2 - z_2^2 + \frac{(C_2 z_2 + D_2) G_2 + F_2}{E_2} \right]^{1/2} \tag{7}$$

$$z_2 = \frac{-L_2 \pm \sqrt{Q_2}}{K_2 C_2} - \frac{D_2}{C_2}$$

where

$$\begin{aligned} A_2 &= a_2^2 + r_3^2 + (z_3 + d_3)^2 \\ B_2 &= -4 \ a_2^2 \ r_3^2 \\ C_2 &= 2 \ a_2 \ / \ s\alpha_2 \\ D_2 &= -2 \ a_2 \ (z_3 + d_3) \ c\alpha_2 \ / \ s\alpha_2 \end{aligned} \tag{8}$$

The quantities r_3 and z_3 are the coordinate distances of H from Z_3 axis and they are evaluable for each value of θ_4 from the geometry of the manipulator chain, Fig.1 or Eqs.(1)-(3), as

$$\begin{aligned} r_3 &= [(a_4 \ c\theta_4 + a_3)^2 + (a_4 \ s\theta_4 \ c\alpha_3 + d_4 \ s\alpha_3)^2]^{1/2} \\ z_3 &= d_4 \ c\alpha_3 - a_4 \ s\theta_4 \ s\alpha_3 \end{aligned} \tag{9}$$

The remaining coefficients can be computed in the form

$$\begin{aligned} E_2 &= -2 \, a_4 (d_3 \ s\alpha_3 \ c\theta_4 + a_3 \ s\theta_4) \\ F_2 &= 4 \ a_2^2 \ a_4 (a_4 \ s\alpha_3^2 \ s\theta_4 \ c\theta_4 + a_3 \ s\theta_4 - d_4 \ s\alpha_3 \ c\alpha_3 \ c\theta_4) \\ G_2 &= 2 \, a_2 \ a_4 \ c\alpha_2 \ s\alpha_3 \ c\theta_4 \ / \ s\alpha_2 \\ K_2 &= G_2^2 + E_2^2 \\ L_2 &= F_2 \ G_2 \\ Q_2 &= L_2^2 - K_2 \left(F_2^2 + B_2 \ E_2^2 \right) \end{aligned} \tag{10}$$

Hence, the workspace $W_{4R}(H)$ can be evaluated by means of Eqs.(5) to (10) as the union of all the toroidal surface workspaces represented by Eq.(5) and due to the joint angles θ_1 and θ_2, assuming θ_4 to be variable from 0 to 2π. Consequently, the boundary envelope of a 4R manipulator workspace can be thought as the geometrical envelope of the toroidal surface θ_4-family. An analytical expression can be obtained by solving Eq.(5) and its derivative with respect to θ_4. Hence, the axial z_1 and the radial r_1 coordinates of the $W_{4R}(H)$ envelope contour points, after some algebra, can be expressed in the form

$$r_1 = \left[A_1 - z_1^2 + \frac{(C_1 \ z_1 + D_1) G_1 + F_1}{E_1} \right]^{1/2} \tag{11}$$

$$z_1 = \frac{-L_1 \pm \sqrt{Q_1}}{K_1 C_1} - \frac{D_1}{C_1}$$

where

$$
\begin{aligned}
E_1 &= T_2 + V_2 \\
F_1 &= -2\ a_1^2\ T_2 \\
G_1 &= -2\,a_1\ z_2'\ c\alpha_1\ /\ s\alpha_1 \\
K_1 &= G_1^2 + E_1^2 \\
L_1 &= F_1\ G_1 \\
Q_1 &= L_1^2 - K_1\left(F_1^2 + B_1\ E_1^2\right)
\end{aligned}
\tag{12}
$$

in which the sign ' is the derivative operator with respect to θ_4. The derivatives z_2' and r_2' can be explicitly calculated from Eqs.(7) to give the expressions

$$
\begin{aligned}
z_2' &= \frac{(\pm\ 0.5\ Q_2'\ /\ \sqrt{Q_2} - L_2')\ K_2 + (L_2 \mp \sqrt{Q_2})\ K_2'}{C_2 K_2^2} - \frac{G_2}{C_2}, \\
T_2 &= \frac{(C_2 z_2 + D_2)(E_2 G_2' - G_2 E_2') + F_2' E_2 - F_2 E_2' + G_2 E_2 (C_2 z_2' + G_2)}{E_2^2} \\
&\qquad + E_2 - 2\ z_2\ z_2' \\
V_2 &= 2(z_2 + d_2)\,z_2'
\end{aligned}
\tag{13}
$$

where the derivatives of coefficients are calculated by means of Eqs.(8) and (10) as

$$
\begin{aligned}
E_2' &= -2\,a_4(a_3\ c\theta_4 - d_3\ s\alpha_3\ s\theta_4) \\
F_2' &= 4\ a_2^2\ a_4\left[a_4\ s\alpha_3^2\ (c^2\theta_4 - s^2\theta_4) + a_3\ c\theta_4 + d_4\ s\alpha_3\ c\alpha_3\ s\theta_4\right] \\
G_2' &= -2\,a_2\ a_4\ c\alpha_2\ s\alpha_3\ s\theta_4\ /\ s\alpha_2 \\
K_2' &= 2(G_2 G_2' + E_2 E_2') \\
L_2' &= F_2'\ G_2 + F_2\ G_2' \\
Q_2' &= 2\,L_2 L_2' - K_2'\left(F_2^2 + B_2\ E_2^2\right) - 2\,K_2\left[F_2\left(F_2' + E_2'\right) + B_2 E_2 E_2'\right]
\end{aligned}
\tag{14}
$$

Eqs.(11) through Eqs.(12) to (14) and (6) to (10) provide an algorithm for determining the $W_{4R}(H)$ workspace contour of a general 4R manipulator as a function of only the θ_4 angle.
It is to note that the boundary envelope equations do not represent the $W_{4R}(H)$ workspace contour when $\alpha_j = k\ \pi/2$, $k=1,...,4$ ($j=1,2$) since Eqs.(5) and (7) are deduced with the hypotheses, respectively, $s\alpha_1 \neq 0$ and $s\alpha_2 \neq 0$. In addition, for $c\alpha_j = 0$ $j=1,2$ the generating toroidal surface envelopes are contained inside the workspace contour whose cross-section is given by the largest and the smallest parallel circles of the generating tori. For $s\alpha_3 = 0$ the hollow ring is generated by a thin circular disk and for $c\alpha_3 = 0$ the workspace ring $W_{4R}(H)$ is generated by a circular torus workspace, which is due to the last two revolute joints in the chain and which is still tracing an envelope contour.

WORKSPACE CHARACTERISTICS AND ILLUSTRATIVE EXAMPLES

A cross-section can be determined with respect to the base frame through the expressions of r_1 and $z = z_1 + d_1$ by using Eqs.(11).
In Figs.2, 3 and 4 are reported some examples illustrating the characteristical geometry of the 4R manipulator workspace envelope contour through cross-sections. From Eq.(15) it is observable that d_1 does affect the workspace only shifting it up or down. Therefore it has been assumed to be zero in the examples.
Fig.2 illustrates the generation of a $W_{4R}(H)$ boundary contour as the envelope of toroidal surfaces which are traced by H for different θ_4 values. Some of these tori are drawn in the figure.

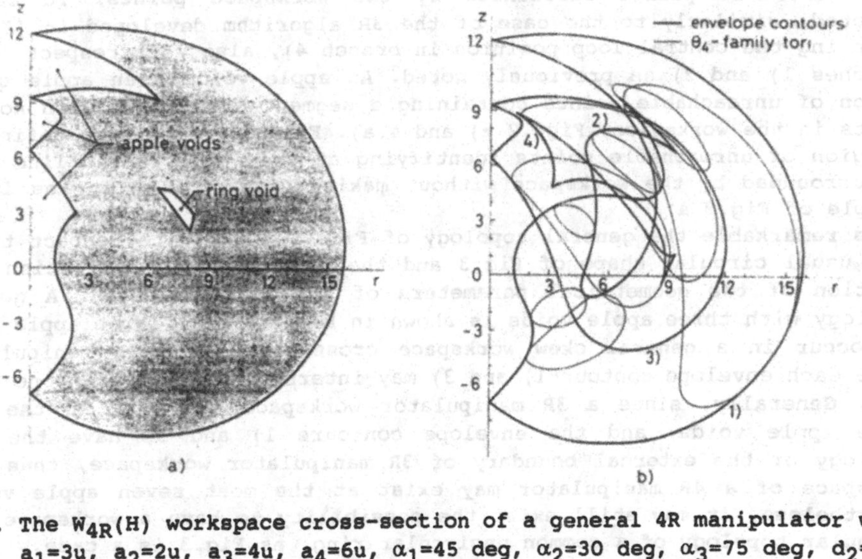

Fig.2. The $W_{4R}(H)$ workspace cross-section of a general 4R manipulator with a_1=3u, a_2=2u, a_3=4u, a_4=6u, α_1=45 deg, α_2=30 deg, α_3=75 deg, d_2=2u, d_3=1u, d_4=3u, where u is a unit length:
a) the workspace boundary contour;
b) all the envelope contours together with the toroidal surface θ_4-family of the generating $W_{3R}(H)$ boundary points.

Since Eqs.(11) show the same form of Eqs.(7), which express the $W_{3R}(H)$ boundary point coordinates, the contour envelope of 4R manipulator workspace shows geometric properties that are similar to those of a 3R manipulator workspace boundary contour. In particular, the envelope, which is represented in a cross-section plane in Figs.2 b), 3 b) and 4 b), consists of four branches, since each one of the two branches of the generating $W_{3R}(H)$ boundary contour, [7], generates two envelopes. This fact can be explained by taking into account that each point of the $W_{3R}(H)$ contour will trace a toroidal surface which is enveloped into two wrapping surfaces, that are 1) and 2) or 3) and 4) in the shown examples. Therefore, the $W_{4R}(H)$ boundary contour is given as the union of the branch segments delimiting unreachable and reachable regions of points, Figs.2 a), 3 a) and 4 a). In addition, the envelope branches 1) and 2) in Fig.2 b) do not delimit a void but they concern with a solid volume of reachable points since they contain the toroidal surfaces. On the contrary, the branches 3) and 4) in Fig.2 b) are generated in such a way that only the envelope contour 4) individuates an effective void when the central loop of the branch does not superimpose on a region of points which has been previously identificated as reachable by means of the branches 1) and 2). The central loop of this branch is the only one that can identify a void, since the other two are particular subregions whose geometry greatly differs from the typical "banana shape" of a ring void cross section, [7]. By means of the same interactive computer graphics reasoning the boundary contour of a workspace cross-section has been generated in the examples of Figs.3 and 4.

Two types of voids may exist in a manipulator workspace: a ring void and an apple void, as shown in Fig.2 a). A ring void is a ring shaped region of unreachable points surrounded by the workspace points. It can be detected, similarly to the case of the 3R algorithm developed in [7], by observing the central loop position in branch 4), also with respect to the branches 1) and 2) as previously noted. An apple void is an apple shaped region of unreachable points containing a segment of Z_1 axis when no hole exists in the workspace, Figs.2 a) and 4 a). Finally, a hole is defined as a region of unreachable points identifying at least a straight line which is surrounded by the workspace without making contact with it, as in the example of Fig.3 a).

It is remarkable the general topology of Figs.2 and 4 with respect to the more usual circular shape of Fig.3 and the great shape modification as a function of the geometrical parameters of the 4R open chain. A general topology with three apple voids is shown in Fig.2 and 4. Three apple voids may occur in a general skew workspace cross-section of 4R manipulators since each envelope contour 1) and 3) may intersect Z_1 axis with one apple void. Generally, since a 3R manipulator workspace may have at the most three apple voids, and the envelope contours 1) and 3) have the same topology of the external boundary of 3R manipulator workspace, thus in a workspace of a 4R manipulator may exist at the most seven apple voids. Nevertheless, it may still exist the possibility to have a workspace with a regular topology of a common uncircular ring, as Fig.3 is a case.

The effect of the manipulator chain parameters on the $W_{4R}(H)$ workspace, on its shape and characteristics, can be investigated directly by means of the proposed algorithm by simply changing the desired parameter and observing the envelope contour, as in the examples of Figs. 2, 3 and 4.

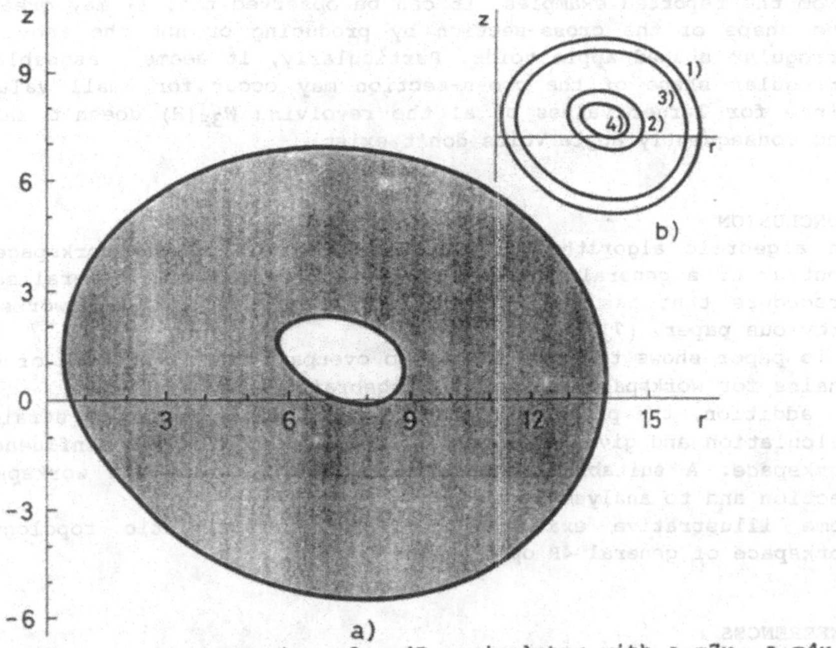

Fig.3. The workspace cross-section of a 4R manipulator with a_1=7u, a_2=4u, a_3=2u, a_4=0.5u, α_1=60 deg, α_2=60 deg, α_3=60 deg, d_2=d_3=d_4=1u, where u is a unit length:
a) the workspace boundary contour;
b) the envelope contours.

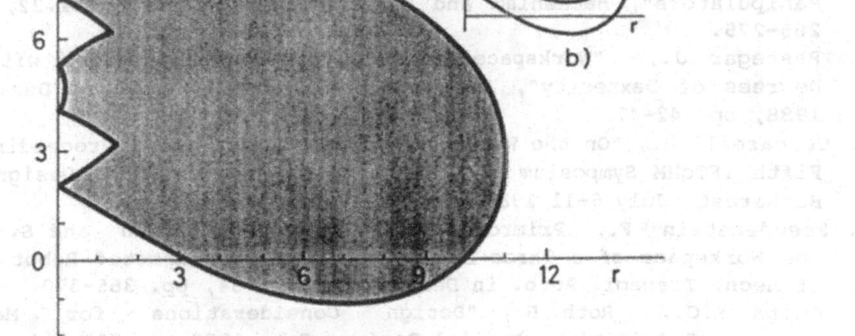

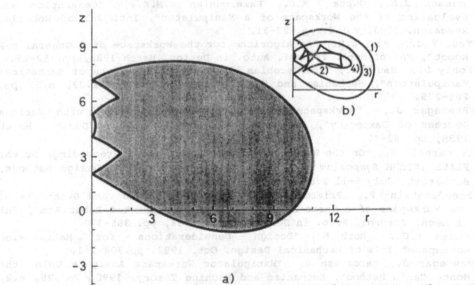

Fig.4. The workspace cross-section of a 4R manipulator with a_1=4u, a_2=3u, a_3=2u, a_4=1u, α_1=30 deg, α_2=60 deg, α_3=30 deg, d_2=3u, d_3=2u, d_4=1u, where u is a unit length:
a) the workspace boundary contour;
b) the envelope contours.

From the reported examples, it can be observed that a_1 may greatly affect the shape of the cross-section by producing or not the above mentioned irregular shaped apple voids. Particularly, it seems reasonable that the irregular shape of the cross-section may occur for small values of a_1, since for larger values of a1 the revolving $W_{3R}(H)$ doesn't intersect Z_1 and consequently apple voids don't exist.

CONCLUSION

An algebraic algorithm for the determination of the workspace boundary contour of a general 4R manipulator is proposed as a generalisation of a procedure that has been adopted for the 3R manipulator workspace in a previous paper, [7].

This paper shows the possibility to overpass the limit case of general 3R chains for workspace analysis by algebraic means, [8].

In addition, the proposed algebraic formulation permits a straightforward calculation and gives a direct insight of the link size influences on the workspace. A suitable computation allows to trace the workspace cross-section and to analyse directly its characteristics.

Some illustrative examples show the characteristic topology of the workspace of general 4R open chains.

REFERENCES

1. Kumar A., Waldron K.J., "The Workspace of a Mechanical Manipulator", Jnl of Mechanical Design, July 1981, pp. 665- 672.
2. Lee T.W., Yang D.C.H., "On the Evaluation of Manipulator Workspace", Jnl of Mech. Transmi. Auto. in Design, March 1983, pp. 70-77.
3. Hansen J.A., Gupta K.C., Kazerounian S.M.K., "Generation and Evaluation of the Workspace of a Manipulator", Int. Jnl of Robotics Research, 1983, Vol.2, pp. 22-31.
4. Tsai Y.C., Soni A.H., "An Algorithm for the Workspace of a General n-R Robot", Jnl of Mech. Transmi. Auto. in Design, March 1983, pp. 52-57.
5. Kohly D., Hsu M.S., "The Jacobian Analysis of Workspace of Mechanical Manipulators", Mechanism and Machine Theory, 1987, Vol.22, n.3, pp. 265-275.
6. Rastegar J., "Workspace Analysis of 4R Manipulators with Various Degrees of Dexterity", Jnl of Mech. Transmi. Auto. in Design, March 1988, pp. 42-47.
7. Ceccarelli M., "On the Workspace of 3R Robot Arms", Proceedings of the Fifth IFToMM Symposium on Linkages and Computer Aided Design Methods, Bucharest, July 6-11 1989, Vol.II-1, pp. 37-46.
8. Freudenstein F., Primrose E.J.F., "On the Analysis and Synthesis of the Workspace of a Three-Link, Turning-Pair Connected Robot Arm", Jnl of Mech. Transmi. Auto. in Design, Sept. 1984, pp. 365-370.
9. Gupta K.C., Roth B., "Design Considerations for Manipulator Workspace" Jnl. of Mechanical Design, Oct. 1982, pp.704-711.
10. Rastegar J., Fardanesh B., "Manipulator Workspace Analysis Using the Monte Carlo Method", Mechanism and Machine Theory, 1990, Vol.25, n.2, pp.233-239.

DISCRETE CALCULATION OF SET POINTS IN MECHANISM MOTION CONTROL

K. Nevala
VTT Automation, Oulu, Finland
and
T. Leinonen
University of Oulu, Oulu, Finland

ABSTRACT

Set points calculation of the mechanism described in the paper are based on a formal method in which the target points of the sequences of motions, the profile functions between the target points and the time of execution between the target points are determined off-line. The internal task of the calculation is to produce the set points between the target points of the motional sequences in either on-line or off-line calculation. Sudden recoil movements at great mass loads result in injurious vibration in the actuators to be controlled. To prevent such vibrations a profile function has been derived which provides a smooth path. The continuous polynomial function was formed of an n-th degree polynomial in which the degree of the polynomial together with the time of execution determines the maximum acceleration of the movement.

1. INTRODUCTION

In a paper center winder, the motion of the winding arms is controlled by means of the hydraulic cylinders attached to the arm mechanism, see Fig 2. Winder control based on a model requires knowledge of the behavior of the winding mechanism, since control accuracy depends on how well the control model describes the behavior of the equipment. From the viewpoint of control, the most important models describing behavior are the kinematic model and the dynamic model. The kinematic model of the winding mechanism describes the movement of the winding arms of the paper roll, when the travel of the piston of the hydraulic cylinder moving the arms is known. The inverse kinematic model is needed when the midpoint of the roll attached to the transfer arms needs to be moved according to the desired profile function. Target angles are determined for the sequences of arm movements and profile functions are fitted between them, after which inverse kinematics is used to solve the set points for the position, velocity and acceleration of the hydraulic cylinder piston [1, 2, 3].

2. DEFINITION OF COUNTERS IN DISCRETE CALCULATION

Time is described by counters in discrete presentation. The set point counter k_{as}, control interval counter k_s and control cycle counter k are defined. By using a notation similar to Fig. 1, the value of the set point counter in the part performance j can be calculated,

$$k_{as} = \frac{k - k_s}{n_k} - \sum_{l=1}^{j-1} (2N_l) \tag{1}$$

in which

$$n_k = \frac{T_{as}}{T_s}, \; n_k \in Z_+, \; T_{as} \in R, \; T_s \in R \tag{2}$$

$$\sum_{l=1}^{j-1} (2N_l) = 2N_1 + 2N_2 + ... + 2N_{j-1}, \; k \in Z_+, \; k \in \left[0, n_k \sum_{j=1}^{J} (2N_j)\right]$$

$$k_s \in Z_+, \; k_s \in [0, n_k - 1], \; k_{as} \in Z_+, \; k_{as} \in [0, 2N_j - 1], \; N_j \in Z_+$$

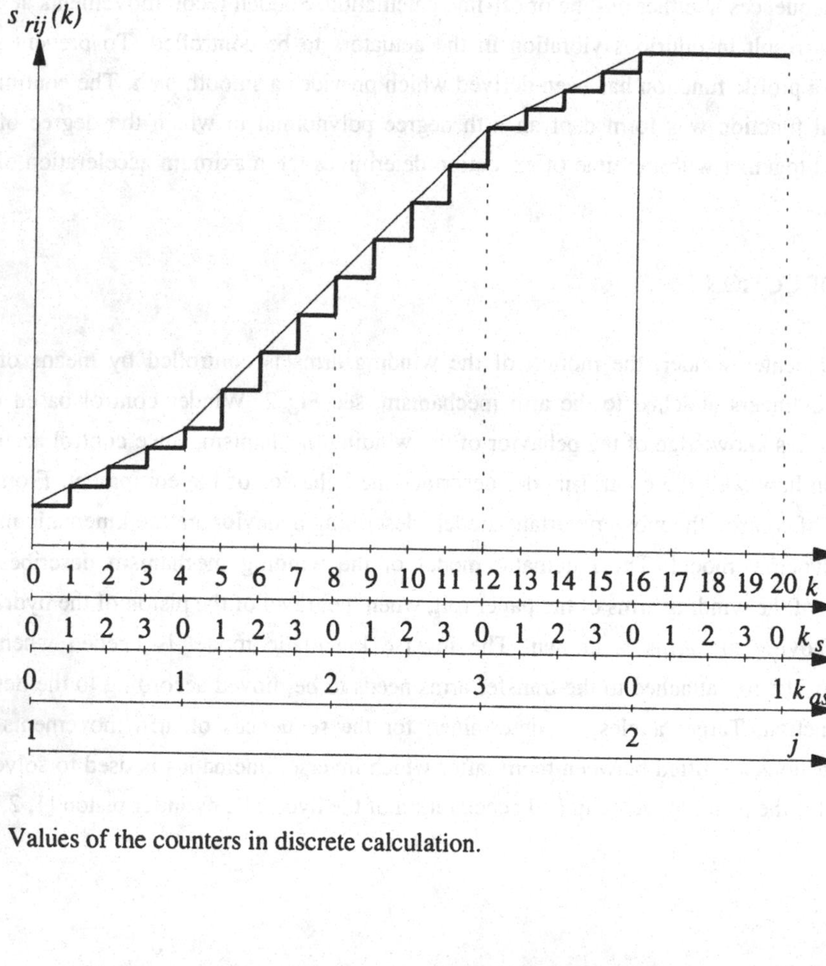

Fig. 1. Values of the counters in discrete calculation.

With the formula (1) it is possible to solve the value for the control cycle counter k in the part performance j,

$$k = \left[k_{as} + \sum_{l=1}^{j-1} (2N_l) \right] n_k + k_s \ . \tag{3}$$

The total time of execution t_t of the sequence of motions can be calculated,

$$t_t = T_{as} \sum_{j=1}^{J} \left(2N_j \right) . \tag{4}$$

The timings of the set point and regulation calculation required by the control are based on the use of counters which have been synchronized to function at the clock frequency of an external interrupt. The clock frequency determines which control interval T_s is used, while the length T_{as} of the set point interval is determined by multiplying the length of the control interval by the empirically chosen integer n_k. The control takes care of activating the set point calculation before the previously set point is achieved. In this way the functions of the set point calculation and of the counters used by the control are linked together. The set point counter k_{as} is incremented according to the time of execution of the function fitted between the target points. Tabulation of the target points of the motional sequence, calculation of the performance index j as well as the selection of the set point function on the basis of the function fitted between the performance index and the target points are connected to the functions of the set point counter. The control period counter k_s is incremented according to the ratio n_k of the set point interval T_{as} and control interval T_s, and the control cycle counter k is incremented according to the total execution time of the sequence of motions

3. SET POINT CALCULATION FOR THE MECHANISM

The cycle that corresponds to paper winding consists of a control sequence which involves the control of the movements of the winding arms and nip force control either by regulating the pressures of the hydraulic cylinders fitted to the arms or the force acting on the arms due to nip force between the rolls. In the mechanism shown in Fig. 2, the number of hydraulic cylinders is $I = 3$ and the number of part performances is $J = 6$ in the cycle corresponding to paper winding. Of the hydraulic cylinders, $i = 1, 2$ are to be synchronized and the cylinder $i = 3$ produces the movement which corresponds to the growth of the paper roll during nip load control.

The set point angles $\theta_{rij}(k_{as})$ of the winding arms for the arm i during the paper winding cycle in the part performance j are calculated,

$$\left[\theta_{rij}(k_{as})\right]_{\substack{i=1\ldots3\\ j=1\ldots6}} = \left[\overline{\theta}^{T}_{rij}(k_{as})\right]_{i=1\ldots3} \tag{5}$$

In the formula (5),

$$\overline{\theta}_{rij}(k_{as}) = \overline{\theta}_{i(j-1)} + diag\left[\overline{F}_{ij}(k_{as})\right]\left(\overline{\theta}_{ij} - \overline{\theta}_{i(j-1)}\right) \tag{6}$$

in which

$$diag\left[\overline{F}_{ij}(k_{as})\right] = diag\left[\overline{F}_{jO}(k_{as})\right]\Psi_{iO} + diag\left[\overline{F}_{jC}(k_{as})\right]\Psi_{iC} + diag\left[\overline{F}_{jP}(k_{as})\right]\Psi_{iP}\,. \tag{7}$$

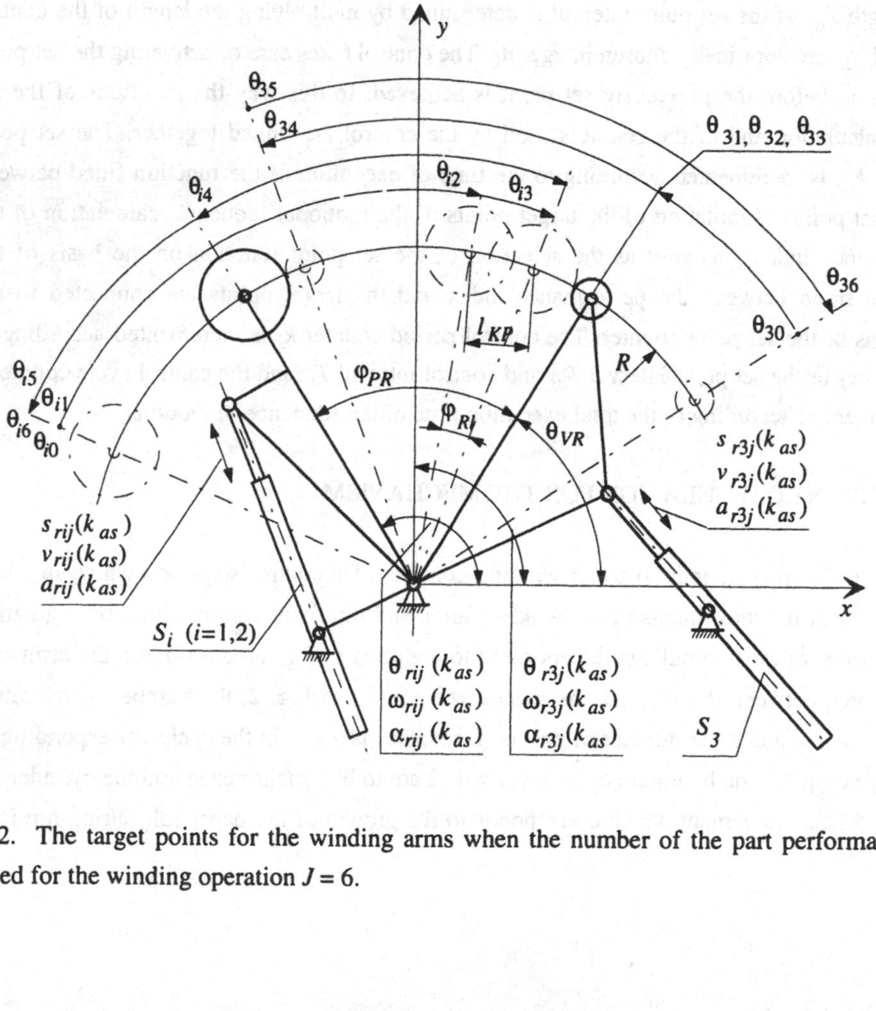

Fig. 2. The target points for the winding arms when the number of the part performances defined for the winding operation $J = 6$.

The set point calculations for the movements of the winding arms are carried out using the n-th degree polynomial. A profile function defined for a constant velocity is used to drive the rolls together. The functions are fitted between the target angles determined for the winding arms, see Fig. 2. The profile functions $F_j(k_{as})$ between the target points are calculated using the formulas [3]

$$\overline{F}_{jO}(k_{as}) = [1]_{j=1\ldots J} \tag{8}$$

$$\overline{F}_{jC}(k_{as}) = \left[\frac{k_{as}}{2N_J}\right]_{j=1\ldots J} \tag{9}$$

$$\overline{F}_{jP}(k_{as}) = \left[\frac{n_{j1}\left(\frac{k_{as}}{N_J}-1\right)^{2n_j+3} - n_{j2}\left(\frac{k_{as}}{N_J}-1\right)^{2n_j+1} + n_{j3}\left(\frac{k_{as}}{N_J}-1\right) + n_{j4}}{2n_{j4}}\right]_{j=1\ldots J} \tag{10}$$

In the formula (10),

$$n_{j1} = n_j(2n_j+1)$$
$$n_{j2} = 2n_j^2 + 5n_j + 3$$
$$n_{j3} = 4n_j^2 + 8n_j + 3$$
$$n_{j4} = 4n_j(n_j+1)\ .$$

The selection matrix for the profile function is determined as follows:

$$\Psi_{ix} = diag\left\{[\psi_j]_{j=1\ldots J}\right\}\ ,\ x \in \{O, C, P\} \tag{11}$$

The selection matrices of the profile function for the winding arms $i = 1,2$
to be synchronized are

$$\Psi_{iO} = diag[1\ \ 0\ \ 0\ \ 0\ \ 0\ \ 1], \Psi_{iC} = diag[0\ \ 0\ \ 1\ \ 0\ \ 0\ \ 0], \Psi_{iP} = diag[0\ \ 1\ \ 0\ \ 1\ \ 1\ \ 0]$$

The selection matrices of the profile function for the winding arm $i = 3$ are

$$\Psi_{3O} = diag[0\ \ 1\ \ 1\ \ 0\ \ 1\ \ 0], \Psi_{3C} = diag[\overline{0}], \Psi_{3P} = diag[1\ \ 0\ \ 0\ \ 1\ \ 0\ \ 1]$$

When the kinematics of the winder mechanism are known and the initial values given are the winding start angle θ_{VR}, the winding angle φ_{PR} and the drive-together length between the rolls l_{KP}, see Fig. 2, the target points θ_{ij} for the arms of the mechanism can be calculated:

$$\left[\theta_{i(j-1)}\right]_{\substack{i=1...3\\j=1...6}} = \begin{bmatrix} \theta_{1\max} & \theta_{1\max} & \theta_{KP} & \theta_{RI} & \theta_{PR} & \theta_{1\max} \\ \theta_{2\max} & \theta_{2\max} & \theta_{KP} & \theta_{RI} & \theta_{PR} & \theta_{2\max} \\ \theta_{3\min} & \theta_{VR} & \theta_{VR} & \theta_{VR} & \theta_{PI} & \theta_{PI} \end{bmatrix} \tag{12}$$

$$\left[\theta_{ij}\right]_{\substack{i=1...3\\j=1...6}} = \begin{bmatrix} \theta_{1\max} & \theta_{KP} & \theta_{RI} & \theta_{PR} & \theta_{1\max} & \theta_{1\max} \\ \theta_{2\max} & \theta_{KP} & \theta_{RI} & \theta_{PR} & \theta_{2\max} & \theta_{2\max} \\ \theta_{VR} & \theta_{VR} & \theta_{VR} & \theta_{PI} & \theta_{PI} & \theta_{3\min} \end{bmatrix}. \tag{13}$$

In the formulas (12) and (13),

$$\theta_{KP} = \theta_{VR} + \varphi_{RI} + \varphi_{KP},\ \theta_{RI} = \theta_{VR} + \varphi_{RI},\ \theta_{PR} = \theta_{VR} + \varphi_{PR},\ \theta_{PI} = \theta_{VR} + \varphi_{PR} - \varphi_{RI} \tag{14}$$

in which

$$\varphi_{KP} = 2\arcsin\left(\frac{l_{KP}}{2R}\right),\ \varphi_{RI} = 2\arcsin\left(\frac{d_{RI} + D_{VR}}{4R}\right) \tag{15}$$

The number of set point intervals $2N_j$ and the degree of the polynomial n_j in the part performance j,

$$2\bar{N}_j = 2\begin{bmatrix} N_1 \\ N_2 \\ \cdot \\ \cdot \\ \cdot \\ N_6 \end{bmatrix},\quad \bar{n}_j = \begin{bmatrix} n_1 \\ n_2 \\ \cdot \\ \cdot \\ \cdot \\ n_6 \end{bmatrix}. \tag{16}$$

When the set point angles $\theta_{rij}(k_{as})$ of the winding arms are known, the set points $s_{rij}(k_{as})$ for the position of the piston of the hydraulic cylinder can be calculated using the conversion formula [3]

$$\left[s_{rij}(k_{as})\right]_{\substack{i=1...3\\j=1...6}} = \left[\sqrt{p_i - b_i\cos\left[\beta_i - \theta_{rij}(k_{as})\right]} - l_{i2}\right]_{\substack{i=1...3\\j=1...6}} \tag{17}$$

in which

$\theta_{rij}(k_{as})$ is an element of the matrix $\left[\theta_{rij}(k_{as})\right]$.

The target points s_{ij} of the piston position that correspond to the arm angles θ_{ij} can be calculated as

$$\left[s_{ij}\right]_{\substack{i=1...3\\ j=1...6}} = \left[\sqrt{p_i - b_i \cos\left[\beta_i - \theta_{ij}\right]} - l_{i2}\right]_{\substack{i=1...3\\ j=1...6}} \tag{18}$$

in which

θ_{ij} is an element of the matrix $\left[\theta_{ij}\right]$.

Figure 3 presents the set points for cylinder piston position, velocity and acceleration in the cycle corresponding to paper winding, in which the number of part performances is $J = 6$.

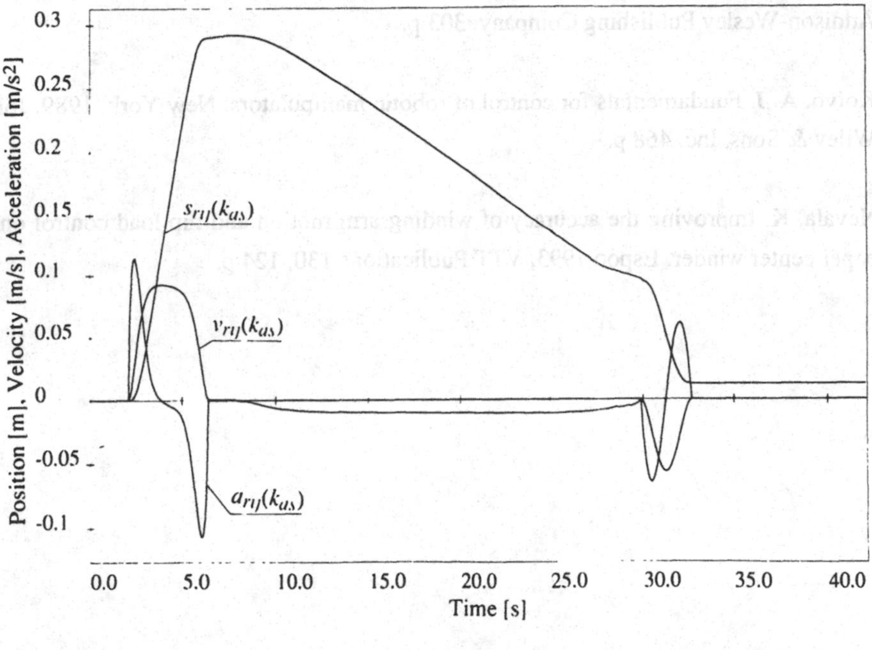

Fig. 3. The cycle corresponding to paper winding, in which the number of part performances is $J = 6$. The figure presents the set points for cylinder piston position $s_{rij}(k_{as})$, velocity $v_{rij}(k_{as})$ and acceleration $a_{rij}(k_{as})$.

4. CONCLUSIONS

The set point calculation of motion control means the calculation of target points for position, velocity and acceleration in accordance with functions determined for the motion. In digital control the calculation is performed in a discrete form. Sudden recoil movements at great mass loads result in injurious vibration in the actuators to be controlled. To prevent such vibrations a profile function has been derived which provides a smooth path. The smoothness of the path requires a function which has a continuous derivative with relation to time. The continuous polynomial function was formed of an n-th degree polynomial in which the degree of the polynomial together with the time of execution determines the maximum acceleration of the movement. To optimize the performance of the microprocessor used in the digital control, the set point calculation and control calculation are assigned priorities on various levels and their functioning is synchronized by means of counters.

REFERENCES

[1] Craig, J. J. Introduction to robotics, mechanics & control. Massachusetts, 1986, Addison-Wesley Publishing Company. 303 p.

[2] Koivo, A. J. Fundamentals for control of robotic manipulators. New York, 1989. John Wiley & Sons, Inc. 468 p.

[3] Nevala, K. Improving the accuracy of winding arm motion and nip load control on a paper center winder. Espoo 1993, VTT Publications 130. 124 p.

Part II
Mechanics II

ANALYSIS OF THE ACCELERATION CHARACTERISTICS OF MANIPULATORS

A. Bowling and O. Khatib
Stanford University, Stanford, CA, USA

Abstract

The study of the acceleration properties at the end effector is important in the analysis, design, and control of robot manipulators. In previous efforts aimed at addressing this problem, the end-effector acceleration has been treated as a vector combining both the linear and angular accelerations. The methodology presented in this article provides characterizations of these two different types of accelerations and describes the relationship between them. This work is an extension of our previous studies on manipulator inertial and acceleration properties. The treatment relies on the *ellipsoid expansion* model, a simple geometric approach to efficiently analyze end-effector accelerations. Results of the application of this analysis to the PUMA 560 manipulator are discussed.

1 Introduction

The end-effector acceleration properties are important performance indicators for manipulator systems. Several frameworks have been proposed for the analysis of end-effector accelerations resulting in different measures of acceleration characteristics. One such measure is the isotropic acceleration: the amount of acceleration achievable in or about every direction.

When determining this measure for systems with many degree of freedom (DOF) it is necessary to account for linear as well as angular accelerations. The non-homogeneity between linear and angular accelerations has been often ignored and these two quantities have been treated as component of a single vector.

Yoshikawa [1] proposed the Dynamic Manipulability Ellipsoid (DME). Here linear and angular accelerations are analyzed as components of a single vector. Scaling factors are used to bring the differing magnitudes into range with each other. In practice these scaling factors are difficult to determine for many DOF systems and their selection is somewhat arbitrary. In addition, the method yields a conservative estimate of the isotropic acceleration.

In an earlier study, we have proposed [2] the hyper-parallelepiped of acceleration as a description of end-effector accelerations. The exact isotropic acceleration can be found by inscribing a circle/sphere within the acceleration hyper-parallelepiped. However, this process proves difficult when analyzing systems with many DOF. In addition, linear and angular accelerations have been also treated as components of the same vector and weighting factors have been used to deal with their different nature.

The work presented in this article is a pursuit of the previous study with the aim of explicitly addressing the differences between linear and angular acceleration. This work has resulted in the *ellipsoid expansion* approach. In this approach linear and angular accelerations are considered as two separate entities.

2

The ellipsoid expansion method provides a simple geometric solution to the system of inequalities associated with the bounds on actuator torques. It allows an exact determination of the isotropic linear acceleration, the angular acceleration, and the relationship between them. The results from application of this approach to the PUMA 560 manipulator are presented.

2 Torque/Acceleration Relationship

In this section the model which is the basis for the *ellipsoid expansion* model is derived. The derivation begins with the joint space equations of motion,

$$A(\mathbf{q})\ddot{\mathbf{q}} + \mathbf{b}(\mathbf{q}, \dot{\mathbf{q}}) + \mathbf{g}(\mathbf{q}) = \tau. \tag{1}$$

In equation(1), $\mathbf{q}$ is the vector of n joint coordinates, $A(\mathbf{q})$ is the kinetic energy matrix, $\mathbf{b}(\mathbf{q}, \dot{\mathbf{q}})$ is the centrifugal and Coriolis force vector, $\mathbf{g}(\mathbf{q})$ is the gravity force vector, and τ is the vector of joint torques. The end-effector linear and angular velocities $\mathbf{v}$ and ω are related to the joint velocities $\dot{\mathbf{q}}$ by

$$\vartheta \triangleq \begin{bmatrix} \mathbf{v} \\ \omega \end{bmatrix} = J_0(\mathbf{q})\dot{\mathbf{q}}; \tag{2}$$

where J_0 is the basic Jacobian matrix[1]. Differentiating equation (2) with respect to time yields, $\ddot{\mathbf{q}} = J_0^{-1}\dot{\vartheta} - J_0^{-1}\dot{J}_0\dot{\mathbf{q}}$, which is substituted into equation (1) to obtain

$$AJ_0^{-1}\dot{\vartheta} \; - \; AJ_0^{-1}\dot{J}_0\dot{\mathbf{q}} \; + \; \mathbf{b} \; + \; \mathbf{g} \; = \; \tau. \tag{3}$$

The bounds on τ can be written as

$$-\tau_{bound} \leq \tau \leq \tau_{bound}. \tag{4}$$

To normalize the bounds on τ, we introduce the diagonal matrix N with components $N_{ii} = \frac{1}{\tau_{bound(i)}}$. Now using equations (3) and (4) yields,

$$-\mathbf{1} \; \leq \; N(AJ_0^{-1}\dot{\vartheta} \; - \; AJ_0^{-1}\dot{J}_0\dot{\mathbf{q}} \; + \; \mathbf{b} \; + \; \mathbf{g}) \; \leq \; \mathbf{1}; \tag{5}$$

where $\mathbf{1}$ is a vector of length n with each element equal to one. Using equation (2) the above equation can be rewritten as,

$$\tau_{lower} \leq [E_v \; E_\omega] \begin{bmatrix} \dot{\mathbf{v}} \\ \dot{\omega} \end{bmatrix} + \begin{bmatrix} \vartheta^T M_1 \vartheta \\ \vdots \\ \vartheta^T M_n \vartheta \end{bmatrix} \leq \tau_{upper}; \tag{6}$$

where M_i are symmetric matrices and,

$$[E_v \; E_\omega] \; = \; NAJ_0^{-1}; \tag{7}$$

$$\begin{bmatrix} \vartheta^T M_1 \vartheta \\ \vdots \\ \vartheta^T M_n \vartheta \end{bmatrix} \; = \; N(\mathbf{b} \; - \; AJ_0^{-1}\dot{J}_0\dot{\mathbf{q}}); \tag{8}$$

$$\tau_{upper} \; = \; \mathbf{1} - N\mathbf{g}; \tag{9}$$

$$\tau_{lower} \; = \; -\mathbf{1} - N\mathbf{g}. \tag{10}$$

[1]for notational simplicity $(\mathbf{q})$ and $(\mathbf{q}, \dot{\mathbf{q}})$ will be omitted in subsequent equations

3

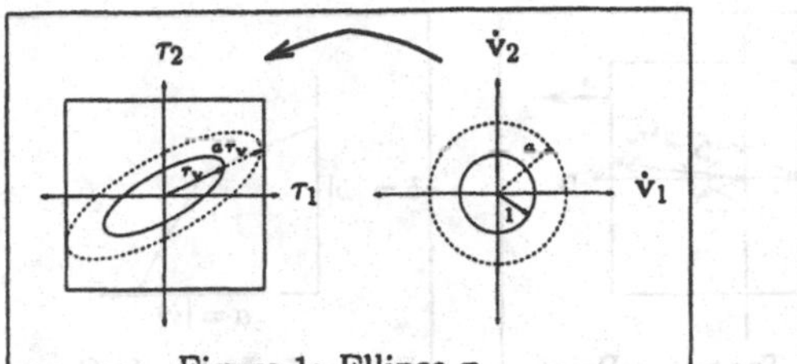

Figure 1: Ellipse τ_v.

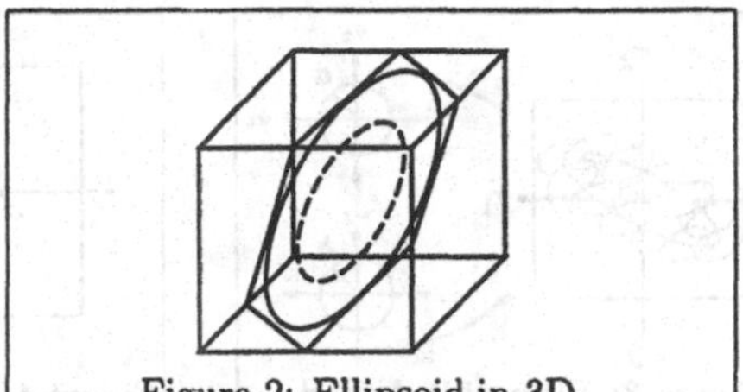

Figure 2: Ellipsoid in 3D.

Finally, the governing equation for this analysis is

$$\tau_{lower} \leq E_v\dot{v} + E_\omega\dot{\omega} + \begin{bmatrix} \vartheta^T M_1 \vartheta \\ \vdots \\ \vartheta^T M_n \vartheta \end{bmatrix} \leq \tau_{upper} \tag{11}$$

The separation of linear and angular accelerations in equation (11) is motivated by the need to analyze each of them independently. The Coriolis/centrifugal terms can also be analyzed to determine the general effect end-effector velocities have on the isotropic accelerations.

3 Ellipsoid Expansion Model

3.1 Acceleration Terms

Consider a manipulator where $\dot{\omega} = \dot{q} = 0$. Equation (11) becomes,

$$\tau_{lower} \leq E_v\dot{v} \leq \tau_{upper}. \tag{12}$$

Geometrically isotropic linear acceleration can be represented as a sphere with some radius a, $\dot{v}^T\dot{v} = a^2$. The value of a represents the magnitude of achievable linear acceleration in every direction. The relationship between acceleration and torque is $\tau_v = E_v\dot{v}$. With respect to $\dot{v}$, the above relationship is an over constrained system of equations, whose unique solution is given by

$$\dot{v} = E_v^+ \tau_v; \tag{13}$$

where E_v^+ is the left inverse[2] of E_v. Using equation (13), the sphere of acceleration is transformed into,

$$\tau_v^T (E_v E_v^T)^+ \tau_v = a^2. \tag{14}$$

Since E_v is at most of rank three, this surface is a hyper-cylinder in n-dimensions. However, we may eliminate the cylindrical portion of this surface because we only want to consider torques that contribute to end-effector acceleration. This corresponds to excluding vectors in the null space of E_v^+. The remaining surface is an ellipsoid in three or less dimensional subspace.

The isotropic acceleration is determined by expanding/contracting the ellipsoid (14), changing a, until it lies within and is tangent to the torque bounds. The bounds are represented as an n-dimensional hypercube whose center is shifted from the origin by the gravity effect, i.e. Ng. In Figure 1 this process is shown for a simple case and Figure 2 shows a more general case. The dashed ellipse in both figures corresponds to $a = 1$. Note that only the vectors associated with the tangent points, $2n$ points, need to be examined.

[2] given by the left pseudo-inverse $E_v^+ = (E_v^T E_v)^{-1} E_v^T$.

4

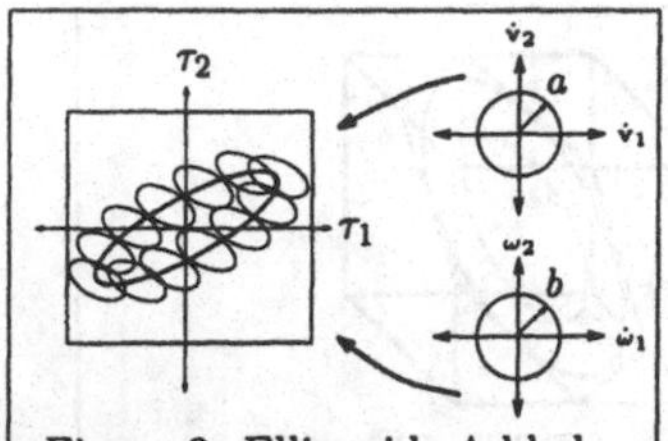

Figure 3: Ellipsoids Added.

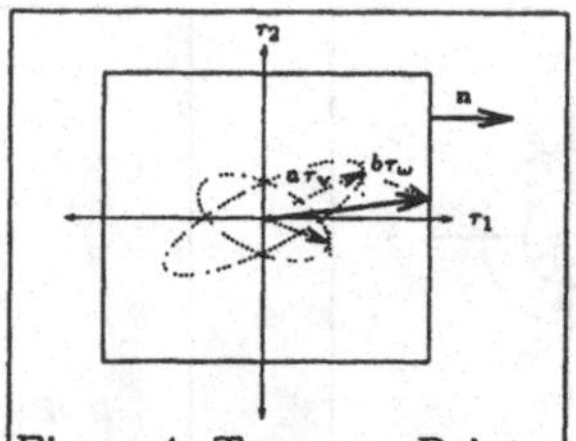

Figure 4: Tangency Point.

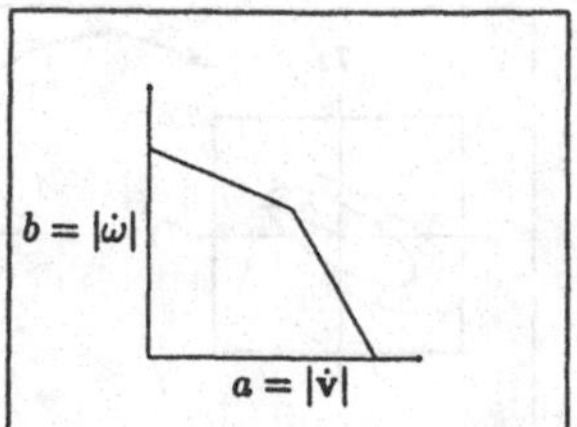

Figure 5: Isotropic Curve

Let us now consider a manipulator where $\dot{\mathbf{q}} = 0$; $\tau_{lower} \leq E_v\dot{\mathbf{v}} + E_\omega\dot{\omega} \leq \tau_{upper}$. Just as in the linear case, the isotropic angular acceleration is represented as a sphere with some radius b, $\dot{\omega}^T\dot{\omega} = b^2$. This sphere is mapped into torque space using the relationship $\tau_\omega = E_\omega\dot{\omega}$, resulting in a second ellipsoid,

$$\tau_v^T(E_vE_v^T)^+\tau_v = a^2 \quad \text{and} \quad \tau_\omega^T(E_\omega E_\omega^T)^+\tau_\omega = b^2. \tag{15}$$

To insure that the sum of $a\tau_v$ and $b\tau_\omega$ remains within the bounds, all possible combinations of these two vectors must be considered. This process is represented as a mapping of one ellipsoid onto every point on the surface of the other ellipsoid, graphically illustrated in Figure 3. As before it is only necessary to examine the vectors representing the tangency points between the resulting surface and the bounding box. These points are found by adding vectors representing the tangency point on each surface corresponding to the same boundary; Figure 4. The possible combinations of $a\tau_v$ and $b\tau_\omega$ yielding a resultant which touches a bound is described by the following equation;

$$\mathbf{n}_i \cdot (a\tau_{v_i} + b\tau_{\omega_i}) = \tau_{upper_i} \text{ or } \tau_{lower_i} \tag{16}$$

where n_i is the vector normal to the boundary plane i; Figure 4. $2n$ of these relationships are found for the $2n$ boundaries. The equation from this set which yields the smallest value for a given b or vice versa determines a curve which is piecewise linear, Figure 5. A discontinuity in the curve occurs when the limiting boundary changes.

3.2 Centrifugal and Coriolis Terms

Another aspect of the analysis involves determining the effect of end-effector velocities on the isotropic accelerations. A specific velocity is analyzed as a boundary shift, just as the gravity terms were. For more general information, surfaces representing all possible velocities, $\mathbf{v}^T\mathbf{v} = c^2$ and $\omega^T\omega = d^2$, can be mapped through the Coriolis/centrifugal terms yielding a hypothetical surface. However, determining an analytic expression for this mapping is difficult. Thus we resort to scaling to try and obtain information from the Coriolis centrifugal terms,

$$\vartheta^T W \vartheta = e^2 \tag{17}$$

where W is an invertible diagonal positive definite matrix used to scale linear and angular velocities. A reasonable scaling factor can be found in the a-b isotropicity curve as the slope of one of the line segments. A change of variables, $\mathbf{y} = W^{1/2}\vartheta$ yields,

$$\mathbf{y}^T\mathbf{y} = e^2 \qquad \text{and} \qquad \tau_c = \begin{bmatrix} \mathbf{y}^T(W^{-1/2})^T M_1 W^{-1/2}\mathbf{y} \\ \vdots \\ \mathbf{y}^T(W^{-1/2})^T M_n W^{-1/2}\mathbf{y} \end{bmatrix}. \tag{18}$$

This mapping produces a surface in torque space, Figure 6. The vectors of the resulting surface representing the tangency point to a particular boundary has the largest component in the direction

5

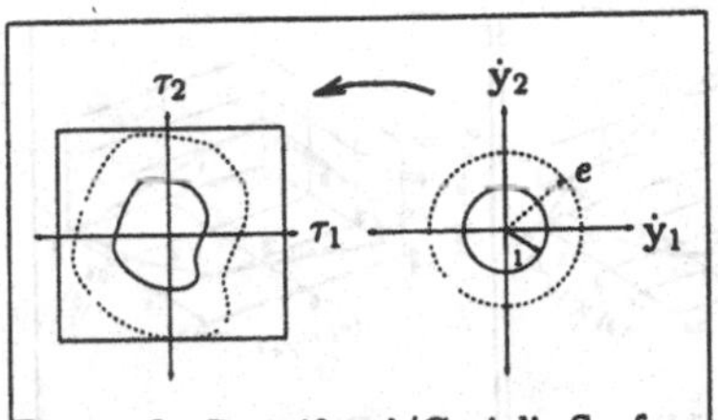

Figure 6: Centrifugal/Coriolis Surface

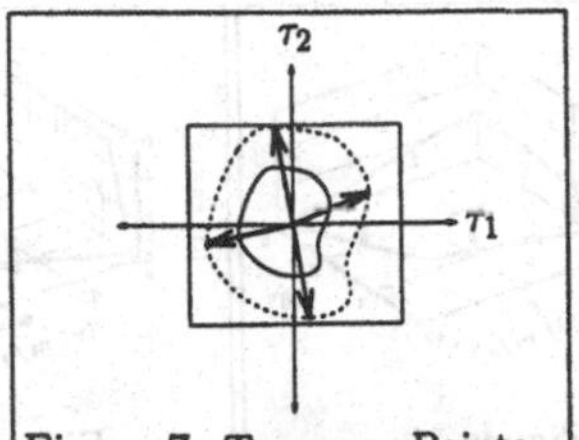

Figure 7: Tangency Points.

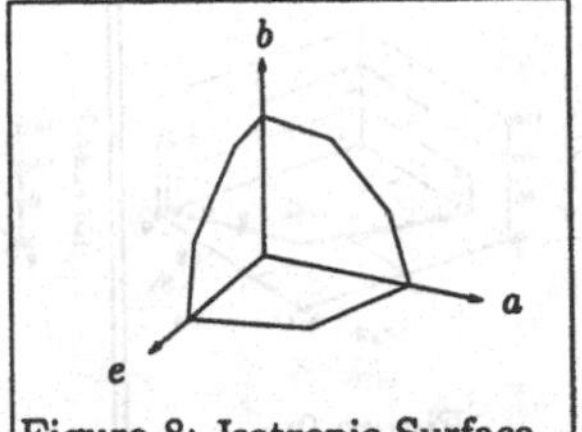

Figure 8: Isotropic Surface.

of the vector normal to that boundary plane; Figure 7. By analyzing each element of the centrifugal/Coriolis we can easily determine which **y** yields the largest, (most positive), and smallest, (most negative), values for that component. However, the details of this process will not be discussed here. These vectors are added to the acceleration torque vectors yielding a relationship between the three magnitudes;

$$\mathbf{n}_i \cdot (a\tau_{v_i} + b\tau_{\omega_i} + e\tau_{c_i}) = \tau_{upper_i} \text{ or } \tau_{lower_i}. \tag{19}$$

The representation of this information is a three-dimensional surface with planar facets. In Figure 8 only the intersection of the surface with the three coordinate planes is shown because most of the interesting information lies in these planes. In the a-b plane the information shown in Figure 5 is duplicated; i.e. the relationship between isotropic linear and angular acceleration at zero end-effector velocity. Also consider the point along the e axis where a and b are driven to zero. This point gives the state of end-effector velocity when, in some direction, all actuator torque is being used to compensate for Coriolis/centrifugal and gravity forces. Given that the velocity must lie on the surface $\vartheta^T W \vartheta = e^2$, this point marks the first instance when the manipulator will exhibit poor response to a controller commanded acceleration.

3.3 Other Information

Information concerning how much achievable end-effector accelerations vary in different directions at a given configuration can be obtained as the condition numbers of the matrices E_v and E_ω, $\kappa(E_v)$ and $\kappa(E_\omega)$ respectively. Information about the limiting actuator(s), the actuator limiting the isotropic accelerations, is also readily available. Depending on the manipulator, increasing the most limiting actuator torque(s) may increase the isotropic accelerations.

4 Application to the PUMA 560

The approach presented above is illustrated here on the PUMA 560. The motor torque bounds are found by subtracting the breakaway torques from the maximum torques as given in [3];

$$\tau_{bound} = \{1.45Nm, 1.70Nm, 1.62Nm, 0.30Nm, 0.27Nm, 0.26Nm\}.$$

Figures 9 through 12 show the results of analysis of the PUMA 560 for different manipulator parameters. For all cases the weighting matrix, W, is identity. In each figure the label beside each line segment indicates the limiting actuator for that segment. Many comparisons can be made between these figures but only a few will be discussed below.

In Figure 9 isotropicity information for the wrist point of the PUMA 560 at a well conditioned configuration in the workspace is shown. The condition numbers for this configuration are, $\kappa(E_v) = 2.8$ and $\kappa(E_\omega) = 1.1$. The shape of the curve in the $|\dot{\mathbf{v}}|$-$|\dot{\omega}|$ plane indicates that linear and angular accelerations are nearly decoupled at the wrist point; i.e. Λ_0 is nearly block diagonal. If the two

6

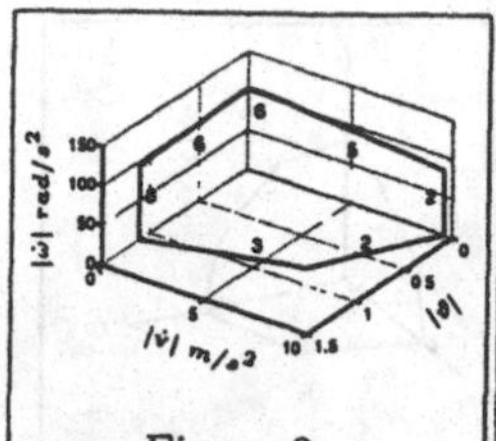

Figure 9:

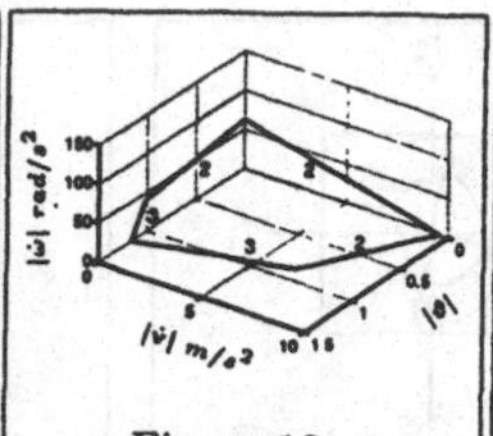

Figure 10:

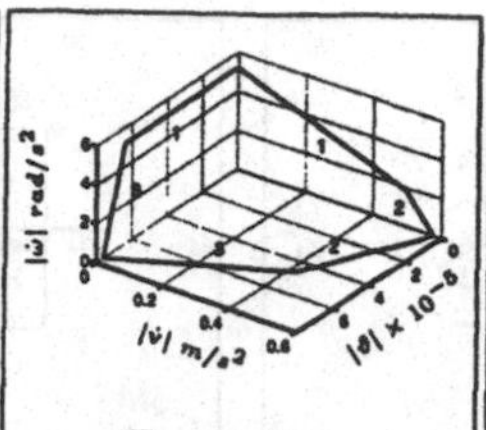

Figure 11:

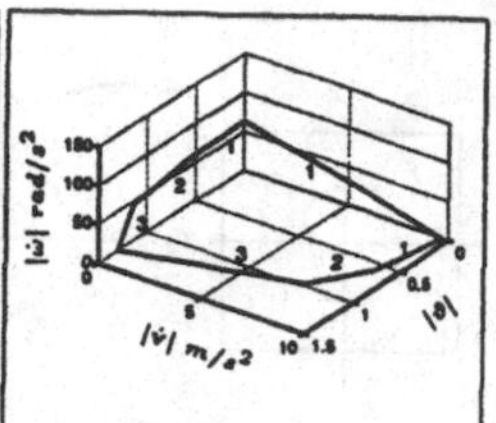

Figure 12:

accelerations were completely decoupled the curve would form a rectangle with the two coordinate axes. The same behavior is apparent in the $|\dot{\omega}|$-$|\vartheta|$ plane.

Figure 10 shows the isotropicity information for a different operational point about 6 inches out from the wrist point at the same configuration as in Figure 9. The shape of the curves show much more coupling between each of the elements. The condition numbers for this configuration are $\kappa(E_v) = 2.8$ and $\kappa(E_\omega) = 2.1$. Figure 11 shows the operational point of Figure 10 for a poorly conditioned configuration. Note the decrease in magnitudes of the isotropic accelerations. The condition numbers for this configuration are $\kappa(E_v) = 54$ and $\kappa(E_\omega) = 26$. Finally, Figure 12 shows information for the same operational point and configuration as in Figure 10 with the maximum torques on joints 2 and 3 increased by 15%. In Figure 10 actuators 2 and 3 limit the isotropic accelerations. Note a small increase in the isotropic acceleration and a much larger sustainable end-effector velocity before the isotropic accelerations are driven to zero, as compared to Figure 10. Also notice that joint 1 has become the limiting actuator in the $|\dot{v}|$-$|\dot{\omega}|$ plane which explains the small increase in the accelerations for that plane. Next one would try also increasing the joint 1 actuator torque. It is also interesting to note that increasing the torques worsened the condition numbers; $\kappa(E_v) = 3.1$ and $\kappa(E_\omega) = 2.1$.

5 Conclusion

We have presented an approach based on the *ellipsoid expansion* model for the analysis of linear and angular accelerations, given the limits on actuator torque capacities, at a given configuration. This model has been shown to provide a simple geometric solution for determining these accelerations. The result of the analysis is a surface which completely describes the dependency between the isotropic accelerations associated with linear and angular motions and end effector velocities. Condition numbers have been used to provide a measure of the extent of magnitude variations for the linear and angular accelerations in different directions. The results obtained with the PUMA 560 manipulator, illustrate the effectiveness of this approach for acceleration analysis. This approach is being extended to redundant manipulators.

References

[1] Yoshikawa T.; *Dynamic Manipulability of Robot Manipulators*, Proc. 1985 IEEE International Conference on Robotics and Automation, St. Louis, 1985, pp. 1033-1038.

[2] Khatib, O. and Burdick, J.; *Dynamic Optimization in Manipulator Design: The Operational Space Formulation*, The International Journal of Robotics and Automation, vol.2, no.2, 1987, pp. 90-98.

[3] Armstrong, B., Khatib, O., and Burdick, J.; *The Explicit Model and Inertial Parameters of the PUMA 560 Arm*, Proceedings IEEE Intl. Conference on Robotics and Automation, vol.1, pp. 510-518, 1986.

DYNAMICS OF ELASTIC ROBOT ARMS WITH VARYING LENGTH

SYSTEMATIC MODELLING, SIMULATION AND CONTROL

D. Söffker
University of Wuppertal, Wuppertal, Germany

INTRODUCTION

In the last years, modeling, simulation, and control of flexible structures have made an essential progress. This is especially stimulated by the requirements of space operations. For this working field, flexible lightweight robots can enhance the range of the work space of space robots very well if the length of the robot arm is variable. The space requirements can only be fulfilled by flexible structures, but the speed of motion induces large inertia forces leading to strong vibrations and stability problems of the flexible structure. A lot of theoretical and practical contributions concerning modeling, control and realization of elastic robots with rotatory joints have appeared until now, cf. [1,2,3,4,5]. Only a few works deal with an elastic robot of varying length. Chalhoub and Ulsoy discuss modeling, control, and experimental setup of a telescopic robot [6], Wang and Wei consider elastic vibrations of a very flexible translating beam and examine stabilizing and destabilizing effects of the varying length [7]. Pan introduces a general method building up finite prismatic beams using external constraints and adding stiffness terms to linear equations considering effects of axial forces [8]. He tests his formalism using the experimental setup according to Chalhoub. Concerning this topic actual works deal with control design for the principal (mostly simplified) structural model, cf. [14], or regard higher nonlinear effects of the mechanical model [13,14]. In this contribution both aspects are examined. The developed beam element modules can easily be used to build up elastic links of varying length. Here a controller structure is extended, which is well-known for its robust and stationary accurate behaviour because of using integral parts [15]. Considering that the end-effector position is not measurable, an extended state observer scheme is used to reconstruct the positions and deflections using the deflections measured by strain-gauges. This design concept is tested out first with a simulation example of an inverted but flexible pendulum [10], and therefore will only be shortly introduced. The observer-based control strategy does not use the developed model consciously.

CONCEPTIONAL ASPECTS

Concerning technical realizations two different kinds of robot constructions with links of varying length can be distinguished: robots with translating links, where only the part between joint and end-effector is of interest [7], and robots with telescopic links, where the elements are sliding into one another. As far as modeling is concerned, principally three different kinds of modeling and discretization procedures are well known [12]: Method of global (assumed) modes, Rigid Finite Element Method (RFE), and the Finite Element Method (FEM). With the requirements of modeling both technical realizations with one concept, considering frictional effects in the prismatic links also of the telescopic robot possibly, modeling the links as a beam of higher order, considering geometric nonlinear stiffening and weakening effects, considering the influence of the reference motion to the elastic vibrations of the beam, the FEM is chosen, using a cartesian joint based coordinate system. Beam element modules represent the mechanical elements of every desired modeling task: the link of the translating beam is modeled as a sequence of a few equal beam modules, and the link of the telescopic robot by using two different modules for the link parts and the prismatic joint part.

MODELING

Using the definitions for the joint-based coordinate system (G) as depicted in fig. 1, the kinematics of an infinitesimal mass element in the infinitesimal rigid disc can be described by

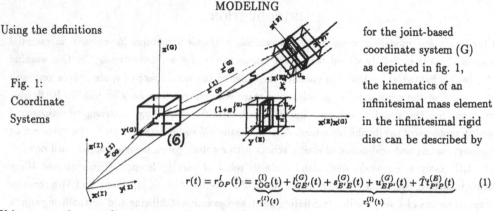

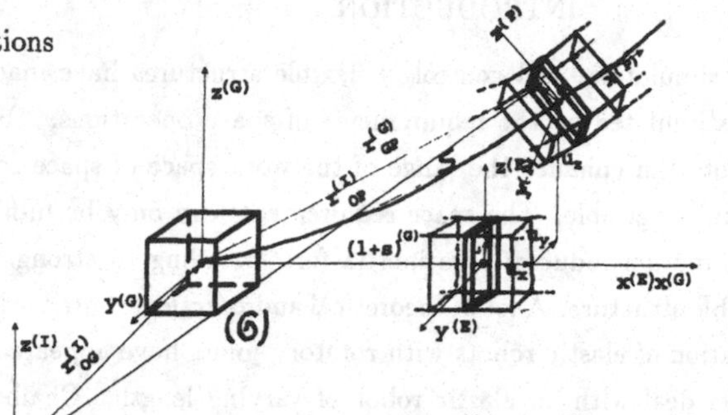

Fig. 1: Coordinate Systems

$$r(t) = r^I_{OP}(t) = \underbrace{\mathrm{r}^{(\mathrm{I})}_{\mathrm{OG}}(\mathrm{t})}_{r_1^{(I)}(t)} + \underbrace{l^{(G)}_{GE'}(t) + s^{(G)}_{E'E}(t) + u^{(G)}_{EP'}(t) + \hat{T}t^{(E)}_{P'P}(t)}_{\mathrm{r}_2^{(\mathrm{I})}(\mathrm{t})} \tag{1}$$

Using a cartesian coordinate system, the strain of the neutral axis is

$$\epsilon = \mathrm{u}'_\mathrm{x} + \frac{1}{2}(\mathrm{u}'^2_\mathrm{y} + \mathrm{u}'^2_\mathrm{z}) \tag{2}$$

with the elastic displacements $\mathrm{u_x, u_y, u_z}$, assuming

$$\mathrm{u}'_\mathrm{x} \approx \mathrm{u}'^2_\mathrm{y} + \mathrm{u}'^2_\mathrm{w} \quad \longrightarrow \quad \mathrm{u}'^2_\mathrm{x} \approx 0 \ , \tag{3}$$

and neglecting also other terms of third and higher order. Using the skrew symmetric relation between the curvatures $\kappa_1, \kappa_2, \kappa_3$ and $T^T T'$, $T' = \frac{\partial}{\partial_s} T$ using the rotational matrix of Cardanus angles $T = T(\phi, \psi, \theta)$, the nonlinear expressions of second order for the curvatures are

$$\kappa_\mathrm{x} = -\alpha' + \mathrm{u}'_\mathrm{y}\mathrm{u}'_\mathrm{z} \quad \kappa_\mathrm{y} = \mathrm{u}''_\mathrm{y} - \mathrm{u}'_\mathrm{y}\mathrm{u}''_\mathrm{x} - \mathrm{u}''_\mathrm{y}\mathrm{u}'_\mathrm{x} - \alpha'\mathrm{u}'_\mathrm{z} \quad \kappa_\mathrm{z} = -\mathrm{u}''_\mathrm{z} + \mathrm{u}''_\mathrm{z}\mathrm{u}'_\mathrm{x} + \mathrm{u}'_\mathrm{z}\mathrm{u}''_\mathrm{x} - \alpha'\mathrm{u}'_\mathrm{y} \quad , \tag{4}$$

where α represents the angle of torsion of the beam. Calculating $\delta\epsilon, \delta\kappa_x, \delta\kappa_y, \delta\kappa_z$ by the rules of variation theory the virtual potential work of the deformed beam results to

$$\delta \mathrm{V_e} = \int_0^l (\delta\epsilon \mathrm{N} + \delta\tau_{\mathrm{xy}}\mathrm{Q_y} + \delta\tau_{\mathrm{xz}}\mathrm{Q_z} + \delta\kappa_\mathrm{x}\mathrm{M_x} + \delta\kappa_\mathrm{y}\mathrm{M_y} + \delta\kappa_\mathrm{z}\mathrm{M_z})\mathrm{dx} \tag{5}$$

with the normal load N, the shears $\mathrm{Q_x, Q_y}$, the moments $\mathrm{M_x, M_y, M_z}$, the strain ϵ, the shear strains τ_{xy}, τ_{xz}, and the curvatures κ_x, κ_y und κ_z. The virtual work of a mass element of the beam is

$$\delta W_m = \int\int\int \rho(\delta r^{(I)}_{OP})^T \ddot{r}^{(I)}_{OP} dm \tag{6}$$

and can be calculated using (1) and derivates. Here $\tilde{T}$ denotes the relation of the coordinate systems (I) and (G) by $r_2^{(I)} = \tilde{T} r_2{}^{(G)}$. Using the above mentioned variations, neglecting shear deformation, assuming a stiff disc perpendicular to the neutral axis (fig. 1) and homogenious material with $\rho = const.$ the eq.(6) results to

$$\delta W_m = \int\int\int \rho \left(\delta r_{OG}{}^{(I)^T} + (\delta l^{(G)^T} + \delta s^{(G)^T} + \delta u^{(G)^T} + \delta t^{(G)^T} \hat{T}^T) T^T\right) \tag{7}$$

$$\left(\ddot{r}^{(I)}_{OG} + \ddot{T}(l^{(G)} + s^{(G)} + u^{(G)} + \hat{T}t^{(G)}) + 2\dot{T}(\dot{l}^{(G)} + \dot{s}^{(G)} + \dot{u}^{(G)} + \dot{\hat{T}}t^{(G)}) + T(\ddot{l}^{(G)} + \ddot{s}^{(G)} + \ddot{u}^{(G)} + \ddot{\hat{T}}t^{(G)}\right)$$

The main idea is to arrange the terms in the way that similar to the usual structural dynamics matrices mechanical element matrices M_e, D_e, K_e, f_e can be built, which additionally depend on the nodes variables. To make the previous equation integrable, terms of the kind $\hat{T}t^{(E)}, \dot{\hat{T}}t^{(E)}, \ddot{\hat{T}}t^{(E)}$ have to be reformulated using only kinematic dependent matrices N_i, as

$$\hat{T}t^{(E)} = N_0 u' + N_{1a} u' + N_{1b}\beta_\alpha + N_{2a} u'^2 + N_{2b}\beta_\alpha{}^2 + N_3 u'_m \quad fg.. \tag{8}$$

Details to this extensive procedure are given in [9]. Knowing the principal geometry of the beam, this integrations can be done resulting to only length-dependent geometry matrices. Ordering the appearing scalar expressions and separating the variables by using a set $f(x)$ of Hermité polynomials for the bending coordinates of the beam and a set $g(x)$ of linear shape functions for the torsion and the axial displacement the coupled variables for the free-free supported beam are decoupled, the equations are discretized as usual. Normally the time-independent shape function products can be integrated, here because of the higher order state dependent product terms like

$$\int_0^l (\delta u_y^T(x) u_z(x,t), u_x(x,t))\, dx = \delta u_y^T \underbrace{\int_0^l (f_y(x) f_z^T(x) u_z(t) g_x^T(t))\, dx}_{\tilde{F}}\, u_x \tag{9}$$

arise, for a coupling matrix $\tilde{F}$ as one part of the element matrices. Redefining the products [9], it is also possible to preintegrate the time independent parts to get state independent products, which must be multiplied by states

$$\tilde{F} = u_{z1}\int_0^l f_{z1} f_y g_x^T\, dx + u_{z2}\int_0^l f_{z2} f_y g_x^T\, dx + u_{z3}\int_0^l f_{z3} f_y g_x^T\, dx + u_{z4}\int_0^l f_{z4} f_y g_x^T\, dx \tag{10}$$

$$\tilde{F} = \sum u_{zi} \underbrace{\int (f_{zi}(t) f_y(x) g_x(x))\, dx}_{D_{f_i fg}} = \sum (u_{zi} D_{f_i fg}) \quad , fg. \tag{11}$$

In this way for all coupling terms of the elastic coordinates and combinations of shape functions and node variables typical products arise, which can be easily implemented in a systematic scheme. Using this length-dependent matrices as a library, stiffness, damping and mass matrices for a beam element can be built up in every time step of numerical calculation due to the nonlinear consideration. The global system matrices are found by adding the specified element matrices in the usual way, so the global system arises as

$$M_g \ddot{u}_g + D_g \dot{u}_g + (K_g + K_{gg}) u_g = f_{g1} + f_{g2} + f_{gg} \tag{12}$$

with the mass matrix $M_g(u, l_{ei})$, the damping matrix $D_g(u, \omega, l_{ei})$, the stiffnes matrix $K_g(u, \omega, l_{ei})$ and vectors f_{g1}, f_{g2} resulting from translational and rotational motion of the joint, and the matrices of gravitational effects f_{gg} (K_{gg}). The integrated matrices and the complete scheme can be found in [9]. The advantage of the developed modeling procedure is, that the modeling can be easily done by handling only a few modules, organizing all terms and products by a given scheme [9].

OBSERVER-BASED CONTROL

The control task is to damp out motion induced elastic vibrations. Under the assumption that the kinematics of the considered elastic beam resulting from the motion of the preceding link of the robot is given, this motion can also be used for control of the elastic vibrations by changing the desired reference motion. Here special care must be taken in order not to compensate the rigid motion.

Based on a simplified linear joint-based mechanical description of the beam

$$M_{red}\ddot{u}_g + D_{red}\dot{u}_g + K_{red}u_g = f_{red}(t) + Nn(t) + b_{red,fic} \tag{13}$$

with reference motion independent matrices $M_{red}, D_{red}, K_{red}, f_{red}$ all unmodeled dynamics, coupling terms of the reference motion etc. are interpreted as external disturbances $Nn(t)$ to the beam which excite the system. The task of the control scheme is to compensate this 'disturbances' with respect to the end-effector displacements [10]. As a control input the angular acceleration will be used, so a fictitious input vector $b_{red,fic}$ has to be built describing the effects of the additional kinematic control input to the reduced system. The main idea is the extension of the reduced model by a fictitious subsystem, to consider these external disturbances. For the linear observer design, the extended system-description

$$\begin{bmatrix} \dot{x} \\ \dot{v} \end{bmatrix} = \begin{bmatrix} A_{red} & N_1 H_1 \\ 0 & 0 \end{bmatrix} \begin{bmatrix} x \\ v \end{bmatrix} + \begin{bmatrix} b_{red,fic} \\ 0 \end{bmatrix}, A_{red} = \begin{bmatrix} 0 & I \\ -M_{red}^{-1}K_{red} & -M_{red}^{-1}D_{red} \end{bmatrix} \tag{14}$$

in state-space with the state vector x and an extension v is used. A_{red} represents the linear reduced mechanical model and N_1H_1 couples a subsystem to the model. Calculating observer gain matrices L_x, L_v usual methods of control theory, Ricatti design eg., can be used. The loop corresponding to the subsystem represents an additional integral feedback loop. Details are given in [10]. For control the same procedure is used calculating the controller gains K_x, K_v. Here the available information of the measurements and the observer estimates is used exciting the integral subsystem as long as estimations x, v are not equal zero. As a base calculating control gains the description

$$\begin{bmatrix} x \\ v \end{bmatrix} = \begin{bmatrix} A_{red} & 0 \\ Q & 0 \end{bmatrix} \begin{bmatrix} x \\ v \end{bmatrix} + \begin{bmatrix} b_{red,fic} \\ 0 \end{bmatrix} \tag{15}$$

is used, where Q represents an input matrix for excitation of the extended controller by the estimates x in the observer-based closed loop [10]. In this way the well known disturbance rejection philosophy of Davision [15] is enlarged by using the estimations of the introduced extended observer. Extended applications of the observer technique are given in [11], the observer technique combined with static compensation in [4,5], and combined with dynamic compensation in [10].

DYNAMIC STUDIES

The quality of the model will be shown by a simulation of a very flexible telescopic beam of three elements for space applications. The geometric parameters of the system are: $I_1 = 6.110^{-6}m^4, I_2 = 5.710^{-6}m^4, I_3 = 5\,310^{-6}m^4, E = 7\cdot 10^{10}\frac{N}{m^2}, A_1 = 0.124410^{-3}m^4, A_2 = 0.124410^{-3}m^4, A_3 = 0.124410^{-3}m^4, \rho = 2.610^{-3}\frac{Ns^2}{m^2}$. The prismatic joints are considered as a link built up of the two link elements, so the geometry is built up with five finite beam modules of the lengths $l_1 = 3.6m, l_2 = 1.3m, l_3 = 2.0m, l_4 = 1\,6m, l_5 = 3.3$. The whole length is $11\,8m$, the weight is $48kg$.

Without reference motion, the coupling effects are stimulated by the following simulations: *a)* a0) No load, a1) Constant axial load $F_{ax1} = 10000N$, a2) $F_{ax2} = -10000N$, (End-effector initial conditions: $u_{x,ef}(t=0) = -5\ 10^{-3}m, u_{y,ef}(t=0) = u_{z,ef}(t=0) = 10\cdot 10^{-3}m$) *b)* Rotating eccentricity at the end-effector $m_{ec}\epsilon = 2\ 10^3 Ns^2, \Omega_{rot} = 0.1\frac{rad}{s}$, (End-effector initial conditions: $u_{x,ef}(t=0) = u_{y,ef}(t=0) = u_{z,ef}(t=0) = 0m$) *c)* Varying length of the beam c1) $v_{ef} = 0.5\cdot 10^{-3}\frac{m}{s}$, c2) $v_{ef} = -0.5\cdot 10^{-3}\frac{m}{s}$ (End-effector initial conditions: $u_{x,ef}(t=0) = -2\cdot 10^{-3}m, u_{y,ef}(t=0) = u_{z,ef}(t=0) = 20\cdot 10^{-3}m$):

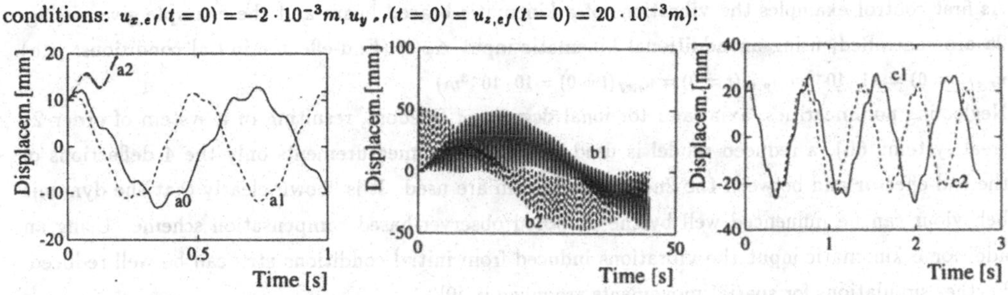

Fig. 2 a,b,c: Coupling effects: a) Axial load effects b) Coupled bendings c) Length Influence

Fig. 2a demonstrates the effects of axial loads, buckling (a1) and stiffening (a2). Fig. 2b demonstrates the coupling between the bendings and it shows an energy exchange between both bendings (b1,b2). The in- and decreasing influence of the length variability is shown in Fig. 2c. With the planar reference motion $\varphi(t)$ of the rotating joint given by

$$\dot{\varphi}(t) = \frac{\Omega}{T_0}\left(t - \frac{T_0}{2\pi}\sin(2\pi\frac{t}{T_0})\right), t \leq T_0, \dot{\varphi}(t) = \Omega_{max}, t > T_0, \Omega_{max} = 0.1\frac{rad}{s}, T_0 = 10s, v_{ef} = 0.1\frac{m}{s} \quad (16)$$

combined with the variable length the end-effector displacement results (End-effector initial conditions: $u_{x,ef}(t=0) = u_{y,ef}(t=0) = u_{z,ef}(t=0) = 0$) as

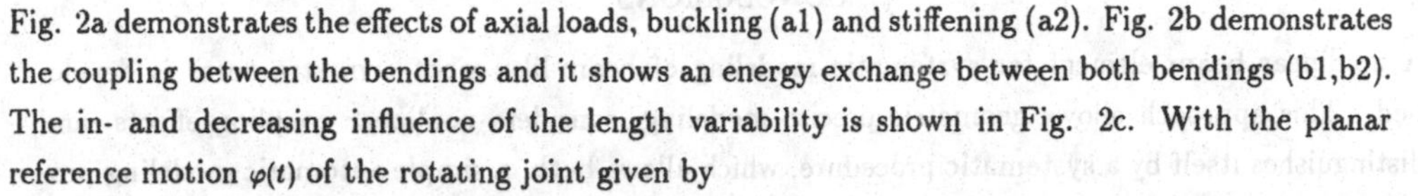

Fig. 3 a,b,c: End-effector displacement

with the cases of *a)* No payload (a1) linear, a2) nonlinear), *b)* Payload = 100*kg* (b1) linear, b2) nonlinear). *c)* Axial displacement, Payload = 100*kg* (c1) linear, c2) nonlinear).

In the figures it is shown, that for this example the destabilizing effects resulting from nonlinear modeling are important. In case 3a the relatively small static axial displacement results from stiffening effects, in case of 3b the additonal large dynamic bending displacement is mainly induced by weakening effects.

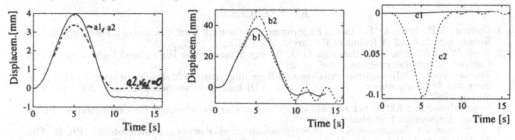

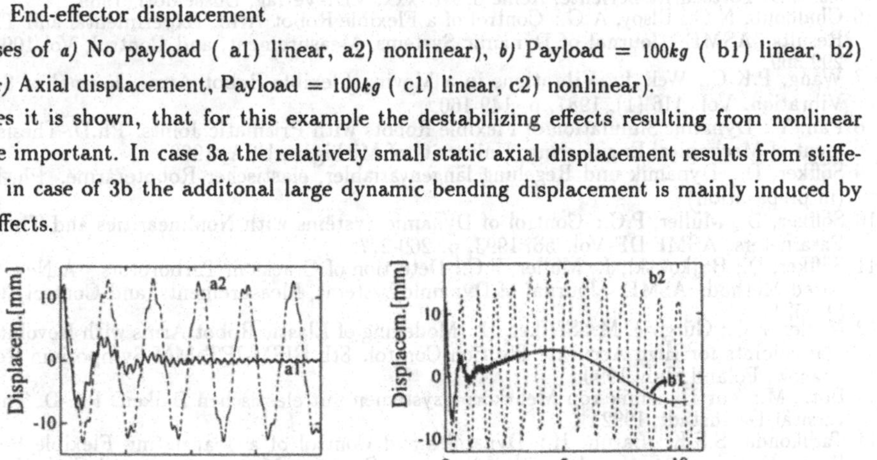

Fig. 4 a,b: Controlled end-effector displacement a) Braced beam b) Rotating beam

As first control examples the vibrations of a kinematic braced beam and the example given in fig. 2a are controlled, using an additional kinematic input $\Delta\dot{\varphi}(t)$ (End-effector initial conditions: a,b) $u_{x,ef}(t=0) = -1 \cdot 10^{-3}m, u_{y,ef}(t=0) = u_{z,ef}(t=0) = 10 \cdot 10^{-3}m$).

Neglecting nonlinearities, axial and torsional degrees of freedom, resulting in a system of order 24 (real system: 60), a reduced model is used therefore. As measurements only the 4 deflections of the end-effector and between the 2nd and 3rd beam are used. It is shown clearly that the dynamic behaviour can be influenced well by the proposed observer-based compensation scheme. Using an additional kinematic input the vibrations induced from initial conditions etc. can be well reduced. Further simulations for spatial movements are given in [9].

CONCLUSIONS

A nonlinear beam element for systematic modeling of beam-like robot arm structures is developed. This approach allows geometric precise modeling, considers nonlinear coupling effects, and distinguishes itself by a systematic procedure, which allows both, a simple automatic modeling procedure combined with modeling of higher order terms and modular procedure. The simulations demonstrate the representation of coupling effects by the developed modules and the importance of nonlinear modeling concerning weakening effects, which lead to instability. With the developed extended observer-based compensation scheme it is possible to reduce elastic vibrations.

REFERENCES

1 Cannon, R.H.; Schmitz, E.: Initial Experiments on the End-Point Control of a Flexible One-Link Robot. Int. Journal of Robotics Research. Vol. 3, No. 3, 1985, p.62-75.

2 Book, W.J.; Alberts, T.E.; Hastings, G.G.: Design Strategies for High-Speed Lightweight Robot. Comput. Mech. Engineering., 1986, p. 26-33.

3 Truckenbrodt, A.: Bewegungsverhalten und Regelung hybrider Mehrkörpersysteme mit Anwendung auf Industrieroboter. Ph.-D. Thesis. VDI-Fortschrittsbericht, Reihe 8, Nr. 33, VDI-Verlag, 1980.

4 Henrichfreise, H.: Aktive Schwingungsdämpfung an einem elastischem Knickarmroboter. Ph.-D. Thesis. Universität Paderborn, 1989.

5 Ackermann, J.: Positionsregelung reibungsbehafteter elastischer Industrieroboter. Ph.-D. Thesis. VDI-Fortschrittsberichte, Reihe 8, Nr. xxx, VDI-Verlag, Düsseldorf, 1989.

6 Chalhoub, N.G.; Ulsoy, A.G.: Control of a Flexible Robot Arm: Experimental and Theoretical Results. ASME - Journal of Dynamic Systems, Measurement, and Control, Vol 109, 1987, p. 299-309.

7 Wang, P.K.C.; Wei, J.: Vibrations in a Moving Flexible Robot Arm. Journal of Sound and Vibration. Vol. 116 (1), 1987, p. 149-160.

8 Pan, Y.: Dynamic Simulation of Flexible Robots with Prismatic Joints. Ph.D.-Thesis, Department of Mechanical Engineering, University of Michigan, USA, 1988.

9 Söffker, D.: Dynamik und Regelung längenvariabler, elastischer Roboterarme. Ph.-D. Thesis (in preparation).

10 Söffker, D.; Müller, P.C.: Control of Dynamic Systems with Nonlinearities and Time-Varying Parameters. ASME DE-Vol. 56, 1993, p. 269-277.

11 Söffker, D.; Bajkowski, J.; Müller, P.C.: Detection of Cracks in Turborotors - A New Observer-based Method. ASME - Journal of Dynamic Systems, Measurements, and Control, 3, 1993, p. 518-524.

12 Müller, P.C.; Gürgöze, M.; Söffker, D.: Modelling of Elastic Robot Arms with Revolute or Prismatic Joints for High Accurate Position Control. 8th CISM-IFToMM Symposium Ro.Man.Sy, Cracow, Poland, July 1990.

13 Botz, M.: Zur Dynamik von Mehrkörpersystemen mit elastischen Balken. Ph.-D. Thesis, Universität Darmstadt, 1992.

14 Tadikonda, S.S.K.; Baruh, H.: Dynamics and Control of a Translating Flexible Beam With Prismatic Joint. ASME - Journal of Dynamics Systems, Measurement, and Control. Vol. 114, Sept. 1992, p. 424-427.

15 Davison, E.J.: The output control of linear time-invariant multivariable systems with unmeasurable arbitrary disturbances. IEEE Transact. Automatic Control, AC-17, 1972, p. 561-573.

Part III
Mechanics III

AN APPLICATION OF THE RIGID FINITE ELEMENT METHOD TO MODELLING OF FLEXIBLE STRUCTURES

E. Wittbrodt
Technical University of Gdansk, Gdansk, Poland
and
S. Wojciech

Technical University of Lodz in Bielsko-Biala, Bielsko-Biala, Poland

Summary: An application of the rigid finite element method (RFEM) to modelling of spatial continuous and discrete systems is presented in the work. It is assumed that the base motion causing vibrations of a flexible system is known. The linear and nonlinear models of a system are described. The nonlinear model enables large displacements to be taken into account.

1. Introduction

In many problems of dynamic analysis and synthesis it is necessary to take into account the flexibility of a system. The problem is difficult to solve especially when changing configuration occurs. Such problems can often be met in dynamics of manipulators and spacecrafts [1,2,3]. The finite element method is the method most often used in order to model flexible systems [4,5]. If a structure of the system is complex the procedure reducing the number of degrees of freedom it is carried out usually using the modal method In this work an application of the RFEM [6,7] to the modelling of the flexible systems with a moving base is presented. At first the rigid finite element method was formulated in order to analyse linear vibrations of complex continuous and discrete systems [6]. Its applications for the vibration analysis of ships, cranes and other machines have proved usefulness and numerical effectiveness of the method. This work presents the further extension of the RFEM for spatial systems with a moving base.

2. The rigid finite element method

It is assumed that the considered system (Fig. 1a) is a system of n+1 of rigid bodies called rigid finite elements (RFE) which are connected by m spring-damping elements (SDE). Each of RFEs is characterised by its mass and mass moment of inertia whereas SDEs are massless and nondimensional and their characteristics are linear The procedure enabling continuos systems such as plates or beams to be replaced by a suitable system of RFEs and SDEs is presented in [6].

Fig 1b shows an example of a discretization of a prismatic beam into the rigid finite elements The essential feature of RFEM is the possibility of substituting the continuous system into the system of rigid bodies connected by springs. It allows the complex continuous and discrete systems, which are often met in practical problems of machine dynamics, to be treated equally

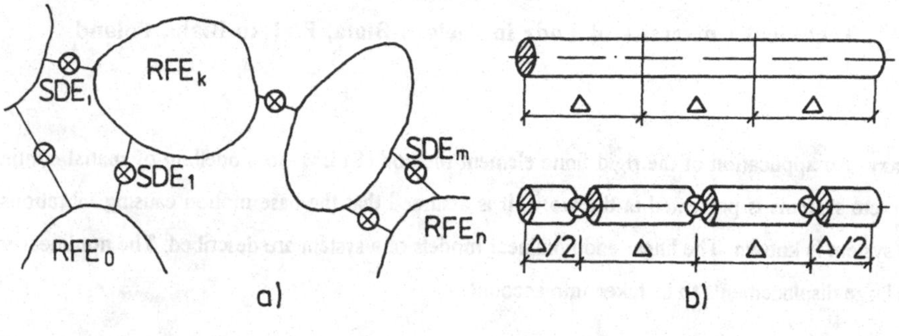

Fig 1 The Rigid Finite Element Method

a) the system of RFEs connected by SDEs, b) the continuous system replaced by RFEs and SDEs.

3. Mathematical model of the system

The considered system is shown in Fig 2 It is assumed that the motion of the RFE 0, to which the coordinate system {b} is fixed, is known. A local coordinate system {i} is fixed to the *i*th RFE. The origin of the coordinate system is placed in the mass center of the RFE and its axes cover with principal axes of inertia of the element

If the considered system is not deformed the transformation of coordinates from the system {i} to the inertial system {0} is executed using transformation matrices [8] as follows

$$\mathbf{A}_{i0}=\mathbf{A}_{i0}(t)=\mathbf{A}_0(t)\bullet\mathbf{A}_{bi} \tag{1}$$

where $\mathbf{A}_0(t)={}^0_b\mathbf{A}$ is the transformation matrix from the coordinate system {b} to the inertial coordinate system,

$\mathbf{A}_{bi}$= const is the transformation matrix from the coordinate system {i} to the system {b}.

If the system is flexible a relative motion of each element is described using components of the following vectors

$$\mathbf{q}_i=\left[x_i,\ y_i,\ z_i,\ \varphi_{i1},\ \varphi_{i2},\ \varphi_{i3}\right]^* \tag{2}$$

where * indicates the transpose of a vector or matrix,

x_i, y_i, z_i are relative displacements of the *i*th RFE with respect to the system {i},

φ_{i1}, φ_{i2}, φ_{i3} are rotary angles of the *i*th RFE about axes of the system {i}

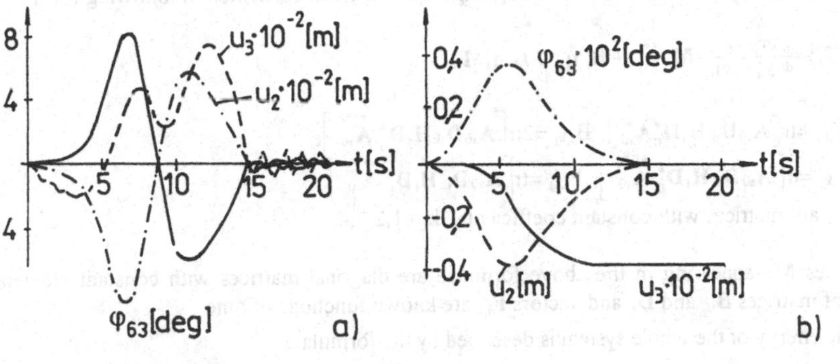

Fig 4 The results corresponding to a) Fig. 3 b, b) Fig. 3 c.

5. Conclusions

In the work the application of the RFEM to modelling of flexible systems with changing configuration is presented In order to derive the equations of motion the formalism for rigid manipulators is used. The presented procedure is effective especially when analyzing the continuous and discrete systems. The nonlinear model enables the large deflections to be analyzed The simple form Eqs (11) allows to hope for using the RFEM in control

6. References

[1] Cetinkunt S, Book W J. Performance limitations of joint variable - feedback controllers due to manipulator structural flexibility. IEEE Transactions on Robotics and Automation. Vol. 6, No 2, April 1990, pp 219-231.
[2] Vincent T L, Lin Y Ch, Joshi S.P.: Positioning and active damping of flexible beams. Journal of Guidance, Control and Dynamics. Vol. 13, No 4, July - August 1990, pp. 714-724
[3] Meirovitch L, Kwak M K.: Dynamics and control of spacecraft with retargeting flexible antennas Journal of Guidance, Control and Dynamics Vol. 13, No 2, March - April 1990, pp. 241-248.
[4] Kane T R, Ryan R.R, Banerjee A.K.: Dynamics of beam attached to a moving base. Journal of Guidance, Control and Dynamics. Vol. 10, No2, March-April 1987, pp. 1-36..
[5] Du H, Hitching P., Davies G A.O.: A finite element structural dynamics model of beam with an arbitrary moving base Part I and II Finite Element in Analysis and Design Vol 12, pp. 117-150
[6] Kruszewski J, Gawroński W, Wittbrodt E., Najbar E, Grabowski E The rigid finite element method Arkady, Warsaw, 1974 (in Polish).
[7] Wittbrodt E Dynamics of the systems with changing in time configuration using the finite element method Papers of TU Gdańsk, Mechanika XLVI, Gdańsk 1983 (in Polish).
[8] Craig J J Introduction to robotics. Addison Wesley Publ. Comp Massachusetts, 1988.

After some necessary transformations the Lagrange operators can be written in following form

$$\varepsilon(T_i)=\frac{d}{dt}\frac{\partial T_i}{\partial \dot{q}}-\frac{\partial T_i}{\partial q_i}=\mathbf{M}_i\ddot{\mathbf{q}}_i+\mathbf{B}_i\dot{\mathbf{q}}_i+\mathbf{D}_i\mathbf{q}_i+\mathbf{F}_i \tag{6}$$

where $M_{i,kj}=\mathrm{tr}\left[\mathbf{A}_{i0}\mathbf{D}_{ik}\mathbf{H}_i\mathbf{D}_{ij}^{*}\mathbf{A}_{i0}^{*}\right]$, $B_{i,kj}=2\mathrm{tr}\left[\mathbf{A}_{i0}\mathbf{D}_{ik}\mathbf{H}_i\mathbf{D}_{ij}^{*}\ddot{\mathbf{A}}_{i0}\right]$,
$D_{i,kj}=\mathrm{tr}\left[\mathbf{A}_{i0}\mathbf{D}_{ik}\mathbf{H}_i\mathbf{D}_{ij}^{*}\dddot{\mathbf{A}}_{i0}\right]$, $F_{i,kj}=\mathrm{tr}\left[\mathbf{A}_{i0}\mathbf{D}_{ik}\mathbf{H}_i\mathbf{D}_{ij}^{*}\dddot{\mathbf{A}}_{i0}\right]$,
$\mathbf{D}_{ik}$ are matrices with constant coefficients, $k,j=1,2,3,4$

The matrices $\mathbf{M}_i$ appearing in the above formulae are diagonal matrices with constant elements while elements of matrices $\mathbf{B}_i$ and $\mathbf{D}_i$ and vectors $\mathbf{F}_i$ are known functions of time
The kinetic energy of the whole system is described by the formula

$$T=\sum_{i=1}^{n}T_i \tag{7}$$

hence, the Lagrange operator of the system is as follows

$$\varepsilon(T)=\frac{d}{dt}\frac{\partial T}{\partial \dot{q}}-\frac{\partial T}{\partial q}=\mathbf{M}\ddot{\mathbf{q}}+\mathbf{B}\dot{\mathbf{q}}+\mathbf{D}\mathbf{q}+\mathbf{F} \tag{8}$$

where $\mathbf{q}=\left[\mathbf{q}_1^{*}\ \mathbf{q}_2^{*}\ \ \mathbf{q}_n^{*}\right]^{*}$ is the vector of the generalized coordinates of the system,
$\mathbf{M}=\mathrm{diag}\{\mathbf{M}_1\ \mathbf{M}_2\ \ \mathbf{M}_n\}$, $\mathbf{B}=\mathrm{diag}\{\mathbf{B}_1\ \mathbf{B}_2\ \ \mathbf{B}_n\}$,
$\mathbf{D}=\mathrm{diag}\{\mathbf{D}_1\ \mathbf{D}_2\ \ \mathbf{D}_n\}$, $\mathbf{F}=[\mathbf{F}_1^{*}\ \mathbf{F}_2^{*}\ \ \mathbf{F}_n^{*}]^{*}$

B. The potential energy of SDE's strain

If the spring-damping element connects the RFEs l and p the potential energy the SDE e is described by the formula

$$V^e=\tfrac{1}{2}k_x^e\left[x_{bp}^e-x_{bl}^e\right]^2+\tfrac{1}{2}k_y^e\left[y_{bp}^e-y_{bl}^e\right]^2+\tfrac{1}{2}k_z^e\left[z_{bp}^e-z_{bl}^e\right]^2+$$
$$+\tfrac{1}{2}k_1^e[\varphi_{p1}-\varphi_{l1}]^2+\tfrac{1}{2}k_2^e[\varphi_{p2}-\varphi_{l2}]^2+\tfrac{1}{2}k_3^e[\varphi_{p3}-\varphi_{l3}]^2 \tag{9}$$

where k_x^e, k_y^e, k_z^e are coefficients of the translation rigidity of the SDE e,
k_1^e, k_2^e, k_3^e are coefficients of the rotary rigidity of the SDE e,
$\mathbf{r}_{bp}^e=\left[x_{bp}^e\ y_{bp}^e\ z_{bp}^e\ 1\right]^{*}={}_p^b\mathbf{A}\mathbf{A}_p\bullet\mathbf{r}_p^e$,
$\mathbf{r}_{bl}^e=[x_{bl}^e\ y_{bl}^e\ z_{bl}^e\ 1]^{*}={}_l^b\mathbf{A}\mathbf{A}_l\bullet\mathbf{r}_l^e$,
$\mathbf{r}_p^e$, $\mathbf{r}_p^e$ are coordinate vectors of the SDE e with respect to the local coordinate systems {p} and {l} respectively

It is necessary to calculate $\frac{\partial V^e}{\partial q_p}$ and $\frac{\partial V^e}{\partial q_l}$ since these are components of the Lagrange equations The potential energy of the whole system is defined in the form.

$$V=\sum_{e=1}^{m}V^e \tag{10}$$

Further on the following two cases are considered·

I· The linear model. In this case the expressions (9) are calculated using the simple form of matrices $\mathbf{A}_i$ defined (4) and after that they are differentiated

II The nonlinear model. In order to calculate (9) matrices $\mathbf{A}_i$ defined in (3) are used Only after differentiations mentioned above the trigonometrical functions are linearized. It causes that nonlinear components of the second order occur in the equations of motion.

Thus, assuming that damping forces are linear functions of the generalized velocities, the equations of the system are written in the form

$$\mathbf{M}\ddot{\mathbf{q}}+(\mathbf{B}+\mathbf{L})\dot{\mathbf{q}}+(\mathbf{D}+\mathbf{K})\mathbf{q}+\mathbf{C}(\mathbf{q})+\mathbf{F}=\mathbf{Q} \qquad (11)$$

where **M, B, D, F** are defined in (8),

K, L are stiffness and damping matrices respectively, defined in [6],

C(q) is the vector of nonlinear components; if linear model is considered this expression has to be neglected,

Q is the vector of the generalized coordinates

4. A numerical example

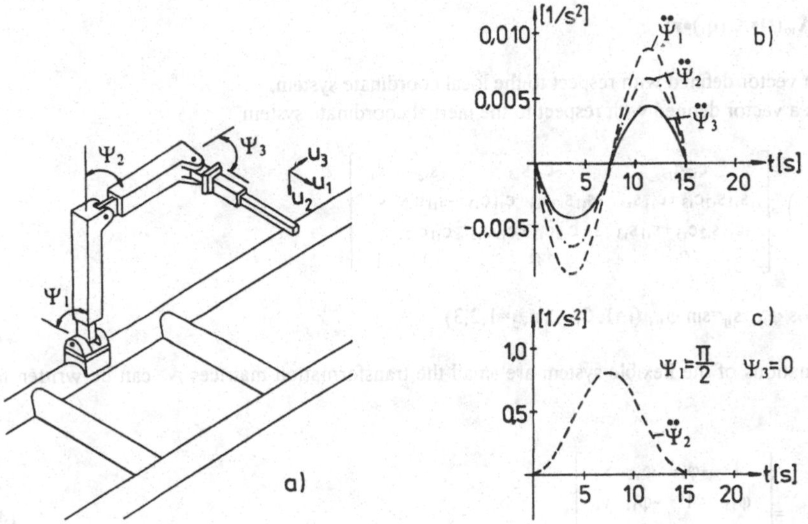

Fig 3 The example: a) the considered manipulator, b,c) the accelerations of joint angles.

In order to check the correctness of the presented models the calculations for a manipulator from Fig. 3 a have been carried out using parameters given in [4].

Assuming n=6 (the third link is divided into 7 of RFEs and 6 of SDEs) and the base motion defined as in Fig 3 b,c the results shown in Fig. 4 a,b have been obtained.

The different results obtained by the RFEM and their comparison with the results obtained by finite element method [4], [5], will be presented at the Conference

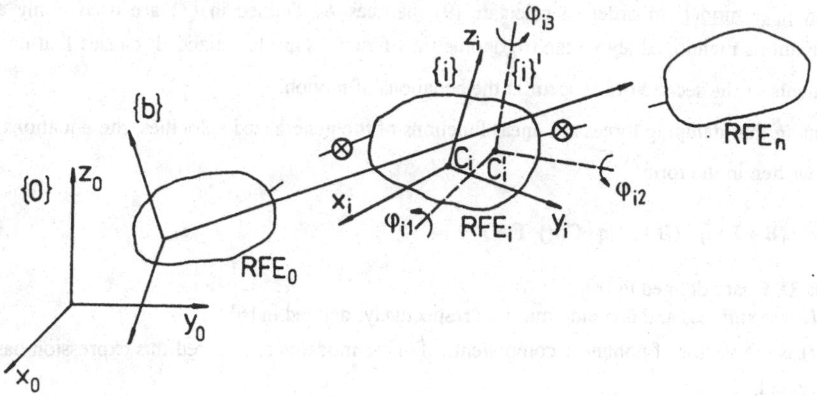

Fig 2 The considered system

The coordinates of any vector $\mathbf{r}_i'$ defined with respect to the local coordinate system can be defined with respect to the inertial coordinate system as follows

$$\mathbf{r}_{i0}=\mathbf{A}_{i0}(t)\bullet\mathbf{A}_i(\mathbf{q}_i)\bullet\mathbf{r}_i' \tag{3}$$

where $\mathbf{r}_i'$ is a vector defined with respect to the local coordinate system,
$\mathbf{r}_{i0}$ is a vector defined with respect to the inertial coordinate system

$$\mathbf{A}_i\left(\mathbf{q}_i\right)=\begin{bmatrix} c_{i2}c_{i3} & -c_{i2}s_{i3} & s_{i2} & x_i \\ s_{i1}s_{i2}c_{i3}+c_{i1}s_{i3} & -s_{i1}s_{i2}s_{i3}+c_{i1}c_{i3} & -s_{i1}c_{i2} & y_i \\ -c_{i1}s_{i2}c_{i3}+s_{i1}s_{i3} & s_{i1}c_{i2}c_{i3}+c_{i1}c_{i3} & c_{i1}c_{i2} & z_i \\ 0 & 0 & 0 & 1 \end{bmatrix}$$

$c_{ij}=\cos\varphi_{ij}$, $s_{ij}=\sin\varphi_{ij}$, $(i=1, 2, \ldots, n, j=1,2,3)$

If the deformations of the flexible system are small the transformation matrices $\mathbf{A}_i$ can be written in the form

$$\mathbf{A}_i\left(\mathbf{q}_i\right)=\begin{bmatrix} 1 & -\varphi_{i3} & \varphi_{i2} & x_i \\ \varphi_{i3} & 1 & -\varphi_{i1} & y_i \\ -\varphi_{i2} & \varphi_{i1} & 1 & z_i \\ 0 & 0 & 0 & 1 \end{bmatrix} \tag{4}$$

The Lagrange equations of the second order are used in order to derive the equations of the motion of the considered system Thus the kinetic and potential energies of the system have to be defined

A. The kinetic energy of the RFE

The kinetic energy of the ith RFE can be presented in the form

$$T_i=\frac{1}{2}\,\mathrm{tr}\left[\dot{\mathbf{A}}_{i0}\,\mathbf{A}_i\mathbf{H}_i\mathbf{A}_i^T\,\dot{\mathbf{A}}_{i0}^T\right]+\mathrm{tr}\left[\dot{\mathbf{A}}_{i0}\,\mathbf{A}_i\mathbf{H}_i\,\dot{\mathbf{A}}_i^T\,\mathbf{A}_{i0}^T\right]+\frac{1}{2}\mathrm{tr}\left[\mathbf{A}_{i0}\,\dot{\mathbf{A}}_i\,\mathbf{H}_i\,\dot{\mathbf{A}}_i^T\,\mathbf{A}_{i0}^T\right] \tag{5}$$

where $\mathrm{tr}[\mathbf{A}]$ indicates the trace of the matrix $\mathbf{A}$, $\mathbf{H}_i$ is an inertia matrix of the ith element

SENSITIVITY ANALYSIS OF MULTIBODY SYSTEMS

A NEW APPROACH BASED ON THE CONCEPT OF GLOBAL INERTIA

F. Pfister, M. Fayet and A. Jutard

National Institute of Applied Sciences of Lyon, Villeurbanne, France

ABSTRACT: The paper presents a unified formalism for sensitivity model construction of open-loop mechanical systems. Using *global-inertia tensors* in conjunction with *basic-kinetic tensors* as an "unifying vehicle" it is shown that the sensitivity model with respect to any system-parameter (kinematic, geometric, inertia) can be formulated in one coherent and economical framework. The integral formulation of Christoffel-Symbols, not much used hitherto, now comes into its own. The procedure involves vector algebraic manipulations, but not numerical differentation. The evaluation of the obtained expressions is computational efficient and naturally follows a recursive procedure. Beyond the advantages of representation, the theory presented herein stresses the common, fundamental structure of the elements that make up motion equations and provides significant physical insight.

I. INTRODUCTION

Analyzing the dynamic of multibody-systems (MBSs) is an important and complicated interdisciplinary problem. Various techniques are available that permit the automated equation formulation, their numerical integration and computer animation. Symbolic and parallel computing become increasingly popular for real-time simulation.

The subject of this paper is the *sensitivity analysis* of MBSs. The need for efficent techniques for sensitivity analysis is well established. It has arisen in the construction of linearized or partially linearized dynamic models around a given trajectory, for stability investigations and the development of control algorithms [VUKOB.&STOKIC,1982], [MILLS&GOLD.,1994]. In order to study systems with variable or unknown inertia characteristics (masses, moments of inertia) one is driven to a derivation of *inertia sensitivity models* (see [GAUT.,1990], [GAUT.&KHALIL,1992], [SEEGER,1992] and the literature cited there). In addition, the determination of dependent inertial parameters provides a method of simplifying the motion-equations. Other applications are design optimization[YOUC.&ASADA,1986] and optimal tolerance problems[YANG&TZENG,1986].

Within the past few years a number of methods and algorithms have been proposed to obtain sensitivity models computationally. They include *inter alia*[VUCOB.&KIRC.,1982] and [LI,1990] for variations of joint parameter and [VUCOB.&KIRC.,1984] and[ATERSON et al.,1986] for variations of inertia parameters .

The essential aim of this paper is to show that the sensitivity with respect to the *totality of parameters* of the MBS (inertia, geometric and kinematic parameters) can be found *from one kind of tensor*, namely the *basic kinetic-tensor*. This route towards sensitivity modelling is suggested as an alternative to the methods already refered to and is believed to show some improvements with respect to clarity and concision. The approach (in contrast to e.g. [LI,1990]) is a pure *tensor-approach*.. The results may be seen as the extension of[FAYET&RENAUD,1989] and[FAYET&PFISTER,i.p.]. Light is shed on certain other matters in passing.

II. GENERAL CONSIDERATIONS

Topology: The numerotation of the N rigid bodies $\mathcal{B}_i$ of the open-loop multibody system $\Sigma := \{\mathcal{B}_1, \cdots, \mathcal{B}_N\}$ is considered to be *regular*. The *interconnection structure* is described with the help of the matrices $\mathbf{a}^q$, $\mathbf{D}^q$ and $\mathbf{P}^q$, whose elements are defined as

(2.3a-c) $$a^q_j := \begin{cases} i: \text{ if } \mathcal{B}_j \text{ is a direct successor of } \mathcal{B}_i \\ 0: \text{ otherwise} \end{cases}, \quad A^q_{ij} := \begin{Bmatrix} 1: \text{ if } a_j = i \\ 0: \text{ otherwise} \end{Bmatrix} \text{ and } P^q_{ij} := \begin{Bmatrix} 1: \text{ if } \mathcal{B}_i \text{ is on the path between } \mathcal{B}_j \text{ and } \mathcal{B}_0 \\ 0: \text{ otherwise} \end{Bmatrix}.$$

The superscript q, though of no immediate importance, will be found useful further on.

Motion-equations:

(2.1) $$Q_i = E^i(T), \text{ where } Q_i \equiv G^i + Q_i^{\Delta}, \; E^i(T) \equiv \left. \sum_{j=1}^{N} a^{ij} \ddot{q}_j + \sum_{j,k=1}^{N} \Gamma^{i,jk} \dot{q}_j \dot{q}_k \right\} \forall\, i=1 \ldots N$$

(q_i: generalized coordinates; Q_i: generalized force; $G^i := -g\,\vec{Z}_0 \cdot \int_\Sigma \partial\overrightarrow{OM}/\partial q_i \, dm$: generalized gravity term.) The coefficents

(2.2a,b) $$a^{ij} = \int_\Sigma \frac{\partial\overrightarrow{OM}}{\partial q_i} \cdot \frac{\partial\overrightarrow{OM}}{\partial q_j} dm \qquad \text{and} \qquad \Gamma^{i,jk} = \int_\Sigma \frac{\partial\overrightarrow{OM}}{\partial q_i} \cdot \frac{\partial^2\overrightarrow{OM}}{\partial q_j \partial q_k} dm$$

are respectively the elements of the mass-matrix and the Christoffel-Symbols. $\overrightarrow{OM}$, $\partial\overrightarrow{OM}/\partial q_i$ and $\partial^2\overrightarrow{OM}/\partial q_j\partial q_k$ are the position vector and the first and second order partial derivatives of the (mass) particle (M,dm) as viewed by a spectator on a galilean frame of reference $F_0 \text{–} (O, \vec{X}_0, \vec{Y}_0, \vec{Z}_0)$. F_0 is fixed on body $\mathcal{B}_0 := \mathcal{B}_{a_1}$. The coefficents G^i, a^{ij} and $\Gamma^{i,jk}$ are, in general, complex functions of the arguments $q := (q_1, \cdots, q_N)^T$ and the position independent parameters (link lengths, twists, and offsets) $d := \{d_1, \cdots, d_{dN}\}^T$. They are, however, linearly dependent on the inertia parameters $\theta := (\theta_1, \cdots, \theta_{10\times N})^T$.

In the authors opinion, the formulation (2.2b) of Christoffel-Symbols has recieved less attention thus far than it deserves. In this paper it will be shown, how it can be used to full advantage. In most publications it is written in the form $\Gamma^{i,jk} = \frac{1}{2}(\partial a^{ij}/\partial q_k + \partial a^{ki}/\partial q_j - \partial a^{jk}/\partial q_i)$.

Problem statement: The problem discussed in this paper is the development of an algorithm for the formulation of the three Jacobians L_q, L_d, L_θ whose ij-entry is given by $\frac{\partial L^i}{\partial q_j}, \frac{\partial L^i}{\partial d_j}, \frac{\partial L^i}{\partial \theta_j}$ respectively, $L^i := E^i(T) - G^i$.

Due to the fact that $\dot{q}_i$ and $\ddot{q}_i$ figure *explicitly* in (2.1), it is natural to think of the problem of determining the Jacobians as falling into *two stages*. First the evaluation of a^{ij}, $\Gamma^{i,jk}$ and G^i and their partial derivatives with respect to the parameters of the system and second the problem of summing up the obtained terms. The second stage is simple, at least in theory, and not nearly as interesting as the first. Therefore, we shall in what follows concentrate on the first.

III. ON TENSOR FORMULATIONS OF PARTIAL DERIVATIVES

The derivatives of a^{ij}, $\Gamma^{i,jk}$ and G^i with respect to any $p_i \in p := \{q_1, \cdots, q_N, d_1, \cdots, d_{dN}\}$ lead to integrals of the form

(3.1a-d) $$\left| \begin{array}{l} (i\,|\,j) := \int_\Sigma \frac{\partial\overrightarrow{OM}}{\partial p_i} \cdot \frac{\partial\overrightarrow{OM}}{\partial p_j} dm; \quad (i\,|\,j,l) := \int_\Sigma \frac{\partial\overrightarrow{OM}}{\partial p_i} \cdot \frac{\partial^2\overrightarrow{OM}}{\partial p_j \partial p_l} dm; \\ (i,l\,|\,j,k) := \int_\Sigma \frac{\partial^2\overrightarrow{OM}}{\partial p_i \partial p_l} \cdot \frac{\partial^2\overrightarrow{OM}}{\partial p_j \partial p_k} dm; \quad (i\,|\,j,k,l) := \int_\Sigma \frac{\partial\overrightarrow{OM}}{\partial p_i} \cdot \frac{\partial^3\overrightarrow{OM}}{\partial p_j \partial p_k \partial p_l} dm \end{array} \right.$$

With the help of the definition $\partial\overrightarrow{OM}/\partial q_0 := \vec{Z}_0 \; \forall\, M \in \Sigma$, we have $G^i = -g\,(0\,|\,i)$ and $\partial G^i/\partial q_j = -g\,(0\,|\,j,l)$.

In this section we consider the question of the how partial dervatives $\partial\overrightarrow{OM}/\partial p_i$, ..., $\partial^3\overrightarrow{OM}/\partial p_j\partial p_k\partial p_l$ can be "algebrized". *This provides the key to determining the integrals* (3.1a-d). The means of determining the partial derivatives of (3.1) with respect to mass parameters appears later. The first partial derivative with respect to q_i can be written as

(3.2a) $$\frac{\partial\overrightarrow{OM}}{\partial q_i} = P^q_{iM}\left(\vec{\varepsilon}^{\,q}_i \vec{Z}^q_i + \varepsilon^q_i \hat{\vec{Z}}^q_i \cdot \overrightarrow{O_i{}^q M}\right), \quad \text{where } \hat{\vec{Z}} \cdot \vec{v} := \vec{Z} \times \vec{v} \;\; \forall \vec{v}.$$

$\vec{Z}_i^q$ is the unit vector along the joint axis and $\overrightarrow{O_i^qM}$ is the position vector of M from any point O_i^q on the joint axis. Moreover,

$$\varepsilon_i^q=1,\ \bar{\varepsilon}_i^q=0,\ \text{if } J_i^q \text{ is revolute and } \varepsilon_i^q=0,\ \bar{\varepsilon}_i^q=1\ ,\ \text{if } J_i^q \text{ is prismatic}\ . \tag{3.3}$$

We can extend the result (3.2a) to include a relationship for $\partial\overrightarrow{OM}/\partial d_i$ by introducing an *artificial joint* J_i^d, so that the geometric parameter d_i becomes a kinematic one. This *"new" joint corresponds to a "new" body* $\mathcal{B}_i^d$ *without mass*. With this artifice all our results are also valid for the geometric model:

$$\frac{\partial\overrightarrow{OM}}{\partial d_i}=P_{iM}^d\left(\bar{\varepsilon}_i^d\,\vec{Z}_i^d+\varepsilon_i^d\,\hat{\vec{Z}}_i^d\cdot\overrightarrow{O_i{}^dM}\right)\ . \tag{3.2b}$$

The different quantities figuring in this expression should be self explanatory. For the sake of unification we write

$$\frac{\partial\overrightarrow{OM}}{\partial p_i}=\bar{\varepsilon}_iP_{iM}\ \vec{Z}_i+\varepsilon_i\ \hat{\vec{Z}}_i\cdot P_{iM}\ \overrightarrow{O_iM}\ \ \forall\ p_i\in \mathbf{p}\ . \tag{3.2c}$$

Taking the partial derivative of (3.3) we obtain the following important relation

$$P_{ij}=1 :\Rightarrow \frac{\partial^2\overrightarrow{OM}}{\partial p_i\partial p_j}=P_{iM}\ \varepsilon_i\ \vec{Z}_i\times\frac{\partial\overrightarrow{OM}}{\partial p_j}\ \forall\ p_i,p_j\in \mathbf{p}\ . \tag{3.4}$$

Since the proof of (3.4) is given in [FAYET&RENAUD,1989], we do not reproduce it here. *Warning:* (3.4) is *not* true for $\left(P_{ji}=1\ i\neq j\right)$ and serious fallacies often arise from this fact. (3.4) can be used for a recursive formulation of *any higher-order partial derivative*. For the third order partial derivative we write

$$P_{ij}P_{ik}=1 :\Rightarrow \frac{\partial^3\overrightarrow{OM}}{\partial p_i\partial p_j\partial p_k}=P_{iM}\varepsilon_i\ \vec{Z}_i\times\frac{\partial^2\overrightarrow{OM}}{\partial p_j\partial p_k}\ \ \forall\ p_i,p_j,p_k\in \mathbf{p}\ . \tag{3.5}$$

IV. GLOBAL INERTIA AND BASIC KINETIC TENSORS

Now that the partial derivatives with respect to the geometric and kinematic parameters have been determined, they may be substituted into (3.1). Before doing so, it is convenient to introduce the following notation:

$$\text{(4.1a-d)}\quad \left|\begin{array}{ll} \langle i,j\rangle := \displaystyle\int_\Sigma \frac{\partial\overrightarrow{OM}}{\partial p_i}\cdot\frac{\partial\overrightarrow{OM}}{\partial p_j}\,dm; & \langle i,jkl\rangle := \displaystyle\int_\Sigma \frac{\partial\overrightarrow{OM}}{\partial p_i}\cdot\vec{Z}_j\times\!\left(\vec{Z}_k\times\frac{\partial\overrightarrow{OM}}{\partial p_l}\right)dm \\ \langle i,jk\rangle := \displaystyle\int_\Sigma \frac{\partial\overrightarrow{OM}}{\partial p_i}\cdot\vec{Z}_j\times\frac{\partial\overrightarrow{OM}}{\partial p_k}\,dm; & \langle ij,kl\rangle := \displaystyle\int_\Sigma \vec{Z}_i\times\frac{\partial\overrightarrow{OM}}{\partial p_j}\cdot\vec{Z}_k\times\frac{\partial\overrightarrow{OM}}{\partial p_l}\,dm \end{array}\right.$$

It is easy to confirm that $\langle i,jk\rangle+\langle k,ji\rangle=0$ and $\langle ij,kl\rangle+\langle j,ikl\rangle=0$. If we insert (3.2-4) into (3.1a-d), we obtain from (4.1) and the preceeding properties

$$\text{(4.2a-c)}\quad \left\{\begin{array}{l}
(i|j)=\langle i',j'\rangle \quad \text{with}\begin{cases} i'=\min(i,j)\\ j'=\max(i,j)\end{cases}\\
(i\,|\,j,l)=\varepsilon_{j'}\,P_{ij'}\,\langle i,j'k'\rangle \quad \text{with}\begin{cases} j'=\min(j,k)\\ k'=\max(j,k)\end{cases}\\
(i\,|\,j,k,l)=\varepsilon_{j'}\varepsilon_{k'}\,P_{j'k'}\,P_{k'l'}\,\langle i,j'k'l'\rangle \quad \text{with}\begin{cases} j'=\min(j,k,l)\quad k'=\mathrm{mid}(j,k,l)\\ l'=\max(j,k,l)\end{cases}\\
(i,l\,|\,j,k)=-\varepsilon_{i'}\varepsilon_{k'}P_{i'j'}P_{k'l'}\ \langle j',i'k'l'\rangle \ \text{with}\begin{cases} i'=\min(i,j,k,l\ ) & k'=\begin{cases}\min(k,l)\text{ if } i'\in\{i,j\}\\ \min(i,j)\text{ if } i'\in\{k,l\}\end{cases}\\ j'=\begin{cases}\max(i,j)\text{ if } i'\in\{i,j\}\\ \max(k,l)\text{ if } i'\in\{k,l\}\end{cases} & l'=\begin{cases}\max(k,l)\text{ if } i'\in\{i,j\}\\ \max(i,j)\text{ if } i'\in\{k,l\}\end{cases}\end{cases}
\end{array}\right.$$

In order to shorten the calculation of these expressions and to get a better symmetry in all formulas, we introduce the *basic-kinetic-tensor*

(4.3) $$\vec{\vec{K}}^{il} := \int_\Sigma \frac{\partial \overrightarrow{OM}}{\partial q_i} \otimes \frac{\partial \overrightarrow{OM}}{\partial q_l}\, dm$$

and thus obtain

(4.4a-c) $$\langle i,k\rangle = \vec{\vec{I}} \cdot\cdot \vec{\vec{K}}^{ik} \;;\; \langle i,jk\rangle = \hat{\vec{\vec{Z}}}_j \cdot\cdot \vec{\vec{K}}^{ik} \text{ and } \langle i,jkl\rangle = \vec{\vec{A}}_{jk} \cdot\cdot \vec{\vec{K}}^{il}, \text{ where } \vec{\vec{A}}_{jk} := \hat{\vec{\vec{Z}}}_j \cdot \hat{\vec{\vec{Z}}}_k \text{ and } (\vec{a}\otimes\vec{b})\cdot\cdot(\vec{c}\otimes\vec{d}) = (\vec{a}\cdot\vec{c})(\vec{b}\cdot\vec{d}).$$

Replacing (3.2) in (4.3) and introducing

(4.5a-c) $$M^{ij} := \int_\Sigma P_{iM}P_{jM}\,dm,\quad \vec{U}^{ij} := \int_\Sigma P_{iM}P_{jM}\overrightarrow{O_iM}\,dm \text{ and } \vec{\vec{\Pi}}^{ij} := \int_\Sigma P_{iM}P_{jM}\,\overrightarrow{O_iM}\otimes\overrightarrow{O_jM}\,dm$$

the basic kinetic tensor may be expressed in the form

(4.6) $$\vec{\vec{K}}^{ij} = \bar{\varepsilon}_i\bar{\varepsilon}_j\vec{Z}_i\otimes\vec{Z}_j M^{ij} + \bar{\varepsilon}_i\varepsilon_j\,\vec{Z}_i\otimes\left(\hat{\vec{\vec{Z}}}_j\cdot\vec{U}^{ji}\right) + \varepsilon_i\bar{\varepsilon}_j\left(\hat{\vec{\vec{Z}}}_i\cdot\vec{U}^{ij}\right)\otimes\vec{Z}_j - \varepsilon_i\varepsilon_j\hat{\vec{\vec{Z}}}_i\cdot\vec{\vec{\Pi}}^{ij}\cdot\hat{\vec{\vec{Z}}}_j\ .$$

Several points concerning the tensors (4.5a-c) are worthy of comment. First, they are, in contrast to those proposed by other authors, defined *relative to two points*. Secondly, as we see from (4.6), they completely *seperate terms that depend on the Euclidean vector space (the unit vectors) from those that depend on the affine space and mass*. Finally, they contain a *complete* description of the mass distribution of the system. This explains their name *global inertia tensors* (GITs).

V. PROPERTIES AND STRUCTURES

The coefficents (4.1) have interesting properties, which we shall now study.

1. From (4.4), (4.5) we realize with *direct arguments* the well known functional (in)dependence:

(5.1) $$\partial\langle i,\cdots j\,|k,\cdots,l\rangle/\partial q_h = 0,\quad \text{if } \mathcal{B}_h \text{ is "inboard" of } \mathcal{B}_p\ \forall\, p\in\{i,\cdots,l\}.$$

2. Analyzing (4.2), we see that the following integrals (4.1) need to be calculated:

(5.2) $$\langle i,j\rangle\ \forall\ P_{ij}=1\ ;\qquad \langle i,jk\rangle\ \forall\ P_{ik}P_{jk}=1\ ;\qquad \langle i,jkl\rangle\ \forall\ P_{jk}P_{kl}=1\ ;\ \forall\, i\ .$$

3. As a kind of computational bonus we have

(5.3) $$\langle i,jk\rangle = \langle i,j'k\rangle \text{ and } \langle i,jkl\rangle = \langle i,j'kl\rangle = \langle i,jk'l\rangle,\ \text{if } \vec{Z}_j\,||\,\vec{Z}_{j'} \text{ or/and } \vec{Z}_k\,||\,\vec{Z}_{k'}.$$

4. Moreover, $\langle i,jkl\rangle$ can be deduced from $\langle l,jki\rangle$ by the relation

(5.4) $$\langle i,jkl\rangle - \langle l,jki\rangle = \vec{Z}_k\cdot\left(\mathrm{vec}\left(\vec{\vec{K}}^{il}\right)\times\vec{Z}_j\right),\ \text{where } \mathrm{vec}(\vec{a}\otimes\vec{b}) := 1/2\,\vec{b}\times\vec{a}$$

5. In order to gain an overall picture the integrals (4.1) may be expressed in terms of GITs. This can be done from (4.5) and (5.3) using vector algebra idendities. The results are summarized in Table 1.

	$\langle i,l\rangle$	$\langle i,jl\rangle$	$\langle i,jkl\rangle$
$\bar{\varepsilon}_i\bar{\varepsilon}_l=1$	$\vec{Z}_i\cdot\vec{Z}_l\,M^{il}$	$[\vec{Z}_i,\vec{Z}_j,\vec{Z}_l]\,M^{il}$	$(\vec{Z}_i\times\vec{Z}_k)\cdot(\vec{Z}_j\times\vec{Z}_l)\,M^{il}$
$\bar{\varepsilon}_i\varepsilon_l=1$	$[\vec{Z}_i,\vec{U}^{li},\vec{Z}_l]$	$[\vec{Z}_j,\vec{Z}_i\times\vec{U}^{li},\vec{Z}_l]$	$[\vec{Z}_l,\vec{U}^{li},\vec{Z}_k\times(\vec{Z}_j\times\vec{Z}_i)]$
$\varepsilon_i\bar{\varepsilon}_l=1$	$[\vec{Z}_i,\vec{U}^{il},\vec{Z}_l]$	$[\vec{Z}_i,\vec{Z}_l\times\vec{U}^{il},\vec{Z}_j]$	$[\vec{Z}_i,\vec{U}^{il},\vec{Z}_j\times(\vec{Z}_k\times\vec{Z}_l)]$
$\varepsilon_i\varepsilon_l=1$	$\vec{\vec{I}}\cdot\cdot\left(\hat{\vec{\vec{Z}}}_i\cdot\vec{\vec{\Pi}}^{il}\cdot\hat{\vec{\vec{Z}}}_l\right)$	$2\,\vec{Z}_j\cdot\mathrm{vec}\left(\hat{\vec{\vec{Z}}}_i\cdot\vec{\vec{\Pi}}^{il}\cdot\hat{\vec{\vec{Z}}}_l\right)$	$(\vec{Z}_k\times\vec{Z}_i)\cdot\vec{\vec{\Pi}}^{il}(\vec{Z}_j\times\vec{Z}_l) - (\vec{Z}_j\cdot\vec{Z}_k)\,\langle i,l\rangle$

Table 1: The integrals (4.1) as functions of GITs

As an example, two properties that can be directly obtained from Table 1 are stated:

(5.5a,b) $$\begin{cases} \langle i,jkl\rangle = 0 \text{ if } (\vec{Z}_l\,||\,\vec{Z}_k \text{ or } \vec{Z}_j\,||\,\vec{Z}_i) \text{ and } J_i,J_l\in P \\ \langle i,jkl\rangle = \vec{Z}_j\cdot\vec{Z}_k\,\langle i,l\rangle \text{ if } (\vec{Z}_i\,||\,\vec{Z}_k \text{ or } \vec{Z}_j\,||\,\vec{Z}_l) \text{ and } J_i,J_l\in R \end{cases}$$

It should be stressed that all these properties do *not* depend on the modelling algorithm but present *fundamental properties* of the mathematical models of tree-structured mechanisms.

VI. RECURSIVE EVALUATION OF GITs

Techniques to compute global inertia tensors $\vec{U}^{ij}$ and $\overrightarrow{\overrightarrow{\Pi}}^{ij}$ for spatial systems where first proposed by [FISCHER, 1905] (for $\vec{U}^{ij}$) and [RENAUD, 1980] (for $\overrightarrow{\overrightarrow{\Pi}}^{ij}$). They consist of the calculation of the mass moments of the *generalized body* $\mathcal{G}_i := \bigcup_{\{j \mid P_{ij}=1\}} \mathcal{B}_j$ as a function of the mass moments of $\{\mathcal{G}_j \mid D_{ij}=1\}$. To this end $\mathcal{B}_j^{+} := \mathcal{B}_j \bigcup_{\{k \mid D_{jk}=1\}} (O_k, M^k)$, the *augmented body* provides an excellent means. Let $\vec{u}_+^j$ and $\overrightarrow{\overrightarrow{\pi}}_+^j$ be the first and second moment of mass of $\mathcal{B}_j^+$ with respect to O_j:

(6.1a)
$$\vec{u}_+^j := \vec{u}^j + \sum_k D_{jk}\overrightarrow{O_jO_k}\, M^k \text{ , where } \vec{u}^j := \int_{\mathcal{B}_j} \overrightarrow{O_jM}dm \text{ and } M^k := \int_{\mathcal{G}_j} dm \text{ ,}$$

(6.1b)
$$\overrightarrow{\overrightarrow{\pi}}_+^j := \overrightarrow{\overrightarrow{\pi}}^j + \sum_k D_{jk}\left\{\overrightarrow{O_jO_k}\otimes\overrightarrow{O_jO_k}\, M^k\right\} \text{, where } \overrightarrow{\overrightarrow{\pi}}^j := \int_{\mathcal{B}_j} \overrightarrow{O_jM}\otimes\overrightarrow{O_jM}dm \text{ .}$$

Therefore,

(6.2 a,b)
$$\vec{U}^{jj} = \vec{u}_+^j + \sum_k D_{jk}\vec{U}^{kk} \quad \text{and} \quad \overrightarrow{\overrightarrow{\Pi}}^{jj} = \overrightarrow{\overrightarrow{\pi}}_+^j + \sum_k D_{jk}\left\{\overrightarrow{O_jO_k}\otimes\vec{U}^{kk} + \vec{U}^{kk}\otimes\overrightarrow{O_jO_k} + \overrightarrow{\overrightarrow{\Pi}}^{kk}\right\}$$

and

(6.3a,b)
$$\vec{U}^{ij}{}_{i\neq j} = P_{ij}\left(\overrightarrow{O_iO_j}\, M^{jj} + \vec{U}^{jj}\right) + P_{ji}\,\vec{U}^{ii};\ \overrightarrow{\overrightarrow{\Pi}}^{ij}{}_{i\leq j} = P_{ij}\left(\overrightarrow{O_iO_j}\otimes\vec{U}^{jj} + \overrightarrow{\overrightarrow{\Pi}}^{jj}\right);\ \overrightarrow{\overrightarrow{\Pi}}^{ji} = \overrightarrow{\overrightarrow{\Pi}}^{ijT} \text{ .}$$

With this, the sensitivity-model with respect to kinematic and geometric parameters is constructed.

Historical aside concerning augmented bodies: Otto FISCHER from Leipzig did more than any man before him, or than any one would do for several decades afterwards, on multibody dynamics. It appears to be one of those peculiar facets of history that O. FISCHER was too far ahead of his time and his work, although appreciated by leading mechanicians early this century (including HEUN, GRAMMEL and WINKELMANN), has essentially been lost after World War two. In 1893 he recognized the augmented body, which he termed the "reduziertes System", as an important construct in dynamics, but he was not the first. Twenty years before, a german schoolteacher by the name of Wilhelm August DUMAS derived and discussed the motion-equations of a special planar N-body pendulum,[DUMAS,1874]. Albeit for the simple case under consideration, he observed that the augmented body (he did not give a special name to it) significantly simplifies the description. W. A. DUMAS work contains the *earliest explicit statement of augmented bodies* that we have been able to find. This work seems never to have become widely known. Some more details are given in [PFISTER, 1994].

VII. INERTIA SENSITIVITY

We now turn to the study of inertia parameters $u_k^p(p)$, $\pi_{jk}^p(p)$ which, together with M^i (not m^i, the mass of $\mathcal{B}_i$!) make up the vector θ. $u_k^p(j)$ and $\pi_{jk}^p(j)$ are the components of the inertia tensors $\vec{u}^p$ and $\overrightarrow{\overrightarrow{\pi}}^p$. With respect to any basis $E_j := \{\vec{e}_1(j), \vec{e}_2(j), \vec{e}_3(j)\}$, they take the form

(7.1a,b)
$$\vec{u}^p = \sum_{k=1}^{3} u_k^p(j)\,\vec{e}_k(j) \text{ and } \overrightarrow{\overrightarrow{\pi}}^p = \sum_{j,k=1}^{3} \pi_{jk}^p(i)\,\vec{e}_j(i)\otimes\vec{e}_k(i) \text{ .}$$

The first and obvious remark is that the formulation of (4.1), provided by (4.4) and (4.6), expresses these integrals as linear functions of the GITs. Moreover, as we have alredy noted these tensors contain *all the inertia characteristics of the system.*. The upshot of this matter is that *the problem of inertia sensitivity-model construction is reduced to the analysis of* GITs. The recursive formulations of Chapter VI provide a convenient and straightforward means of evaluating the individual coefficients. The results are given in Table 2.

This approach to inertia sensitivity model construction conforms most naturally with the preceding Chapters, but other approaches are possible. We can construct the inertia sensitivity matrix taking the idea of the LUH-WALKER-PAUL method. This may involve fewer numerical calculations. On the other hand, our approach is of particular interest, because it offers deep insight into the structure of the coefficents:

(a) Most of the *analytic* methods proposed in [DOMBRE & KHALIL, 1988] for finding inertia parameters that are irrelevant to for the motion of the system, or others that can be grouped together can directly be seen from the results given in Table 2 , the definition (4.6) and the recursive relations of Chapter VI.

(b) The procedure directly allows one to write Lagrangian L in the form $L=\psi^T\theta$ and therefore applies to *energy-based identification methods*.

(c) As the joint velocities and accelerations figure explicitely in our model, it is an adequate tool if we are interested in special test motions, based on moving only one or two axes at a time [SEEGER,1992].

(d) A third advantage results from one's ability to construct, with few additional computations, the *inertia sensitivity of the linearized model*. This in turn allows one to apply existing *linear-system identification theories*.

	M^{ij}	$\vec{U}^{ij}$	$\vec{\vec{\Pi}}^{ij}$, $P_{ij}=1$
$\frac{\partial}{\partial M^p}$	$= \begin{cases} 1 \text{ if } p=\max(i,j) \\ 0 \text{ otherwise} \end{cases}$	$= \begin{cases} P_{ji}\,\overrightarrow{O_iO_p}, \text{ if } j \le i = a_p \\ P_{ji}\sum_l D_{il}\frac{\partial \vec{U}^{il}}{\partial M^p}, \text{ if } j \le i \ne a_p \\ P_{ip}\left(\overrightarrow{O_iO_j}\frac{\partial M^{ij}}{\partial M^p} + \frac{\partial \vec{U}^{jj}}{\partial M^p}\right), \text{ if } j > i \end{cases}$	$= \begin{cases} \overrightarrow{O_jO_p}\otimes\overrightarrow{O_jO_p}, \text{ if } i=j=a_p \\ \sum_l D_{jl}\left(\overrightarrow{O_jO_l}\otimes\frac{\partial \vec{U}^{il}}{\partial M^p} + \frac{\partial \vec{U}^{il}}{\partial M^p}\otimes\overrightarrow{O_iO_l} + \frac{\partial \vec{\vec{\Pi}}^{il}}{\partial M^p}\right), \text{ if } i=j\ne a_p \\ P_{ij}\left(\overrightarrow{O_iO_j}\otimes\frac{\partial \vec{U}^{jj}}{\partial M^p} + \frac{\partial \vec{\vec{\Pi}}^{jj}}{\partial M^p}\right), \text{ if } i<j \end{cases}$
$\frac{\partial}{\partial u^p_k(m)}$	$=\vec{0}\quad \forall k,l,m,p$	$= P_{j'p}\vec{e}_k(m)$ where $j'=\max(i,j)$	$= \begin{cases} \sum_l D_{jl}\left(\overrightarrow{O_jO_l}\otimes\frac{\partial \vec{U}^{il}}{\partial u^p_k(m)} + \frac{\partial \vec{U}^{il}}{\partial u^p_k(m)}\otimes\overrightarrow{O_iO_l} + \frac{\partial \vec{\vec{\Pi}}^{il}}{\partial u^p_k(m)}\right), \text{ if } i=j \\ \overrightarrow{O_iO_j}\otimes\frac{\partial \vec{U}^{jj}}{\partial u^p_k(m)} + \frac{\partial \vec{\vec{\Pi}}^{jj}}{\partial u^p_k(m)}, \text{ if } i<j \end{cases}$
$\frac{\partial}{\partial \pi^p_{kl}(m)}$	$=\vec{0}\quad \forall k,l,m,p$	$=\vec{0}\quad \forall k,l,m,p$	$= P_{jp}\,\vec{e}_k(m)\otimes\vec{e}_l(m)$

Table 2: Partial derivatives of GITs with respect to inertia parameters.

REFERENCES

ATKESON,C.&AN,C.&HOLLERBACH,J.: Estimation of Inertial Parameters of Manipulator Loads and Links, *Int. J Rob Res.*, Vol.5, No.3, 1986.

BENNIS,F.&KHALIL,W.&GAUTIER,M.: Calculation of the base inertial Parameters of Closed-loops robots,IEEE Conference on Robotics and Automation, Nice, France, 1992.

DOMBRE, M. & KHALIL, W.: *Modélisation et commande des robots*. Paris, Hermes, 1988.

DUMAS, W.A. Über Schwingungen verbundener Pendel. *Festschrift zur dritten Säcularfeier des Berlinischen Gymnasiums zum grauen Kloster*. Berlin, 1874.

FAYET, M. & PFISTER, F.: Analysis of multibody systems with indirect coordinates and global inertia tensors. *Europ. J. of Mech*, in press.

FAYET, M. & RENAUD, M.: Quasi-Minimal Computation under an Explicit Form of the Inverse Dynamic Model of a Robot-Manipulator. *Mech. Mach Theory*, 1989, Vol 24, No. 3 pp.165-174.

FISCHER, O.: Die Arbeit der Muskeln und die lebendige Kraft des menschlichen Körpers. *Abh. der math.-phys Klasse d Königl Saechs. Gesellsch d Wissensch*, 1893, Vol 20, No.1.

FISCHER, O.: Über die Bewegungsgl. räuml Gelenksysteme. *Abh der math -phys. Klasse d Königl Sächs. Gesellsch d. Wissensch*, 1905, Vol.29, No.1.

GAUTIER, M.: *Contribution à la modélisation et à l'identification des robots*. Thèse de Doctorat d'état, E.N.S.M., 1990, 265 p..

GAUTIER, M &KHALIL, W ·Exciting Trajectories for the Identification of Base Inertial Parameters of Robots, Int. J. Rob. Res., Vol. 11, No.4, 1992.

LI, C. J. et al.: A New Computational Method for Linearized Dynamic Models for Robot Manipulators. Int. J. of Rob. Res., Vol.9, No. 1, 1990.

MAYEDA, H., YOSHIDA, K. & OSUKA, K.: Base parameters of manipulator dynamics. *Proc IEEE Conf on Rob and Autom*, Philadelphia, 1988.

MILLS, J. & ANDREW, A. K.· Constrained motion task of robotic manipulators. *Mech Mach Theory*, Vol. 29, No. 1, pp. 95-114, 1994.

PFISTER, F.: Die Bewegungsgleichungen der Kreiselketten, Dem Erweiterten Körper zum hundertsten Geburtstag. *GAMM-Jahrestagung*. Braunschweig, 1994.

SEEGER, G.· Selbsteinstellende, modellgestützte Regelung eines Industrieroboters. Braunschweig: Viehweg, 1992.

VUCOBRATOVIC, M. & KIRCANSKI, N.:Computer-oriented Method for Linearization of Dynamic Models of Active Spatial Mechanisms. *Mech and Mach Theory*, Vol. 17, No. 1, pp.21-32, 1982.

VUCOBRATOVIC, M. & KIRCANSKI, N.: Computer assisted Sensitivity Model Generation in Manipulation Robots Dynamics.*Mech and Mach Theory*, Vol. 19,No.2, pp 223-233, 1984.

VUCOBRATOVIC, M. & STOKIC,D.: Control of Manipulation Robots. Berlin: Springer, 1982.

YOUCEF-TOUMI, K. & ASADA, H.: The Design of Open-Loop Manipulator Arms with Decoupled and Configuration-Invariant Inertia Tensors, *Proc. of IEEE International Conference on Robotics and Automation*, San Francisco, pp. 2018-2026, 1986.

YANG, D.C.H. & TZENG, S.W.: Simplification and Linearization of Manipulator Dynamics by the Design of Inertia Distribution, *Int. J. Rob. Res*, Vol.5, No. 3, pp.261-268, 1989.

SINGULARITY ANALYSIS OF SPHERICAL PARALLEL MANIPULATORS

H.R.M. Daniali, P.J. Zsombor-Murray and J. Angeles
McGill University, Montreal, Quebec, Canada

Abstract. The Jacobian matrices of two spherical manipulators of the parallel type, with three degrees of freedom, are derived. One is a spherical 3-legged parallel manipulator; the other, a spherical double-triangular manipulator. A general classification of parallel-manipulator singularities into three groups, which relies on the properties of the Jacobian matrices of the manipulator, is described. Finally, the three types of singularity are identified for the two manipulators.

1. Introduction

Manipulator accuracy can be characterized using some properties of the associated Jacobian matrices. Germane to accuracy is the concept of singularities, which have been extensively studied in the realm of serial manipulators (Sugimoto et al., 1982; Litvin et al., 1986; Lai and Yang, 1986; Angeles et al., 1988). On the other hand, with regard to manipulators with kinematic loops, works found in the literature are more recent (Merlet, 1989; Gosselin and Angeles, 1990; Ma and Angeles, 1992; Gosselin and Sefrioui, 1992; Mohammadi, Zsombor-Murray and Angeles, 1993). In the last reference, a general classification of the singularities pertaining to planar parallel manipulators into three main groups has been suggested. This classification is based on the singularities of the two Jacobian matrices that relate the input rates to the Cartesian speeds. Here, we recall the above-mentioned classification of singularities encountered in parallel manipulators. The three types of singularities will be illustrated with some examples. These are a spherical 3-dof three-legged (3L) parallel manipulator, Fig. 1a, and a spherical 3-dof double-triangular (DT) parallel manipulator, Fig. 1b.

2. Jacobian Matrices

The general relationship between actuator coordinates, expressed by the 3-dimensional vector $\boldsymbol{\theta}$, and the Cartesian coordinates, grouped in vector $\mathbf{x}$, in parallel mechanisms of the spherical type can be written as:

$$\mathbf{f}(\boldsymbol{\theta}, \mathbf{x}) = \mathbf{0} \tag{1}$$

where $\mathbf{f}$ is a vector function of vectors $\boldsymbol{\theta}$ and $\mathbf{x}$. We assume here that an invariant representation of rotations has been adopted, based on either linear or quadratic invariants (Angeles, 1988), which resort to a 4-dimensional vector representation, and hence, $\mathbf{x}$ is 4-dimensional. The underlying differential kinematic relations can be derived for the case at hand from simple kinematic relations between the joint rates and the angular velocity of the end-effector (EE). The resulting differential relations take the general form

$$\mathbf{J}\dot{\boldsymbol{\theta}} + \mathbf{K}\boldsymbol{\omega} = \mathbf{0} \tag{2}$$

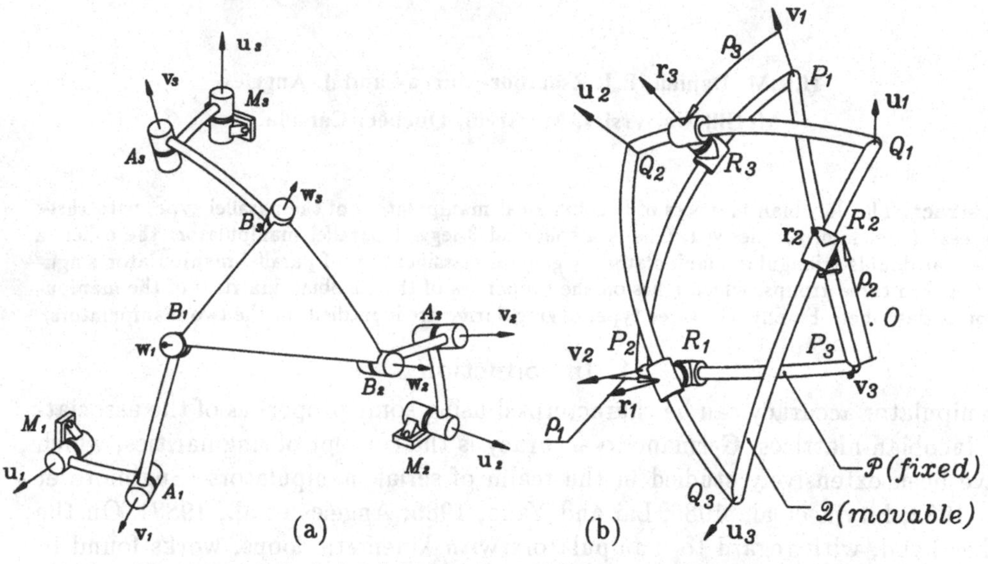

Fig. 1. a) Spherical 3-L Manipulator; b) Spherical DT Manipulator

where **J** and **K** are the 3×3 Jacobian matrices whose properties are of paramount interest in this study. Moreover, $\dot{\boldsymbol{\theta}}$ is the 3-dimensional joint-rate vector, while $\boldsymbol{\omega}$ is the 3-dimensional angular-velocity vector, representing the Cartesian velocity.

Below we derive expressions for **J** and **K** for two types of spherical 3-dof parallel manipulators.

2.1. 3L Manipulator

A spherical parallel manipulator is depicted in Fig. 1a, that we term, *faute de mieux*, 3-legged (3L) manipulator. All the joints of this manipulator are revolutes and the three motors, M_1, M_2 and M_3, are fixed to the base. The angular velocity $\boldsymbol{\omega}$ of the EE can be written as

$$\dot{\theta}_i \mathbf{u}_i + \dot{\alpha}_i \mathbf{v}_i + \dot{\gamma}_i \mathbf{w}_i = \boldsymbol{\omega}, \quad i = 1,2,3 \tag{3}$$

where $\mathbf{u}_i$, $\mathbf{v}_i$ and $\mathbf{w}_i$ are the unit vectors pointing from the center of the sphere to points M_i, A_i and B_i, respectively. Moreover, $\dot{\theta}_i$, $\dot{\alpha}_i$ and $\dot{\gamma}_i$ are the rates of the joint attached to the base, the intermediate joint and the joint attached to EE, as pertaining to the ith leg, respectively. Below we eliminate the rates of the unactuated joints by dot-multiplying both sides of the foregoing equation by $\mathbf{v}_i \times \mathbf{w}_i$, thereby obtaining

$$\dot{\theta}_i \mathbf{u}_i \cdot (\mathbf{v}_i \times \mathbf{w}_i) = \boldsymbol{\omega} \cdot (\mathbf{v}_i \times \mathbf{w}_i), \quad i = 1,2,3 \tag{4}$$

which can be written in turn as

$$\dot{\theta}_i (\mathbf{u}_i \times \mathbf{w}_i) \cdot \mathbf{v}_i + (\mathbf{v}_i \times \mathbf{w}_i)^T \boldsymbol{\omega} = 0, \quad i = 1,2,3 \tag{5}$$

The above equation for $i = 1,2,3$, are now assembled in the form of eq.(2) in which **J** and **K** are defined as

$$\mathbf{J} \equiv \begin{bmatrix} (\mathbf{u}_1 \times \mathbf{w}_1) \cdot \mathbf{v}_1 & 0 & 0 \\ 0 & (\mathbf{u}_2 \times \mathbf{w}_2) \cdot \mathbf{v}_2 & 0 \\ 0 & 0 & (\mathbf{u}_3 \times \mathbf{w}_3) \cdot \mathbf{v}_3 \end{bmatrix}, \quad \mathbf{K} \equiv \begin{bmatrix} (\mathbf{v}_1 \times \mathbf{w}_1)^T \\ (\mathbf{v}_2 \times \mathbf{w}_2)^T \\ (\mathbf{v}_3 \times \mathbf{w}_3)^T \end{bmatrix} \tag{6}$$

2.2. DT Manipulator

A spherical parallel manipulator of a different class is depicted in Fig. 1b; it consists of two spherical triangles, $\mathcal{P}$ and $\mathcal{Q}$, with vertices P_1, P_2, P_3 and Q_1, Q_2, Q_3, respectively. Because of its special architecture, we call this class double-triangular (DT). Triangle $\mathcal{P}$ is designated to be the *fixed triangle*, while $\mathcal{Q}$ is the *movable triangle*. The motion of triangle $\mathcal{Q}$ can be described through the arc lengths ρ_i, or *actuator coordinates*, for $i = 1, 2, 3$.

We introduce the normalized vectors $\mathbf{w}_i$ and $\mathbf{t}_i$, for $i = 1, 2, 3$, which are perpendicular to the planes of arcs $P_{i+1}P_{i+2}$ and $Q_{i+1}Q_{i+2}$, respectively, i.e.,

$$\mathbf{w}_i = (\mathbf{v}_{i+1} \times \mathbf{v}_{i+2})/\|\mathbf{v}_{i+1} \times \mathbf{v}_{i+2}\|, \quad \mathbf{t}_i = (\mathbf{u}_{i+1} \times \mathbf{u}_{i+2})/\|\mathbf{u}_{i+1} \times \mathbf{u}_{i+2}\| \tag{7}$$

where $\mathbf{v}_i$ and $\mathbf{u}_i$ are both unit vectors directed from O to P_i and Q_i, respectively.

The angular velocity $\boldsymbol{\omega}$ of the EE can now be written as

$$\dot{\theta}_i \mathbf{w}_i + \dot{\alpha}_i \mathbf{r}_i + \dot{\gamma}_i \mathbf{t}_i = \boldsymbol{\omega}, \quad i = 1, 2, 3 \tag{8}$$

where $\mathbf{r}_i$ is the unit vector directed from the center of the sphere to R_i. Moreover, α_i is the angle between planes of P_{i+1}, P_{i+2} and Q_{i+1}, Q_{i+2}, while γ_i is the angle between $\mathbf{u}_{i+1}$ and $\mathbf{r}_i$. Furthermore, we have $\dot{\theta}_i = \dot{\rho}_i$, for $i = 1, 2, 3$, in which the radius of the sphere is assumed to be of unit length, without loss of generality.

The inner product of both sides of eq.(8) with $\mathbf{r}_i \times \mathbf{t}_i$, upon simplification, leads to an equation free of rates of unactuated joints, namely,

$$\dot{\theta}_i(\mathbf{r}_i \times \mathbf{t}_i) \cdot \mathbf{w}_i - (\mathbf{r}_i \times \mathbf{t}_i) \cdot \boldsymbol{\omega} = 0, \quad i = 1, 2, 3 \tag{9}$$

The above equations, for $i = 1, 2, 3$, are now assembled in the form of eq.(2) in which $\mathbf{J}$ and $\mathbf{K}$ are defined as

$$\mathbf{J} \equiv \begin{bmatrix} (\mathbf{r}_1 \times \mathbf{t}_1) \cdot \mathbf{w}_1 & 0 & 0 \\ 0 & (\mathbf{r}_2 \times \mathbf{t}_2) \cdot \mathbf{w}_2 & 0 \\ 0 & 0 & (\mathbf{r}_3 \times \mathbf{t}_3) \cdot \mathbf{w}_3 \end{bmatrix}, \quad \mathbf{K} \equiv \begin{bmatrix} -(\mathbf{r}_1 \times \mathbf{t}_1)^T \\ -(\mathbf{t}_2 \times \mathbf{t}_2)^T \\ -(\mathbf{t}_3 \times \mathbf{t}_3)^T \end{bmatrix} \tag{10}$$

3. Singularity Analysis

In parallel manipulators, singularities occur whenever $\mathbf{J}$ or $\mathbf{K}$ becomes singular. Thus, for these manipulators, a distinction can be made among three types of singularities, which have different kinematic interpretations, namely,

1) The first type of singularity occurs when $\mathbf{J}$ becomes singular but $\mathbf{K}$ is invertible. This type of singularity consists of the set of points where different branches of the inverse kinematic problem meet. Since the nullity of $\mathbf{J}$ is not zero, we can find a set of actuator velocity vectors $\dot{\boldsymbol{\theta}}$ for which the Cartesian velocity vector $\boldsymbol{\omega}$ is zero. Then, some Cartesian velocity vectors $\mathbf{K}\boldsymbol{\omega}$, those lying in the nullspace of $\mathbf{J}^T$, cannot be produced.

2) The second type of singularity, occurring only in closed kinematic chains, arises when $\mathbf{K}$ becomes singular but $\mathbf{J}$ is invertible. This type of singularity consists of a point or a set of points whereby different branches of the direct kinematic problem meet, the latter being understood here as the computation of the values of the Cartesian variables from given values of the driving-joint variables. Since the nullity of $\mathbf{K}$ is not zero, we can find a set of nonzero Cartesian velocity vectors

ω for which the actuator velocity vector $\dot{\boldsymbol{\theta}}$ is zero. Then, the mechanism gains one or more degrees of freedom or, equivalently, cannot resist moments in one or more directions, even if all the actuators are locked.
3) The third type of singularity occurs when both **J** and **K** are simultaneously singular, while none of the rows of **K** vanishes. Under a singularity of this type, configurations arise for which the EE of the manipulator can undergo finite motions even if the actuators are locked or, equivalently, it cannot resist moments in one or more directions over a finite portion of the workspace, even if all the actuators are locked. As well, a finite motion of the actuators produces no motion of the EE and some of the Cartesian velocity vectors cannot be produced.

Furthermore, depending on the formulation, it can happen that one or more rows of **K** vanish. It turns out, then, that the corresponding rows of **J** vanish as well, **J** and **K** thus becoming singular simultaneously. In other words, the formulation leads to the third type of singularity. In this case, it is possible to reformulate the problem and the new formulation may lead to any of the three types of singularities. If this is not the case, we do not have a singular configuration at all. Therefore, this type of singularity, which arises merely from the way in which the kinematic relations are formulated, is in fact a *formulation singularity*.

3.1. 3L PARALLEL MANIPULATOR

In this subsection the three types of singularities discussed above are investigated for the manipulator of Fig. 1a. It is recalled that the first type of singularity occurs when the determinant of **J** vanishes. From eq.(6) this condition yields

$$(\mathbf{u}_i \times \mathbf{w}_i) \cdot \mathbf{v}_i = 0 \quad \text{for } i = 1 \text{ or } 2 \text{ or } 3 \tag{11}$$

This type of configuration is reached whenever $\mathbf{u}_i$, $\mathbf{w}_i$ and $\mathbf{v}_i$, for $i = 1$ or 2 or 3, are coplanar, which means that one or some of the legs are fully extended or folded. At each of these configurations the motion of one actuator, that corresponding to the fully extended or folded leg, does not produce any motion of the EE.

The second type of singularity occurs when the determinant of **K** vanishes, which occurs when the rows or columns of **K** are linearly dependent. By inspection of eq.(6), we claim that this type of singularity occurs when the three planes defined by the axes of the revolutes parallel to the unit vectors $\{\mathbf{v}_i, \mathbf{w}_i\}_1^3$ intersect at a common line. This can be readily seen by noting that the three vectors $\mathbf{v}_i \times \mathbf{w}_i$, for $i = 1, 2, 3$, which are perpendicular to the plane of $\mathbf{v}_i$ and $\mathbf{w}_i$, are perpendicular to the intersection line. Then, these vectors are coplanar and each of them, representing a row of **K**, can be written as a linear combination of the other two, the claim thus being made apparent. This type of singularity is depicted in Fig. 2a.

The third type of singularity occurs when the determinants of **J** and **K** both vanish. We have this type of singularity whenever the two foregoing singularities occur simultaneously. In this case $\mathbf{k}_i \neq \mathbf{0}$, where $\mathbf{k}_i^T$, for $i = 1, 2, 3$, is the ith row of **K**. At this type of singularity the motion of at least one actuator does not

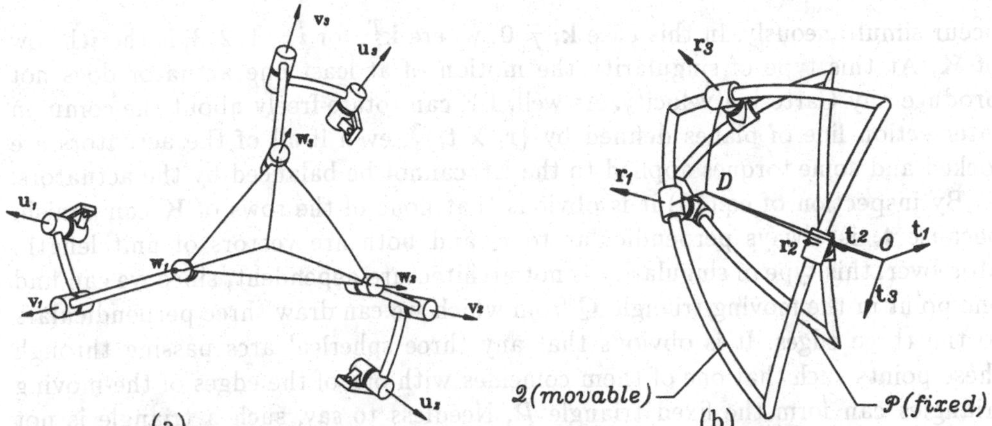

Fig. 2. The Second Type of Singularity of a) a 3L Manipulator; b) a DT Manipulator

produce any Cartesian velocity. As well, the gripper can rotate freely about the common intersection line of planes defined by the axes of the revolutes parallel to the unit vectors $\{\mathbf{v}_i, \mathbf{w}_i\}_1^3$, even if all of the actuators are locked and some torques applied to the EE cannot be balanced by the actuators.

This type of singularity is not architecture-dependent, because we can change the link lengths M_iA_i and A_iB_i, for $i = 1, 2, 3$, correspondingly, while maintaining the third type of singular posture. Moreover, by inspection of eq.(6) it is apparent that the ith row of $\mathbf{K}$ vanishes only if $\mathbf{v}_i = \pm\mathbf{w}_i$. In this case we have a degenerate case of a 3L manipulator with one leg of zero or π length. Such a manipulator is irrelevant for our study and is thus left aside.

3.2. DT Manipulator

In this subsection the three types of singularities are investigated for the manipulator of Fig. 1b. It is recalled that the first type of singularity occurs when the determinant of $\mathbf{J}$ vanishes. From eq.(10) this condition yields

$$(\mathbf{r}_i \times \mathbf{t}_i) \cdot \mathbf{w}_i = 0 \quad \text{for } i = 1 \text{ or } 2 \text{ or } 3 \tag{12}$$

This type of configuration is reached whenever $\mathbf{w}_i$ is perpendicular to $\mathbf{r}_i \times \mathbf{t}_i$, but $\mathbf{r}_i$ lies in the plane whose normal is $\mathbf{w}_i$. Then, this type of singularity occurs whenever $\mathbf{t}_i$ and $\mathbf{w}_i$ coincide. In other words, each pair of two sides of the two triangles lies in the same plane.

The second type of singularity occurs when the determinant of $\mathbf{K}$ vanishes, which occurs in turn when the rows or columns of $\mathbf{K}$ are linearly dependent. By inspection of eq.(10), we claim that this type of singularity occurs when the three planes containing vectors $\mathbf{r}_i$ and $\mathbf{t}_i$ intersect at a common line. This can be readily seen by noting that the three vectors $\mathbf{r}_i \times \mathbf{t}_i$, for $i = 1, 2, 3$, which are perpendicular to the planes, are perpendicular to the intersection line as well. Then, these vectors are coplanar and each of them, which represents a row of $\mathbf{K}$, can be written as a linear combination of the other two, the claim thus being made apparent. This type of singularity is depicted in Fig. 2b.

The third type of singularity occurs when the determinants of $\mathbf{J}$ and $\mathbf{K}$ both vanish. We have this type of singularity whenever the two foregoing singularities occur simultaneously. In this case $\mathbf{k}_i \neq \mathbf{0}$, where $\mathbf{k}_i^T$, for $i = 1, 2, 3$, is the ith row of $\mathbf{K}$. At this type of singularity the motion of at least one actuator does not produce any Cartesian velocity. As well, EE can rotate freely about the common intersection line of planes defined by $\{\mathbf{r}_i \times \mathbf{t}_i\}_1^3$, even if all of the actuators are locked and some torques applied to the EE cannot be balanced by the actuators.

By inspection of eq.(10) it is obvious that none of the rows of $\mathbf{K}$ can vanish, because $\mathbf{t}_i$ is always perpendicular to $\mathbf{r}_i$ and both are vectors of unit length. Moreover, this type of singularity is not architecture-dependent, since we can find one point in the moving triangle $\mathcal{Q}$ from which we can draw three perpendiculars to the three edges. It is obvious that any three spherical arcs passing through these points such that one of them coincides with one of the edges of the moving triangles can form the fixed triangle $\mathcal{P}$. Needless to say, such a triangle is not unique. In other words, we can choose the fixed and moving triangles arbitrarily.

4. Conclusion

A general classification of singularities of parallel manipulators into three groups was given. The three types of singularities, which have different kinematic interpretations for these manipulators, were identified. Moreover, it has been shown that the third type of singularity is not necessarily architecture-dependent.

Acknowledgements

This research work was supported by NSERC (Natural Sciences and Engineering Research Council, of Canada) under Grants OGP0004532, OGP0139964, STRGP 205, and EQP00-92729. Support from IRIS, the Institute for Robotics and Intelligent Systems, a network of Canadian centres of excellence, is also acknowledged. The first author was funded by the Ministry of Culture and Higher Education of the Islamic Republic of Iran.

References

Angeles, J.: 1988, *Rational Kinematics*, Springer-Verlag, New Yory.

Angeles, J., Anderson, K., Cyril, X. and Chen, B.: 1988, 'The Kinematic Inversion of Robot Manipulators in the Presence of Singularities', *ASME J. of Mechanisms, Transmissions, and Automation in Design* **vol. 110**, pp. 246–254.

Gosselin, C. and Angeles, J.: 1990, 'Singularity Analysis of Closed-Loop Kinematic Chains', *IEEE Transactions on Robotics And Automation* **vol. 6, No. 3**, pp. 281–290.

Gosselin, C. M. and Sefrioui, J.: 1992, 'Determination of the Singularity Loci of Spherical Three-Degree-of-Freedom Parallel Manipulators', *Proc. of the ASME Mechanisms Conf., Phoenix*, pp. 329–335.

Lai, Z. C. and Yang, D. C. H.: 1986, 'A New Method for the Singularity Analysis of Simple Six-Link Manipulators', *Int. J. of Robotics Research* **vol. 5, No. 2**, pp. 66–74.

Litvin, F. L., Yi, Z., Parenti-Castelli, V. and Innocenti, C.: 1986, 'Singularities, Configurations, and Displacement Functions for Manipulators', *Int. J. of Robotics Research* **vol. 5, No. 2**, pp. 52–65.

Ma, O. and Angeles, J.: 1992, 'Architecture Singularities of Parallel Manipulators', *The Int. J. of Robotics and Automation* **vol. 7, No. 1**, pp. 23–29.

Merlet, J-P.: 1989, 'Singular Configurations of Parallel Manipulators and Grassmann Geometry', *Int. J. of Robotics Research* **vol. 8, No. 5**, pp. 45–56.

Mohammadi Daniali, H. R., Zsombor-Murray, P. J. and Angeles, J.: 1993, 'On Singularities of Planar Parallel Manipulators', *Proc. ASME Conf. on Design Automation, Albuquerque*, pp. 593–599.

Sugimoto, K., Duffy, J. and Hunt, K. H.: 1982, 'Special Configurations of Spatial Mechanisms and Robot Arms', *Mechanism and Machine Theory* **vol. 17, No. 2**, pp. 119–132.

DYNAMICS OF EVOLVING SYSTEMS

A. Barraco and B. Cuny

Ecole Nationale Supérieure d'Arts et Métiers, Paris, France

2 2-02-1995 RO.MAN.SY. 10

1- Introduction.

Writing and solving the set of equations describing the dynamic behavior of spatial complex systems built of rigid bodies is now a resolved problem. The same problem for mechanisms composed of both rigid and flexible members is quite solved, and many references are available for both subjects.

The problem we adress in this paper is the treatment of evolving systems. We know how to write the equations for a given system with constant mass and permanent boundary conditions. If one or both of these properties change, do we have to treat this problem as a completely new problem or can we formulate the problem in such a manner that the amount of computational time is minimized ?

The greatest part of the total computational time is devoted to calculate theinertial, Coriolis and centrifugal terms. If our goal is to minimize the total time, we will first have to minimize the change, or to keep constant the inertial, Coriolis and centrifugal matrices.

A walking machine is a typical problem with changing boundary conditions, especially if we suppose that we do not know which point will be the next contacting point. Two robots grasping the same load, that each of them can not bear, is an example in which three independent open loop systems become a single closed loop system and in which the natural treatment of each independent system would lead, if applied to the closed loop system, to a very complex formulation.

In this paper we develop an approach to treat evolving systems. It is based on first computing once the inertial, Coriolis and centrifugal matrices and then introducing the effects of topological changes by complementary equations. In the next section we begin first to summarize the solution of what we may call the "standard problem", constant mass and permanent boundary conditions and then we present a methodology to treat the evolving systems. In the last section we show two examples of simple applications.

2- Solution for the "standard problem".

We define the "standard problem" as a problem with constant mass and permanent boundary conditions. The complete solution for such a problem is formed by two parts : first the equations of motion the solution of which gives the kinematic parameters as functions of the external given generalized forces then the remaining equations of

Newton and Euler to retrieve the unknown internal forces and torques at each link. It is well known that the first part of the solution does always exist and that for open loop systems the second part of the solution is always complete while for closed loop systems it may be complete or incomplete. Therefore for any open or closed loop system the set of equations describing the dynamic solution is :

$$\mathbf{A}(\mathbf{q}).\ddot{\mathbf{q}}+\mathbf{B}(\mathbf{q},\dot{\mathbf{q}})=\tau$$
$$\mathbf{X}_l=\mathbf{X}_l(\mathbf{q},\dot{\mathbf{q}},\ddot{\mathbf{q}},\tau) \qquad (1)$$

where $\mathbf{q}$ is the column matrix containing the " m " independent parameters sufficient to describe at any time the geometric position, velocities and accelerations. $\mathbf{A}(\mathbf{q})$ is the square symmetricinertial matrix, $\mathbf{B}(\mathbf{q},\dot{\mathbf{q}})$ is the column vector containing the Coriolis and centrifugal effects, τ is the column vector of the given external generalized forces, $\mathbf{X}_l$ is the matrix containing the internal unknown forces and torques expressed as a function of the parameters and the external given forces. If $\mathbf{X}_l$ is a subset of the complete set of all the internal forces the solution is not complete.

For a spatial closed loop system (S), equations (1) are very difficult or almost impossible to obtain, and we prefer to solve an associated larger problem obtained by defining an associate open loop system (So) with constraint equations which force the solution of problem (So) to be identical to the solution of the initial problem (S). The set of equations (1-1) is now written as :

$$\mathbf{A}_o.\ddot{\mathbf{q}}_o+\mathbf{B}_o=\mathbf{H}_o^T.\lambda_o+\tau_o$$
$$\mathbf{J}_o(\mathbf{q}_o)=0 \ , \ \mathbf{H}_o.\dot{\mathbf{q}}_o=0 \ , \ \mathbf{H}_o.\ddot{\mathbf{q}}_o-\varphi_o=0 \qquad (2).$$

The first equation of (2) has the same meaning as equation (1), except for the term $\mathbf{H}_o^T\,\lambda$ which contains the Lagrange multipliers associated with the links that were opened. The second equation, written in the three forms presented, is called the constraint equation and forces the solution of (So) to be identical with that of problem (S).

The equations (2) are written in a single matricial equation as :

$$\begin{bmatrix}\mathbf{A}_o & -\mathbf{H}_o^T\\ -\mathbf{H}_o & 0\end{bmatrix}\begin{bmatrix}\ddot{\mathbf{q}}_o\\ \lambda_o\end{bmatrix}+\begin{bmatrix}\mathbf{B}_o\\ 0\end{bmatrix}=\begin{bmatrix}\tau_o\\ -\varphi_o\end{bmatrix} \qquad (3)$$

Once the solution of (3) is established, the equation (1-2) is used to express the unknown internal general forces. The solution of (3) is now well known and references can be found in Ref. [1].

Of particular interest, for the treatment of evolving system, is the evaluation of the different costs required to establish each of the terms of (3) and to solve the matricial equation. It is obvious that the main cost lies in computing matrices $\mathbf{A}_o$, $\mathbf{B}_o$ and φ , while $\mathbf{H}_o$ matrix is quite easy to establish. Several articles deal with the best method to obtain these matrices with the minimum number of operations, Ref. [2]. If the set of parameters, $\mathbf{q}_o$, becomes inadequate to describe the position all the matrices have to be computed again and their sizes are modified.

3- The evolving systems.

We consider three kinds of instantaneous changes for a mechanical system : the mass changes, the boundary conditions changes, the velocities and accelerations discontinuities. These changes may occur each alone, independently, or with a combination of two or three of them. But we will consider, in the next sections, the effect of each of them independently of the others.

3-1 Non-permanent boundary conditions.

Two cases must be distinguished. The first one is the case of one hinge which was active and becomes inactive : for instance two bodies previously tied together are separated and behave like two independent bodies. The second one is the creation of one or more hinges, which did not exist beforehand and become now active : for instance two robots grasping the same load.

This second case is easy to treat. As the set of kinematic parameters does not change, all matrices, except matrix H, are kept and one or more new constraint equations are added so that equations (2-2) are updated.

The treatment of the first case depends on whether or not the hinge, which was active and becomes inactive, belongs to the set of hinges previously opened to transform the closed loop system into an open loop system. If the hinge, which becomes inactive, belongs to this set we simply have to remove the corresponding constraint equation from equation (2-2) and follow the normal procedure. Conversely if the hinge which becomes inactive did not belong to this set then the set of kinematics parameters is no more sufficient to describe the position of the mechanism and one or more parameters must be added. In this case the matrices have to be computed again.

3-2 The mass is modified.

The groups of terms with geometry and mass dependency appearing in equation (2) are time invariant if each individual body's mass is constant.

A robot leaving or grasping a load has its last member's mass changing, and while it is still described by the same sets of parameters and boundary conditions, although most of the terms of equations (2) are changed. We have to update the groups in which the new terms appear. If the problem is described by equation (1) this procedure is not possible, whereas if the problem is expressed by equation (2), a lot of references in Ref. [3] are available.

For instance two robots grasping the same body may be considered as one single completely new mechanical system or as a collection of three independent systems linked by constraint equations. In the first modeling all the terms of the different matrices are expressed in terms of all body masses, and there is no way to deduce them from the previous terms calculated for each of the three independent systems. In the second case all the matrices calculated are still valid and the different constraint equations are easy to compute.

3-3 The impact problem.

When an impact occurs, velocities, accelerations and internal forces change instantaneously, while the position and external forces remain constant. Several cases of impact may be considered : one body of one system undergoes an impact with some external moving or fixed object, or two bodies of two different systems come into contact with impact. If we suppose that the impact occurs at one single point, the different equations available are the equation related to the conservation of the kinetic quantities and the local equation related to the change of the normal velocity at the tangent plane of contact. This local equation implies a coefficient, e , of energy restitution. The set of equations is now :

$$\begin{aligned} &\mathbf{A}_0 . \Delta\dot{\mathbf{q}}_0 = \mathbf{H}_0^T . \hat{\lambda}_0 + \mathbf{H}_i \hat{\lambda}_i \\ &\mathbf{H}_0^T . \Delta\dot{\mathbf{q}}_0 = 0 \qquad (4) \\ &\Delta\vec{\mathbf{V}}_i^{+} . \vec{\mathbf{n}} = -e . \Delta\vec{\mathbf{V}}_i^{-} . \vec{\mathbf{n}} \end{aligned}$$

where Δ stands for variation during impact, the hat symbol for impulses and the subscript i for the point where impact occurs.

When an impact is detected, these equations are used to compute the new velocities and accelerations at the end of the impact and then the normal procedure restarts.

3-4 The methodology for evolving systems.

We will consider the case of systems with predictable evolution. We suppose that we know what will happen and where it will happen but not when. This type of problem covers a large field of applications, including walking machines and co-operation of robots.

The previous chapters lead us to conclude that the worst case is when the set of kinematic parameters has to be changed. In order to avoid this flaw we define the total system by a set of subsystems (Sa) of less or equal size to (S) which are described by a subset of parameters which remains unaffected by all changes occurring during the total time of analysis. Each of these subsystems (Sa) may be composed of closed loop systems in which case an associated open loop system is used, but we have to be sure that this subset of equations will always use the same subset of parameters. Then the equations for (Sa) are :

$$\begin{aligned} &\mathbf{A}_a . \ddot{\mathbf{q}}_a + \mathbf{B}_a = \mathbf{H}_{aa}^T . \lambda_{aa} + \tau_a + \sum_b \mathbf{H}_{ab}^T . \lambda_{ab} \qquad (5) \\ &\mathbf{H}_{aa}^T . \dot{\mathbf{q}}_a = 0 \quad , \quad \mathbf{H}_{ab}^T . \dot{\mathbf{q}}_a + \mathbf{H}_{ba}^T . \dot{\mathbf{q}}_b = 0 \end{aligned}$$

where λ_{ab} represents the Lagrange multipliers, which appear at the links between subsystem (Sa) and all other subsystem (Sb), λ_{aa} represents the Lagrange multipliers at internal opened hinges to obtain the associate open loop. The number of multipliers λ_{aa} is time independent, but λ_{ab} is evolving with time according to the status of the different hinges, wether active or inactive.

The constraint equations take the two forms presented in (5-2).

The dynamic behavior of the system is described by equations (5). Any modification of the mass or the boundary conditions affects equation (5-2) and the right hand size of (5-1), but leaves the main matrices of (5-1) unchanged.

In order to know when a change occurs, we monitor some functions whose changes in values indicate either a contact, with or without impact, or the loss of contact between two specified points.
The following two examples illustrate the use and the possibilities of the proposed method.

5- Examples.
The first example shows the different situations encountered in the treatment of a walking machine. The mechanism, represented figure n°1, is constituted by one mass and two bodies, moving in a plane.

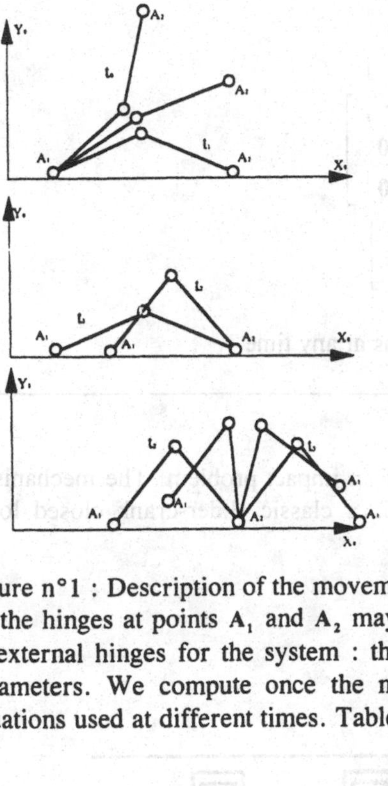

For time $t \in [t_0, t_1]$ the revolute joint at point A_1 is free to move until it comes into contact without impact with the $\bar{X}_0$ axis. Then for time $t \in [t_1, t_2]$ point A_2 is fixed and point A_1 slides on the $\bar{X}_0$ axis. At time t_1 this point leaves the $\bar{X}_0$ axis, A_2 remains fixed and for $t \in [t_2, t_3]$ A_1 is free until it comes into contact with some point of the $\bar{X}_0$ axis.
For $t \in [t_0, t_1]$ and $t \in [t_2, t_3]$ the mechanism is simply a reverse double pendulum, and for $t \in [t_1, t_2]$ it becomes a slider-crank mechanism, which is a closed loop system.

Figure n°1 : Description of the movement.
As the hinges at points A_1 and A_2 may be active or inactive we must select these points as external hinges for the system : the system is represented on Figure n°2 with four parameters. We compute once the matrices and then construct the different sets of equations used at different times. Table n°1 shows these sets.

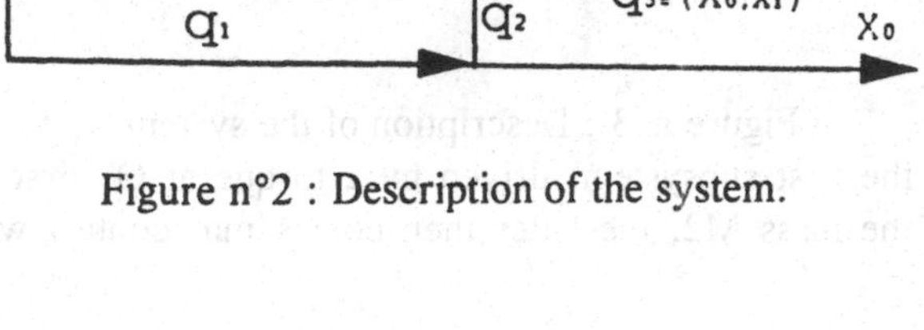

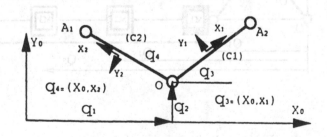

Figure n°2 : Description of the system.

$$t \in [t_0, t_1] \qquad A.\ddot{q} + B = \tau + H_1^T.\lambda_1$$
$$H_1.\dot{q}_1 = 0$$
$$t \in [t_1, t_2] \qquad A.\ddot{q} + B = \tau + H_2^T.\lambda_2$$
$$H_2.\dot{q}_2 = 0$$
$$t \in [t_2, t_3] \qquad A.\ddot{q} + B = \tau + H_3^T.\lambda_3$$
$$H_3.\dot{q}_3 = 0$$

With the different matrices :

$$t \in [t_0, t_1] \qquad H_1 = \begin{bmatrix} 1 & 0 & 0 & -ls_4 \\ 0 & 1 & 0 & lc_4 \end{bmatrix}$$

$$t \in [t_1, t_2] \qquad H_2 = \begin{bmatrix} 0 & 1 & 0 & lc_4 \\ 1 & 0 & -ls_3 & 0 \\ 0 & 1 & lc_3 & 0 \end{bmatrix}$$

$$t \in [t_2, t_3] \qquad H_3 = \begin{bmatrix} 1 & 0 & -ls_3 & 0 \\ 0 & 1 & lc_3 & 0 \end{bmatrix}$$

Table n°1 : Sets of equations at any time.

The second example, Figure n°3, is an example of an impact problem. The mechanism is composed of three sub-systems, the first one is a classic slider-crank closed loop mechanism, the other two being simply two masses.

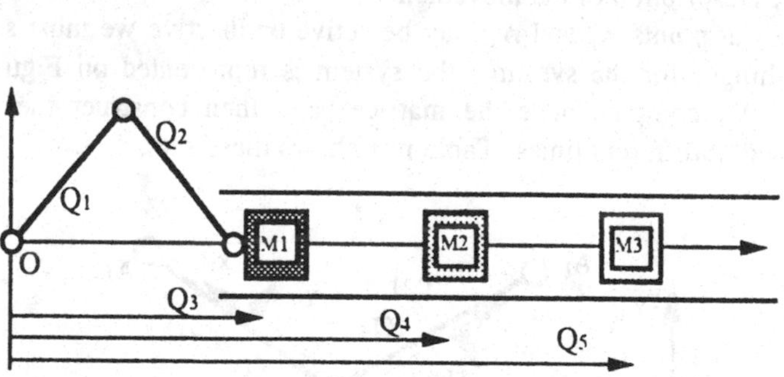

Figure n°3 : Description of the system.

The mass M1 of the first subsystem, driven by a torque at O, first come into contact with impact with the mass M2, the latter then comes into contact with mass M3, and

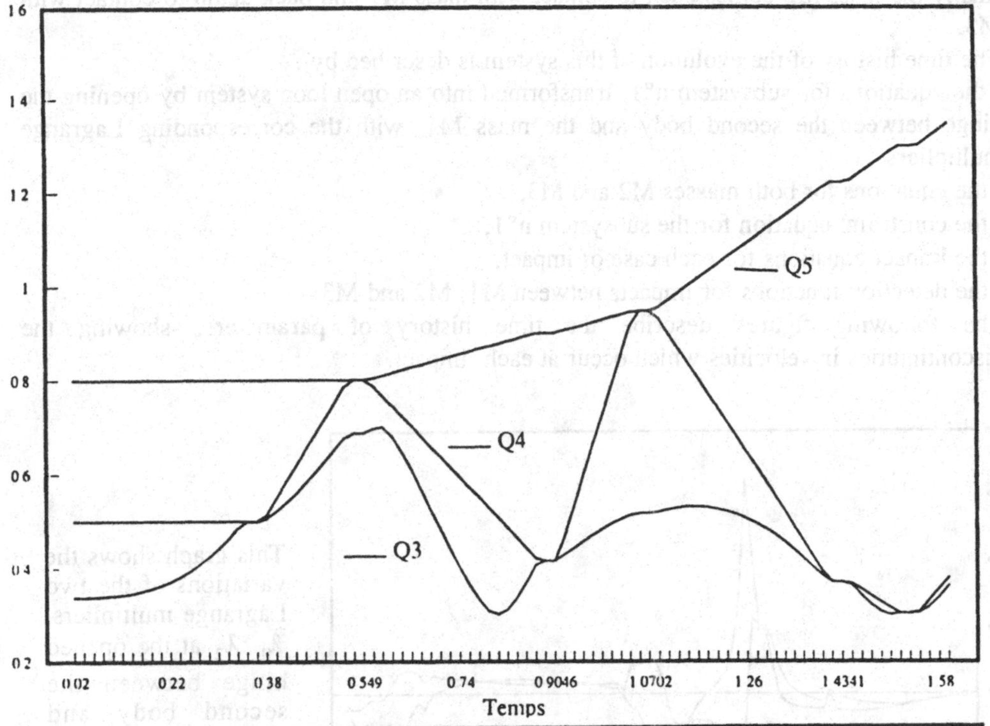
1 6
1 4
1 2
1
0 8
0 6
0 4
0 2
Q5
Q4
Q3
0 02
0 22
0 38
0 549
0 74
0 9046
1 0702
1 26
1 4341
1 58
Temps

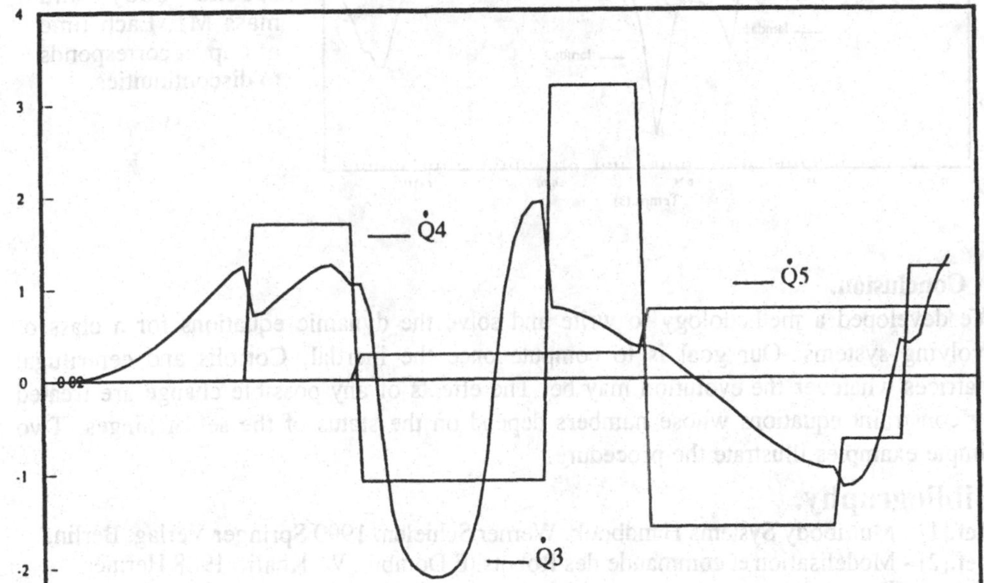
4
3
2
1
0
-1
-2
-3
$\dot{Q}4$
$\dot{Q}5$
$\dot{Q}3$
0 02
0 38
0 74
1 0702
1 4341
Temps (s)

finally the mass M2 comes back to contact with mass M1 and back again to contact with M3.

The time history of the evolution of this system is described by :

- the equations for subsystem n°1, transformed into an open loop system by opening the hinge between the second body and the mass M1, with the corresponding Lagrange multipliers.
- the equations for both masses M2 and M3,
- the constraint equation for the subsystem n°1,
- the impact equations for each case of impact,
- the detection functions for impacts between M1, M2 and M3.

The following figures describe the time history of parameters, showing the discontinuities in velocities which occur at each impact.

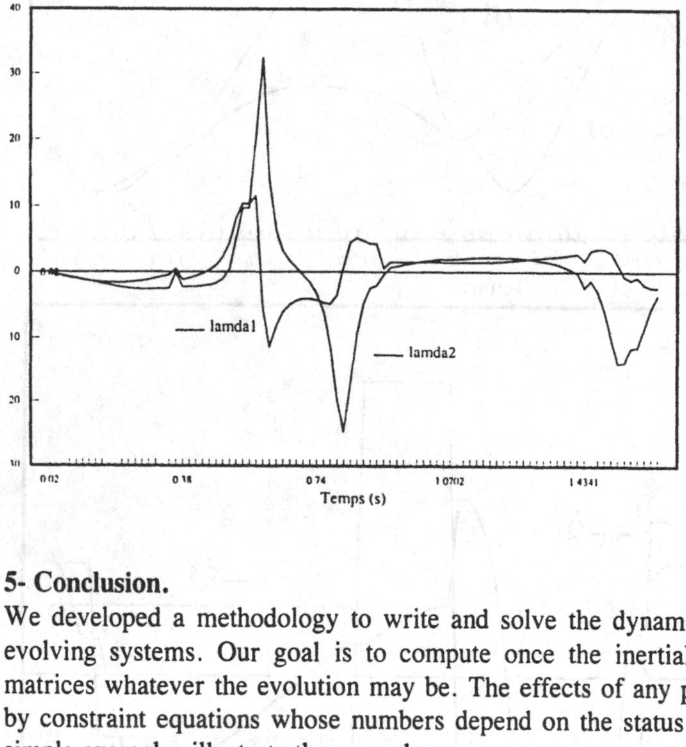

This graph shows the variations of the two Lagrange multipliers, λ_1 λ_2 at the opened hinge between the second body and mass M1. Each time of impact corresponds to discontinuities.

5- Conclusion.

We developed a methodology to write and solve the dynamic equations for a class of evolving systems. Our goal is to compute once the inertial, Coriolis and centrifugal matrices whatever the evolution may be. The effects of any possible change are treated by constraint equations whose numbers depend on the status of the set of hinges. Two simple examples illustrate the procedure.

Bibliography.

Ref.[1] - Multibody Systems Handbook. Werner Schielen. 1990 Springer Verlag. Berlin.

Ref.[2] - Modélisation et commande des Robots. E Dombre, W. Khalil. 1988 Hermès. Paris.

Ref.[3] - Computer aided kinematics and dynamics of mechanical system. E. J. Haug.1989 Allyn and Bacon. Boston.

Ref.[4] - Walking without impacts as a motion/force control problem. W. Blajer, W. Schielen. Transaction of ASME. Vol 114, pp 660-665. Dec 1992.

Part IV
Control of Motion I

ANTHROPOMORPHIC TELEMANIPULATION SYSTEM IN TERMINUS CONTROL MODE

B.M. Jau, M.A. Lewis and A.K. Bejczy
California Institute of Technology, Pasadena, CA, USA

Summary

The paper describes a prototype anthropomorphic kinesthetic telepresence system that is being developed at JPL. It utilizes dexterous terminus devices in form of an exoskeleton force-sensing master glove worn by the operator and a replica four finger anthropomorphic slave hand. The newly developed master glove is integrated with our previously developed non-anthropomorphic six degree-of-freedom (DOF) universal force-reflecting hand controller (FRHC). The mechanical hand and forearm are mounted to an industrial robot (PUMA 560), replacing its standard forearm. The notion of "terminus control mode" refers to the fact that only the terminus devices (glove and robot hand) are of anthropomorphic nature, and the master and slave arms are non-anthropomorphic. The system is controlled by a high performance distributed controller. Control electronics and computing architecture were custom developed for this telemanipulaton system. The system is currently being evaluated, focusing on tool handling and astronaut equivalent task executions. The evaluation revealed the system's potential for tool handling but it also became evident that hand tool manipulations and space operations require a dual arm robot.

Introduction

This telerobotic system was designed to perform dexterous manipulations in hazardous environments. In order to perform a large variety of tasks, it must be able to use tools. Most tools can only be handled by fingered hands. The robot's fingered hand provides tool handling and other dexterous manipulation capabilities.

An obvious hazardous environment is space. Cost savings might result by performing certain space tasks by robots because they don't need life support systems and can operate for long time periods in hostile environments. For space operations, the robot must be able to perform typical Extra Vehicular Activity (EVA) astronaut tasks which require to use the same EVA tools and equipment that are available and certified for astronaut use [1]. Evaluating the robot's tool and EVA equipment handling skills thus became a major task. In fact, the success of this robot depends to a large degree on its ability to handle EVA equipment.

Besides using the system as a telemanipulator, it is also noted here that there is a strong medical interest for the backdrivable glove as a rehabilitation physical therapy aid for helping patient's recovery from hand

a vertical orientation. It controls the robot's wrist in either position/orientation or force control and provides equivalent force feedback to the operator. The telescoping part of the FRHC is gravity compensated so that the operator does not feel any gravitational effects from the master controller. The operational space at the wrist is a 45 cm cube working area.

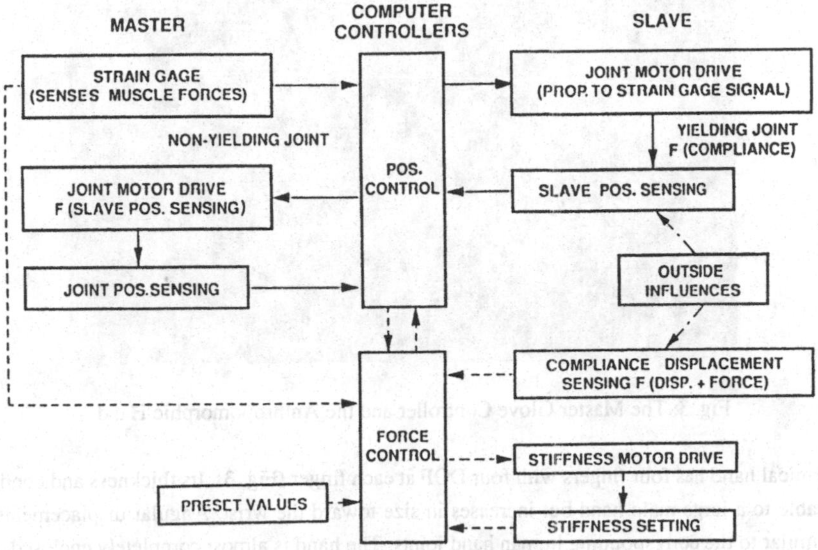

Fig. 2: The Overall Sensing and Control Information Flow Block Diagram

A glove-type device [3] is worn by the operator (Fig. 3). Its force sensors enable hybrid position/force control and compliance control of the mechanical hand. Four fingers are instrumented, each having four DOF. Position feedback from the mechanical hand is providing position control for each of the 16 glove joints. The glove's feedback actuators are remotely located and linked to the glove through flex cables. A one-to-one kinematic mapping exists between master glove and slave hand joints, thus reducing the computational efforts and control complexity of the terminus subsystem. The exceptions to the direct mapping are the two thumb base joints which need kinematic transformations.

Manipulator Arm

The manipulator arm consists of a PUMA 560 robot with its forearm replaced by the new forearm assembly. The forearm weighs approximately 50 kg. A cable links the forearm to an overhead gravity balance suspension system, relieving the PUMA upper arm of this additional weight. The forearm has two sections, a rectangular and a cylindrical. The cylindrical section, extending beyond the elbow joint, contains the wrist actuation system. The rectangular cross section houses the finger drive actuators, all sensors and the local control and computational electronics. The wrist has three DOF with angular displacements similar to the human wrist. The wrist is linked to an AEC system that controls the wrist's stiffness. It is noted that the slave hand, wrist and forearm form a mechanically closed system, that is, the hand cannot be used without its wrist.

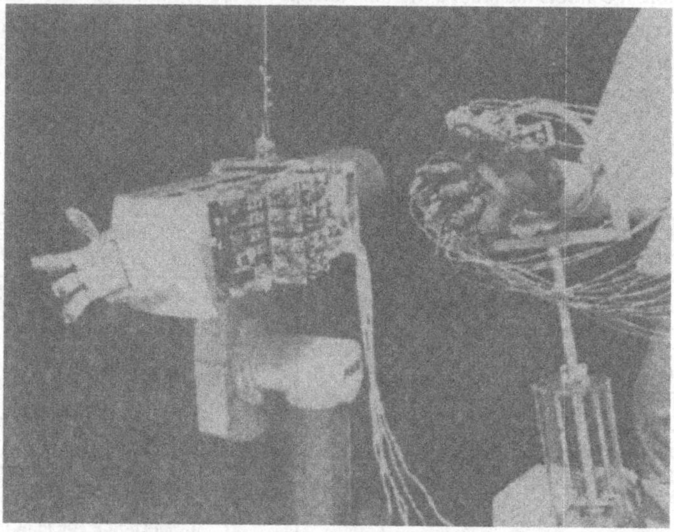

Fig. 3: The Master Glove Controller and the Anthropomorphic Hand

The mechanical hand has four fingers with four DOF at each finger (Fig. 3). Its thickness and configuration is comparable to a large male hand but increases in size toward the wrist. Angular displacements at each joint are similar to the corresponding human hand joints. The hand is almost completely enclosed, preventing object intrusions that could jam its mechanism. All finger joints are linked to the actuating system through flex cables. Each finger is linked to its own AEC system, enabling human-like soft grasping. The hand's kinematics is similar to the human hand, enabling tool manipulations and direct human control through the glove. A more detailed description of the hand can be found in publication [3].

Control Electronics

Early robotic control systems used parallel microprocessors to satisfy computational needs. Typically, one processor was devoted to each joint. In addition, a host processor provided mass storage, a user interface and coordinated joint motion. This configuration provided a cost effective solution to meet the processing requirements.

The dramatic performance increase of modern processors has had a major effect on the computational architecture of modern robots. It is now evident that using a large number of simple processors is not optimal. Today's Digital Signal Processing (DSP) chips represent a cost effective solution to handle all computational needs. The high performance of the DSP chips allows the functionality of a large number of simple processors to be consolidated onto one chip, thus reducing system complexity.

For this anthropomorphic telemanipulation system, the Texas Instrument TMS320C40 (C40) processor was selected to handle the digital control system computation and to provide user services. Key features that make the C40 DSP chip an attractive processor for robotics use are: 1) 40-50 MFLOPS performance 2) Six 20 MBYTES/sec communication ports, each with a separate Direct Memory Access (DMA) processor

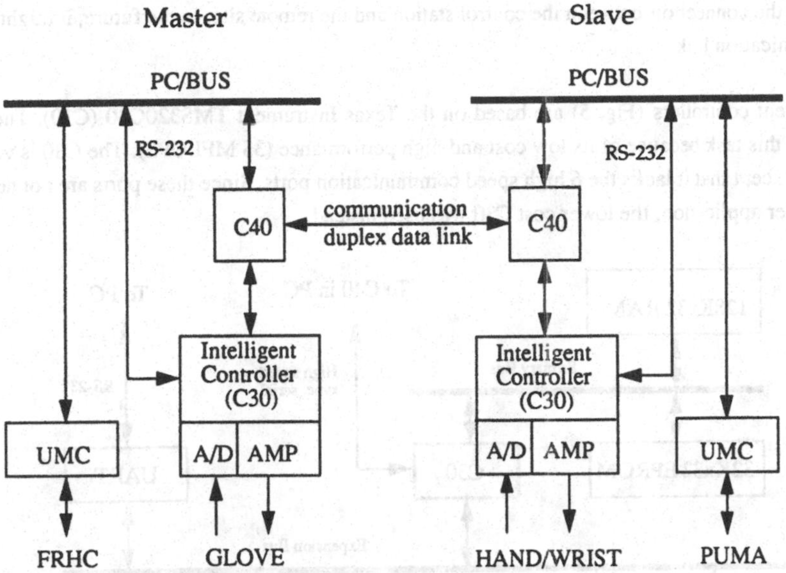

Fig. 4: Control Architecture Overview

capable of communicating with other processors without additional interface logic (i.e. glue logic) 3) Optimized for multiplications and accumulative type operations (i.e. A*B +...) 4) Availability of high performance real-time operating system 5) Low latency interrupt service and 6) Low cost PC based development environment.

The C40 is an attractive alternative to the use of a general purpose processor for robotics control (i.e. Motorola 68040). A consideration in designing a high performance computing architecture is upgradability. The C40 has been designed to be the building block of larger, higher speed computing networks. Its concept it is very similar to the Transputer, manufactured by Inmos and SGS-Thompson, which has been a popular parallel computing building block, particularly in Europe.

The control electronics (Fig. 4) for the master glove and the anthropomorphic hand/wrist are comprised of PC based computational engines, using TMS320C40 (C40) processors and 2 custom designed intelligent controllers. The interface to the FRHC and the PUMA upper arm joints is provided by two separate Universal Motor Controllers (UMC). The UMC has been described previously in [4].

The C40 development system is comprised of Texas Instrument Modules (TIM), daughter boards, a PC and software tools. The C40s are placed on daughter boards that provide communication to a supervisory program on the PC. The development system also provides C source level debugging capabilities. All programs were written in the C language, assembly language programming was not necessary. In this implementation, the SPOX Real-Time Operating System (Spectrum Microsystems) was used to facilitate the development of multi-process programs.

The C40s communicate with each other via a single duplex communication channel. This communication link will be the connection between the control station and the remote site. In the future, it might be a satellite communication link.

The intelligent controllers (Fig. 5) are based on the Texas Instrument TMS320C30 (C30). The C30 was selected for this task because of its low cost and high performance (33 MFLOPS). The C30 is very similar to the C40 except that it lacks the 6 high speed communication ports. Since these ports are not necessary in this controller application, the lower cost C30 was used instead.

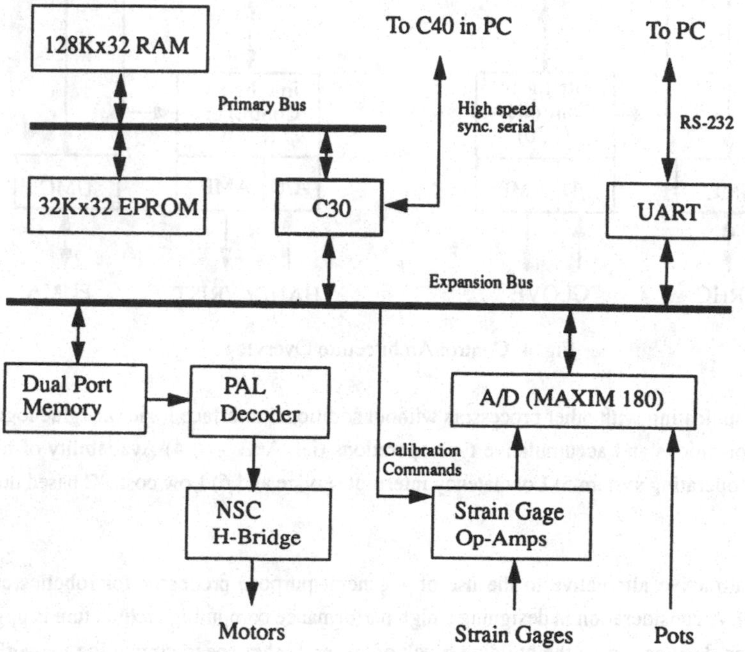

Fig. 5: Schematic of Intelligent Controllers

The two intelligent controllers are placed near the system's sensors, one is near the master glove, the other is near the anthropomorphic hand and wrist. The function of the controllers are to provide sampling of analog signals, filtering of these signals, to provide digital calibration of strain gages, modeling the actuator voltage-velocity curve, the generation of PWM signals and to communicate with the PC based computational engine.

Sensor signals are sampled at 2 kHz using 12-bit, 8 channel A/D converters (MAXIM 180). All strain gage signals are amplified by digitally-calibrated signal stage OP-AMP circuits. The motors are driven by a custom designed PWM circuit, composed of a Dual Ported Memory and several PALs (Programmable Array Logic). The circuit generates the 16 PWM signals needed to backdrive the exoskeleton glove and the 20 signals necessary to drive the anthropomorphic hand, including the four compliance drives (one for each finger) and 4 PWM signals for the 3 DOF wrist and its compliance control. In addition, the controller monitors joint and force limits and can stop the system if preset limits are exceeded. The amplifier drive circuits, based on the National Semiconductor 18201 H-Bridge (PWM amplifier) provides power signals to the motor.

The C30 uses a UART (Universal Asynchronous Receiver/Transmitter) to provide an RS-232 serial line communication to the PC. The RS-232 serial interface between the Intelligent Controller and the PC is used to download programs. Communication of sensor data and actuator commands to the computational engine is via a custom built 4 MHz synchronous serial interface between the C30 and one of the six parallel communication ports of the C40.

A monitor program was written for the C30 and resides in the EPROM. This program boots the computer, provides functions such as memory test, calibration and program downloading. Programs are downloaded via the RS-232 into the RAM memory.

Computing Architecture

The computing architecture (Fig. 4) was designed as an efficient means of handling the computational requirements for this system, currently with 49 DOF. It supports several distinct functions: 1) Filtration of sensed signals 2) Modeling of voltage-velocity curves for motor control 3) Control law implementation and 4) Inverse kinematics.

Filtration of Sensed Signals- Sensed signals include joint encoder signals, compliance deflection signals, compliance setting encoder signals and strain gage signals. Sensed values require filtration to estimate the true signal value. In this system, an Infinite Impulse Response (IIR) filter was used and is given by the following equation:

$$\hat{y}_n = \hat{y}_{n-1} + K \cdot (x_n - \hat{y}_{n-1}) \qquad (1)$$

where $\hat{y}_n$ is an unbiased estimate [5] of the true signal y_n at sample time n, K is a gain factor that depends on the signal noise and x_n is the value of the raw, unfiltered A/D converted sample.

Modeling of Voltage-Velocity Curve- A model of each actuator system is necessary for achieving high performance control. Each actuator system is comprised of a motor, a gear train, flex cables to the joint and the joint itself. For the hand, the joint's position pot, which is driven by the joint actuator cable, is the actual controlled quantity. For the glove, the positional output of the gear train is measured and the position of its joints are inferred from this measurement. Any other variable, i.e. true motor position, motor torque or true joint position are not measured in this system and hence are not considered in the model.

To obtain a first approximation, a voltage-velocity curve can be modelled by using two straight lines interrupted by a discontinuity at the origin. This model is given by the following equation:

$$f_{model}(Y, voltage) = k_1(Y) \cdot voltage + v_0^+(Y) \quad if(voltage) \geq 0 \qquad (2)$$

$$f_{model}(Y, voltage) = k_2(Y) \cdot voltage + v_0^-(Y) \quad if(voltage) < 0 \qquad (3)$$

where f_m is the modelled velocity of the position sensor, k_1 and k_2 are gains, the v_0's are the minimum starting voltages required to move the joint, and Y is an estimated value depending on the hand configuration. The values k_1, k_2, and v_0 are estimated by a calibration procedure. Note that back EMF and other velocity dependent terms are not included. Also note that the k_1, k_2 and v_0 variables are configuration dependent due to friction variations throughout the range of motion of each joint and due to interactions between joints. To simplify the problem, the variables were regarded as configuration independent at this time. In the near future, the calibration procedure will be augmented to generate several configuration dependent models for each actuator in the terminus devices.

Control Law Implementation- There are two modes of operation of the control system. In the first mode, called "free motion", the anthropomorphic hand/wrist assembly moves freely without contacting the environment. Once a finger comes into contact with a surface, its deflection sensors inform the controller of an externally induced finger deflection which causes the controller to automatically switch to the "force control" mode for that finger: Instead of specifying a joint position, joint torques and compliance settings are being servoed (force control of the wrist functions in similar fashion). Experimentation in this mode are still ongoing, the theory behind this control mode will be published at a later date.

In the "free motion" mode, the control laws are PD (positional and derivative) control laws. The input to the control law is the desired position, the current position and its time differential (which approximates velocity). Once a command is computed using the PD control law, the command signal is mapped through an inverse model of the motor. In this way, the motor's behavior is linearized. The equation for the control law is given in equation (4) and the equation for the inverse model is given in equation (5):

$$u = K_p \ (\hat{y}_n - y_d) + K_v \ (\hat{y}_n - \hat{y}_{n-1}) \tag{4}$$

$$u' = f^{-1}{}_{model} \cdot u \tag{5}$$

where u is the control command to the actuator model, $\hat{y}_n$ is the estimated position of the joint being controlled, and y_d is the desired position of the joint. The variables K_p and K_v are gain factors for the position error and the velocity error respectively. In equation (5), u' is the control output after being transformed by the inverse of the motor model f_{model}.

In the "free motion" mode, the desired position command y_d is derived from various sources: For most joints in the anthropomorphic hand, the desired position is given by the integral of the torque, sensed at the corresponding joint in the glove. This torque is a measure of the force exerted by the human operator. The equation for the desired position is given by the equation below:

$$y_{danthro} = \int_0^T K_{sg} \cdot \hat{y}sg_n dt \tag{6}$$

Here $y_{danthro}$ is the set point command to the motor, K_{sg} are gain factors for strain gages at individual joints and $\hat{y}sg_n$ is the estimated strain gage value which is derived from the torque at that joint of the exoskeleton glove.

The lateral motion of the three fingers is position controlled and follows the same concept as described below for the glove.

The position of each joint in the exoskeleton glove is determined by the position of each corresponding joint in the anthropomorphic hand. That is:

$$y_{dglove} = y_{anthro} \tag{7}$$

where y_{anthro} is the actual position of the corresponding anthropomorphic hand joint.

Finally, the position and orientation of the PUMA and the wrist is determined by a kinematic mapping of the FRHC to the PUMA and the wrist. Likewise, the FRHC is backdriven by the PUMA and the wrist in similar fashion.

Inverse Kinematics- The closed form inverse kinematics solution for the 6 dof robot arm (first 3 joints of the PUMA and the 3 dof wrist) are computed by standard methods.

The computation is distributed between the C40s and the C30s. The C40s perform the bulk of the computations which are equations (4) through (7). Equation (1) and the inverses of the motor model, given by equations (2) and (3) are performed by the C30 in the intelligent interface controller. The C40s subdivide the computational load according to the natural division of master and slave, thus minimizing the bandwidth requirements of the communication link between the two C40s. If, in the future, additional processing requirement are desired, 2 or more C40s can be built into the system, thus forming a high performance parallel machine. The closed loop bandwidth of the system is approximately 1 kHz. This will increase as the code is optimized.

Performance Evaluation

The following discussion describes some of the handling skills that were demonstrated in the initial performance evaluation of this telemanipulation system. Testing is still in progress and actual test data will be published at a later date.

Object Grappling- Grasping objects with the hand in compliant mode simplifies the grappling process because compliance enables self-alignment of the hand to the object, easing positional accuracy requirements during final approach and grappling. Compliance also enables multi point object contact because individual fingers self-align to objects, resulting in a tight grip. The hand grasps objects primarily from one side, enabling a better oversight over the hand's activities while requiring less workspace around objects.

Tool Guidance- Tools that need to be guided along linear paths (i.e. knife) can be handled quite well due to the hand's compliance. Tools requiring tightening motions around an axis (i.e. a wrench moving around the screw axis), could also be handled. Two key capabilities make this possible: The hand's articulation enables it to embrace the tool handle, thus holding it without using much clamping force. To tighten the screw, the wrench has to move around the screw axis in a circular path but the hand does not have to follow this path accurately because the hand, formed as a hook around the embraced tool, allows some relative motion between itself and the tool without loosing contact with the tool. Additionally, the wrist's compliance provides some self-alignment of the hand's orientation w.r.t. the tool orientation. Not having to follow the tool path accurately simplifies this tool guidance operation considerably.

In order to use a tool with the hand, it first must be placed into the hand properly. An often overlooked fact is that orienting and placing the tool handles correctly into the robot hand requires assistance from a second hand. Also, tools often need to be regrasped, held and guided or held and manipulated which requires a second hand as well. It was found that EVA-type remote tool operations are only meaningful if a second hand is available to assist the tool handling operation.

Tool Manipulations- Hand tool manipulations are surprisingly difficult to perform, even with the articulation this hand has. However, some tool manipulation tasks have already been demonstrated (i.e. scissor). The lack of tactile sensing was quite evident in tool manipulations: human tactile sensors not only sense the locations of contact but also sense the strength and direction of the applied forces, enabling the hand to exert proper reactive forces. This makes human tool manipulations easy. The lack of tactile sensing in the mechanical hand severely hampers tool manipulations.

Most tool manipulations will require two hands. For instance, a pliers can be held near its hinging point by one hand while the other hand operates the tool at the handles. In essence, tool manipulations will be transformed into two simpler tasks: tool holding and tool actuation.

Number of fingers needed- In most cases, it takes at least three fingers to rigidly hold an object within the hand while a fourth finger performs a manipulaton such as squeezing a trigger. Likewise, object manipulations require three fingers to firmly hold the object while the fourth finger is free to regrasp. Tests with the four fingered hand proved that most hand tools could not be held and operated by using only three fingers. Thus, the minimum number of fingers is four, one being a thumb.

EVA tool evaluation- An evaluation was undertaken [1], analyzing the feasibility of handling astronaut tools by a general purpose space telerobot. Major findings were that of the 195 astronaut EVA items evaluated, only 6 could be handled be an industrial type end effector. A one arm robot with a fingered hand could handle 29 items whereas a dual arm robot with fingered hands would be capable of handling 171 items.

The following listing states the EVA items that can be handled by one fingered hand (excluding the tethering operation): Battery, bolt puller, camera actuator, connector tool, door latch, door stay, drill (on handle), drive unit preload tool, fastener (1/4 turn), force measurement tool, hammer, J-hook, hydrazine brush, loop pin extractor, probe, pry bar and wrenches (open, box ends and allen).

Other EVA items that are being considered for robot testing include: bolt puller, connector demate tools, connector pin straightener, power drive disconnect tool, ratchet wrenches, screwdrivers with shroud, sockets, bags, connector and cap, mirror, tape caddies, tool boards, knobs and switches.

Prerequisite for dual arm manipulations- Redundant 7 DOF arms are needed to reach around obstructions and to properly align the arms w.r.t. each other, avoiding arm interferences. Full arm compliance is needed for tool guidance operations (see above) and for cooperative dual arm manipulations of rigid objects. (Seven DOF anthropomorphic master arm controllers already exist).

Tethering operations- Shuttle safety manifests require that all loose items must be tethered in space to prevent free floating. All tethers securing tools or equipment while being used are lock-lock tethers which require two simultaneous operations to unlock the tether's hook, requiring two handed operations even for the astronaut. Tethering operations with the robot also require dual hands.

Conclusions

The initial evaluation revealed the system's potential for multifunctional operations, including tool handling and manipulation tasks. However, hand tool manipulations and EVA tasks require a dual fingered hand system with at least four fingers and 7 DOF compliant arms. The system can only reach its full potential after its expansion into a dual arm system with active electromechanical compliance, enabling human-equivalent telepresence in space, including tool use.

References

[1] Jau, B. M., "Feasibility Analysis of Performing EVA Tasks with Dexterous Robots", JPL IOM Nr. 3474-94-007, Jet Propulsion Laboratory, Pasadena, CA, Feb. 25, 1994.
[2] Bejczy, A. K. and Salisbury, J. K., "Kinesthetic Coupling between Operator and Remote Manipulator", Proceedings of the Int. Computer Technology Conference, ASME Century 2, Vol. 1, San Francisco, CA, Aug. 1980, pp 197-211.
[3] Jau, B. M., "Man-Equivalent Telepresence Through Four Fingered Human-Like Hand System", IEEE Int. Conference on Robotics and Automation, Nice, France, May 12-14, 1992, pp 843-848.
[4 Bejczy, A. K., Szakaly, Z. F., "Universal Computer Control System (UCCS) for Space Telerobots," Proceedings of the IEEE Int. Conference on Robotics and Automation, Raleigh, NC, March 30-Apr. 3, 1987, pp 318-324.
[5] Mendel, Jerry M., "Lessons in digital estimation theory", Englewood Cliffs, N.J., Printice-Hall, 1987.

Acknowledgment

This work was carried out at the Jet Propulsion Laboratory, California Institute of Technology under a contract by the National Aeronautics and Space Administration.

PATH PLANNING OF REDUNDANT MANIPULATOR AND DETERMINATION OF ITS CONFIGURATION OF BENDING ELASTIC PLATE SPRING

R. Katoh, T. Fujimoto and T. Yamashita
Kyushu Institute of Technology, Kitakyushu, Japan

SUMMARY

In tasks of assembling machines by manipulator, operations such that the manipulator deforms the shape of an object, for example, bending spring or rubber, are often needed, while many studies as for the operations of the manipulator with geometrical constraints, which means the absence of the shape change of the object, have been reported. This paper deals with two problems; a path planning of a manipulator's end-effector and a determination method of its configuration in such a operation that an end-effector of the manipulation is commanded to move one end of an elastic plate spring with the another fixed to a wall to a desired position by bending it.

INTRODUCTION

The robot technology have made it possible of robot manipulators to do various type of tasks, for example, inserting a peg to a hole, rotating a crank, tracing to an object surface, and so on. In the conventional studies, the objects manipulated by robots are assumed to be rigid or the structures consisting of some rigid bodies. However, there are situations such that the flexibility must be considered in manipulations; the flexibility of the manipulators and that of the object manipulated by robot. The former has been studied in many papers[1][2], but the latter has rarely researched yet. The latter type of flexibility is related to the deformation of an object by the manipulator. Indeed in manufacturing process, this type of operations such that the manipulator must deform the shape of the object, for example, bending spring or rubber, are often needed.

This paper deals with two problems; a path planning of a manipulator's end-effector and a determination method of its configuration in such an operation that the effector is commanded to move one end of an elastic plate spring with the another fixed to a wall to a desired position by bending it. It is assumed here the energy can't be regenerated.

COMPLIANCE ELLIPSE

Compliance of a plate spring

Let an elastic plate spring, whose one end is fixed to a wall (x_*, y_*), be deformed by a force f on a x-y plane and the another be set in a point (x,y), as shown in Fig.1. Two components f_x and f_y of the force f are functions of x and y, namely $f_x=f_x(x,y)$ and $f_y=f_y(x,y)$. If the position (x,y) is enough near to a given position (x_0,y_0), the following relations are satisfied;

$$f_x(x,y)-f_x(x_0,y_0)$$
$$=\frac{\partial f_x(x_0,y_0)}{\partial x}(x-x_0)+\frac{\partial f_x(x_0,y_0)}{\partial y}(y-y_0) \equiv a_x(x-x_0)+b_x(y-y_0) \quad (1)$$

$$f_y(x,y)-f_y(x_0,y_0)$$
$$=\frac{\partial f_y(x_0,y_0)}{\partial x}(x-x_0)+\frac{\partial f_y(x_0,y_0)}{\partial y}(y-y_0) \equiv a_y(x-x_0)+b_y(y-y_0) \quad (2)$$

By introducing new variables X, Y, F_x and F_y such that

$$X = x-x_0, \quad Y = y-y_0, \quad F_x = f_x(x,y) - f_x(x_0,y_0), \quad F_y = f_y(x,y) - f_y(x_0,y_0)$$

we have the following relation;

$$F \equiv \begin{bmatrix} F_x \\ F_y \end{bmatrix} = \begin{bmatrix} a_x & b_x \\ a_y & b_y \end{bmatrix}\begin{bmatrix} X \\ Y \end{bmatrix} \equiv A\begin{bmatrix} X \\ Y \end{bmatrix} \quad (3)$$

If a norm of the additional force F applied to one end of the spring has a constant value, i.e. $F^TF=C^2$, then the following equation can be derived from Eq.(3);

$$F^TF = (X,Y)\,A^TA\begin{bmatrix} X \\ Y \end{bmatrix} = C^2. \quad (4)$$

This is an equation of an ellipse, which means the displacement of the spring tip when the force applied to its tip is changed by F. At the same time it means the compliance of the spring, because it is assumed the magnitude of the additional force F is constant. We call this as a "Compliance Ellipse".

Compliance of manipulation

It is assumed that the force generated by the manipulator is balanced to the reaction force caused by the spring. If the torque of the manipulator is changed by $\Delta\tau$, then the following equation must be satisfied;

$$\Delta\tau = J_0{}^TF = J_0{}^T\begin{bmatrix} F_x \\ F_y \end{bmatrix} \quad (5)$$

where J_0 is a Jacobian matrix of the manipulator when the end-effector's position is (x_0,y_0). Eqs.(4) and (5) give the following relation;

$$\Delta\tau^T\Delta\tau = \|\Delta\tau\|^2 = (F_x,F_y)\,J_0J_0{}^T\begin{bmatrix} F_x \\ F_y \end{bmatrix} = (X,Y)(J_0{}^TA)^T(J_0{}^TA)\begin{bmatrix} X \\ Y \end{bmatrix}. \quad (6)$$

If $\|\Delta\tau\|$ is constant, then this is also an ellipse. This ellipse means the displacement of the end-effector, i.e. that of the spring's tip, moved by the additional torque $\Delta\tau$ with constant magnitude. The spring can be easily deformed along the major axis of the ellipse, but not along the minor. This is called here as a compliance ellipse of the spring and manipulator (or a "Compliance Ellipse of Manipulation").

PATH AND CONFIGURATION PLANNING

Path planning

A path planning of the manipulator on the basis of the compliance ellipse of manipulation described above is proposed. Fig.2 shows a concept of this planning method. In this method, it is assumed that a target position and a current end-effector's position can be measured. The method is given by the following procedures.

(A) At first, a compliance ellipse is drawn at the current position of the end-effector for a given value of $\|\Delta\tau\|$.

(B) Next, a circle, which has a center at a position of the target and contacts with this ellipse is drawn and the contact point is determined.

(C) The direction determined by both the contact point and the current end-effector's position is set as a direction which the end-effector should be moved to.

The contact position means the nearest position to the target among the positions that the end-effector can be moved under the condition of a constant $\|\Delta\tau\|$. If the magnitude of end-effector's velocity $(\dot{x}, \dot{y})^T \equiv \dot{p}$ and a sampling time are given, the current reference position of the end-effector is determined.

Determination of manipulator's configuration

Many researches as for the use of the manipulator's redundancy have been reported and many methods for them have been developed [(3)-(5)]. Here the method developed by Liegeois [(3)] is used for determining the configuration of a redundant manipulator shown in Fig.2. The configuration is determined such that the relation (7) is satisfied and that the following criterion L (Eq.(8)) is minimized.

$$\dot{p} = J_o \dot{\theta} \tag{7}$$

$$L = \|\dot{\theta}\|^2 + k_p \|\Delta\tau\|^2 \tag{8}$$

where $\dot{\theta}=(\dot{\theta}_1, \dot{\theta}_2, \dot{\theta}_3)^T$ is an angular velocity vector of the manipulator. According to the Liegeois's method, the angular velocity reference of the manipulator $\dot{\theta}_d$ can be determined as follows;

$$\dot{\theta}_d = J_o^+ \dot{p} + (I - J_o^+ J_o)\,\xi\, k_p, \quad \xi = (\xi_1, \xi_2, \xi_3)^T, \quad \xi_i = \partial L/\partial\theta_i \ (i=1,2,3) \tag{9}$$

where J_o^+ is a pseudoinverse of the Jacobian matrix J_o, and k_p is a weighting parameter.

Eq.(9) gives the value of $\|\Delta\tau\|$ as well as the manipulator's configuration. This value is not same as that given in procedure (A). Then, above procedures (A), (B) and (C) are repeated until the value of $\|\Delta\tau\|$ converges to a constant after resetting the value in (A) to that obtained by Eq.(9).

SIMULATION

Modelling of a plate spring

To execute the present planning method by simulation, it is necessary to make a spring model. This model was experimentally obtained by measuring the forces f_x and f_y shown in Fig.1, and by applying the least square method to these experimental data as follows;

$$f_x = -114.5x^2 - 75.6y^2 + 18.9y + 28.4$$

$$f_y = -250.5x^2 - 526.2y^2 - 286.4y - 11.6$$

In the experiment, a plate spring 500mm long by 40mm wide by 0.8mm thick of phosphor bronze was used, and the position (x_0, y_0) that the spring was fixed was set at (0, 0). This model can't represent the singular state of the spring such that the force f_x or f_y has an infinity value in the finite area of (x,y). However, this model gives the convenient results in the area which the spring will be deformed in usual way except the natural position of the spring. For the bellow simulation, the spring model obtained by transferring the position (x_0, y_0)=(0, 0) to (0.4, 0.6) was used.

Simulation method

Simulations were done under the following conditions.

(1) A velocity of the end-effector and a sampling time are set at 0.1m/s and 0.01s. An initial value of $\|\Delta\tau\|^2$ is given at $10^{-3}N^2m^2$.

(2) All links of the manipulator has the same length 0.2m.

(3) If the end-effector approaches near a target position, i.e. the difference between two positions becomes smaller than 0.02m, then a straight line passing through two points is selected as the path.

Simulation results

Fig. 3 shows a typical simulation result. An initial position of the end-effector and a target position are set at (0.35, 0.103) and (0.2, 0.3), respectively. Configurations of the manipulator are drawn every 0.5 s. Fig. 4 shows the compliance ellipses of manipulation at every configurations. These figures show that the path is reasonable. Fig. 5 shows the time histories of $\|\Delta\tau\|$. The value of $\|\Delta\tau\|$ is very small in the first half. This means the spring can be easily deformed. The larger parameter k_p becomes, the smaller the value of $\|\Delta\tau\|$ in the latter half. But too large value of k_p gives the sudden change of angular velocity $\dot{\theta}$. The curves drawn in the left side of Fig.5 are the time histories of $\|\Delta\tau\|$ in a

case that the end-effector's path is given as a straight line from an initial position to a target position. In this case, larger values of $\|\Delta\tau\|$ are needed compared with the result by the proposed method.

CONCLUSION

This paper treated with a path planning and a configuration determination of a redundant manipulator in such an operation that an elastic plate spring is deformed by the manipulator. For this method a compliance ellipse of manipulation was introduced. Computer simulations make it clear that the method gives a reasonable path of the manipulator's end-effector.

REFERENCE

(1) De Luca A., Dynamic Control of robots with joint elasticy, Pro. IEEE Conf. on Robotics and Automation, pp.152-158, 1988,

(2) D.M.Rovner and R.H.Cannon, Experiments toward on-line identification and control of a very flexible one-link manipulator, Int. J. of Robotics Research, Vol.6, No.4, pp.3-19, 1987.

(3) A. Liegeois, Automatic Supervisory Control of the Configuration and Behavior of Multibody Mechanisms, IEEE Trans. SMC-7, 12, pp.868-871, 1977.

(4) T. Yoshikawa, Analysis and Control of Robot Manipulators with Redundancy, Robotics Research: The First Int. Symp., (M.Brady and R.Paul ed.), MIT Press, Cambridge, Mass. pp.735-747, 1984.

(5) J.M. Hollerbach and KI C. Suh, Redundancy Resolution of Manipulators through Torque Optimization, IEEE Trans. RA-3, 4, pp.308-316, 1987.

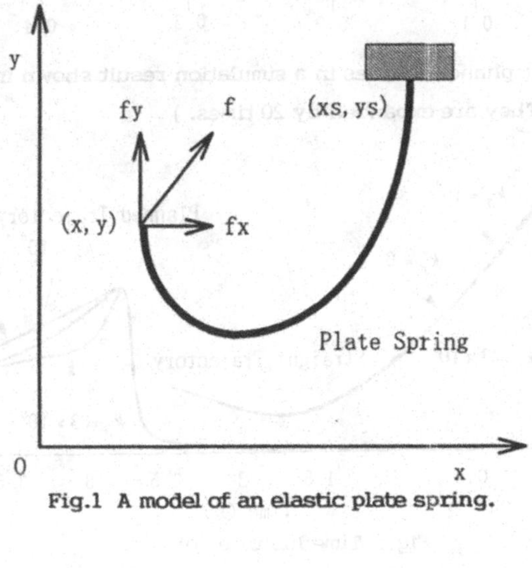

Fig.1 A model of an elastic plate spring.

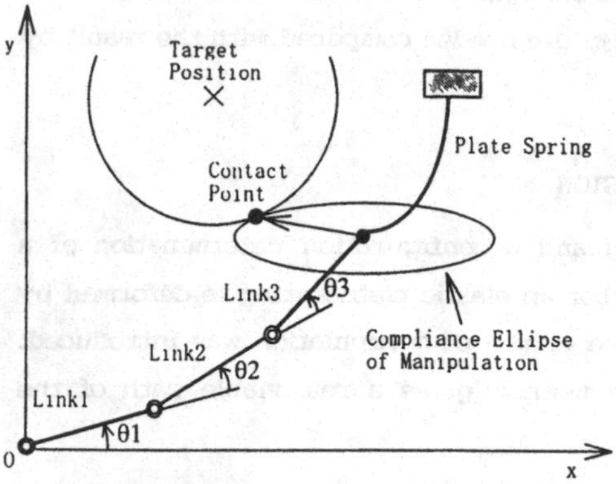

Fig.2 A concept for a path planning.

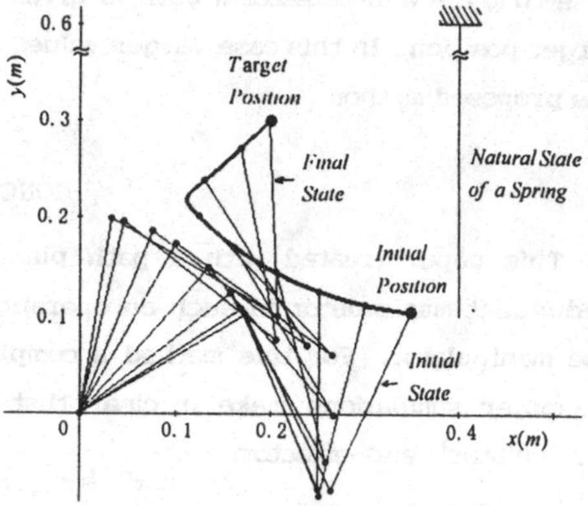

Fig.3 A typical simulation result (k_p=10^9).

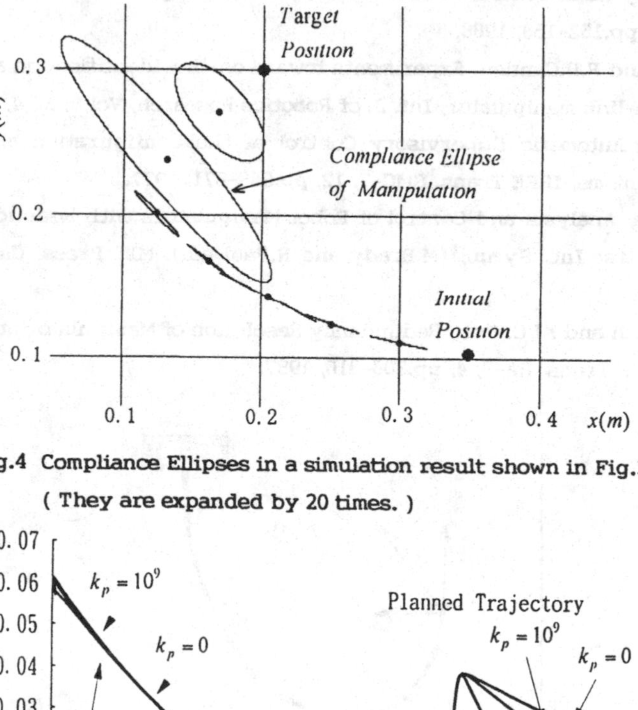

Fig.4 Compliance Ellipses in a simulation result shown in Fig.3.
(They are expanded by 20 times.)

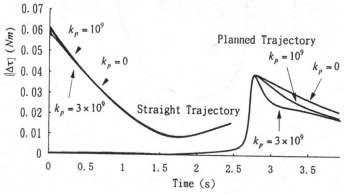

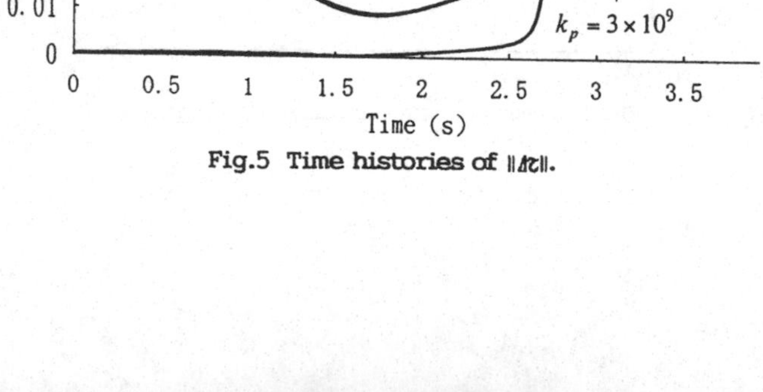

Fig.5 Time histories of $\|\Delta\tau\|$.

REDUNDANCY RESOLUTION FOR TWO COOPERATIVE SPATIAL MANIPULATORS WITH A SLIDING CONTACT

P. Chiacchio, S. Chiaverini and B. Siciliano
Federico II University of Naples, Naples, Italy

INTRODUCTION

Cooperative robot manipulator systems are receiving an increasing interest in the research community in view of their potential over ordinary single robot manipulators. The goal is to achieve coordinated motion of the two robots so that a commonly held object can be effectively manipulated.

The task formulation plays an important role in the kinematics description of this kind of systems. Previous cooperative task descriptions [1,2] have the inconvenience that the interpretation of task variables is not always straightforward, especially for what concerns orientation in spatial manipulators.

Following our earlier results for planar manipulators [3], a cooperative task description has recently been proposed for spatial manipulators [4] that allows the user to specify the motion in terms of meaningful absolute and relative variables in a clear fashion.

In this work a cooperative system of two spatial manipulators is considered where one of them is allowed a sliding contact with the object. Parallel research efforts for the case of rolling contacts are reported in [5,6].

Motion coordination is achieved by adopting a closed-loop inverse kinematics scheme [7] which allows computation of joint trajectories corresponding to given task trajectories. Redundancy resolution is performed in view of the extra degrees of freedom available from the sliding contact.

A system of two PUMA 560 robot manipulators is taken to develop numerical case studies aimed at showing the effectiveness of the approach.

REVIEW OF COOPERATIVE TASK FORMULATION

The typical task of a cooperative two-arm system is to manipulate a common object. This demands for a task description in terms of absolute variables describing the motion of the object and relative variables describing the mutual location between the end effectors which in turn characterize the object grasp. Both absolute and relative variables include position and orientation.

In [4] an effective user-oriented task formulation was established which unambiguously defines the cooperative task for spatial manipulators as well as allows the user to give a direct specification of the task in terms of meaningful variables.

2

The absolute position and orientation are defined by attaching a suitable frame to the object and relating its origin position vector $\boldsymbol{p}_a$ and rotation matrix $\boldsymbol{R}_a$ to the position and orientation of the two manipulators' end effectors, i.e.

$$\boldsymbol{p}_a = \tfrac{1}{2}(\boldsymbol{p}_1 + \boldsymbol{p}_2) \qquad \boldsymbol{R}_a = \boldsymbol{R}_1\, \boldsymbol{R}^1_{k^1_{12}}(\vartheta_{12}/2) \tag{1}$$

where $\boldsymbol{p}_1$, $\boldsymbol{p}_2$ are the end-effectors' position vectors, $\boldsymbol{R}_1$, $\boldsymbol{R}_2$ are the end-effectors' rotation matrices; these can be conveniently expressed as a function of the joint variable vectors $\boldsymbol{q}_1$ and $\boldsymbol{q}_2$. Also in (1) $\boldsymbol{k}^1_{12}$ and ϑ_{12} are respectively the unit vector and the angle realizing the rotation described by $\boldsymbol{R}^1_2$; then the actual rotation is by half the angle needed to align $\boldsymbol{R}_2$ with $\boldsymbol{R}_1$. The relative position and orientation are defined as

$$\boldsymbol{p}_r = \boldsymbol{p}_2 - \boldsymbol{p}_1 \qquad \boldsymbol{R}^1_r = \boldsymbol{R}^1_2. \tag{2}$$

All the above quantities are referred to a common base frame.

The differential kinematics corresponding to (1),(2) are found to be

$$\begin{aligned} \dot{\boldsymbol{p}}_a &= \tfrac{1}{2}(\dot{\boldsymbol{p}}_1 + \dot{\boldsymbol{p}}_2) & \dot{\boldsymbol{p}}_r &= \dot{\boldsymbol{p}}_2 - \dot{\boldsymbol{p}}_1 \\ \boldsymbol{\omega}_a &= \tfrac{1}{2}(\boldsymbol{\omega}_1 + \boldsymbol{\omega}_2) & \boldsymbol{\omega}^1_r &= \boldsymbol{\omega}_2 - \boldsymbol{\omega}_1 \end{aligned} \tag{3}$$

which simply express the linear and angular velocities in the cooperative task space as a function of the differential kinematics of the single manipulators.

SLIDING CONTACT

The foregoing task formulation can be directly applied to specify tasks for a tightly grasped object by assigning suitable values to the absolute and relative variables.

To embed the possibility of handling a *sliding contact* in either of the two manipulators, a *virtual end effector* can be introduced by adding an adequate number of fictitious joints at the actual end effector. The sliding contact is realized if the orientation of the actual and virtual end effectors coincide while their positions differ according to the geometry of the sliding surface.

It can be recognized that a sliding contact requires at most *two degrees of freedom*. In the case of a planar surface, *two virtual prismatic joints* are added at the end effector with their axes realizing two degrees of freedom along the surface; of course, the two axes must not be aligned. The result is an augmented kinematics expressing the location of the virtual end effector.

In the case of a curved surface, sliding contact still requires at most two degrees of freedom. However, three virtual prismatic joints have to be introduced at the end effector with a geometric constraint so as to realize two independent degrees of freedom along the surface.

Consider a system of two cooperative 6-degree-of-freedom manipulators. For each manipulator, let $\boldsymbol{q}_i$ indicate the (6×1) vectors of joint variables. The geometric Jacobian $\boldsymbol{J}_i(\boldsymbol{q}_i)$ is the (6×6) matrix relating the joint velocity vectors $\dot{\boldsymbol{q}}_i$ to the linear and angular end-effector velocities in the base frame as

$$\begin{bmatrix} \dot{\boldsymbol{p}}_i \\ \boldsymbol{\omega}_i \end{bmatrix} = \boldsymbol{J}_i(\boldsymbol{q}_i)\dot{\boldsymbol{q}}_i \qquad i = 1, 2. \tag{4}$$

3

Let q_S denote the (2×1) vector of the additional joint variables. Without loss of generality, manipulator 2 is assumed to make the sliding contact; this implies that p_2 becomes a function of both q_2 and q_S. Hence the virtual end effector still makes a tight grasp with the object and the foregoing task formulation can be retained.

The task space differential kinematics associated to (3) become:

$$\begin{bmatrix} \dot{p}_a \\ \omega_a \end{bmatrix} = J_a \begin{bmatrix} \dot{q}_1 \\ \dot{q}_2 \\ \dot{q}_S \end{bmatrix} \tag{5}$$

for the *absolute* part where

$$J_a(q_1, q_2, q_S) = \left[\tfrac{1}{2} J_1(q_1) \quad \tfrac{1}{2} J_2(q_2, q_S) \right], \tag{6}$$

and

$$\begin{bmatrix} \dot{p}_r \\ \omega_r^1 \end{bmatrix} = J_r \begin{bmatrix} \dot{q}_1 \\ \dot{q}_2 \\ \dot{q}_S \end{bmatrix} \tag{7}$$

for the *relative* part where

$$J_r(q_1, q_2, q_S) = \left[-J_1(q_1) \quad J_2(q_2, q_S) \right]. \tag{8}$$

INVERSE KINEMATICS WITH REDUNDANCY RESOLUTION

The task formulation of the previous section constitutes the basis for a kinematic control problem, that is finding the joint variable trajectories corresponding to given trajectories for the absolute and relative task variables; these trajectories will be the reference inputs to some joint space control scheme.

Define $q = [\, q_1^{\mathrm{T}} \quad q_2^{\mathrm{T}} \quad q_S \,]^{\mathrm{T}}$ and $J = [\, J_a^{\mathrm{T}} \quad J_r^{\mathrm{T}} \,]^{\mathrm{T}}$. The following closed-loop *inverse kinematics* scheme with *redundancy resolution*, originally proposed for single arms [7], can be adopted:

$$\dot{q} = J^{\dagger}(q)(v_d + Ke) + (I - J^{\dagger}J)k_c \left(\frac{\partial c(q)}{\partial q} \right)^{\mathrm{T}} \tag{9}$$

where K is a positive definite diagonal matrix, $e = [\, e_a^{\mathrm{T}} \quad e_r^{\mathrm{T}} \,]^{\mathrm{T}}$ is the task space error, $v_d = [\, v_{ad}^{\mathrm{T}} \quad v_{rd}^{\mathrm{T}} \,]^{\mathrm{T}}$ is the desired feedforward velocity, c is a constraint function of the joint variables that is optimized locally in the null space of J and k_c is a signed constant.

In detail, the absolute error is

$$e_a = \begin{bmatrix} p_{ad} - p_a \\ \tfrac{1}{2}(n_a \times n_{ad} + s_a \times s_{ad} + a_a \times a_{ad}) \end{bmatrix} \tag{10}$$

where p_{ad} is the desired absolute position specified by the user in the base frame, p_a is the actual absolute position that can be computed as in (1), n_{ad}, s_{ad}, a_{ad} are the column vectors of the rotation matrix R_{ad} giving the desired absolute orientation specified by

4

the user in the base frame, and $\boldsymbol{n}_a, \boldsymbol{s}_a, \boldsymbol{a}_a$ are the column vectors of the rotation matrix $\boldsymbol{R}_a$ in (1). The relative error is given by

$$\boldsymbol{e}_r = \begin{bmatrix} \boldsymbol{R}_a \boldsymbol{p}^a_{rd} - \boldsymbol{p}_r \\ \frac{1}{2}(\boldsymbol{n}^1_r \times \boldsymbol{n}^1_{rd} + \boldsymbol{s}^1_r \times \boldsymbol{s}^1_{rd} + \boldsymbol{a}^1_r \times \boldsymbol{a}^1_{rd}) \end{bmatrix}. \tag{11}$$

The rotation $\boldsymbol{R}_a$ is aimed at expressing the desired relative position $\boldsymbol{p}^a_{rd}$, assigned by the user in the object frame, in the base frame; in this way, if an error occurs on the object frame orientation this does not affect the specification of the desired relative position between the two end-effectors. Further in (11), $\boldsymbol{p}_r$ can be computed as in (2), $\boldsymbol{n}^1_{rd}, \boldsymbol{s}^1_{rd}, \boldsymbol{a}^1_{rd}$ are the column vectors of the rotation matrix $\boldsymbol{R}^1_{rd}$ giving the desired relative orientation specified by the user in the end-effector frame of the first manipulator, and $\boldsymbol{n}^1_r, \boldsymbol{s}^1_r, \boldsymbol{a}^1_r$ are the column vectors of the rotation matrix $\boldsymbol{R}^1_r$ in (2). The absolute velocity term is given by

$$\boldsymbol{v}_{ad} = [\dot{\boldsymbol{p}}^{\mathrm{T}}_{ad} \quad \boldsymbol{\omega}^{\mathrm{T}}_{ad}]^{\mathrm{T}} \tag{12}$$

where $\dot{\boldsymbol{p}}_{ad}$ and $\boldsymbol{\omega}_{ad}$ are respectively the desired absolute linear and angular velocities specified by the user in the base frame. The relative velocity term is given by

$$\boldsymbol{v}_{rd} = \begin{bmatrix} \boldsymbol{R}_a \dot{\boldsymbol{p}}^a_{rd} + \boldsymbol{\omega}_a \times \boldsymbol{R}_a \boldsymbol{p}^a_{rd} \\ \boldsymbol{\omega}^1_{rd} \end{bmatrix} \tag{13}$$

where $\dot{\boldsymbol{p}}^a_{rd}$ is the desired relative linear velocity specified by the user in the object frame and $\boldsymbol{\omega}^1_{rd}$ is the desired relative angular velocity specified by the user in the end-effector frame of the first manipulator. Notice that the expression of the translational part of the relative velocity presents an additional term which is a consequence of having assigned the relative position in the object frame.

NUMERICAL CASE STUDIES

A system of two cooperative PUMA 560 robot manipulators has been considered to work out numerical case studies. The base of manipulator 1 is located at $(0, -0.1501, 0)$ and the initial configuration places the end effector at $\boldsymbol{p}_1 = [0.4 \quad 0 \quad 0.5]^{\mathrm{T}}$; a constant rotation matrix has been used so that $\boldsymbol{R}_1 = \boldsymbol{I}$. The base of manipulator 2 is located at $(1, 0.1501, 0)$ and the initial configuration places the end effector at $\boldsymbol{p}_2 = [0.6 \quad 0 \quad 0.5]^{\mathrm{T}}$ with $\boldsymbol{R}_2 = \boldsymbol{I}$. Manipulator 2 is allowed to make a sliding contact with the object.

Using (1),(2), the initial values for the task variables are computed:

$$\boldsymbol{p}_a = [0.5 \quad 0 \quad 0.5]^{\mathrm{T}} \quad \boldsymbol{R}_a = \boldsymbol{I} \quad \boldsymbol{p}_r = [0.2 \quad 0 \quad 0]^{\mathrm{T}} \quad \boldsymbol{R}^1_r = \boldsymbol{I}.$$

A sketch of the initial configuration is depicted in Fig. 1. For clarity of illustration, the dimensions of the object have been enlarged.

The task is to move the object to the absolute location

$$\boldsymbol{p}_a = [0.5 \quad 0 \quad 0.7]^{\mathrm{T}} \qquad \boldsymbol{R}_a = \begin{bmatrix} \cos\pi/4 & -\sin\pi/4 & 0 \\ \sin\pi/4 & \cos\pi/4 & 0 \\ 0 & 0 & 1 \end{bmatrix}.$$

5

A rectlinear path is assigned from the initial to the final position, whereas an angular motion about a fixed axis in space is assigned to pass from the initial to the final orientation. Smooth trajectories are imposed using 5th-order interpolating polynomials with null initial and final velocities and accelerations, and a time duration of 1 s.

As for the relative variables, it is opportune to define the task with reference to the absolute frame. The task is to keep $\boldsymbol{p}_r^a$ and $\boldsymbol{R}_r^1$ constant; note the simplicity of task specification when it is referred to the absolute frame.

The closed-loop inverse kinematics scheme based on (9), at first without the null space term, has been implemented in MATLAB at 1 ms sampling time; the gain matrix has been chosen as $\boldsymbol{K} = \text{block diag}\{500\boldsymbol{I}_6, 1000\boldsymbol{I}_6\}$. The resulting final system configuration is also shown in Fig. 1. As can be recognized, the object is taken to the desired final position whereas the final orientation of end effector 2 accounts for the sliding contact. The time history of the norm of position and orientation components of both absolute and relative errors (Fig. 2) confirm the good tracking capabilities of the inverse kinematics scheme.

Another numerical case study has been worked out where the kinematic redundancy introduced by the sliding contact is exploited to minimize the constraint $c = 0.5(q_{2,1}(t) - q_{2,1}(0))^2$, i.e. to keep the base revolute joint of manipulator 2 constant. The initial configuration is the same as before (Fig. 1) and the same task has been assigned for the absolute and relative variables. Further, in the discrete-time implementation of (9) it has been chosen $k_C = 3000$. The final configuration in Fig. 3 shows the correct execution of the task while the base revolute joint of manipulator 2 is remarkably kept close to the initial value as shown by the time history of the constraint value.

CONCLUSIONS

An inverse kinematics scheme for a system of two spatial manipulators holding a common object with a sliding contact has been presented. The scheme is based on a cooperative task formulation that allows the user to give a straightforward description of coordinated motions in terms of absolute and relative variables. Two numerical case studies have demonstrated the good performance of the scheme also when the redundant degrees of freedom introduced by the sliding contact are effectively exploited.

ACKNOWLEDGEMENTS

The research reported in this work was supported by *Consiglio Nazionale delle Ricerche* under contract n. 93.00946.PF67.

REFERENCES

[1] J.Y.S. Luh and Y.F. Zheng, "Constrained relations between two coordinated industrial robots for motion control," *Int. J. of Robotics Research*, vol. 6, n. 3, pp. 60–70, 1987.

[2] M. Uchiyama, and P. Dauchez, "A symmetric hybrid position/force control scheme for the coordination of two robots", *Proc. 1988 IEEE Int. Conf. on Robotics and Automation*, Philadelphia, PA, pp 350–356, 1988

[3] P. Chiacchio, S. Chiaverini, and B. Siciliano, "Kineto-static analysis of cooperative robot manipulators achieving dexterous configurations," *9th CISM-IFToMM Symp. on Theory and Practice of Robots and Manipulators*, Udine, I, Sep. 1992, in *RoManSy 9*, Lecture Notes in Control and

6

Information Sciences 187, A. Morecki, G. Bianchi, K. Jaworek (Eds.), Springer-Verlag, Berlin, D, pp. 93–100, 1993.

[4] P. Chiacchio, S. Chiaverini, and B. Siciliano, "Task-oriented kinematic control of two cooperative 6-dof manipulators," *Proc. 1993 American Control Conf.*, San Francisco, CA, pp. 336–340, 1993.

[5] P. Chiacchio, S. Chiaverini, and B. Siciliano, "User-oriented task description for cooperative spatial manipulators: One-degree-of-freedom rolling grasp," *Proc. 32nd IEEE Conf. on Decision and Control*, San Antonio, TX, pp. 1126–1127, 1993.

[6] P. Chiacchio and S. Chiaverini, "User-oriented task description for cooperative spatial manipulators: Rolling grasp," *5th Int. Symp. on Robotics and Manufacturing*, Maui, HI, Aug. 1994.

[7] P. Chiacchio, S. Chiaverini, L. Sciavicco, and B. Siciliano, "Closed-loop inverse kinematics schemes for constrained redundant manipulators with task-space augmentation and task-priority strategy", *Int. J. of Robotics Research*, vol. 10, pp. 410–425, 1991.

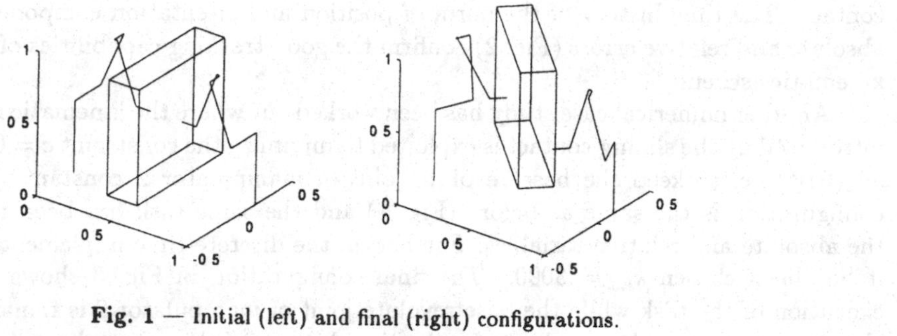

Fig. 1 — Initial (left) and final (right) configurations.

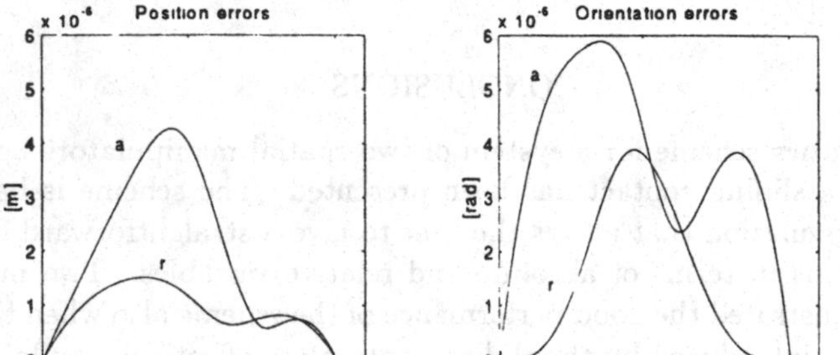

Fig. 2 Time history of norm of position and orientation errors: a—absolute; r—relative.

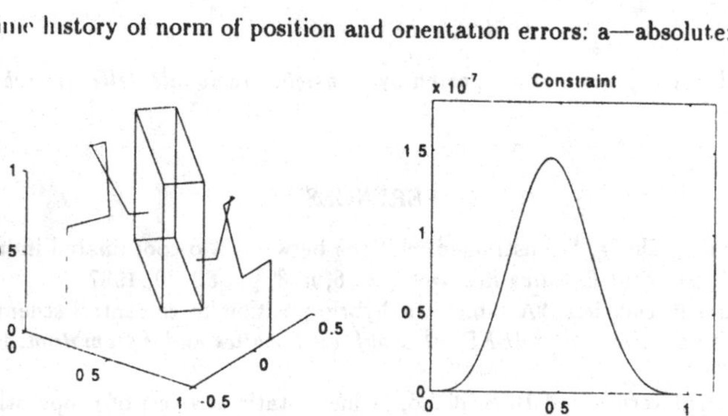

Fig. 3 Final configuration and time history of constraint value in case of redundancy resolution.

COMPLIANCE CONTROL OF DIRECT DRIVE MANIPULATOR USING ULTRASONIC MOTOR

A. Kato, N. Kondo and H. Narita

Aichi Institute of Technology, Toyota, Japan

and

K. Ito and Z.W. Luo

Toyohashi University of Technology, Toyohashi, Japan

ABSTRACT: A compact size light weight direct drive(DD) manipulator with traveling wave ultrasonic motors (USMs) as actuators was developed. Contact tasks[1)] were realized by adapting an adjustable compliance control system of the USM. In the system, a torque feedback control by interaction force to environment is not used. Experiments on contact tasks like crank rotation and cooperative motion of two manipulators were observed.

1. Introduction

The absence of any reduction gear system between the actuators and the joints gives the DD manipulator no backlash and low non linear friction characteristics. Therefore it is good for realizing the compliant motion of the manipulators. The DD actuators that rotate low speed and high torque have large size and heavy weight, so the DD manipulators also become large and heavy. The compact size, light weight, low speed rotation, high speed response and high torque weight ratio characteristic of the USM make desirable actuators for DD manipulators.

2. Adjustable Compliant Motion of the Traveling Wave Ultrasonic Motor

As previously mentioned, the USM has good characteristics which make it excellent DD manipulators. On the other hand, since the motor generates torque by the Coulomb's frictional force between its rotor and stator, the torque and speed control is not easy to use as a servo motor. Thus, a new method of rotation speed and torque control by the phase-difference control of the two-phase driving signals of the motor was developed.[2)]The method realized the adjustable compliant motion and damped frictional motion using the output shaft angle feedback and the angular velocity feedback of the shaft, respectively. It can be regard as the PD control system. The stator of the USM is driven by two out-of-phase sine-wave electric signals generating two out-of-phase periodic bending waves are on the stator. The bending waves composed of nine periodic traveling waves is propagated along the stator ring. The traveling waves generate ellipsoidal motion on the stator surface. Equations (1) (2) are the perpendicular and circumference components of the ellipsoidal motion.

for perpendicular component to the stator flat surface,

$$\xi = \xi_0 \sin(nX)\sin(\omega t) + \xi_0 \cos(nX)\sin(\omega t + \phi) \quad (1)$$

for circumference component along the stator ring.

$$\zeta = \zeta_0 n(H + T/2)\sin(nX)\sin(\omega t) + \zeta_0 n(H + T/2)\cos(nX)\sin(\omega t + \phi) \quad (2)$$

where,

ξ perpendicular component of bending wave
ξ_0 amplitude of perpendicular
ζ circumference component of bending wave
ζ_0 amplitude circumferential component
ϕ phase difference of driving signals.
ω angular frequency of bending vibration.
n number of resonant bending wave period
H height of teeth.
T thichness of stator ring.
X position in circumference.
t time

Fig.1 presents a representative torque-phase difference characteristic. It is almost coincide with the result of the simulated value.[3)]

Figure 1:

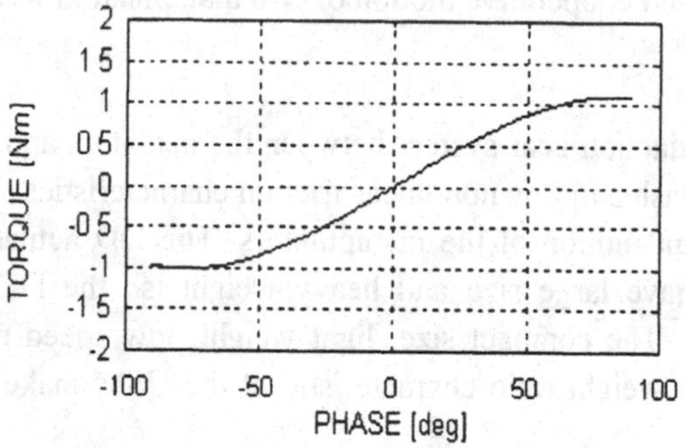

Output Torque Change with Phase Difference of the Two Driving Signals.

2.2 Adjustable Compliance Control

Fig.2 shows the PD control system. The phase control signal is supplied to the torque and speed control system by the feedback of the shaft angle. It is supposed that the torque is generated in the opposite direction of the external torque.

Figure 2:

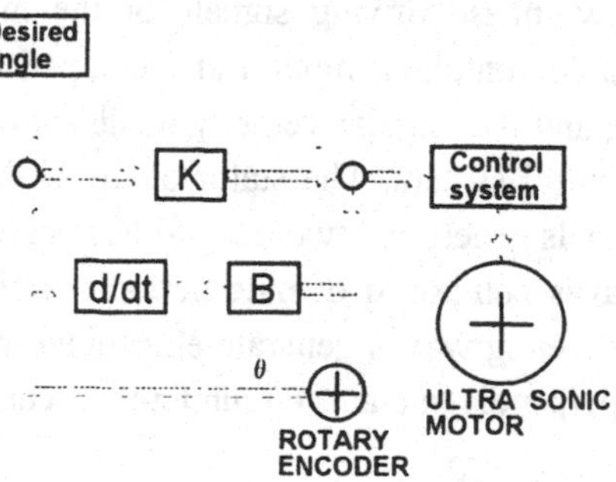

Compliance Control System of the USM

Figure 3;

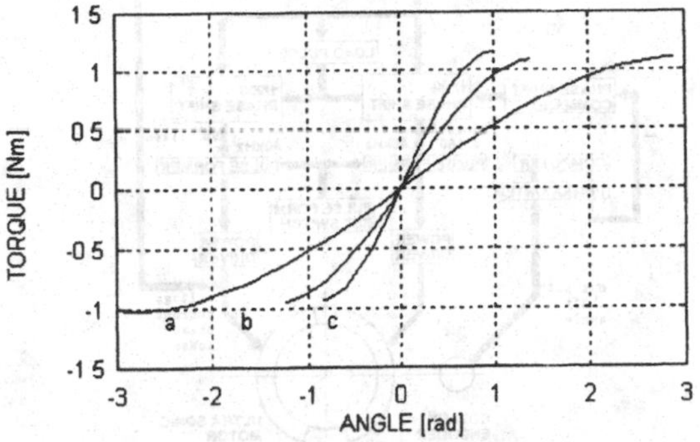

Elastic Torque Responses.

Fig.3 shows the characteristics elastic torque - output shaft angle. The shaft rotates compliantly against the external torque, when the D factor B is zero and the P factor K is non zero. It shows that the system works like a spring whose spring constant can be regulated freely.

Figure 4;

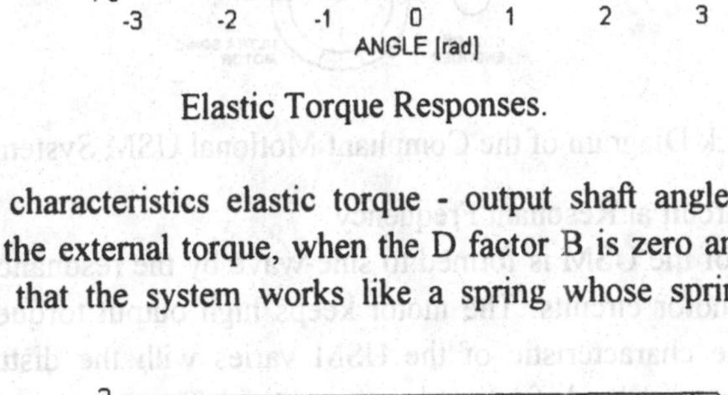

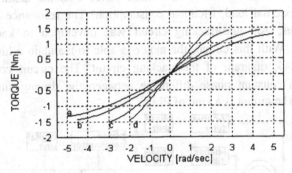

Viscous Frictional Torque Responses.

Fig.4 shows the characteristics of the viscous frictional torque to the angular velocity of the shaft. When the P factor is zero and the D factor is non zero, the motor rotates with viscous damping. When both K and B are non zero, the motor has the characteristics of both elasticity and viscous damping.

3. Implementation of the Control System

3.1 Basic Composition

Fig.5 shows the block diagram of the control system consisting of analog and digital circuit. The output shaft angle is detected by a rotary encoder. The computer composes the phase shift command based on the deflection angle of the shaft from the desired angle and angular velocity of the shaft. Then, the command is supplied to the variable phase shifter. Thus, the gain adjustable PD control system is made for the adjustable compliant-damped motion on the shaft.

Figure 5;

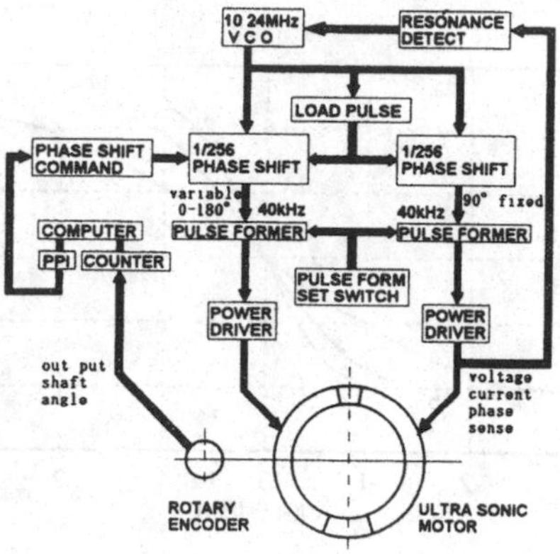

Block Diagram of the Compliant Motional USM System.

3.2 Auto-tracking Circuit at Resonant Frequency

The driving signal of the USM is formed to sine-wave by the resonance characteristic of the transformer and motor circuits. The motor keeps high output torque at the resonance frequency. The torque characteristic of the USM varies with the disturbances.(e.g., the stator temperature, external load, frictional consumption) The resonance frequency varies with the disturbances. An auto-tracking circuit was developed to keep track of the resonance frequency. With this circuit, the motor is able to maintain a high output torque. **Fig.6** shows the block diagram of the tracking circuit. The circuit controls a driving frequency to keep a constant phase difference between the driving sine wave electric voltage and current. The circuit is a kind of PLL.

Figure 6;

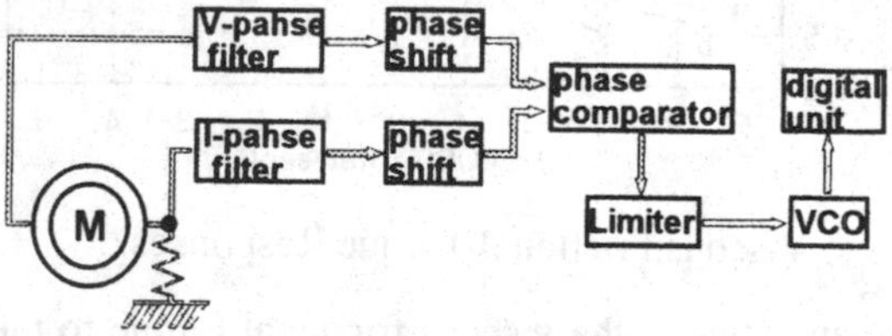

Block Diagram of the USM Auto Frequency Tracking Circuit.

4. SCARA Type Two-Vertical Axis Joints Direct Drive Manipulator

4.1 Structure

Two compact size DD manipulators were made. Each one has two vertical axis joints and moves in the horizontal plane. The developed compliant motional USM systems are used as actuators. The mechanical structure is shown in **Fig.7**.

Figure 7;

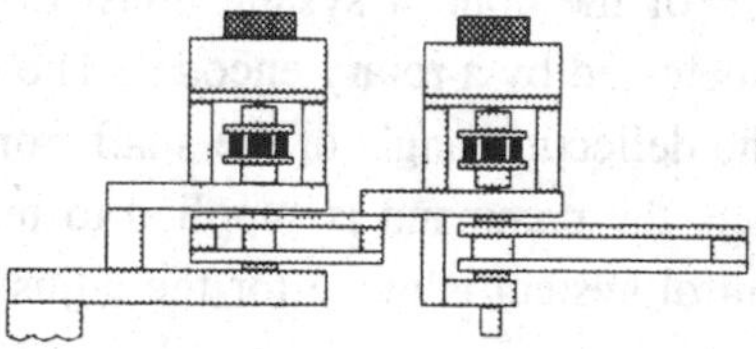

Structure of the Compact Size DD Manipulator.

4.2 Non-linearity Compensation of the Manipulator

The kinematic equation of two vertical axis joints manipulator is given as follows,

$$\tau_0 = \tau - M(\theta)\ddot{\theta} - h(\theta,\dot{\theta}) \qquad (3)$$

where,

$\tau_0 = [\tau_{01}, \tau_{02}]^T$: output torque $\quad M(\theta)$: 2×2 matrix of inertia

$\tau = [\tau_1, \tau_2]^T$: joint torque $\quad h(\theta,\dot{\theta})$: 2×1 vector of nonlinear forces

$\theta = [\theta_1, \theta_2]^T$: joint variable

The non-linear terms $M(\theta)\ddot{\theta}$ and $h(\theta,\theta)$ depend on the posture of the manipulator. So, the $\dot{\theta}$ and the $\ddot{\theta}$ are computed from the joint angle θ and the interrupt interval. The terms of $M(\theta)\ddot{\theta}$ and $h(\theta,\dot{\theta})$ are computed with the parameters of the manipulator.

Fig.8 shows the model of the manipulator and its parameter table.

Figure 8;

	LINK1	LINK2
m (kg)	1.0785	0.0625
l (m)	0.12	0.12
lg (m)	0.102	0.089
I (kgm²)	0.015	6.6×10^{-4}

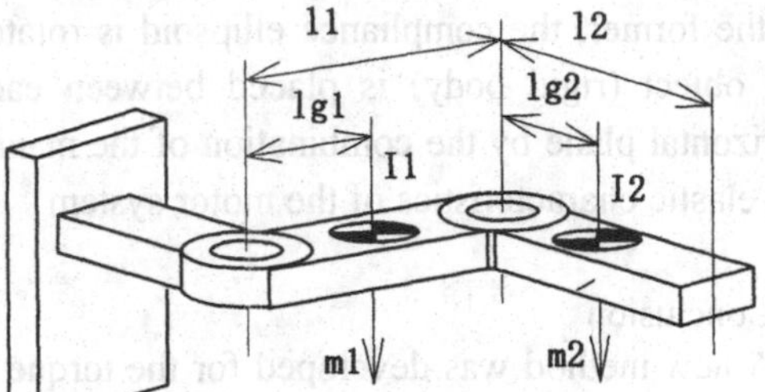

Model and Parameter Table of the DD Manipulator.-

The manipulator has the characteristics of the compliance and damping giving the output torque as follows,

$$\tau_0 = K(\theta_d - \theta) - B\theta \qquad (4)$$

From Eq.(3) and Eq.(4) the compensated joint torque is

$$\tau = M(\theta)\ddot{\theta} + h(\theta,\dot{\theta}) - (B - B_m)\dot{\theta} + K(\theta_d - \theta) \qquad (5)$$

where,

$K = [K_1 \;, K_2]^T$. elastic coefficient $\quad B_m = [B_{m1}, B_{m2}]^T$: natural viscosity of the motor

$B = [B_1 \;, B_2]^T$: viscous coefficient $\quad \theta_d = [\theta_{d1}, \theta_{d2}]^T$: desired angle of joints

4.3 Step Responses

Fig.9 shows the step responses of the manipulator shoulder arm. **(a)** is the natural response and **(b)** is the response with compensated non-linearity.

Figure 9;

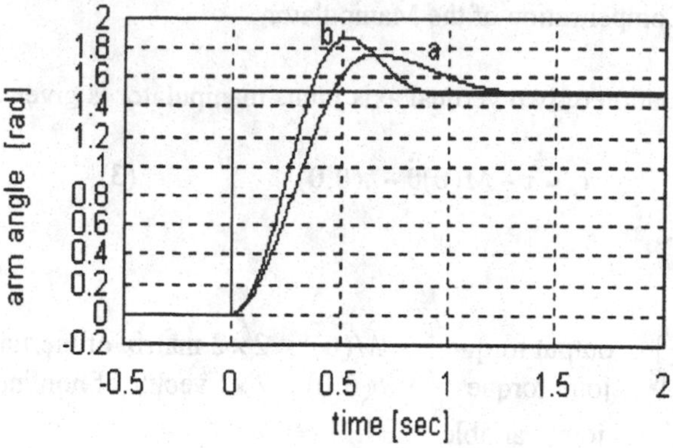

Step Responses of the Manipulator.

5. Contact Tasks by the Manipulator

Some contact tasks were carried out using the developed DD manipulator. For example, the crank rotation tasks by one manipulator and the cooperative tasks by two manipulators. In the former, the compliance ellipsoid is rotated along the crank trajectory. In the latter, the object (rigid body) is placed between each manipulator end-effector moved in the horizontal plane by the combination of the manipulators. The holding force is generated by the elastic characteristics of the motor system.

6. Conclusion

A new method was developed for the torque and speed control of the USM. The torque and speed were controlled by the phase difference of the two-phase driving signals of the motor. The adjustable compliant motion was realized by the feedbacks of the shaft angle and angular velocity of the shaft. With the compliance control, some contact tasks were carried out.

References;

1) N. Hogan Proc. of IEEE int. conf. Robotics and Information Vol.2 1047-1054.(1987)
2) A. Kato, K. Ito, and M. Ito, "Compliance Control of Circular Traveling Wave Motor", Proc. of IEEE IECON'91, 538-542,(1991).
3) A. Kato, K. Ito, and M. Ito, "Adjustable Compliant Motion of Ultrasonic Motor", Journal of Robotics and Mechatronics Vol.5 No.5,(1993).

OPTIMAL EFFICIENCY OF A ROBOT ENVIRONMENT INTERACTION TASK IN A MATCHING IMPEDANCE APPROACH

G.A. Ombede, J.P. Simon, M. Betemps and A. Jutard
National Institute of Applied Sciences of Lyon, Villeurbanne, France

Abstract : This paper investigates a matched impedance approach using scattering waves in robot manipulators task in contact with the environment. Given that the dynamic performance of robot in constrained situations is very dependent on the environment parameters, matched conditions are established to optimize the power transferred on a fixed configuration. Simulations show that on matched conditions, the coupled robot-environment consume maximum power supplied.

Nomenclature

A_c	Controller-actuator gain (diagonal matrix)
B	Force / velocity relation
B_c	Damping matrix of controller actuator
B_e	Damping matrix of environment
B_r	Damping matrix of the robot
C_c	Input torque
C_e	Environment torque
F	Contact force
F_e	Environment contact force
$g(\theta)$	Robot gravity terms
$I(\theta)$	Inertia tensor in actuator coordinates
$J(\theta)$	Jacobian
K	Force / displacement relation
K_c	Stiffness matrix of controller actuator
K_e	Stiffness matrix of environment
K_r	Stiffness matrix of the robot
L_1, L_2	Link lengths
M	Inertia tensor in end point coordinates
M_c	Inertia matrix of controller actuator
M_e	Inertia matrix of environment
M_r	Inertia matrix of the robot
Rc	real part of Zc

Re	real part of Ze
Rr	real part of Zr
Rre	real part of Zre
Rp	Real part of Zp
S	Power wave scattering matrix
V_e	Environment velocity
V_{in}	Nominal velocity input
X	End point position
Xc	Imaginary part of Zc
Xe	Imaginary part of Ze
Xr	Imaginary part of Zr
Xre	Imaginary part of Zre
Z_c	Controller-actuator impedance
Z_e	Environment impedance
Zp	Diagonal matrix of internal impedance of the circuit
Zr	Robot impedance
Zre	coupling robot-environment impedance
θ	Actuator position or angle
θ_1, θ_2	Absolute joint angle
θ_{in}	Desired joint position
θ_e	Environment equilibrium position in joint frame
Ω_{in}	Joint velocity input
Ω_e	Environment velocity
τ_{int}	Interface torque

I - Introduction

Most assembly operations and manufacturing tasks require mechanical interactions with the environment where the manipulated object is in a constrained space[2], [3], [6]. As the robot comes into contact with a semi-rigid or rigid environment, the coupling between the robot and environment affects the behaviour of the overall system.
There is currently extensive research underway on force and control impedance. The goal is the implementation of a target impedance i.e an equivalent dynamics which supersedes the nominal robot dynamics. According to Hogan [2], the robot behaves as a mass-spring-dashpot system whose parameters (Inertia-Stiffness-Damping) can be specified arbitrarily. The objective of matched impedance control is to optimize the exchange power between the robot and environment in spite of the force of interaction which varies due to incertainty on the location of the point of contact and environment properties. Hogan [2] also suggests to use mass-spring-dashpot as the target impedance model knowing that typically the contact force and the resulting motion are generally related by a second order differential equation $M\ddot{X}+B\dot{X}+KX=F$.

II - Problem definition

We are interesting given the closed loop system control configuration proposed by Goldenberg [1] (fig1-a) to implement a matched target impedance of a robot coupled with the environment. The methodology is based on replacing the robot-environment model by an analogous equivalent electrical circuit to analyse his comportment in a frequency domain, in the way of wave scattering approach [7], [8], [9].

III - Model of robot and environment

For impedance control, Goldenberg [1] suggests the following diagram:

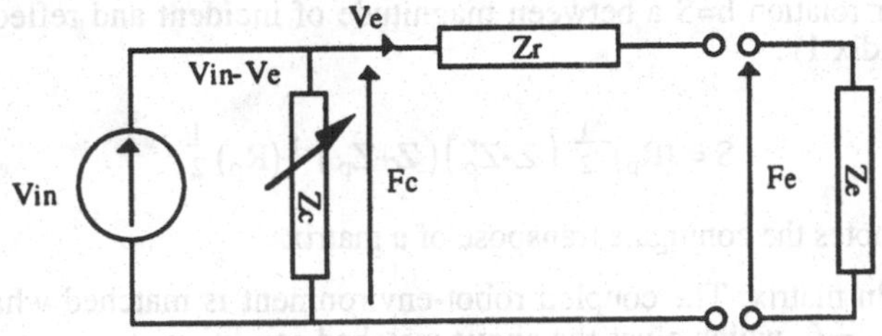

Fig 1-a:

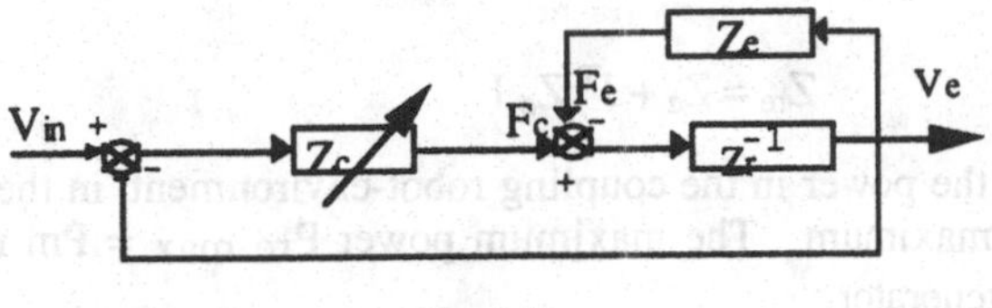

Fig 1-b:

The required diagram in actuator coordinates is:

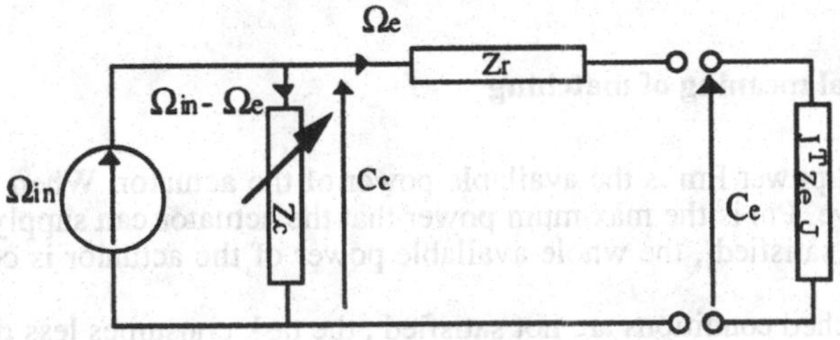

Fig 2-a

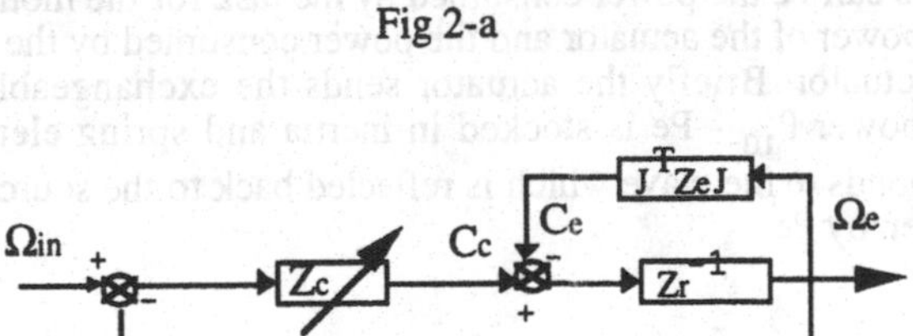

Fig 2-b

IV - Modelization in terms of waves scattering

IV.1 - Matching conditions

In the scattering description, the interaction between the robot and environment during the task is modelized considering an incident power wave emitted by a source, splited in a transmitted power wave absorbed by the task and a reflected power wave not used up by the task. According Simon [4], [5], with the equations of robot-environment model :

$$\frac{1}{2}(R_p)^{-\frac{1}{2}}\left(C+Z_p\Omega\right)=a \quad (1)$$

$$\frac{1}{2}(R_p)^{-\frac{1}{2}}\left(C-Z_p^*\Omega\right)=b \quad (2)$$

and the linear relation b=S a between magnitude of incident and reflected power waves,we obtain (appendix 1) :

$$S=(R_p)^{-\frac{1}{2}}\left(Z-Z_p^*\right)\left(Z+Z_p\right)^{-1}(R_p)^{\frac{1}{2}} \quad (3)$$

where "*" denotes the conjugate transpose of a matrix

S is a 2n x 2n matrix. The coupled robot-environment is matched whatever Ω_{in} when the submatrice $S_{11}=0$ which gives the above matched condition :

$$Z_c = Z_{re}^* \quad (4)$$

where

$$Z_{re} = Z_c + J^T Z_e J$$

Let's call P_{re} the power in the coupling robot-environment, in the matching case, the power P_{re} becomes maximum . The maximum power $P_{re\ max}$ = Pm is called the exchangeable power of the generator.

In order to realize the maximum power consumption, (4) has to be satisfied . This condition $Z_c = Z_{re}^*$ is called the matching condition .If (4) is not satisfied, the task consumes less power than the power available.

IV.2 - Physical meaning of matching

The maximum power Pm is the available power of the actuator. When the real part of Zc is definite positive, Pm is the maximum power that the actuator can supply. When the matched conditions are satisfied , the whole available power of the actuator is consumed by the task for motion.

When the matched conditions are not satisfied , the task consumes less power than the power available. Let's call Pe the power consumed by the task for the motion, the difference between the available power of the actuator and the power consumed by the task is the power reflected back to the actuator. Briefly the actuator sends the exchangeable power Pm to the task , however the power P_m - Pe is stocked in inertia and spring elements of the system. This power corresponds to the wave which is reflected back to the source . Hence the net power to the task is given by Pe

V - Implementation of a matched target impedance

The elements of the target impedance correspond to elements of a PID controller actuator impedance. The advantage of a matched target impedance is that in spite of the change of environment structural elements, the target impedance takes off values which satisfy matching conditions and then assure that the task consumes the maximum power.

Given $Z_c = R_c + j\,X_c$ the controller actuator impedance and $Z_{re} = Z_r + J^T Z_e J = R_{re} + j\,X_{re}$ the load impedance, equation (4) gives:

$$R_c = R_{re} \quad \text{and} \quad X_c = -X_{re} \quad (5)$$

The dynamic model of an n-DOF rigid manipulator is non linear and expressed by the relation:

$$I(\theta)\ddot{\theta} + f(\theta,\dot{\theta}) + g(\theta) = \tau_{int} \quad (6)$$

where $I(\theta)$ is the mass inertia matrix, $f(\theta,\dot{\theta})$ is the rate and configuration dependent terms of the moment equation, $g(\theta)$ is the gravity and τ_{int} is the actuator input.
We linearize the dynamic model around a functional state $H=[C_c, C_e, \Omega_e, \Omega_{in}]_0^T$ to obtain a variational linear model.
The robot impedance Zr becomes (appendix 2) :

$$Z_r = M_r\, s + B_r + \frac{K_r}{s}$$

Let's assume that the controller-actuator model and the environment model can be represented by a combination of spring-mass-dashpot; then their impedances are given respectively by:

$$Z_c = A_c(s)\left[M_c\, s + B_c + \frac{K_c}{s}\right]$$
$$Z_e = M_e\, s + B_e + \frac{K_e}{s}$$

In the frequency domain, s = j ω, we have:

$$Z_e = B_e + j\left(M_e\,\omega - K_e\,\omega^{-1}\right)$$
$$Z_r = B_r + j\left(M_r\,\omega - K_r\,\omega^{-1}\right)$$
$$Z_c = A_c(j\omega)\, B_c + j\, A_c(j\omega)\left(M_c\,\omega - K_c\,\omega^{-1}\right)$$

Replacing on equation (5), matched target impedance values must satisfy equations:

$$\left(J^T M_e\, J + M_r + A_c(j\omega)\, M_c\right)\omega_0^2 = \left(J^T K_e\, J + K_r + A_c(j\omega)\, K_c\right)$$
$$A_c(j\omega)\, B_c = J^T B_e\, J + B_r$$

V.1 - Compute motion

Given fig 2-a , let's write :

$$C_c = Z_c\left(\Omega_{in} - \Omega_e\right) \quad (7)$$

and :

$$C_c = \left(Z_r + J^t\, Z_e\, J\right)\Omega_e \quad (8)$$

(7) and (8) gives :

$$Z_c\, \Omega_{in} = \left(Z_r + Z_c + J^t\, Z_e\, J\right)\Omega_e \quad (9)$$

For a fixed Ω_{in} , we have to determine the output motion Ω_e by solving in time domain the above equation:

$$\left(M_r + A_c M_c + J^t M_e J\right)\ddot{\theta}_e + \left(B_r + A_c B_c + J^t B_e J\right)\dot{\theta}_e + \left(K_r + A_c K_c + J^t K_e J\right)\theta_e = M_c\ddot{\theta}_{in} + B_c\dot{\theta}_{in} + K_c\theta_{in}$$

For

$$M = M_r + A_c\, M_c + J^t\, M_e\, J$$
$$B = B_r + A_c\, B_c + J^t\, B_e\, J$$
$$K = K_r + A_c\, K_c + J^t\, K_e\, J$$

We have :

$$\theta_e + M^{-1}B\,\theta_e + M^{-1}K\,\theta_e = M^{-1}M_c\,\theta_{in} + M^{-1}B_c\,\theta_{in} + M^{-1}K_c\,\theta_{in} \qquad (10)$$

On harmonic excitation, $\theta_{in} = a \sin \omega t$, $\theta_{in}(0) = \theta_{in}(0) = O_n$ where $a = [a_1, a_2, \ldots\ldots, a_n]^T \in R^n$ denotes the vector of amplitude and O_n denotes the zero vector in R^n ; the mass matrix $M \in R^n \times R^n$, the damping matrix $B \in R^n \times R^n$ and the stiffness matrix $K \in R^n \times R^n$ are all symmetric and positive definite. The equation (10) is solved for $\theta_e(t) = \Omega_e(t)$ by direct numerical integration.

V.2 - Compute powers

The environment torque value is $C_e = Z_e\,\dot{\theta}_e$, then the environment power value is given by :

$$P_e = C_e^T\,\dot{\theta}_e = \dot{\theta}_e^T\,Z_e^T\,\dot{\theta}_e$$

This power represents the net power consumes in the environment. With fig 2-a, let's write :

$$C_c - C_e = Z_r\,\Omega_e$$

where $$Z_r = M_r\,s + B_r + \frac{K_r}{s}$$

we obtain : $$C_c - C_e = M_r\,\ddot{\theta}_e + B_r\,\dot{\theta}_e + K_r\,\theta_e$$

The exchangeable power value is:

$$P_T = (C_c - C_e)^T\,\dot{\theta}_e$$

with $C_c - C_e$ the physical resultant coupling robot-environment torque on the driving shaft of the link.

VI - Application to a scara mechanism in interaction with the environment : simulations and results.

We conduct simulation using a 2-DOF horizontaly articulated manipulator. The first axis (θ_1) and the second axis (θ_2) are active joint driven by DC servomotors with gear, fig-4 shows the model of manipulator.

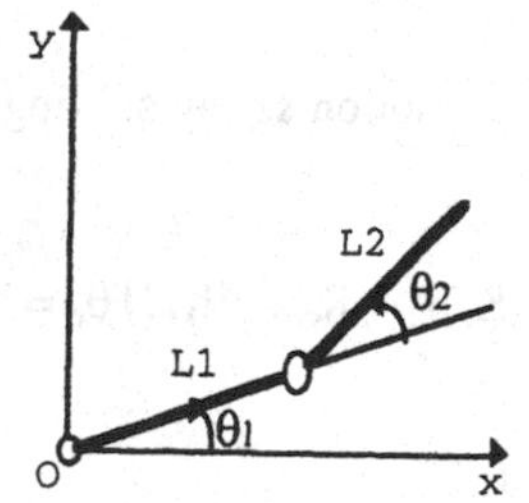

Fig-4 model of robot.

The dynamic parameters $I(\theta)$, $B(\theta)$ and $C(\theta)$ of (6) are defined in appendix (2). The table shows parameters of the model of manipulator :

Mass of link 1	m1= 1 Kg
Mass of link 2	m2= 1 Kg
Length of link 1	L1= 0,3 m
Length of link 2	L2= 0,3 m
Inertia moment of actuator 1	Jm1=0,03 Kg m^2
Inertia moment of actuator 1	Jm1=0,03 Kg m^2

In simulations, the environment is modelized by a mass-spring-dashpot with :

$$M_e=\begin{pmatrix} 3 & 0 \\ 0 & 3 \end{pmatrix},\ B_e=\begin{pmatrix} 50 & 0 \\ 0 & 50 \end{pmatrix},\ K_e=\begin{pmatrix} 200 & 0 \\ 0 & 200 \end{pmatrix}$$

Graph 1 to 5 show the average net power $P_{Eav}=\frac{1}{T_{i+1}-T_i}\int_{T_i}^{T_{i+1}} C_e^T\,\dot{\theta}_e\,dt$

to the environment, function of parameter α with $R_C=\alpha\,R_L$

P_{Eav} is the energy consumed in the environment to generate movement per period. For matched conditions, i.e $\alpha = 1$, P_{Eav} is maximum for each period Ti i=1...5.The whole power supplied by the source is consumed by the environment.

For unmatched conditions, i.e a = 0, 0.1, 0.6, 1.5, 2, P_{Eav} is not maximum and then there is power which is reflected back to the source.

Graph 6 shows the average exchangeable power $P_{Tav}=\frac{1}{T_{i+1}-T_i}\int_{T_i}^{T_{i+1}} (C_c-C_e)^T\,\theta_e\,dt$

function of the pulsation. P_{Tav} is the energy needed for the movement during one period. For $\alpha = 1$, and $\omega = \omega_0$, (matching conditions) , the curve has a minimum which represents a reduction of 85% of P_{Tav}. It means that the robot and environment are accommodated around $\dot{\omega}0$.

For $\alpha = 6$, the coupling robot-environment is out of matching and graph 6 has not a significant reduction of power.

VII - Conclusion.

A method to implement a matched impedance control in joint space is proposed. This method takes account of the dynamic caracteristics of the coupled robot-environment mechanism. The optimization of the net power gives the conditions to generate a matched impedance robot-environment using scattering approach. This fact is imaged by simulations in which the net power of the environment is maximum at the matching robot-environment interaction.

Appendix 1: Expression of scattering matrix S of the circuit.

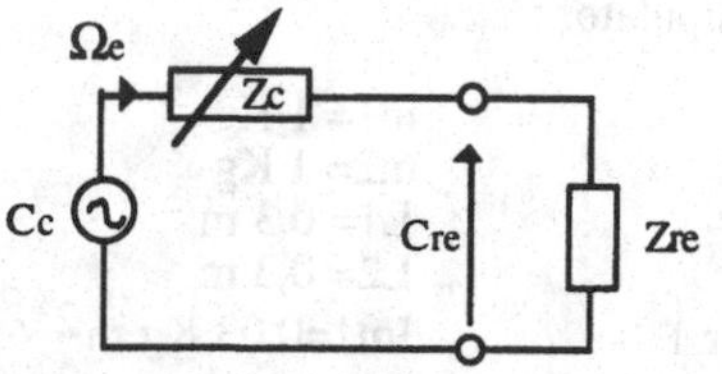

To define the scattering matrix, let's consider the linear circuit above. We have the incident power wave a and the reflected power wave b defined in section IV by :

$$\frac{1}{2}(R_p)^{-\frac{1}{2}}\left(C+Z_p\Omega\right)=a \qquad \frac{1}{2}(R_p)^{-\frac{1}{2}}\left(C-Z_p^*\Omega\right)=b$$

where $C=[C_{re}\ C_c]^T$ and $\Omega=\left[\Omega_e\ \Omega_{in}\right]^T$

Since there is a linear relation between C and Ω given by :

$$C=Z\,\Omega$$

then there is a linear relation between a and b : $b=S\ a$

Elimination of a, b and C in above equations gives:

$$S=(R_p)^{-\frac{1}{2}}\left(Z-Z_p^*\right)\left(Z+Z_p\right)^{-1}(R_p)^{\frac{1}{2}}$$

Given our circuit, let's write :

$$C_{re}=Z_{re}\,\Omega_e \text{ and } C_c=(Z_{re}+Z_c)\,\Omega_e$$

with :

$$Z=\begin{pmatrix} Z_{re} & 0 \\ Z_c+Z_{re} & 0 \end{pmatrix} \qquad Z_p=\begin{pmatrix} Z_c & 0 \\ 0 & Z_{re} \end{pmatrix}$$

Then the following expression of S is obtained by :

$$S=\begin{pmatrix} (R_p)^{-\frac{1}{2}}\left(Z_{re}-Z_c^*\right)\left(Z_{re}+Z_c\right)^{-1}(R_p)^{\frac{1}{2}} & 0 \\ (R_p)^{-\frac{1}{2}}\left(I+Z_{re}^*\,Z_{re}^{-1}\right)(R_p)^{\frac{1}{2}} & -(R_p)^{-\frac{1}{2}}Z_{re}^*\,Z_{re}^{-1}(R_p)^{\frac{1}{2}} \end{pmatrix}$$

Matched condition is obtained when $Z_c=Z_{re}^*$. With this condition, the reflected wave b is always nulle whatever the incident wave a, then the net power consumed by the task is maximal.

Appendix 2: Explicit expression of dynamic coefficients of the 2-dof SCARA manipulator.

The dynamic model of the manipulator expressed explicitely is:

$$I(\theta)\,\ddot{\theta}+B(\theta)\,\dot{\theta}\dot{\theta}+C(\theta)\,\dot{\theta}^2=\tau_{int}$$

with :

$I(\theta)$ Inertia matrix of the robot.

$B(\theta)$ Matrix of coriolis terms.

$C(\theta)$ Matrix of centrifigural terms.

In our model case ,we have:

$$I(\theta)=\begin{pmatrix} \frac{1}{3}m_1L_1^2+\frac{1}{3}m_2L_2^2+m_2L_1^2+m_2L_1L_2\cos\theta_2+J_{m1} & \frac{1}{2}m_2L_1L_2\cos\theta_2+\frac{1}{3}m_2L_2^2 \\ \frac{1}{2}m_2L_1L_2\cos\theta_2+\frac{1}{3}m_2L_2^2 & \frac{1}{3}m_2L_2^2+J_{m2} \end{pmatrix}$$

$$B(\theta)=\begin{pmatrix} m_2L_1L_2\cos\theta_2 \\ 0 \end{pmatrix}$$

$$C(\theta)=\begin{pmatrix} 0 & -\frac{1}{2}m_2L_1L_2\sin\theta_2 \\ \frac{1}{2}m_2L_1L_2\sin\theta_2 & 0 \end{pmatrix}$$

Appendix 3 : Linearization of the dynamic model of the 2-dof SCARA manipulator.

Given the dynamic model of the manipulator expressed explicitely in (6):

$$I(\theta)\ddot{\theta}+B(\theta)\dot{\theta}\dot{\theta}+C(\theta)\dot{\theta}^2=f(\theta,\dot{\theta},\ddot{\theta})$$

with $I(\theta)$, $B(\theta)$ and $C(\theta)$ defines in appendix 2.

To obtain the variational model,we linearize the dynamic model around a functionnal state. Then , we can write:

$$f(\theta,\dot{\theta},\ddot{\theta})-f_0(\theta_0,\dot{\theta}_0,\ddot{\theta}_0)=\left(\frac{\partial f}{\partial\ddot{\theta}}\right)_{\theta_0}(\ddot{\theta}-\ddot{\theta}_0)+\left(\frac{\partial f}{\partial\dot{\theta}}\right)_{\theta_0}(\dot{\theta}-\dot{\theta}_0)+\left(\frac{\partial f}{\partial\theta}\right)_{\theta_0}(\theta-\theta_0)$$

Resolving, we obtain :

$$f(\theta,\dot{\theta},\ddot{\theta})-f_0(\theta_0,\dot{\theta}_0,\ddot{\theta}_0)=\left(M_r s+B_r+\frac{K_r}{s}\right)(\dot{\theta}-\dot{\theta}_0)=Z_r(\dot{\theta}-\dot{\theta}_0)$$

Where :

$$M_r=I(\theta)$$

$$B_r=m_2L_1L_2\begin{pmatrix} \dot{\theta}_{20}\cos\theta_{20} & (\dot{\theta}_{10}\cos\theta_{20}-\dot{\theta}_{20}\sin\theta_{20}) \\ \dot{\theta}_{10}\sin\theta_{20} & 0 \end{pmatrix}$$

$$K_r=\frac{1}{2}m_2L_1L_2\begin{pmatrix} 0 & \dot{\theta}_{20}^2\cos\theta_{20}-(2\dot{\theta}_{10}\dot{\theta}_{20}+2\ddot{\theta}_{10}+\ddot{\theta}_{20})\sin\theta_{20} \\ 0 & -\ddot{\theta}_{10}\sin\theta_{20}+\dot{\theta}_{10}^2\cos\theta_{20} \end{pmatrix}$$

References

[1] Goldenberg A. A. ," Implementation of force and impedance control in Robot manipulators", IEEE Int. Conf. on Robotic and Automation, Philadelphia,1988.,

[2] HOGAN N. " Impedance Control : An Approach to Manipulation", ASME Journal of Dynamic Systems, Measurement and Control, Vol 107, 1985, pp 1- 24

[3] Raibert M. H. , Craig J. J. " Hybrid Position/ Force Control of Manipulators" , ASME Journal of Dynamic Systems, Measurement and Control, Vol 102, 1981, pp 418- 432

[4] Simon J.P. , Betemps M., Jutard A., "Matching Impedance Model of a constrained robot-environment task using scattering S matrix", IFACS-IMACS-IEEE International Workshop on Motion Control for Intelligent Automation, Vol 2, pp 31- 37 , Perugia Italy 27-29 October 1992.

[5] Simon J.P. , Betemps M., Jutard A., "Application to the Wave Scatter Theory to the Impedance Model of a Robot ", INRIA - SIAM Int. Conf. on Mathematical and Numerical Aspect of propagation Phenomena, Strasbourg, April 23-26, 1991

[6] Simon J.P. , Betemps M. "An Active Compliant Parallel Link Manipulator" IEEE - IES Int Workshop on Sensorial Integration for Industrial Robots, Zaragoza, Nov 22-24 1989

[7] Kurokawa K. ," Power Waves and Scattering Matrix" , IEEE on Transaction Microwaves Theory and Techiques. March 1965.

[8] Oswald J. " Sur la repartition de l'énergie dans les réseaux linéaires" Câbles et Transmissions, Octobre 1958, pp 303-326

[9] Rivier E. , Sardos R. "La matrice S" , Edition Masson 1982

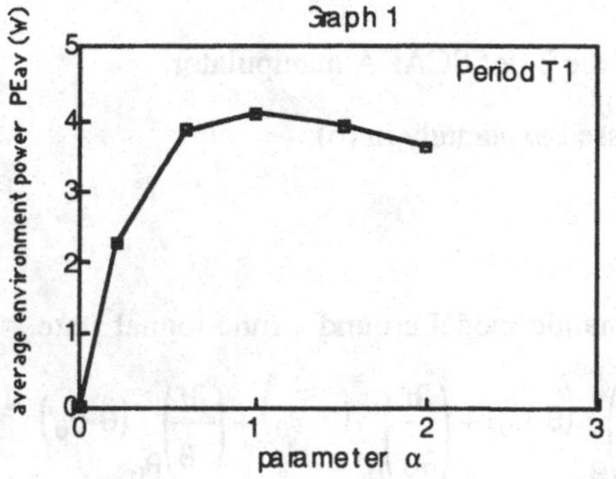

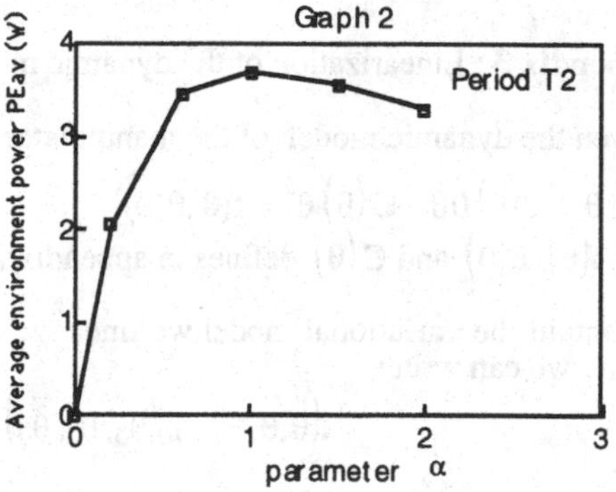

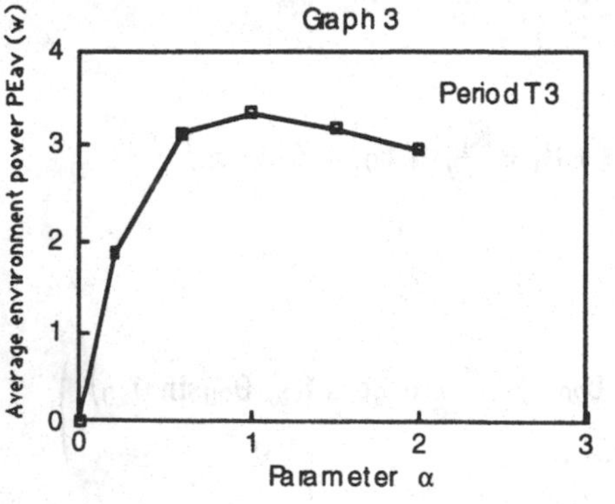

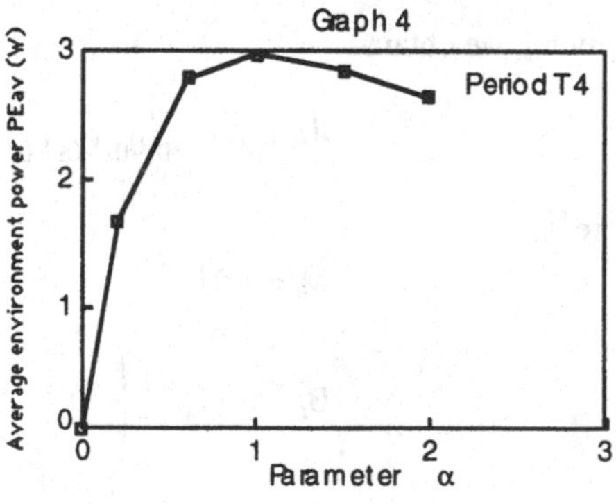

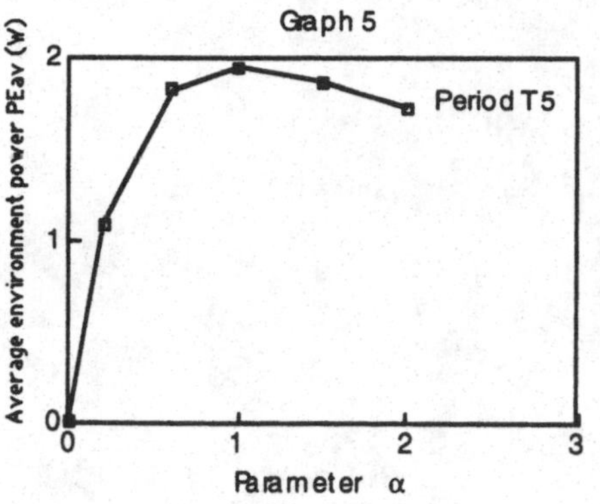
Graph 5
Average environment power PEav (W)
2
1
0
0
1
2
3
Period T5
Parameter α

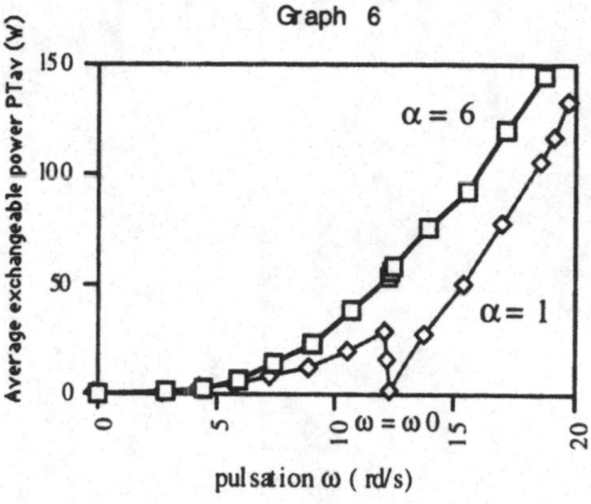
Graph 6
Average exchangeable power PTav (W)
150
100
50
0
0
5
10
15
20
ω = ω0
α = 6
α = 1
pulsation ω (rd/s)

MODELLING AND CONTROL OF CONSTRAINED ROBOTS

K.A. Tahboub
Hebron Technical Engineering College, Hebron, Palestine
and
P.C. Müller
University of Wuppertal, Wuppertal, Germany

1 Abstract

This paper deals with modeling and control of constrained industrial robots. The considered constraints arise during the contact of a robot with an external object in its environment. A novel method that is based on task-space formulations and on the modes-of-motion approach is developed to represent constraints and constrained tasks. It is shown that this methodology can handle constrained motions generally without having any of the substantial limitations of other proposed methods. For control purposes. a novel method is developed. The key feature of this method is to transform nonlinear effects (disturbances) acting on the system to where they can be estimated and statically compensated. For these purposes. only a rough model knowledge and only the position and force measurements are required. Simulation results show the superiority of the proposed method.

2 Introduction

Successful completion of robot applications (like deburring. grinding. and assembly where parts with low clearance should be mated) calls for monitoring and controlling contact forces (forces and torques) between the end effector and the environment. During the conduction of any of these examples. the end effector approaches an external object and maintains contact with its surface. This implies that the end effector loses at least one of its positional degrees of freedom. in the sense that a motion in a certain direction is either not possible because of the external body. or causes a loss of contact and thus leads to a deficiency. The end effector is. then. said to be constrained. A constrained end effector simply indicates a constrained robot. since the motion of the different joints is constrained to achieve end-effector motion which complies with a desired task.

To have an adequate model describing the situation mathematically. we have some choices. The natural way is to consider the dynamics of two subsystems separately and then bind them together through a kinematical model. The two subsystems. of course. define the robot and environment dynamics.

The first choice. i.e.. subsystems and binding technique. leads to a descriptor model or in other words to a set of differential and algebraic equations related to the dynamics and binding of the subsystems respectively. These models are singular. and thus require special control tools [4].

The other possibility is to modify the model describing the free robot to cope with the constraints. Since we control only the robot. we are interested only in the interaction between it and the environment. This includes a specific motion in the workplace and reaction and surface forces arising due to the contact with the environment. To end up with a regular (nonsingular) model including all necessary information. a model reduction has to be performed.

Special attention should be paid to what we call surface forces, these are all contact forces between the robot and its environment except the normal reaction forces. For example. in assembly. if the mating surface is not smooth. then we should expect some friction forces opposing the motion. In machining. on the other hand. we need some specified cutting forces to remove metal. for example. from the cutting surface as desired. These forces are usually related to the normal reaction forces and when included in the model cause extra difficulties as will be shown in the coming sections.

Due to space limitations. this paper is intended only to give a glance at a more detailed work documented in [4].

3 Constrained Robot Spaces

To model the constraints and the dynamics of constrained robots, it is natural to consider four spaces and frames: The joint space, the inertial reference frame, the end-effector space, and the task space.

Dynamics of free robots is usually described and modelled in terms of joint space variables as

$$\mathbf{M}(\mathbf{q})\ddot{\mathbf{q}} + \mathbf{C}(\mathbf{q}, \dot{\mathbf{q}}) = \mathbf{\Gamma}, \tag{1}$$

where $\mathbf{M}(\mathbf{q})$ is the mass matrix, $\mathbf{C}(\mathbf{q}, \dot{\mathbf{q}})$ is the vector of nonlinear gravity, centrifugal, and Coriolis forces. $\mathbf{\Gamma}$ is the vector of generalized input forces.

The motion and contact forces of the end effector, or a point at its tip, are naturally described in terms of Cartesian variables. So, for a free-robot task for example, the planner determines time history of the desired Cartesian motion of the end effector, i.e., the velocity, acceleration, etc. If any constraints are imposed, then relevant contact forces should be specified. All of these variables can be expressed in an inertial Cartesian frame fixed usually to the robot base with a reasonable orientation. From now on, it will be called the reference frame.

The relation between the velocity of the end effector expressed in the reference frame and the joint velocities is given as

$$\begin{bmatrix} \mathbf{V}_0 \\ \mathbf{\Omega}_0 \end{bmatrix} = \mathbf{J}_0(\mathbf{q})\dot{\mathbf{q}}, \tag{2}$$

where the subscript 0 stands for the reference frame. In light of Equation (2), it is possible to give the relation between the input forces at the joints and the acceleration of the end effector as

$$\mathbf{M}_{x0} \begin{bmatrix} \dot{\mathbf{V}}_0 \\ \dot{\mathbf{\Omega}}_0 \end{bmatrix} + \mathbf{C}_{x0} = \mathbf{\Gamma}, \tag{3}$$

with

$$\mathbf{M}_{x0}(\mathbf{q}) = \mathbf{M}(\mathbf{q})\mathbf{J}_0^{-1}, \tag{4}$$

and

$$\mathbf{C}_{x0}(\mathbf{q}, \dot{\mathbf{q}}) = \mathbf{C}(\mathbf{q}, \dot{\mathbf{q}}) - \mathbf{M}(\mathbf{q})\mathbf{J}_0^{-1}\dot{\mathbf{J}}_0\dot{\mathbf{q}}. \tag{5}$$

When external forces corresponding to contact constraints act on the end effector, they should be additionally considered in the model. The end-effector forces are, as known, related to the joint torques by the transpose of the Jacobian. Then the dynamics equation (3) after including the effect of the contact forces becomes

$$\mathbf{M}_{x0} \begin{bmatrix} \dot{\mathbf{V}}_0 \\ \dot{\mathbf{\Omega}}_0 \end{bmatrix} + \mathbf{C}_{x0} = \mathbf{\Gamma} + \mathbf{J}_0^T \begin{bmatrix} \mathbf{F}_{s0} \\ \mathbf{\Gamma}_{s0} \end{bmatrix}. \tag{6}$$

For simple tasks, it may be possible to express the motion and contact force variables directly in the reference frame. However, when the task becomes more involved, then one notices, that it is not an easy job anymore. Moreover, for model reduction and control purposes, one would like to see the relationship between contact forces and motion in its most obvious form. This can be achieved by defining a task frame. A task frame is a Cartesian coordinate system selected in such a way to make the description of the constrained task as easy and clear as possible. By not limiting the task frame to be stationary, the range of applications that robots may perform will certainly increase. Previous research work in the force-position control area ignored this important issue and thus, the developed control strategies were handicapped to handle certain tasks such as tracking a three-dimensional contour.

Usually, the task frame is associated with the frame fixed to the tip of the end effector and moves with it: here it is called the end-effector frame $\mathbf{e}^7$ (assuming a six-degrees-of-freedom robot). The two frames may coincide, but in general, the relation between them is constant or a simple one. Any way, one can always choose the task frame to have identical origin with that of the end-effector frame. The dynamics of constrained robots can be expressed, as well, in the end-effector frame. The form of the model remains as defined by Equation (6) while changing the subscript 0 to 7.

4 Constraints and Constrained Tasks

In 1985, McClamroch published a paper [2] to introduce the idea of modeling constrained robots as singular systems of differential equations. That was the first time, that constraints imposed on the motion of robots are stated explicitly in an equation form.

The obtained model in the joint space is composed of a differential equation that rehearses the dynamics of the arm

$$\mathbf{M}(\mathbf{q})\ddot{\mathbf{q}} + \mathbf{C}(\mathbf{q}.\dot{\mathbf{q}}) = \mathbf{\Gamma} + \mathbf{J}^{\mathbf{T}}(\mathbf{q})\mathcal{D}^{\mathbf{T}}(\mathbf{X})\bar{\lambda}, \tag{7}$$

and an algebraic equation corresponding to m imposed constraints

$$\mathbf{\Theta}(\mathbf{X}) = 0, \tag{8}$$

where $\mathbf{X} \in \mathbf{R}^6$ denotes the position (position and orientation) of the end effector expressed in the reference frame; $\mathbf{\Theta} : \mathbf{R}^6 \rightarrow \mathbf{R}^m$ is twice continuously differentiable, and the constraints Jacobian is defined by

$$\mathcal{D}(\mathbf{X}) = \frac{\partial \mathbf{\Theta}(\mathbf{X})}{\partial \mathbf{X}}. \tag{9}$$

The Lagrange multipliers *lambd* $\in \mathbf{R}^m$ are in this case the magnitudes of the contact or reaction forces between the end effector and the constraining environment. This art of modeling describes in a useful and correct form the robot interaction with its environment. It stresses the role of writing down constraint equations in finding the contact forces. The differential and the algebraic equations can be cast together in a descriptor form, where the contact force is then considered as a state. For a detailed discussion, the interested reader is referred to the dissertation [4].

Geometric-constraint equation (8) may be differentiated once with respect to the time to get

$$\frac{d\mathbf{\Theta}(\mathbf{X})}{dt} = \frac{\partial \mathbf{\Theta}(\mathbf{x})}{\partial \mathbf{X}}\frac{d\mathbf{X}}{dt} = 0, \tag{10}$$

or

$$\mathcal{D}(\mathbf{X})\dot{\mathbf{X}} = 0. \tag{11}$$

The more general form of this equation

$$\mathcal{D}(\mathbf{X})\dot{\mathbf{X}} = \mathbf{f}(t), \tag{12}$$

is considered in this work following the same philosophy of McClamroch but more applicable techniques are brought in. In general, task-frame models and kinematical constraints defined by "modes of motion" form the core of the proposed method, which is claimed to cross out the drawbacks of geometric-constraints methods.

4.1 Modes of Motion

First, the general kinematic constraint equation on the motion of the end effector in the task frame can be given by

$$\mathbf{W}^{\mathbf{T}}\begin{bmatrix} \mathbf{V} \\ \Omega \end{bmatrix} = \zeta(t), \tag{13}$$

where $\mathbf{W} \in \mathbf{R}^{6\times m}$, *zet* $\in \mathbf{R}^m$, and m denotes the number of constraints. The general solution of Equation (13) in terms of a fewer number of velocity state variables $\dot{\mathbf{P}}$ is

$$\begin{bmatrix} \mathbf{V} \\ \Omega \end{bmatrix} = \mathbf{\Phi}\dot{\mathbf{P}} + \xi, \tag{14}$$

with $\mathbf{\Phi} \in \mathbf{R}^{6\times(6-m)}$, $\dot{\mathbf{P}} \in \mathbf{R}^{6-m}$, and $\boldsymbol{x} \in \mathbf{R}^6$. The interpretation of Equation (13) and Equation (14) depends strongly on the method used for the reduction of the dynamical equations. Here, the interpretation is given in terms of modes of motion which keep geometrical characteristics of the constrained motion clearly in view.

The velocity variables $\begin{bmatrix} \mathbf{V^T} & \mathbf{\Omega^T} \end{bmatrix}^T$ form an element of the six-dimensional vector space $\mathbf{R}^6$. A set of spanning vectors for this space is collected as columns of a matrix

$$\hat{\mathbf{\Phi}} = [\ \hat{\mathbf{\Phi}}_i\ ] \qquad i = 1, 2, ..., 6. \tag{15}$$

Any vector in the space can be represented as a linear combination of these spanning vectors. So, vector $\begin{bmatrix} \mathbf{V^T} & \mathbf{\Omega^T} \end{bmatrix}^T$ can be represented in the form

$$\begin{bmatrix} \mathbf{V} \\ \mathbf{\Omega} \end{bmatrix} = \hat{\mathbf{\Phi}}\hat{\mathbf{g}}, \tag{16}$$

where $\hat{\mathbf{g}} \in \mathbf{R}^6$ and is split into

$$\hat{\mathbf{g}} = \begin{bmatrix} \dot{\mathbf{P}} \\ \kappa(t) \\ 0 \end{bmatrix}. \tag{17}$$

Equation (17) states that some of the velocity variables vanish identically, some are known functions of time ***kapp***(t), and the others may be unknown variables (state velocity variables) $\dot{\mathbf{P}}$.

Selecting the elements of $\hat{\mathbf{g}}$ this way, the columns of $\hat{\mathbf{\Phi}}$, the so-called mode vectors, represent three different types of modes of motion. Locked modes characterized by $\hat{\mathbf{g}}_\mathbf{i} = \mathbf{0}$ and kinematically excited modes characterized by $\hat{\mathbf{g}}_\mathbf{i} =$ ***kapp***$(\mathbf{t})$ result from constraints, their equations can be given without considering the system dynamics. They are summarized as constrained modes of motion. By contrast, the motions represented by $\hat{\mathbf{g}}_\mathbf{i} = \dot{\mathbf{P}}$ cannot be determined by the constraints. These are called the free or dynamic modes of motion.

It is possible to partition matrix $\hat{\mathbf{\Phi}}$ to three submatrices corresponding to three modes of motion as

$$\hat{\mathbf{\Phi}} = [\ \mathbf{\Phi} \quad \bar{\mathbf{\Phi}}_\kappa \quad \bar{\mathbf{\Phi}}_l\ ] = [\ \mathbf{\Phi} \quad \bar{\mathbf{\Phi}}\ ], \tag{18}$$

where $\mathbf{\Phi}$ contains the free modes, $\bar{\mathbf{\Phi}}_\kappa$ contains the kinematically constrained modes, and $\bar{\mathbf{\Phi}}_l$ contains the locked modes. $\bar{\mathbf{\Phi}}_\kappa$ and $\bar{\mathbf{\Phi}}_l$ are collected together into $\bar{\mathbf{\Phi}}$ which contains all constrained modes.

Accordingly, Equation (14) can be rewritten in terms of these submatrices as

$$\begin{bmatrix} \mathbf{V} \\ \mathbf{\Omega} \end{bmatrix} = \mathbf{\Phi}\dot{\mathbf{P}} + \bar{\mathbf{\Phi}}_\kappa \kappa(\mathbf{t}) + \bar{\mathbf{\Phi}}_l[0]. \tag{19}$$

This is the solution of Equation (13), where the term $\mathbf{\Phi}\dot{\mathbf{P}}$ is the solution of the homogeneous part

$$\mathbf{W^T}\begin{bmatrix} \mathbf{V} \\ \mathbf{\Omega} \end{bmatrix} = 0, \tag{20}$$

which implies that

$$\mathbf{W^T}\mathbf{\Phi}\dot{\mathbf{P}} = 0. \tag{21}$$

Since Equation (21) must be satisfied identically for any choice of $\dot{\mathbf{P}}$, it is ready to conclude that

$$\mathbf{W^T}\mathbf{\Phi} = 0. \tag{22}$$

The columns of $\mathbf{\Phi} \in \mathbf{R}^{6\times(6-m)}$ span a $(6-m)$-dimensional subspace of $\mathbf{R}^6$. Adding the m additional base vectors collected in $\bar{\mathbf{\Phi}} \in \mathbf{R}^{6\times m}$, one obtains a complete basis for space $\mathbf{R}^6$. These additional base vectors are selected in a way to define the locked and kinematically-excited modes. This implies the choice

$$\xi = \bar{\mathbf{\Phi}}_\kappa \kappa(\mathbf{t}) + \bar{\mathbf{\Phi}}_l[0]. \tag{23}$$

Because x is a particular solution of Equation (13) for a right hand side equal to ***zet***, the constrained-mode vectors must satisfy

$$\mathbf{W^T}\bar{\mathbf{\Phi}}\begin{bmatrix} \kappa(t) \\ 0 \end{bmatrix} = \zeta. \tag{24}$$

Now, from $\hat{\Phi}$ a second basis for $\mathbf{R}^6$ is constructed, called the dual basis and denoted $\hat{\Psi}$. It is defined by

$$\hat{\Psi}^T = \hat{\Phi}^{-1}. \tag{25}$$

Matrix $\hat{\Psi}$ is partitioned in the same way as $\hat{\Phi}$ into

$$\hat{\Psi} = [\ \Psi \quad \bar{\Psi}_\kappa \quad \bar{\Psi}_l\ ] = [\ \Psi \quad \bar{\Psi}\ ], \tag{26}$$

where the columns of $\hat{\Psi}$ are called the dual-mode vectors.

The definition given in Equation (25) implies that

$$\Phi^T\Psi = I \qquad \Phi^T\bar{\Psi} = 0 \qquad \bar{\Phi}^T\Psi = 0 \qquad \bar{\Phi}^T\bar{\Psi} = I. \tag{27}$$

By comparing Equation (22) with Equation (27), it is seen that the choice

$$\bar{\Psi} = W, \tag{28}$$

makes Equation (22) satisfied automatically. Premultiplying Equation (19) by Ψ^T and $\bar{\Psi}^T$ one obtains

$$\Psi^T \begin{bmatrix} V \\ \Omega \end{bmatrix} = \dot{P}. \tag{29}$$

and

$$\bar{\Psi}^T \begin{bmatrix} V \\ \Omega \end{bmatrix} = \begin{bmatrix} \kappa(t) \\ 0 \end{bmatrix} = \zeta. \tag{30}$$

These are important results. They state that the free and constrained mode variables $\mathbf{P}$ and ***zeta*** being the coordinates of velocity $\begin{bmatrix} V^T & \Omega^T \end{bmatrix}^T$ in basis Φ and $\bar{\Phi}$ respectively, appear when projecting $\begin{bmatrix} V^T & \Omega^T \end{bmatrix}^T$ onto the dual base vectors Ψ and $\bar{\Psi}$.

The contact-force vector, on the other hand, can be represented using basis $\hat{\Psi}$. Considering the partitioning of $\hat{\Psi}$ according to Equation (26), the contact force can be written as:

$$\begin{bmatrix} F_s \\ \Gamma_s \end{bmatrix} = \Psi\lambda + \bar{\Psi}\bar{\lambda}. \tag{31}$$

The elements of ***lambd*** $\in \mathbf{R}^{6-m}$ are projection of the contact force onto the free modes. These are exactly the vectors on which the free-mode velocity variables act. Such surface forces arise due to cutting or friction for example. The elements of ***lambd*** $\in \mathbf{R}^m$ are projections of contact forces onto constrained modes, which is to say the unknown generalized constraint forces. Here, it is important to note that ***lambd*** and ***lambd*** are projections onto basis $\hat{\Phi}$ and not onto basis $\hat{\Psi}$, this can be seen following the same reasoning followed for the interpretation of the velocity variables.

In the case, where there are m constraints and $6 - m$ free modes, a general model for surface forces may be given [4] by

$$\lambda = \Pi\bar{\lambda}, \tag{32}$$

where $\Pi \in \mathbf{R}^{(6-m)\times m}$ is a coefficients matrix.

Kinematically-excited modes, explained previously, can be used to handle dynamic contacts. For example, if the constraining object should be modelled as an elastic system, then the relation between the contact force and the resulting deflection is known. In this case, the contact force is controlled by commanding the motion (in that direction). For moving constraining objects with known motions, it is possible to perform desired tasks if the known motion is modelled as a kinematical excitation.

4.2 Minimal model reduction and state space representation

Due to the m constraints on the motion of the end effector, there are only $(6 - m)$ independent variables in the dynamic model given in Equation (6) (or its equivalent in the task space) which are needed to describe the motion of the constrained system.

Controlling the motion and contact forces of a constrained robot explicitly calls for measuring and commanding them independently. For analysis and synthesis, a reduced model for the independent motion variables apart from the contact forces is needed. In other words one would like to realize a state space representation. Moreover, an explicit solution for contact forces is required if classical (regular) control methods are used.

The velocity and contact-force vectors of the end effector as found earlier in this section are

$$\begin{bmatrix} \mathbf{V} \\ \mathbf{\Omega} \end{bmatrix} = \mathbf{\Phi}\dot{\mathbf{P}} + \bar{\mathbf{\Phi}}_\kappa \kappa(t). \tag{33}$$

and

$$\begin{bmatrix} \mathbf{F_s} \\ \mathbf{\Gamma_s} \end{bmatrix} = \mathbf{\Psi}\boldsymbol{\lambda} + \bar{\mathbf{\Psi}}\bar{\boldsymbol{\lambda}}. \tag{34}$$

where a generalized model for ***lambd*** is given as

$$\boldsymbol{\lambda} = \mathbf{\Pi}\bar{\boldsymbol{\lambda}}. \tag{35}$$

then

$$\begin{bmatrix} \mathbf{F_s} \\ \mathbf{\Gamma_s} \end{bmatrix} = \mathbf{\Psi}\mathbf{\Pi}\bar{\boldsymbol{\lambda}} + \bar{\mathbf{\Psi}}\bar{\boldsymbol{\lambda}}. \tag{36}$$

To have a solution for the acceleration of the end effector, Equation (33) is differentiated once, this yields

$$\begin{bmatrix} \dot{\mathbf{V}} \\ \dot{\mathbf{\Omega}} \end{bmatrix} = \mathbf{\Phi}\ddot{\mathbf{P}} + \bar{\mathbf{\Phi}}_\kappa \dot{\kappa}(t) + \dot{\mathbf{\Phi}}\dot{\mathbf{P}} + \dot{\bar{\mathbf{\Phi}}}_\kappa \kappa(t). \tag{37}$$

Substituting for the acceleration from Equation (37) and for the contact force from Equation (36) into the task-space equivalent of equation (6), one obtains

$$\mathbf{M_X}\mathbf{\Phi}\ddot{\mathbf{P}} + \mathbf{M_X}\bar{\mathbf{\Phi}}_\kappa \dot{\kappa}(t) + \mathbf{M_X}\dot{\mathbf{\Phi}}\dot{\mathbf{P}} + \mathbf{M_X}\dot{\bar{\mathbf{\Phi}}}_\kappa \kappa(t) + \mathbf{C_X} = \mathbf{\Gamma} + \mathbf{J^T}\mathbf{\Psi}\mathbf{\Pi}\bar{\boldsymbol{\lambda}} + \mathbf{J^T}\bar{\mathbf{\Psi}}\bar{\boldsymbol{\lambda}}. \tag{38}$$

Defining the total nonlinear forces vector as

$$\mathbf{C_{XX}} = \mathbf{C_X} + \mathbf{M_X}\bar{\mathbf{\Phi}}_\kappa \dot{\kappa}(t) + \mathbf{M_X}\dot{\mathbf{\Phi}}\dot{\mathbf{P}} + \mathbf{M_X}\dot{\bar{\mathbf{\Phi}}}_\kappa \kappa(t). \tag{39}$$

Equation (38) becomes

$$\mathbf{M_X}\mathbf{\Phi}\ddot{\mathbf{P}} + \mathbf{C_{XX}} = \mathbf{\Gamma} + \mathbf{J^T}\mathbf{\Psi}\mathbf{\Pi}\bar{\boldsymbol{\lambda}} + \mathbf{J^T}\bar{\mathbf{\Psi}}\bar{\boldsymbol{\lambda}}. \tag{40}$$

This is a general dynamical model of constrained robots in the task frame which corresponds to the selected modes of motion.

Now it is ready to eliminate contact forces from Equation (40). By premultiplying this equation with $\mathbf{J^{-T}}$ (which is assumed to be regular), a fully "task-frame" model is obtained

$$\mathbf{J^{-T}}\mathbf{M_X}\mathbf{\Phi}\ddot{\mathbf{P}} + \mathbf{J^{-T}}\mathbf{C_{XX}} = \mathbf{J^{-T}}\mathbf{\Gamma} + \mathbf{\Psi}\mathbf{\Pi}\bar{\boldsymbol{\lambda}} + \bar{\mathbf{\Psi}}\bar{\boldsymbol{\lambda}}. \tag{41}$$

To solve for the constraint forces ***lam̄bd***, the dynamics is projected onto the constrained modes by premultiplying with $\bar{\mathbf{\Phi}}^{\mathbf{T}}$, while recalling that $\bar{\mathbf{\Phi}}^{\mathbf{T}}\mathbf{\Psi} = 0$, this yields

$$\bar{\boldsymbol{\lambda}} = \bar{\mathbf{\Phi}}^{\mathbf{T}}\mathbf{J^{-T}}\mathbf{M_X}\mathbf{\Phi}\ddot{\mathbf{P}} + \bar{\mathbf{\Phi}}^{\mathbf{T}}\mathbf{J^{-T}}\mathbf{C_{XX}} - \bar{\mathbf{\Phi}}^{\mathbf{T}}\mathbf{J^{-T}}\mathbf{\Gamma}. \tag{42}$$

This is an expected result, which says that the constraint forces are projections of direct input forces $\mathbf{\Gamma}$ to the system, inertia forces $\mathbf{M_X}\mathbf{\Phi}\ddot{\mathbf{P}}$, and other forces $\mathbf{C_{XX}}$, onto the constrained modes.

To solve for the state accelerations $\ddot{\mathbf{P}}$, the dynamics of the constrained system is projected onto the free modes by premultiplying with $\mathbf{\Phi}^{\mathbf{T}}$, this yields

$$\mathbf{\Phi}^{\mathbf{T}}\mathbf{J^{-T}}\mathbf{M_X}\mathbf{\Phi}\ddot{\mathbf{P}} + \mathbf{\Phi}^{\mathbf{T}}\mathbf{J^{-T}}\mathbf{C_{XX}} = \mathbf{\Phi}^{\mathbf{T}}\mathbf{J^{-T}}\mathbf{\Gamma} + \mathbf{\Pi}\bar{\boldsymbol{\lambda}}. \tag{43}$$

It can be seen that, the solution for $\ddot{\mathbf{P}}$ is a function of counter surface forces (e.g. friction forces) as should be expected. To eliminate these forces from the solution, Equation (42) is substituted in Equation (43) to become

$$\begin{aligned} \mathbf{\Phi}^{\mathbf{T}}\mathbf{J^{-T}}\mathbf{M_X}\mathbf{\Phi}\ddot{\mathbf{P}} + \mathbf{\Phi}^{\mathbf{T}}\mathbf{J^{-T}}\mathbf{C_{XX}} &= \mathbf{\Phi}^{\mathbf{T}}\mathbf{J^{-T}}\mathbf{\Gamma} \\ &+ \mathbf{\Pi}\bar{\mathbf{\Phi}}^{\mathbf{T}}\mathbf{J^{-T}}\mathbf{M_X}\mathbf{\Phi}\ddot{\mathbf{P}} + \mathbf{\Pi}\bar{\mathbf{\Phi}}^{\mathbf{T}}\mathbf{J^{-T}}\mathbf{C_{XX}} - \mathbf{\Pi}\bar{\mathbf{\Phi}}^{\mathbf{T}}\mathbf{J^{-T}}\mathbf{\Gamma}, \end{aligned} \tag{44}$$

or

$$(\mathbf{\Phi}^T\mathbf{J}^{-T}\mathbf{M_x}\mathbf{\Phi} - \mathbf{\Pi}\bar{\mathbf{\Phi}}^T\mathbf{J}^{-T}\mathbf{M_x}\mathbf{\Phi})\ddot{\mathbf{P}} + (\mathbf{\Phi}^T - \mathbf{\Pi}\bar{\mathbf{\Phi}}^T)\mathbf{J}^{-T}\mathbf{C_{xx}} = (\mathbf{\Phi}^T - \mathbf{\Pi}\bar{\mathbf{\Phi}}^T)\mathbf{J}^{-T}\mathbf{\Gamma}. \quad (45)$$

A unique solution for $\ddot{\mathbf{P}}$ is then possible if and only if

$$rank(\mathbf{\Phi}^T\mathbf{J}^{-T}\mathbf{M_x}\mathbf{\Phi} - \mathbf{\Pi}\bar{\mathbf{\Phi}}^T\mathbf{J}^{-T}\mathbf{M_x}\mathbf{\Phi}) = 6 - m. \quad (46)$$

This is an important result which is least expected. If the coefficients matrix, $\mathbf{\Pi}$, is zero, i.e., no surface forces are present, then a solution exists always since $\mathbf{\Phi}^T\mathbf{J}^{-T}\mathbf{M_x}\mathbf{\Phi}$ is necessarily regular (by noting that $\mathbf{M_x} = \mathbf{M(q)J}^{-1}$). So, surface forces may lead to the unsolvability of the system.

If a solution exists, then it must be given by

$$\begin{bmatrix} \ddot{\mathbf{P}} \\ \bar{\lambda} \end{bmatrix} = \begin{bmatrix} \mathbf{M_x}\mathbf{\Phi} & -\mathbf{J}^T(\mathbf{\Psi}\mathbf{\Pi} + \bar{\mathbf{\Psi}}) \end{bmatrix}^{-1} (\mathbf{\Gamma} - \mathbf{C_{xx}}). \quad (47)$$

5 Control

The feedback-control problem for constrained robots is the problem of determining the input voltages at the motors, in real time, according to available measurements to fulfill a desired task as accurate as possible. Different controllers and different implementations can be chosen in conformity with design objectives. If accessible knowledge about the system to be controlled is used in the determination of the inputs, then the controller is said to be model based.

Two model-based controllers for constrained robots are developed and introduced in [4]. The first is derived from the feedback linearization method, while the second is built around the reverse linearization approach [5] which is proved to recover from catches of the feedback linearization method. Here only the application of the later to our problem will be presented.

The motors dynamics is approximated with the first order model

$$\mathbf{L_m}\dot{\mathbf{m}} = \mathbf{R_m}\mathbf{m} + \mathbf{K_m}\mathbf{u}, \quad (48)$$

where $\mathbf{L_m}$, $\mathbf{R_m}$, and $\mathbf{K_m}$ are motor inductivity, resistance, and torque constants matrices respectively, whereas $\mathbf{m}$ and $\mathbf{u}$ are the motor torques and input voltages. Thus $\mathbf{\Gamma} = \mathbf{G_r}\mathbf{m}$, where $\mathbf{G_r}$ is the gear-reduction ratio matrix.

Denoting the matrix $\begin{bmatrix} \mathbf{M_x}\mathbf{\Phi} & -\mathbf{J}^T(\mathbf{\Psi}\mathbf{\Pi} + \bar{\mathbf{\Psi}}) \end{bmatrix}$ by $\mathbf{Q}$ and decomposing it into

$$\mathbf{Q} = \bar{\mathbf{Q}} + \tilde{\mathbf{Q}}, \quad (49)$$

where $\bar{\mathbf{Q}}$ is a constant regular average and $\tilde{\mathbf{Q}}$ is the nonlinear remaining part yield the state-space form:

$$\begin{bmatrix} \dot{\mathbf{m}} \\ \dot{\mathbf{P}} \\ \ddot{\mathbf{P}} \end{bmatrix} = \begin{bmatrix} -\mathbf{L_m^{-1}R_m} & \mathbf{0} & \mathbf{0} \\ \mathbf{0} & \mathbf{0} & \mathbf{I} \\ \mathbf{\Delta}\bar{\mathbf{Q}}^{-1}\mathbf{G_r} & \mathbf{0} & \mathbf{0} \end{bmatrix} \begin{bmatrix} \mathbf{m} \\ \mathbf{P} \\ \dot{\mathbf{P}} \end{bmatrix} + \begin{bmatrix} \mathbf{L_m^{-1}K_m} \\ \mathbf{0} \\ \mathbf{0} \end{bmatrix} \mathbf{u} + \begin{bmatrix} \mathbf{0} \\ \mathbf{0} \\ -\mathbf{\Delta}\bar{\mathbf{Q}}^{-1} \end{bmatrix} \mathbf{n} \quad (50)$$

$$\begin{bmatrix} \mathbf{P} \\ \bar{\lambda} \end{bmatrix} = \begin{bmatrix} \mathbf{0} & \mathbf{I} & \mathbf{0} \\ \bar{\mathbf{\Delta}}\bar{\mathbf{Q}}^{-1}\mathbf{G_r} & \mathbf{0} & \mathbf{0} \end{bmatrix} \begin{bmatrix} \mathbf{m_t} \\ \mathbf{P} \\ \dot{\mathbf{P}} \end{bmatrix} + \begin{bmatrix} \mathbf{0} \\ \mathbf{0} \\ -\bar{\mathbf{\Delta}}\bar{\mathbf{Q}}^{-1} \end{bmatrix} \mathbf{n}, \quad (51)$$

where

$$\mathbf{n} = \mathbf{C_{xx}} + \tilde{\mathbf{Q}} \begin{bmatrix} \ddot{\mathbf{P}} \\ \bar{\lambda} \end{bmatrix}, \quad (52)$$

$$\mathbf{\Delta} = \begin{bmatrix} \mathbf{I}_{(6-m)\times(6-m)} & \mathbf{0}_{(6-m)\times m} \end{bmatrix}, \quad (53)$$

and

$$\bar{\mathbf{\Delta}} = \begin{bmatrix} \mathbf{0}_{m\times(6-m)} & \mathbf{I}_{m\times m} \end{bmatrix}. \quad (54)$$

One can see here that the above system becomes linear if we can compensate statically (cancel) for the nonlinearities vector $\mathbf{n}$. Unfortunately, this is not possible in this form because the control input does not influence the nonlinearities directly, so there is no control input that cancels $\mathbf{n}$. Moreover, in order to save the computations necessary in the previous method and avoid the sensitivity to model uncertainties, we would like to estimate the nonlinearities in lieu of computing them. Again, this model form does not facilitate the design of such an observer [3].

Fortunately, it is quite feasible through a simple coordinate transformation to solve both problems simultaneously. Defining the following nonlinear coordinate transformation:

$$\mathbf{m_t} = \mathbf{m} - \mathbf{G_r^{-1}n}, \tag{55}$$

transforms the system to the favorable form

$$\underbrace{\begin{bmatrix} \dot{\mathbf{m}}_t \\ \dot{\mathbf{P}} \\ \ddot{\mathbf{P}} \end{bmatrix}}_{\dot{\mathbf{x}}_t} = \underbrace{\begin{bmatrix} -\mathbf{L_m^{-1}R_m} & \mathbf{0} & \mathbf{0} \\ \mathbf{0} & \mathbf{0} & \mathbf{I} \\ \Delta\bar{\mathbf{Q}}^{-1}\mathbf{G_r} & \mathbf{0} & \mathbf{0} \end{bmatrix}}_{\mathbf{A}} \underbrace{\begin{bmatrix} \mathbf{m_t} \\ \mathbf{P} \\ \dot{\mathbf{P}} \end{bmatrix}}_{\mathbf{x_t}} + \underbrace{\begin{bmatrix} \mathbf{L_m^{-1}K_m} \\ \mathbf{0} \\ \mathbf{0} \end{bmatrix}}_{\mathbf{B}} \mathbf{u} + \underbrace{\begin{bmatrix} -\mathbf{L_m^{-1}} \\ \mathbf{0} \\ \mathbf{0} \end{bmatrix}}_{\mathbf{N_t}} \mathbf{n_t}, \tag{56}$$

$$\underbrace{\begin{bmatrix} \mathbf{P} \\ \bar{\boldsymbol{\lambda}} \end{bmatrix}}_{\mathbf{y}} = \underbrace{\begin{bmatrix} \mathbf{0} & \mathbf{I} & \mathbf{0} \\ \Delta\bar{\mathbf{Q}}^{-1}\mathbf{G_r} & \mathbf{0} & \mathbf{0} \end{bmatrix}}_{\mathbf{C}} \begin{bmatrix} \mathbf{m_t} \\ \mathbf{P} \\ \dot{\mathbf{P}} \end{bmatrix}, \tag{57}$$

with

$$\mathbf{n_t} = \mathbf{L_m G_r^{-1}\dot{n}} + \mathbf{R_m G_r^{-1} n}. \tag{58}$$

This transformation frees the outputs from the nonlinearities and transforms the nonlinearities to where they can be compensated statically to obtain a linear system. It is noted that neither the control variables nor the linear system matrices are affected by this transformation. Moreover, one does not need to compensate for the linear part as is the case in the feedback linearization method.

The main artifice in estimating the nonlinearities is to approximate the time behaviour of the nonlinearity with a fictitious model [3] as

$$\dot{\mathbf{z}}(t) = \mathbf{F}\mathbf{z}(t), \tag{59}$$

$$\mathbf{n_t}(\mathbf{x_t}(t)) \approx \mathbf{H}\mathbf{z}(t), \tag{60}$$

where the matrices $\mathbf{F}$ and $\mathbf{H}$ are of appropriate dimensions. The metasystem comprising the linear-system states and the fictitious-model states is characterized by

$$\begin{bmatrix} \dot{\mathbf{x}}_t \\ \dot{\mathbf{z}} \end{bmatrix} = \underbrace{\begin{bmatrix} \mathbf{A} & \mathbf{N_t H} \\ \mathbf{0} & \mathbf{F} \end{bmatrix}}_{\mathcal{A}} \begin{bmatrix} \mathbf{x_t} \\ \mathbf{z} \end{bmatrix} + \underbrace{\begin{bmatrix} \mathbf{B} \\ \mathbf{0} \end{bmatrix}}_{\mathcal{B}} \mathbf{u}, \tag{61}$$

$$\mathbf{y} = \underbrace{\begin{bmatrix} \mathbf{C} & \mathbf{0} \end{bmatrix}}_{\mathcal{C}} \begin{bmatrix} \mathbf{x_t} \\ \mathbf{z} \end{bmatrix}. \tag{62}$$

An identity observer for the metasystem can be designed as:

$$\begin{bmatrix} \dot{\hat{\mathbf{x}}}_t \\ \dot{\hat{\mathbf{z}}} \end{bmatrix} = \begin{bmatrix} \mathbf{A} - \mathbf{L_x C} & \mathbf{N_t H} \\ -\mathbf{L_n C} & \mathbf{F} \end{bmatrix} \begin{bmatrix} \hat{\mathbf{x}}_t \\ \hat{\mathbf{z}} \end{bmatrix} + \begin{bmatrix} \mathbf{B} \\ \mathbf{0} \end{bmatrix} \mathbf{u} + \begin{bmatrix} \mathbf{L_x} \\ \mathbf{L_n} \end{bmatrix} \mathbf{y}. \tag{63}$$

The asymptotic stability of the observer can be guaranteed by a suitable choice of the gain matrices $\mathbf{L_x}$ and $\mathbf{L_n}$ if the metasystem characterized by Equation (61) and Equation (62) is completely observable. The condition is, cf. [3]:

$$rank \begin{bmatrix} s\mathbf{I} - \mathbf{A} & \mathbf{N_t H} \\ \mathbf{0} & s\mathbf{I} - \mathbf{F} \\ \mathbf{C} & \mathbf{0} \end{bmatrix} = dim(\mathbf{x_t}) + dim(\mathbf{z}), \forall s \in C. \tag{64}$$

An estimate of the nonlinearity can now be reconstructed by employing the estimates generated by the observer as

$$\hat{\mathbf{n}}_t = \mathbf{H}\hat{\mathbf{z}}. \tag{65}$$

Conceding the difficulty of finding a linear model that describes adequately the nonlinearity, one would like to use a simplified fictitious model for it. To avoid any sensitivity of the observability to the chosen model, the condition [3]

$$rank \begin{bmatrix} s\mathbf{I}-\mathbf{A} & \mathbf{N}_t \\ \mathbf{C} & 0 \end{bmatrix} = dim(\mathbf{x}_t) + dim(\mathbf{n}_t), \tag{66}$$

should be satisfied for all complex numbers s. This requires, but not only, that the number of nonlinearities may not exceed the number of measurements.

A practical and convenient fictitious model for the nonlinearity is based on the assumption that the nonlinear effects can be considered approximately as stepwise constant, that is

$$\mathbf{F} = 0. \tag{67}$$

This leads sometimes to high observer gains depending on the rate of change of the nonlinearity. The implementation of such an observer, in this case, calls for a fast measurement update. Real-time computations themselves are quite easy to be performed, since the observer is based on a linear compact model.

Based on the estimates generated by the observer, the nonlinearities can be statically compensated as

$$\mathbf{u} = \mathbf{u}_l + \mathbf{K}_m^{-1}\hat{\mathbf{n}}_t \tag{68}$$

where $\mathbf{u}_l$ is determined to assure robust tracking of a desired trajectory. After a design by Davison [1], the linear control input becomes

$$\mathbf{u}_l = -\mathbf{K}_C\mathbf{x}_C - \mathbf{K}\hat{\mathbf{x}}_t \tag{69}$$

where $\mathbf{x}_C$ is the state of the controller which is characterized by

$$\dot{\mathbf{x}}_C = \mathbf{A}_C\mathbf{x}_C + \mathbf{B}_C\mathbf{e}. \tag{70}$$

The gains $\mathbf{K}_C$ and $\mathbf{K}$ are determined to minimize a quadratic performance index. For more details, the reader is referred to the dissertation of Tahboub [4]. The completed control system is given in Figure 1.

6 Discussion and Conclusions

In this work we have presented a general methodology to describe and represent constraints imposed on industrial robots. The generality of the proposed method is preserved by introducing tools for dealing with surface forces acting on the robot end effector due to its motion on a constraining surface. For control and simulation, a method is given to reduce the constrained system to its state-space form; a general solvability condition is derived.

For control, we have developed a novel method to deal with a certain class of nonlinear systems. In this class, the nonlinearities can be modelled as external disturbances acting on a linear system. These nonlinearities can be sometimes compensated statically by a part of the control input. However, for many systems, this is not possible when the control input does not act on the the nonlinearities directly. Fortunately, we have shown that it is quite possible through a suitable nonlinear coordinate transformation to transform them to where the control input acts directly.

This method is applied to robots with constraints. There, nonlinearities arise due to the configuration-dependent mass and Jacobian matrices and to the vector of nonlinear gravity, centrifugal, and Coriolis forces. It is shown by performing the required transformation that the control variables (position and contact forces) are not affected. Furthermore, the dynamics (**A**, **B**, **C** matrices) remain unchanged.

The transformed nonlinearities, which are complex functions of the original ones, can be estimated together with the states of the linear system via an extended observer that requires a few measurements. Based on the estimates of the nonlinear effects, a disturbance-rejection controller is built to end up with a nominally-linear system.

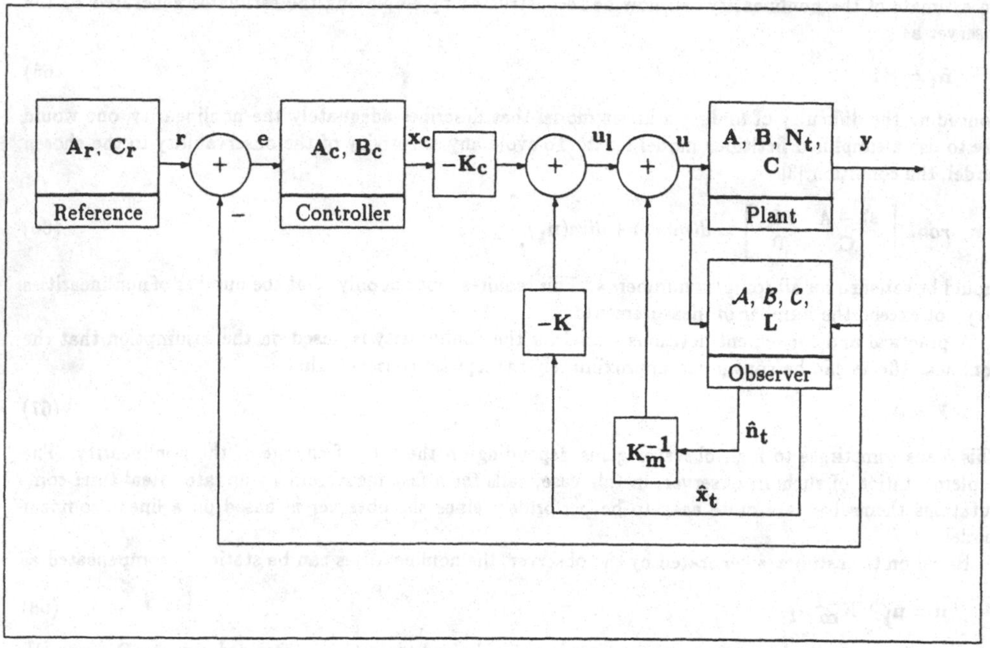

Figure 1: The completed control system

For this kind of linearization, only position and constraint force measurements are required and only the nonlinear term is compensated leaving the original linear part as it is. These are profound advantages over the feedback linearization method which requires a lot of measurements and compensates for the whole dynamics in the system leading to a set of integrators.

Practically, nothing has to be computed online for implementing the proposed method. Specifically, the nonlinear effects are estimated rather than computed which saves a lot of practical problems. The proposed observer and controller are the usual linear ones.

References

[1] Davison, E. J. (1976). The Robust Control of a Servomechanism Problem for Linear Time-Invariant Multivariable Systems. IEEE Transactions on Automatic Control, Vol. AC-21, No. 1, pp. 25-34.

[2] McClamroch, N. H., and H. P. Huang (1985). Dynamics of a Closed Chain Manipulator. American Control Conference, pp. 50-54.

[3] Müller, P. C. (1993). Schätzung und Kompensation von Nichtlinearitäten. VDI Bericht Nr. 1026, VDI-Verlag, Düsseldorf.

[4] Tahboub, K. A. (1993). Modeling and Control of Constrained Robots. Dissertation, VDI-Fortschrittsberichte, Reihe 8, Nr. 361, VDI-Verlag, Düsseldorf.

[5] Tahboub, K. A., and P. C. Müller (1993). A Novel Disturbance-Rejection Control Method Applied to Flexible-Joint Robots. Submitted for Publication in the IEEE Transactions on Robotics and Automation.

IMPEDANCE CONTROL OF AN INDUSTRIAL MANIPULATOR IN A DUAL ARM ASSEMBLY CELL

G. Morel and Ph. Bidaud

University of Versailles St. Quentin, Vélizy, France

Abstract

This paper addresses the problem of the "right hand" control in dual arm assembly cell. While considering the constraints and perturbations that occur during a typical multi-robot assembly task (poorly known and varying environment parameters, mobile obstacles, impacts, friction, noises,...), a controller have been developed in [1]. It is based on an impedance controller which is an intermediate solution between a fully decoupled scheme ([2]) and a linearized scheme. To increase the performances of the resulting controller, a higher control level is added to modify on-line the desired impedance and/or the reference trajectory. This supervisor has been developed using fuzzy logic techniques. The choice of gains to ensure stability is discussed, considering some theorical and practical aspects. Experimental results, obtained on a two-arms assembly system, illustrate the ability of the system to absorb large external impacts.

1 Introduction

Advanced assembly applications sometimes require multiple robotic manipulators to carry out various kinds of tasks in various situations. The management of the kinematical redundancy and the control of the contact between the end-effectors are central problems for the exploitation of these systems and more generally for building dexterous robot systems. This paper concerns both of these issues in the particular context of assembly operations. Figure 1 shows a two-arms assembly system, developed at the *Laboratoire de Robotique de Paris*, and used to address these research topics from an experimental point of view. In this system, the left hand constituted by a PUMA robot is used to optimize the attitude of one of the parts to be mated in the right-hand workspace. The right-hand is a SCARA robot used to realize fine compliant motions during the assembly phase.

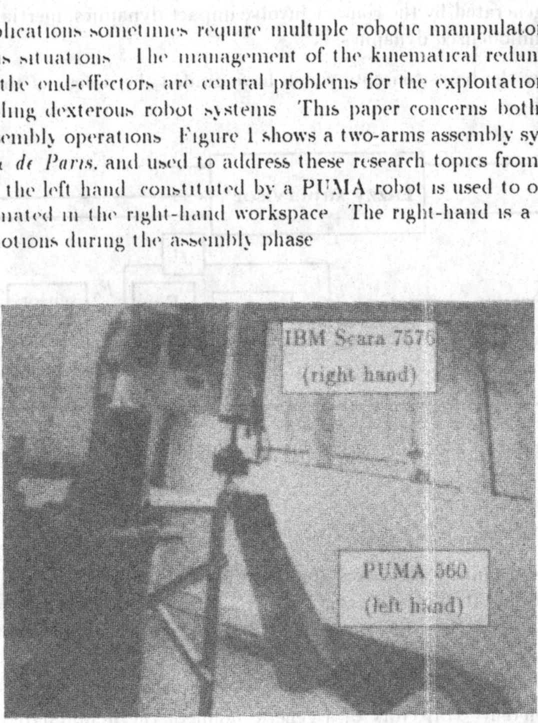

Figure 1: The LRP Two-arms system for dexterous manipulations

The right-hand control scheme must have the ability to establish and maintain a stable contact with a *dynamically complex* environment, indeed:

- The mechanical behavior of the different manipulated components and the manipulation structure, vary within time and space.
- Constraints in the contact are modified while the kinestatics of the contact varies.
- Large impulses can be produced when parts contacting and it is recognized that a simple switch within the transition phase between free and constrained motions creates unstabilities in the controlled system [3].
- Friction can induce jamming, stick-slip or wedging and cause part damages or task failures.
- Models, used to describe the contact and manipulator dynamics are not perfectly known, and the force measurement is generally noisy.

To deal with this, a scheme based on impedance control [2], with a higher level fuzzy supervisor have been developed [1]. In this paper, we just briefly recall its structure and we discuss the choice of the impedance of the system to ensure stability during free and environmental constrained motions and transitions.

2 A supervised impedance controller

Considering problems mentioned above, figure 2 shows the overall structure of the proposed control scheme, which is a kind of adaptive impedance controller composed of two stages :

- At low level, a force controller based on a modified form of the Hogan's impedance controller [2].
- A qualitative supervisor is exploited to modify on line the desired impedance of the controller (i.e. gains) and the reference trajectories, to deal with large unexpected disturbant external forces in the contact (the forces generated by the contact involve impact dynamics, inertia, elastic deformations and friction), and unmodelized dynamics.

In the following, we review the basic concepts which have guided the design of these two stages. More details can be found in [1]

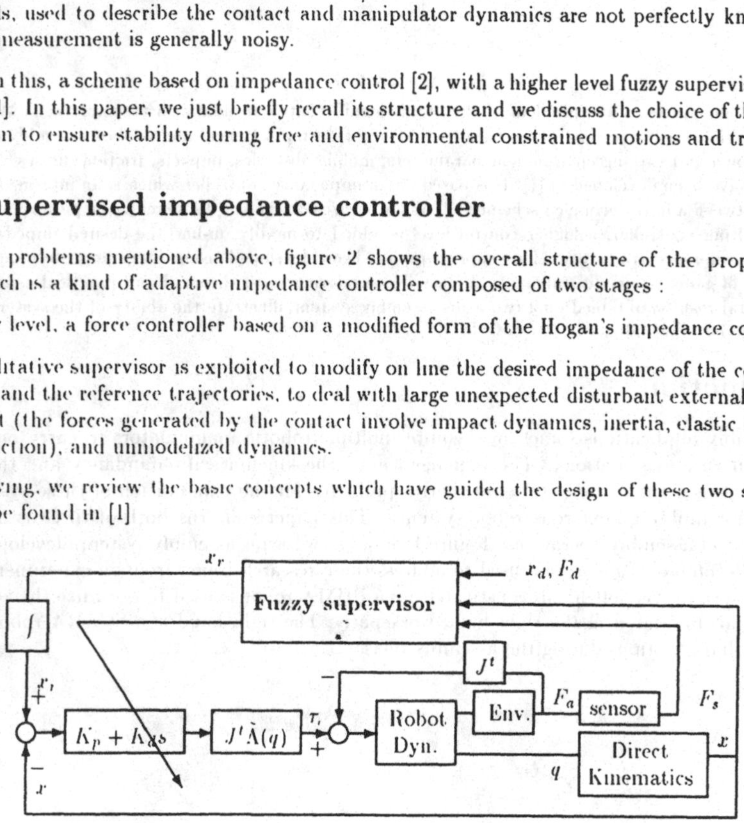

Figure 2. Bloc diagram of the right-hand control system

2.1 Impedance Control

Impedance control provides a unified framework for controlling a manipulator both in free motions and environmental contact motions It allows to program directly the relationship between the velocity $\dot{x}$ of the manipulated object and interaction force F_a as a second-order linear impedance

$$-F_a = M_d.\ddot{x} + B_d.(\dot{x} - \dot{x}_r) + K_d.(x - x_r) \tag{1}$$

In (1) x_r represents the reference trajectory of a generic point of the manipulated object and M_d, K_d and B_d are respectively the desired inertia, stiffness and damping matrices.

Moreover, recent developments have corrected the inherent difficulties in impedance control for task specification :

- Lasky and Hsia [4] proposed to add a trajectory modifying controller in the outer-loop for force tracking ; Seraji and Colbaugh [5] have proposed to provide adaptive properties to this external loop;
- selection matrices as those used in hybrid F/P scheme, which allow the user to easily describe task constraints, have been introduced within an impedance scheme by Anderson and Spong [6].

Basically, two kinds of implementation are proposed in the litterature. The most simple one use a Jacobian transpose PD algorithm to compute consistent command torque τ_c and produce a desired visco-elastic object's behavior in the Cartesian space:

$$\tau_c = J^t.\{B_d.(\dot{x}_r - \dot{x}) + K_d.(x_r - x)\} \tag{2}$$

However, An & Hollerbach [7] have shown clearly the role of the dynamic compensation in cartesian force control. They demonstrated *not only that using a dynamic model leads to more accurate control, but also that not using this model can in certain cases make force control unstable.* The dynamics to be compensated in the control loop can be written in Cartesian space as [8]:

$$F_c - F_a = \Lambda(q)\ddot{x} + \mu(q,\dot{q}) + p(q) \tag{3}$$

$$\Lambda(q) = (J^{-1})^t(q).H(q).J^{-1}(q) \quad ; \quad p(q) = (J^{-1})^t(q).g(q) \quad ; \quad \mu(q,\dot{q}) = (J^{-1})^t(q).b(q,\dot{q}) - \Lambda(q).\dot{J}(q).\dot{q}$$

where q is the joint configuration, $J(q)$ is the Jacobian matrix, F_c is the command force, $\Lambda(q)$ and $H(q)$ are the kinetic energy matrices respectively in Cartesian space and joint space, $\mu(q,\dot{q})$ and $b(q,\dot{q})$ are the vector of Coriolis and centrifugal forces and torques respectively, $g(q)$ the gravity torque vector and $p(q)$ is the gravity force. To deal with this nonlinear coupled dynamics, Hogan has proposed the following decoupling scheme [2] :

$$F_c = (J^{-1})^t(q).\tau_c = \tilde{\Lambda}.M_d^{-1}.\{B_d.(\dot{x}_r - \dot{x}) + K_d.(x_r - x)\} + \tilde{\mu}(q,\dot{q}) + \tilde{p}(q) + (1 - \tilde{\Lambda}.M_d^{-1})F_a \tag{4}$$

where $\tilde{}$ designates estimated values.

It has been shown in [1], that, in practice, errors in force measurements make the choice of the imaginary mass M_d very difficult to keep the system stable throught the non linear force feedback $(1 - \tilde{\Lambda}.M_d^{-1})F_a$. However, the dynamics decoupling is still efficient for free motions, but can be notably simplified in an experimental context as :

$$F_c = \tilde{\Lambda}(q)\{B_d.(\dot{x}_r - \dot{x}) + K_d.(x_r - x)\} \tag{5}$$

To illustrate this discussion, these three controllers (the Hogan's scheme, the Jacobian transpose scheme

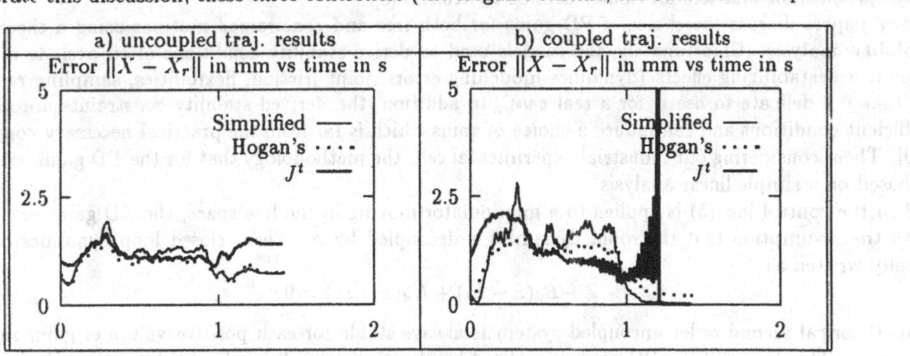

Figure 3: Results for free trajectory tracking

and our simplified impedance controller) has been compared experimentally for free trajectories of a SCARA robot. The purpose of this experiment is to explore the effects and the benefits of a dynamic decoupling in a basic motion (in fig. 3, max. velocity = 0.15 m/s and max. acceleration = 0.5 m/s^2). One can observe that for the first trajectory for which the $\Lambda(q)$ matrix is close to diagonal, the three controllers provide similar results, but the Jacobian transpose scheme is not able to stabilized the manipulator when $\Lambda(q)$ becomes strongly coupled.

2.2 A Fuzzy Supervisor

The higher level of our control scheme is a supervisor, which is used to on-line modify both the reference trajectory x_r and the PD gains of the controller K_d and B_d, from informations on state variables and task specification (x_d, F_d). If we consider realistic experimental conditions, we have to take account of real time constraints (impact process requires some reflex actions), measurement noises, and uncertainties. A complete analytic computation for the supervisor is not reasonable, so we have explored the use of a reactive technique. Since the supervisor must deal with various situations during an assembly sequence, a fuzzy algorithm was prefered to the neuromimetic solution, which is generally used to learn a repetitive task. To compute the desired impedance, we implemented a multi-level fuzzy algorithm, using two kinds of rules :

- **Switching rules** which are used to deal with contact disturbances in transitions between free and constrained motions.
- **Tuning rules** used to modify the nominal impedance parameters when the contact is established, and to control the interaction wrench, throught an outer loop which computes x_r from the force error signal ([4])

3 PD gains tuning using linear analysis

One of the main issue in impedance control is the choice of the target behavior (i.e. controller gains) which ensure stability of the system. It is well known that the stability of an unconstrained manipulator does not guarantee the stability when it becomes in contact with its environment, and PD gains have to be modified between free and compliant motions. In the same way, when a contact is established, target impedance have to be choosen differently with respect to constrained and tangential directions. A solution of this problem is to decompose the desired impedance in two subspaces [6]. The resulting impedance parameters in a reference (fixed) frame K_d and B_d are given by :

$$\left\{ \begin{array}{lcl} B_d & = & b_{d_n}.\Omega + b_{d_t}.\breve{\Omega} \\ K_d & = & k_{d_n}.\Omega + k_{d_t}.\breve{\Omega} \end{array} \right. \text{ with : } \Omega = \begin{pmatrix} R^t S_f R & 0 \\ 0 & R^t S_m R \end{pmatrix} \text{ and : } \breve{\Omega} = \begin{pmatrix} R^t \bar{S}_f R & 0 \\ 0 & R^t \bar{S}_m R \end{pmatrix}$$

where S_f and S_m are the selection matrices for forces and moments respectively [9], R is the rotation matrix between fixed frame and task frame, b_{d_n} and k_{d_n} are scalar parameters in constrained directions (i.e. which have to be choosen according to the environment stiffness) and b_{d_t} and k_{d_t} are also scalars which can be choosen to take friction in the tangential plane into account. Ω and $\breve{\Omega}$ are the ***generalized task specification matrices*** previously defined by Khatib [8].

Many papers discuss the choice of PD gains for both free and constrained motions using a theoretical stability analysis. Unfortunately, the models used to derive stability conditions often exclude a lot of practical destabilizing effects (dynamics modelling errors, joint friction, flexibilities, sampling rate, ...) so that it's delicate to use it for a real case. In addition, the derived stability constraints are usually sufficient conditions and can induce a choice of gains which is far from the practical necessary condition [10]. Then, considering our industrial experimental cell, the methodology that for the PD gains selection is based on a simple linear analysis.

When the control law (5) is applied to a manipulator moving in the free space, the PD gains are tuned with the assumption that the robot dynamics is decoupled by $\tilde{\Lambda}$. Then, closed loop dynamics can be ideally written as :

$$\ddot{x} + B_d(\dot{x} - \dot{x}_r) + K_d(x - x_r) = 0 \tag{6}$$

This theorical second order uncoupled system is always stable for each positive values of gains in each direction The design of the PD gains is realized by choosing a small bandwidth (rarely much than few Hertz for an industrial manipulator) and by overdamping the closed loop dynamics. For the IBM 7576 Scara, we have experimentally set a natural frequency $\omega_0 = 7$ Hz and a damping coefficient $\xi = 1$ to provide a stable closed loop behavior throught the workspace, and acceptable errors.

In the same way, for constrained motions, we assume that the environment, constituted by the (compliant) parts and the left hand, can be modelized by a linear impedance around its equilibrium position x_{r_e} :

$$F_a = M_e \ddot{x}_e + B_e(\dot{x}_e - \dot{x}_{r_e}) + K_e(x_e - x_{r_e}) \tag{7}$$

where M_e , B_e and K_e are the environment mass, damping and stiffness matrices [1]. Then, if we assume that the target impedance is achieved, the bandwidth and damping coefficient in a constrained direction n are :

$$\begin{cases} \omega_0 = \sqrt{\dfrac{k_{d_n}+k_{e_n}}{1+m_{e_n}}} \\ \xi = \dfrac{b_{d_n}+b_{e_n}}{2\omega_0} \end{cases} \quad \text{with :} \begin{cases} k_{e_n} = n^t K_e n \\ b_{e_n} = n^t B_e n \\ m_{e_n} = n^t M_e n \end{cases}$$

4 Impact control

In this section, we will focus to the impact control problem to illustrate how the fuzzy rules can be design to deal with unmodelized effects. When the right hand collides with its environment, the produced impulses depend on velocities and projected masses of the two hands. Considering a normal impact, the impulse P^n is equal to the linear momentum component jump along the normal of contact :

$$P^n = \int_{t_1}^{t_2} F_a^n(t)dt = n^t\Lambda n(\dot{x}_n(t_1) - \dot{x}_n(t_2)) = n^t M_e n(\dot{x}_{e_n}(t_1) - \dot{x}_{e_n}(t_2)) \quad (8)$$

where t_1 and t_2 are the time of the begin and the end of impact. When the velocity is small enough, a simple switch between free motion gains and constrained gains can be sufficient to stabilize the contact (fig 4 - column 1). However, when the relative velocity is increased, a larger impact is produced, and unstability is observed (fig 4 - column 2) [2] A specific strategy has to be developed. Many strategies have been implemented to control impacts. Khatib and Burdick [11] have experimented a method based on maximal damping during impact impulse. Volpe and Kholsa have discussed the role of the desired mass for the impact process control [12] and experimented a method equivalent to an impedance scheme with a large target mass and using the Hogan's force feedback as a feedforward term. The developed method is inspired by those previous works. Both gains and reference velocity are adjusted. To maximize damping, we use an impact detection signal, which is constructed with fuzzy rules, from force measurement. ID is set to 1 when impact is detected and 0 else. Then, if $k^s_{d_n}$ and $b^s_{d_n}$ are the gains of the controller for stable contact (we assume to know them), the actual gains during impact phase are computed by : $\begin{cases} k_{d_n} = (1-0.8ID).k^s_{d_n} \\ b_{d_n} = (1+ID)b^s_{d_n} \end{cases}$ such that the stiffness is deacreased and damping coefficient is increased during impact. To compute the reference velocity, we exploit the following fuzzy table :

$\dot{x}_{r_n}$		dF (normal force variation)				
		NL	NS	Z	PS	PL
ε_{F_n} (force error)	NL	PS	Z	Z	Z	NS
	NS	PS	PS	PS	PS	Z
	Z	PL	PS	PS	PS	PS
	PS	PL	PL	PL	PS	PS
	PL	PL	PL	PL	PL	PS

Since impact impulse is very short, and can not be avoided, the goal of this table is to prepare the end of impact rather than to track the desired force during impact. Thus, during compression, desired penetration is augmented to prepare the restitution phase. We have experimented such a strategy for different configurations and environments. The maximum impact velocity without rebounds is obviously increased (fig. 4 - column 3).

[1] In fact, with our controller, which don't use the Hogan's force feedabck, the environmnet dynamics is coupled through the right hand dynamics, but we can replace M_e, B_e and K_e by linearized forms of $\Lambda^{-1}M_e$, $\Lambda^{-1}B_e$ and $\Lambda^{-1}K_e$ for this analysis

[2] Notice that, when one use "heavy" industrial manipulators, impact impulses can be significantly large (few daN at maximum compression) even for reasonable approach velocities (few cm/s)

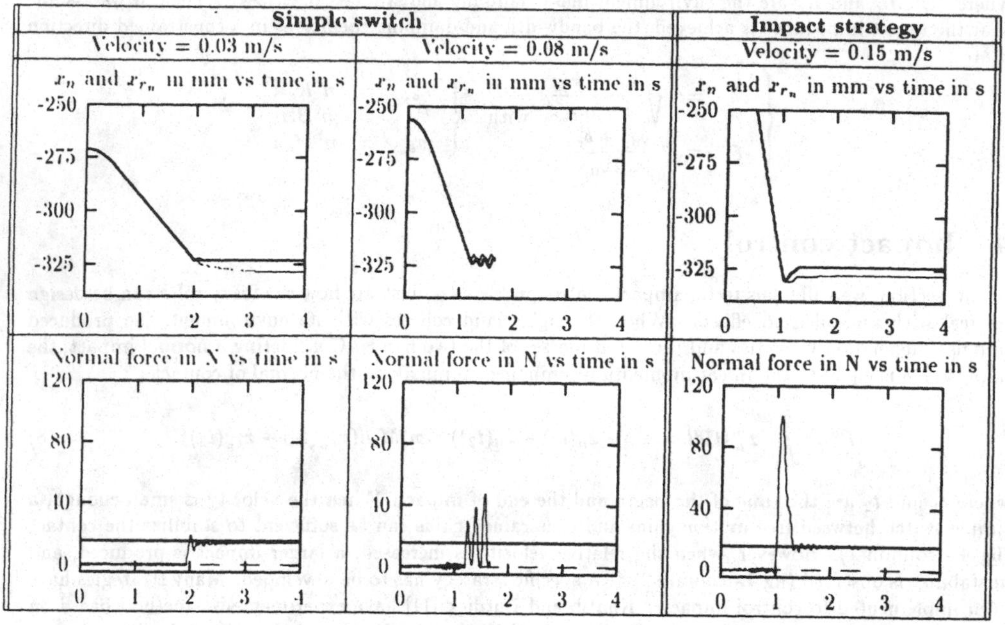

Figure 4. Experimental results on impact control

5 Conclusion

In this paper, we have presented a global control scheme based on impedance control, and a fuzzy supervisor, designed to control a manipulator which have to interact with a complex environment. Some details have been given on the practical choice of the target impedance. In addition, considering the specific problem of transitions between free and constrained motions, we have shown how some simple rules in the supervisor can significantly increase the performances of the system.
Future works concern the development of fuzzy strategies for more complex assembly tasks.

References

[1] G. Morel and P. Bidaud. Experiments on impedance control to derive adaptive strategies. In *Int. Symp. on Experimental Robotics*, pages 418-427, 1993.

[2] N. Hogan. Impedance control : A new approach to manipulation. *ASME J. of Dyn. Sys. Meas. and Control*, 107:1-24, 1985.

[3] J.K Mills. A generalized lyapunov approach to robotic manipulation stability during transition to and from contact tasks. In *Japan-USA Conf. on Flexible Automation*, pages 903–910. ISCIE, 1990.

[4] T.A. Lasky and T.C. Hsia. On force-tracking impedance control of robot manipulators. In *Int. Conf. on Rob. and Automation*, pages 274–280. IEEE, 1991.

[5] H Seraji and R. Colbaugh. Adaptive force-based impedance control. In *Int. Conf. on Intelligent Robots and Systems*, pages 1537-1542. IEEE/RSJ, 1993.

[6] R.J. Anderson and M.W. Spong Hybrid impedance control of robotics manipulators. *IEEE J. of Rob. and Automation*, 4(5):549–556, 1988.

[7] C.H. An and J. M. Hollerbach. The role of dynamic models in cartesian force control of manipulators. *The Int. Jal of Robotics Research*, 8:54-72, 1989.

[8] O. Khatib. A unified approach for motion and force control of robot manipulators: The operational space formulation. *IEEE J. of Rob. and Automation*, 3(1):45–53, 1987.

[9] M.H. Raibert and J.J. Craig. Hybrid position/force control of manipulators. *ASME J. of Dyn. Sys. Meas. and Control*, 102:126–133, 1981.

[10] W. McCormick and H.M. Schwartz. An investigation of impedance control for robot manipulators. *Int. Jal. of Robotics Research*, 12:473–489, 1993.

[11] O. Khatib and J. Burdick. Motion and force control of robot manipulators. In *Int. Conf. on Rob. and Automation*, pages 1381–1386. IEEE, 1986.

[12] R. Volpe and P. Khosla. Experimental verification of a strategy for impact control. In *Int. Conf. on Rob. and Automation*, pages 1854–1860. IEEE, 1991.

Part V
Control of Motion II

MODELLING AND CONTROL SYNTHESIS OF FLEXIBLE MULTIBODY SYSTEMS

J. Haug and W. Schiehlen
University of Stuttgart, Stuttgart, Germany

Abstract

In the last two decades, several models have been developed for the simulation of gross nonlinear motions of multibody systems which incorporate elastic members with small deformations. Although these models represent the real system sufficiently well for simulation, a closed-loop controller designed on the basis of such models may result in instability if applied to the real system.

In this paper, the value of a model consisting of superelements for control synthesis is investigated. The superelements approximating flexible bodies with beam like structures are composed of a series of rigid bodies interconnected by joints and springs. As an example, the nonlinear equations of motion of a plane flexible robot with two links are derived and linearized for control synthesis. A first controller is synthesized by using the pole placement method. Applied to the 'real system' represented by a highly accurate finite element model, the controller leads to instability. Therefore, a second robust controller is synthesized by using left coprime factorization and H_∞ minimization.

1 Introduction

Industrial robot manipulators have to serve conflicting goals: Demands on high velocities lead to large dynamical forces, whereas energy efficiency requires light weight constructions. Both requirements can only be met by allowing flexible deformations at normal working conditions. However, to guarantee safety and accuracy, open and closed loop controllers have to be applied to such systems.

A basic requirement for control synthesis is a mathematical model of the system to be controlled. For the above mentioned structures the method of elastic multibody systems has been established which takes into consideration a gross nonlinear reference motion as well as small deformations. For describing small elastic deformations different approaches are available:

- finite elements,
- continuous systems,
- rigid bodies interconnected by joints and springs.

A controller designed with these models may fail if applied to the real system due to modeling errors and spillover effects, see Engell [1] and Czaijkowski et al. [2].

In this paper, the third approach will be considered in more detail. A plane two-link flexible robot, known from Kleemann [3], Fig. 1, is modeled by 6 rigid bodies interconnected by auxiliary joints and springs. On the basis of this model, called design model in the following, controllers will be designed using two different methods, the pole placement method and H_∞ minimization. To check the influence of modeling errors and the performance of the controlled system, a second, highly accurate model on the basis of a finite element approach is generated. This will be called the plant model and is used as a representation of the real system.

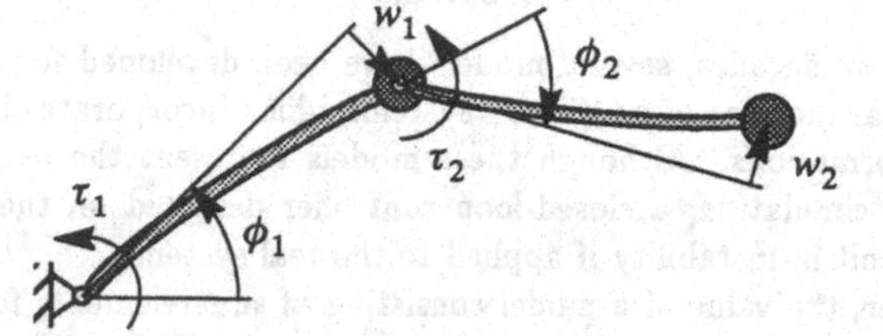

Figure 1: Flexible robot

2 Equations of motion

The equations of motion for the design model can be found from the well-known multibody systems approach. The equations of motion for the plant model can be generated either from a multibody systems approach extended by finite elements with small displacements or from a pure finite element approach including gross nonlinear motion. In the following, the first approach will be used.

2.1 Design Model

In a rigid multibody systems approach, flexible bodies are represented by so-called superelements. A superelement consists of three rigid bodies connected by revolute joints and springs, Fig. 2. For plane motion, a free superelement has two internal and three external degrees of freedom. The parameters for stiffness, inertia and geometry have been identified by Rauh [5].

The model for the plane robot consists of two superelements and two main joints, one for each elastic arm, resulting in 6 degrees of freedom. By formulating Newton's and Euler's equations for each body and applying d'Alembert's principle, the equations of motion can be found as

$$\mathbf{M}(\mathbf{q})\ddot{\mathbf{q}} + \mathbf{k}(\mathbf{q}, \dot{\mathbf{q}}) = \mathbf{h}(\mathbf{q}, \dot{\mathbf{q}}, t) \tag{1}$$

where $\mathbf{M}$ is the mass matrix, $\mathbf{k}$ is the vector of the gyroscopic forces, $\mathbf{h}$ is the vector of generalized applied forces, and $\mathbf{q}$ is the vector of generalized coordinates describing the motion of the robot.

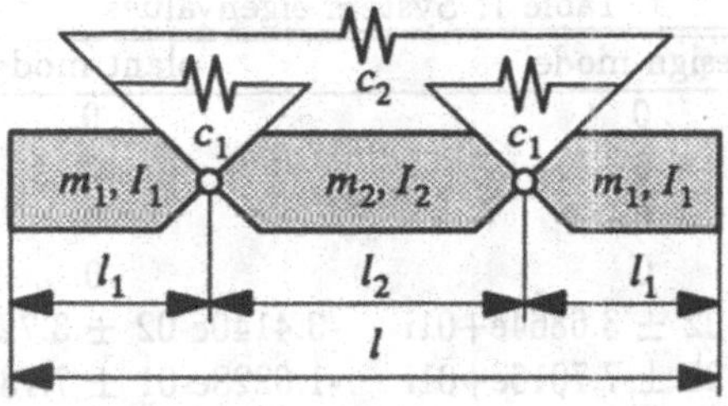

Figure 2: Superelement of a two dimensional beam

For control synthesis, the equations of motion may be linearized with respect to the final position. The resulting equations can be written in state space form:

$$\dot{\mathbf{x}} = \mathbf{A}\mathbf{x} + \mathbf{B}\mathbf{u}\ , \quad \mathbf{y} = \mathbf{C}\mathbf{x} \tag{2}$$

where $\mathbf{x}$ is the state vector summarizing coordinates $\mathbf{q}$ and velocities $\dot{\mathbf{q}}$. The vector $\mathbf{u}$ contains the two inputs, which are the two torques acting at the joint of each body and the vector $\mathbf{y}$ contains the four outputs, assuming that the joint angles and the tip displacements of each body can be measured, Fig. 1:

$$\mathbf{u} = \begin{bmatrix} \tau_1 & \tau_2 \end{bmatrix}^{\mathbf{T}}, \quad \mathbf{y} = \begin{bmatrix} \phi_1 & \phi_2 & w_1 & w_2 \end{bmatrix}^{\mathbf{T}}. \tag{3}$$

2.2 Plant Model

Following the approach used by Melzer [6], the equations of motion for a multibody system with elastic members are derived using d'Alembert's principle, too:

$$\sum_{i=1}^{p} \left(\int_V \delta\mathbf{r}^T(\rho\mathbf{a} - \mathbf{f}^a)dV + \int_V \delta\epsilon^T \sigma^a dV - \int_A \delta\mathbf{r}^T\mathbf{t}^a dA \right)_i = 0 \tag{4}$$

The acceleration vector $\mathbf{a}$, the virtual strain vector $\delta\epsilon$, and the virtual displacement vector $\delta\mathbf{r}$ are found from the position vector of a material point

$$\mathbf{r} = \mathbf{r}(\boldsymbol{\Phi}, \mathbf{q}) \tag{5}$$

using shape functions $\boldsymbol{\Phi}$ and generalized coordinates $\mathbf{q} = [\mathbf{q}_r^T \ \ \mathbf{q}_f^T]^T$ where $\mathbf{q}_r$ describes the rigid body motion and $\mathbf{q}_f$ represents the elastic deformations. The volume integrals in Eq. (4) can be pre-computed using a finite element approach. The resulting equations of motion have the same structure as Eq. (1)

$$\mathbf{M}_{ref}(\mathbf{q}_{ref})\ddot{\mathbf{q}}_{ref} + \mathbf{k}_{ref}(\mathbf{q}_{ref}, \dot{\mathbf{q}}_{ref}) = \mathbf{h}_{ref}(\mathbf{q}_{ref}, \dot{\mathbf{q}}_{ref}, t) \tag{6}$$

where the subscript *ref* denotes the use of Eq. (6) as reference model for the real system. For the plant model of the elastic robot, each arm is discretized by 40 finite elements. After modal condensation we end up with a model with 14 degrees of freedom.

Table 1 shows a comparison of the eigenvalues of the two models for the robot. Obviously, the zero eigenvalues due to the uncoupled rotations of the arms are the same, while the nonzero eigenvalues of the design model deviate from the plant model due to modeling errors. Furthermore, in the design model not all eigenvalues are taken into consideration. Both errors affect the overall behavior of controllers based on the more simple design model.

Table 1: System eigenvalues

design model	plant model
0	0
0	0
0	0
0	0
-3.6744e-02 ± 3.6864e+01i	-3.4140e-02 ± 3.7274e+01i
-1.5428e-01 ± 7.7016e+01i	-1.6228e-01 ± 7.7389e+01i
-8.6413e-01 ± 4.3219e+02i	-7.6067e-01 ± 4.5192e+02i
-3.1699e+00 ± 6.6207e+02i	-3.8222e+00 ± 7.8770e+02i
	-7.3933e+00 ± 1.5741e+03i
	-1.0993e+01 ± 2.5319e+03i
	-3.1359e+01 ± 3.3241e+03i
	-8.5318e+01 ± 5.7518e+03i
	-5.9892e+01 ± 6.0726e+03i
	-2.4022e+02 ± 9.2358e+03i
	-1.7411e+02 ± 1.1411e+04i
	-4.7205e+02 ± 1.8876e+04i

3 Controller Synthesis

To accomplish the global motion illustrated in Fig. 3, a bang–bang open loop controller calculated from rigid body analysis is used. In the final position a closed loop controller is switched on to damp out the vibrations and minimize the position error. The closed loop control design is based on the linearized design model.

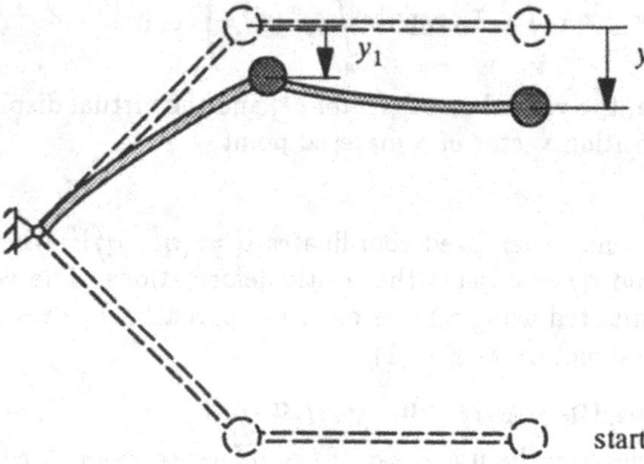

Figure 3: Robot motion

3.1 Pole Placement

For state-variable feedback an observer has to be applied, Fig. 4. The feedback matrices R and E for the controller and observer, respectively, can be calculated independent by the pole placement method from the desired poles of the closed-loop system. This was done by using the software package *MATLAB* [7].

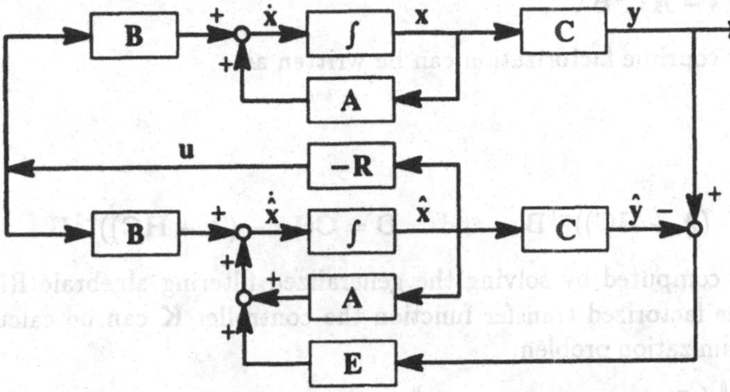

Figure 4: State-variable feedback controller and observer

Applying the resulting controller and observer to the plant model leads to instability due to two positive eigenvalues, Fig. 5. A detailed error investigation shows that this instability effect is not due to truncation of modes but due to errors in the eigenvalues of the design model. Therefore, a robust controller will be designed in the following section.

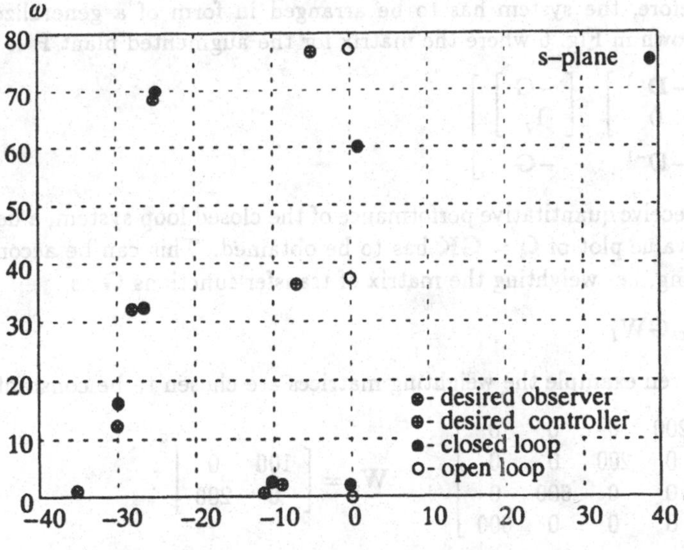

Figure 5: Poles of the open and closed loop system

3.2 Robust Controller

With the normalized left coprime factorization a controller can be designed giving maximal robustness against unstructured model errors, see McFarlain and Glover [8]. The controller is computed by performing H_∞ minimization.

The matrix of transfer functions corresponding to Eq. (2) is

$$\mathbf{G}(s) = \mathbf{C}(\mathbf{I}s - \mathbf{A})^{-1}\mathbf{B}\,. \tag{7}$$

A normalized left coprime factorization can be written as

$$\mathbf{G} = \mathbf{D}^{-1}\mathbf{N} \tag{8}$$

where

$$\mathbf{N} = \mathbf{C}(\mathbf{I}s - (\mathbf{A}+\mathbf{HC}))^{-1}\mathbf{B} \quad \text{and} \quad \mathbf{D} = \mathbf{C}(\mathbf{I}s - (\mathbf{A}+\mathbf{HC}))^{-1}\mathbf{H}\,. \tag{9}$$

The matrix $\mathbf{H}$ is computed by solving the generalized filtering algebraic Riccati equation [8]. With the factorized transfer function the controller $\mathbf{K}$ can be calculated from the following minimization problem

$$\inf_{\mathbf{K}\text{ stabilizing}} \left\| \begin{pmatrix} \mathbf{I}_2 \\ \mathbf{K} \end{pmatrix} (\mathbf{I}_2 + \mathbf{GK})^{-1}\mathbf{D}^{-1} \right\|_\infty = \frac{1}{\rho_{max}} \tag{10}$$

where $\mathbf{I}_2$ is a 2×2 idendity matrix and the maximal stability margin ρ_{max} is calculated as

$$\rho_{max} = \sqrt{1 - \|[\mathbf{N}\ \ \mathbf{D}]\|_H^2} \tag{11}$$

using the Hankel Norm. This problem can be solved by using standard H_∞ minimization [7]. Therefore, the system has to be arranged in form of a generalized feedback arrangement shown in Fig. 6 where the matrix for the augmented plant $\mathbf{P}$ is given as

$$\mathbf{P} = \begin{bmatrix} \begin{bmatrix} -\mathbf{D}^{-1} \\ \mathbf{0} \end{bmatrix} & \begin{bmatrix} -\mathbf{G} \\ \mathbf{I}_q \end{bmatrix} \\ -\mathbf{D}^{-1} & -\mathbf{G} \end{bmatrix}. \tag{12}$$

In order to receive quantitative performance of the closed loop system, a desired shape of the singular value plot of $\mathbf{Q} = \mathbf{GK}$ has to be obtained. This can be accomplished by open loop shaping, i.e. weighting the matrix of transfer functions $\mathbf{G}$ as

$$\mathbf{G}_W = \mathbf{W}_O\mathbf{G}\mathbf{W}_I \tag{13}$$

where for the given example the weighting matrices are chosen to be constant:

$$\mathbf{W}_O = \begin{bmatrix} 200 & 0 & 0 & 0 \\ 0 & 200 & 0 & 0 \\ 0 & 0 & 600 & 0 \\ 0 & 0 & 0 & 600 \end{bmatrix}, \quad \mathbf{W}_I = \begin{bmatrix} 100 & 0 \\ 0 & 200 \end{bmatrix}. \tag{14}$$

After computing the controller $\mathbf{K}_W$ using Eq. (10) on the basis of the weighted plant $\mathbf{G}_W$ the final solution for the original plant can be obtained:

$$\mathbf{K} = \mathbf{W}_I\mathbf{K}_W\mathbf{W}_O\,. \tag{15}$$

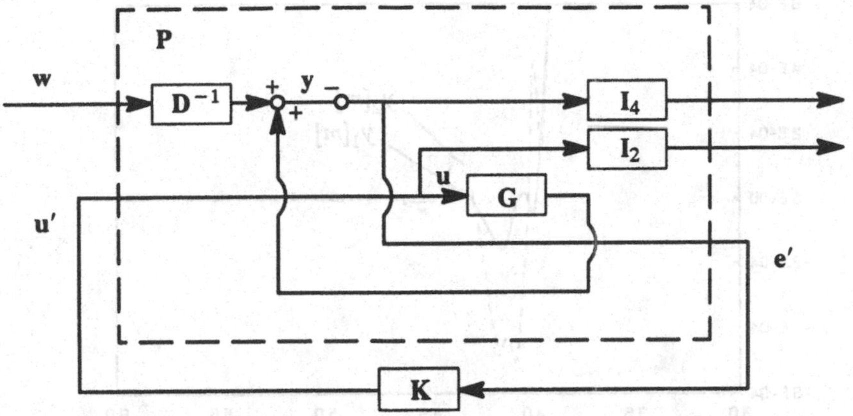

Figure 6: Generalized feedback arrangement

Figure 7 shows the time behavior of the system without closed loop controller, characterized by strong vibrations. Figure 8 represents the time behavior of the closed loop system where the robust controller is applied to the plant model, too. Obviously, the behavior is stable in contrast to the first controller, which has shown some positive eigenvalues, Fig. 5.

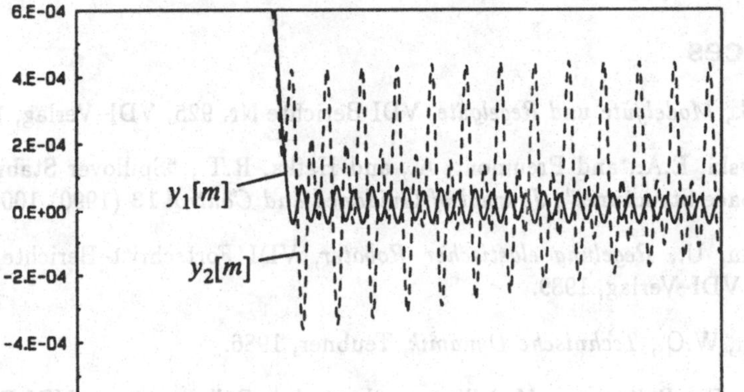

Figure 7: Time response of the displacement of the end effectors after open loop control

4 Conclusions

In this paper, a flexible robot is modeled for control design by a rigid multibody systems approach using superelements for approximating elastic beams. Such a rough model simplifies control design but results in erroneous eigenvalues compared to an accurate

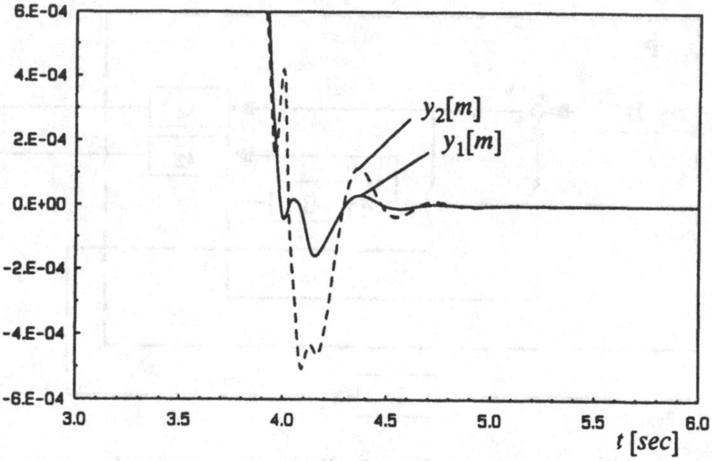

Figure 8: Time response of the displacement of the end effectors for combined open and closed loop control

plant model. A classical state feedback controller designed on the basis of such a simple model may then result in instability if it is applied to the real system, whereas a robust control design yields satisfactory results.

References

[1] Engell, S., *Modelgüte und Regelgüte*, VDI Berichte Nr. 925, VDI–Verlag, 1989.

[2] Czaijkowski, E.A., and Preumont, A., and Hafka, R.T., "Spillover Stabilization of Large Space Structures", *Journal of Guidance and Control* **13** (1990) 1000-1007.

[3] Kleemann, U., *Regelung elastischer Roboter*, VDI Fortschritt–Berichte, Reihe 8, Nr. 191, VDI–Verlag, 1989.

[4] Schiehlen, W.O., *Technische Dynamik*, Teubner, 1986.

[5] Rauh, J., *Ein Beitrag zur Modellierung elastischer Balkensysteme*, VDI Fortschritt–Berichte, Reihe 18, Nr. 37, VDI–Verlag, 1987.

[6] Melzer F., "Semi–Symbolic Equations of Motion for Flexible Multibody Systems", *Proceedings of the 1993 ASME International Computers in Engineering Conference*, San Diego, August 8-12, 1993.

[7] The Math Works Inc., *MATLAB User's Guide*, Natick, MA, 1992.

[8] McFarlain D.C. and Glover K., *Robust Controller Design Using Normalized Coprime Factor Plant Description*, Springer-Verlag, 1990.

CONTRIBUTION TO THE POSITION/FORCE CONTROL OF A ROBOT INTERACTING WITH DYNAMIC ENVIRONMENT IN CARTESIAN SPACE

M. Vukobratović and R. Stojić
M. Pupin Institute, Belgrade, Yugoslavia

Abstract

In this paper the problem of simultaneous stabilization of both the robot motion and interaction force in Cartesian space, based on the unified approach to contact task problem in robotics [1, 2], is considered . This control task is solved under the conditions on environment dynamics which are less restrictive than those in [1, 2] where some particular environment properties are required to ensure overall system stability. The one-to-one correspondence between closed loop motion and force dynamic equations is obtained and unique control law ensuring system stability and preset either motion or force transient response is proposed.

1 Introduction

Based on the stability principle of closed-loop control systems the control laws that simultaneously stabilize both the robot motion and interaction force with the environment has been synthesized in Ref. [1-5] . These control laws as distinct from the control laws synthesized by using the known traditional approaches [6-11], possess the exponential stability of closed-loop systems and ensure the preset quality of transient responses of motion and interaction force.

However, control laws stabilizing desired interaction force with preset quality of transient response are applicable only if the environment posses "internal stability" property [1, 2]. In this paper these restrictive conditions are removed and more general case of dynamic environment is considered. For the case of linear environment dynamics, which is often met in Cartesian space, necessary and sufficient conditions for stability of both motion and force are derived, and corresponding control laws are defined.

2 Task Setting

Consider the dynamics model of the robot interacting with the environment in the form

$$H(q)\ddot{q} + h(q,\dot{q}) = \tau + J^T(q)F \tag{1}$$

where q is the n-dimensional vector of generalized robot coordinates; $H(q)$ is the $n \times n$ positive definite matrix of inertia of the manipulator's mechanism; $h(q,\dot{q})$ is the n-dimensional nonlinear function of centrifugal, Coriollis' and gravitational terms; τ is the n-dimensional vector of the control input; $J^T(q)$ is the $n \times m$ matrix connecting the robot end-effector velocities and the generalized velocities $\dot{q}$; F is the m-dimensional vector of the generalized forces acting on the end-effector from the environment.

Let the dynamics model of the environment be described by the equation

$$M(q)\ddot{q} + L(q,\dot{q}) = -S^T(q)F \tag{2}$$

where $M(q)$ is $n \times n$ matrix, $L(q,\dot{q})$ is n-vector function, and $S(q)$ is $n \times m$ matrix . It is assumed that robot coordinates q and velocities $\dot{q}$ are sufficient to represent state of the dynamic environment. To simplify analysis it will be adopted that $n = m$.

The end-effector equations of motion can be written in the task frame using operational space approach [6].Let p be n-dimensional vector of external coordinates specifying position and orientation in a chosen task frame, and T vector of the control force coordinates. Then the equations (1,2) can be transformed to the form:

$$\Lambda(p)\ddot{p} + \nu(p,\dot{p}) = T + F \tag{3}$$

$$\mathcal{M}(p)\ddot{p} + \mathcal{L}(p,\dot{p}) = -F \tag{4}$$

It is assumed that all the functions in (1,2) i.e. (3,4) are continuously differentiable with respect to all variables, thus ensuring existence and uniqueness of the Cauchy problem solution for (3,4), with the initial conditions $p(t_0) = p_0$, $\dot{p}(t_0) = \dot{p}_0$.

The equation of robot dynamics (3) can be solved with respect to control T:

$$T = U(p,\dot{p},\ddot{p},F) \tag{5}$$

where $U(p,\dot{p},\ddot{p},F) = \Lambda(p)\ddot{p} + \nu(p,\dot{p}) - F$ while the environment dynamics model (4) can be rewritten as:

$$F = f(p,\dot{p},\ddot{p}) \tag{6}$$

where $f(p,\dot{p},\ddot{p}) = -\mathcal{M}(p)\ddot{p} - \mathcal{L}(p,\dot{p})$.

Let $p^0(t)$ be desired motion and $F^0(t)$ be desired interaction force, satisfying eq. (6) The control objective is to realize asymptotically stable $p^0(t), F^0(t)$, i.e. achieve

$$(\ p(t),\dot{p}(t),F(t)\ ) \rightarrow (\ p^0(t),\dot{p}^0(t),F^0(t)\ ) \qquad t \rightarrow \infty$$

It is additionally required that closed loop system possess contact force transient response (force dynamics) specified by equation

$$\ddot{\mu} = Q(\mu,\dot{\mu}) \tag{7}$$

where $\mu(t) = F(t) - F^0(t)$. Alternatively, it may be required that closed loop system posses motion transient response (motion dynamics) specified by equation

$$\ddot{\eta} = P(\eta,\dot{\eta}) \tag{8}$$

where $\eta(t) = p(t) - p^0(t)$. Functions P and Q are assumed to be continuously differentiable with respect to all arguments, such that eq. (7,8) have the asymptotically stable trivial solutions $\eta = 0, \mu = 0$, respectively.

Note that due to relation (6) $Q(\mu(t),\dot{\mu})$ and $P(\eta,\dot{\eta})$ cannot be arbitrary specified at the same time, so that force stabilization task and motion stabilization task should be distinguished.

3 Relation to previous work

Two control laws has been proposed in Ref.[1, 2] as to stabilize the desired motion $p^0(t)$ with a quality of the transient response specified by eq. (8). The first control law utilizes force feedback

$$T = U(p, \dot{p}, \ddot{p}^0 + P(\eta, \dot{\eta}), F) \tag{9}$$

and the second one stabilizes robot motion with no force feedback

$$T = U(p, \dot{p}, \ddot{p}^0 + P(\eta, \dot{\eta}), f(p, \dot{p}, \ddot{p}^0 + P(\eta, \dot{\eta}))) \tag{10}$$

By assumption function $P(\eta, \dot{\eta})$ has the property to assure an asymptotic stability in the whole of the trivial solution $\eta = 0$ of the system (8), i.e. $\eta(t) \rightarrow 0, \ \ \dot{\eta}(t) \rightarrow 0, \ \ t \rightarrow \infty$. Due to continuity of the function P it follows that: $\ddot{p}(t) \rightarrow \ddot{p}^0(t), \ \ t \rightarrow \infty$ and finally, continuity of the function f implies $F(t) \rightarrow F^0(t), \ \ t \rightarrow \infty$.

In this way the control laws (9,10) attain control goal, stabilizing motion with prescribed quality of transient response, as well as stabilizing contact force $F^0(t)$.

Three control laws has been proposed in Ref.[1, 2] as to stabilize desired interaction force with the specified quality of the transient response. However, these control laws enable stable motion of robot in contact only if the environment posesses such properties which ensure fulfilment of the limit relation $p(t) \rightarrow p^0(t), \ \ t \rightarrow \infty$, i.e. the environment stabilizes motion of robot in contact [1, 2].

Obviously, the class of stable motions is narrower than the class of motions corresponding to stable forces. So it is of great practical interest to identify such control laws which enable both stable force with preset transient response and stable motion.

4 Control Laws for Specified Force Dynamics

First, observe the relation between closed loop motion and force dynamics. Assume that the robot is in closed loop, and that corresponding motion and force dynamics are described by eq. (8,7) i.e. by functions $P(\eta, \dot{\eta})$ and $Q(\mu, \dot{\mu})$ respectively. If the motion dynamics is specified, (function $P(\eta, \dot{\eta})$ given), then, due to relation (6) the force dynamics and corresponding function $Q(\mu, \dot{\mu})$ will be uniquely determined. Further, if $\eta = 0, \ \ \dot{\eta} = 0$ then $\mu = 0$. However, in the converse case, having $Q(\mu, \dot{\mu})$ specified, function $P(\eta, \dot{\eta})$ is not uniquely determined in general case. Moreover, $\mu = 0, \ \ \dot{\mu} = 0$ does not imply $\eta = 0$ and therefore force stability does not necessarily imply motion stability.

To obtain correspondence between motion and force dynamics consider the case, often found in Cartesian space, when environment is modelled by a set of linear equations of the type:

$$\mu = M\ddot{\eta} + L_h\dot{\eta} + L_k\eta \tag{11}$$

where M, L_h, L_k are constant matrices of environment inertia, viscous friction and stiffness respectvely.

To simplify derivations, let us focus our attention to a class of control laws enabling linear closed loop motion dynamics:

$$\ddot{\eta} = \Gamma_1\dot{\eta} + \Gamma_2\eta \tag{12}$$

with some $n \times n$ matrices Γ_1, Γ_2.

Observe that force variable μ is linearly dependent on state variables $\eta, \dot{\eta}$:

$$\mu = (L_k + M\Gamma_2)\eta + (L_h M\Gamma_1)\dot{\eta} \tag{13}$$

whereas (13) results from (11) after substituting $\ddot{\eta}$ given by (12).

Time derivatives of variable μ are given as

$$\mu^{(i)} = \frac{d^i}{dt^i}\mu = \alpha_i\eta + \beta_i\dot{\eta} \qquad i = 0,1,2 \tag{14}$$

$$\text{where} \quad \begin{matrix} \alpha_0 = L_k + M\Gamma_2 & \alpha_{i+1} = \beta_i\Gamma_2 \\ \beta_0 = L_h + M\Gamma_1 & \beta_{i+1} = \alpha_i + \beta_i\Gamma_1 \end{matrix} \tag{15}$$

Namely, differentiating (14) for some i and substituting $\ddot{\eta}$ from (12) into obtained relation gives $(i+1)$-th time derivative of μ.

It follows from (14)

$$\begin{bmatrix} \mu \\ \dot{\mu} \end{bmatrix} = \alpha \begin{bmatrix} \eta \\ \dot{\eta} \end{bmatrix} \qquad \text{where} \quad \alpha = \begin{bmatrix} \alpha_0 & \beta_0 \\ \alpha_1 & \beta_1 \end{bmatrix} \tag{16}$$

By taking $i = 2$ into (14) and assuming invertibility of matrix α, it is further obtained

$$\ddot{\mu} = [\alpha_2 \;\; \beta_2] \begin{bmatrix} \eta \\ \dot{\eta} \end{bmatrix} = [\alpha_2 \;\; \beta_2]\,\alpha^{-1} \begin{bmatrix} \mu \\ \dot{\mu} \end{bmatrix} \tag{17}$$

and finally

$$\ddot{\mu} = Q_1\dot{\mu} + Q_2\mu \tag{18}$$

where $[Q_2 \;\; Q_1] = [\alpha_2 \;\; \beta_2]\,\alpha^{-1}$

In this way it is shown that linear closed loop motion dynamics (12) induces linear force dynamics given by (18) .

Consider now the relation between stability characteristics of motion and force dynamics equations (12,18). The following theorem may be formulated.

Theorem

If and only if matrix α is nonsingular, then force and motion dynamics equations (12,18) will have the same characteristic polynomial.

Proof:

Let $\Gamma = \begin{bmatrix} 0 & I \\ \Gamma_2 & \Gamma_1 \end{bmatrix}$, $\quad \Gamma_Q = \begin{bmatrix} 0 & I \\ Q_2 & Q_1 \end{bmatrix}$, then it can be derived from (15)

$$[\alpha_{i+1} \;\; \beta_{i+1}] = [\alpha_i \;\; \beta_i]\,\Gamma \tag{19}$$

and from (12-18)

$$\Gamma_Q = \alpha\Gamma\alpha^{-1} \tag{20}$$

$$det(\Gamma_Q - \lambda I) = det(\Gamma - \lambda I)$$

Above equality is equivalent to

$$det(\lambda^2 I + \lambda Q_1 + Q_2) = det(\lambda^2 I + \lambda\Gamma_1 + \Gamma_2)$$

which proves theorem.

The corollary of the theorem is that if desired force dynamics is specified in the form of eq. (18) and all characteristic roots are stable, then corresponding motion dynamics equation will have stable solution with the same characteristic roots, provided that matrix α is nonsingular.

Finally, answer the question how to find corresponding matrices Γ_1, Γ_2 having specified matrices Q_1, Q_2 which enssure stable force dynamics.

Introduce the matrices

$$c_0 = \begin{bmatrix} L_k & L_h - I \end{bmatrix} \quad c_1 = \begin{bmatrix} I & M \end{bmatrix}$$

Matrix C can be rewritten as $C = c_0 + c_1\Gamma$. It follows from (18) and (19):

$$\ddot{\mu} = [Q_2 \ Q_1] \begin{bmatrix} \mu \\ \dot{\mu} \end{bmatrix} = [Q_2 \ Q_1] \begin{bmatrix} C \\ C\Gamma \end{bmatrix} \begin{bmatrix} \eta \\ \dot{\eta} \end{bmatrix}$$

Comparision of this equation with (14,19) yields the relation

$$C\Gamma^2 = [Q_2 \ Q_1] \begin{bmatrix} C \\ C\Gamma \end{bmatrix} \tag{21}$$

which can be reduced to

$$(c_0 + c_1\Gamma)\Gamma^2 = Q_1(c_0 + c_1\Gamma)\Gamma + Q_2(c_0 + c_1\Gamma) \tag{22}$$

i.e.

$$c_1\Gamma^3 + (c_0 - Q_1c_1)\Gamma^2 - (Q_1c_0 + Q_2c_1)\Gamma - Q_2c_0 = 0 \tag{23}$$

Obtained equation can be solved with respect to unknown matrix Γ i.e. matrices Γ_1, Γ_2 for given environment parameters L_k, L_h, M (i.e. c_0, c_1), and desired (stable) force dynamics specified with Q_1, Q_2, If matrix Γ satisfying eq. (23) exists and if matrix α given by (16) is nonsingular, then desired force dynamics specified by (18) can be realized by control law (9) or (10). At the same time the closed loop robot motion will be stable, with motion dynamics satisfying eq. (12) and force dynamics satisfying eq. (18).

Note that, due to correspondence between closed loop motion and force dynamics, the same control law stabilizes the overall system and also realizes preset transient response of either motion or force.

5 Conclusion

The task of simultaneous stabilization of both the robot motion and the interaction force with closed loop force dynamics specified in advance has been considered in the paper, in the scope of unified approach to control laws synthesis for robotic manipulator in contact with dynamic environment [1, 2]. This task is solved with the conditions set on the environment dynamics which are less restrictive than those in [1, 2] where some particular environment properties are required to ensure overall system stability. Thus, the one-to-one correspondence between the closed loop motion and force dynamic equations is obtained and unique control law ensuring system stability with preset either motion or force transient response is proposed.

Although the environment dynamics is assumed to be linear in Cartesian space, the obtained results can be easily generalized to the case when environment dynamics model is not linear. With minor modifications they can be also extended to the case when $n > m$ (i.e. when dimension of control input is higher than dimension of contact force vector).

References

[1] M. Vukobratović and Y. Ekalo, "Unified approach to control laws synthesis for robotic manipulators in contact with dynamic environment," in *Tutorial S5: Force and Contact Control in robotic systems, IEEE Int. Conf. on Robotics and Automation*, Atlanta , pp. 213–229, 1993.

[2] M. Vukobratović and Y. Ekalo, "New approach to control laws synthesis for robotic manipulators in contact with dynamic environment," *submitted to ASME J. of Dyn. Syst., Meas. and Control.*

[3] Y. Ekalo and M. Vukobratović, "Quality of stabilization of robot interacting with dynamic environment," *Journal of Inteligent and Robotic Systems*, to appear 1994.

[4] Y. Ekalo and M. Vukobratović, "Robust and adaptive position/force stabilization conditions of robotic manipulators in contact tasks," *Robotica*, vol. 11, pp. 373–386, 1993.

[5] M. Vukobratović, R. Stojić, and Y. Ekalo, "Contribution to the correct problem solution of position/force control of manipulaton robots in contact with dynamic environment - a generalization," *submitted to ASME J. of Dyn. Syst., Meas. and Control.*

[6] O. Khatib, "A unified approach for motion and force control of robot manipulators: the operational space formulation," *IEEE Journal of Robotics and Automation*, vol. 3, pp. 43–53, 1987.

[7] M. H. Raibert and J. J. Craig, "Hybrid position/force control of manipulators," *ASME Journal of Dynamic Systems, Measurement and Control*, vol. 103, pp. 126–133, June 1981.

[8] T. M. Mason, "Compliance and force control for computer controlled manipulators," *IEEE Trans. on Systems, Man, and Cybernetics*, vol. SMC-11, pp. 418–432, 1981.

[9] N. Hogan, "Impedance control: an approach to manipulation, part 1.- theory, part 2.- implementation, part 3.- application," *ASME Journal of Dynamic Systems, Measurement and Control*, vol. 107, pp. 1–24, 1985.

[10] T. Yoshikawa, T. Sugie, and M. Tanaka, "Dynamic hybrid position/force control of robot manipulators – control design and experiment," *IEEE Journal of Robotics and Automation*, vol. 4, pp. 699–705, 1988.

[11] A. De Luca and C. Manes, "Hybrid force/position control for robots in contact with dynamic environments," in *Proc. Robot Control, SYROCO '91*, pp. 377–382, 1991.

INVERSE DYNAMICS APPROACH FOR INVARIANT CONTROL OF CONSTRAINED ROBOTS

K.P. Jankowski and H.A. ElMaraghy

McMaster University Hamilton, Ontario, Canada

Abstract

A nonlinear feedback control based on inverse dynamics is proposed for redundant robots with rigid or flexible joints during constrained motion task execution. Based on constrained system formalism, the control scheme presented in the paper achieves simultaneous, independent control of both position and contact force at the robot end-effector. The method is based on the introduction of a set of kinematic parameters, which are defined in a new basis of the working space. In this basis, a general inner product characterized by the unity matrix gives rise to the definition of a new set of metrics for the robot task space. Using these metrics, it becomes possible to decompose the twist and wrench spaces into complementary subspaces. This approach contributes to better understanding of the constrained task decomposition, and provides a consistent interpretation of the analytical procedures used for constraint formulation. An example with a three-link robot operating a moving joystick, with constrained orientation of the end-effector, is presented. The results of numerical simulation are used to show the effectiveness of the proposed controller and its robustness to modeling errors.

1 Introduction

Successful execution of tasks involving planned interactions of the robot end-effector with the environment, such as assembly tasks, edge following or machining operations, requires hybrid position/force control strategies. The basic hybrid control scheme [1] was based on the approach [2] of the task space decomposition onto two orthogonal subspaces. Further developments in this direction appeared, for example, in [3, 4]. Another approach to hybrid control is to consider the conditions of robot contact with the environment as analytical constraints imposed on the considered mechanical system [5-10].

It has been shown in [11] on the basis of classical geometry that the approaches based on the previously proposed idea [2] of the task space decompositon are non-invariant with respect to translating the reference frame origin or changing the length unit. In [11], the "orthogonality" condition was substituted with more appropriate "reciprocity" condition between twists of freedom and wrenches of constraints. Basing on the screw theory, the use of the "sip" matrix (with ones on the opposite diagonal and zeros elsewhere) has been proposed in the literature as a more general approach for the task space decomposition.

As demonstrated in [12], the approach, based on selecting the sip matrix for the pseudo-orthogonality condition, is not general enough to cover all possible task configurations. In particular, it can be applied for five or six dimensional screw spaces only. Moreover, it can be shown that using this matrix does not produce satisfactory results when the constrained motion of the end-effector represents a rotation with respect to an axis which does not pass through the contact point. The analysis presented in this paper provides a more general means of introducing the metrics for the decomposition of the robot working space.

A nonlinear feedback control based on inverse dynamics is proposed in this paper for robot constrained motion task execution. Based on constrained system formalism, the control scheme presented achieves simultaneous, independent control of both position and contact force at the robot end-effector. The method is based on the introduction of a set of kinematic parameters, which are defined in a new basis of the working space. A general inner product is introduced in this basis, characterized by the unity matrix, which gives rise to the definition of a new set of metrics for the robot task space. This approach contributes to better understanding of the constrained task decomposition, and provides a consistent interpretation of the analytical procedures used for constraint formulation. An example with a three-link robot operating a moving joystick, with constrained orientation of the end-effector, is presented. No results of simulation or experimental studies for a similar case have been presented to date in the literature. The results of numerical simulation are used to show the effectiveness of the proposed controller and its robustness to modeling errors.

2 Constraint Formulation

The position and orientation of the robot end-effector are described by the vector $x^T = [x_e^T \ \phi_e^T]$, where x_e denotes a k_1-dimensional vector of Cartesian coordinates of the origin of the frame C_e in the base frame C_o, and ϕ_e represents a k_2-dimensional vector of angular coordinates describing the orientation of the frame C_e with respect to C_o. The dimension of the vector x equals: $k = k_1 + k_2$. The generalized coordinates of the robot are chosen to be the joint coordinates $q \in R^n$. The case of a non-redundant robot is considered with $k = n$, but the approach can be also used for redundant robots. Let the relation between the vectors x and q be given by $x = f(q)$, which results in the velocity relation $\dot{x} = J(q)\dot{q}$, where $J = \partial f/\partial q \in R^{nxn}$. The velocity vector $\vec{v} = \mathrm{d}\vec{r}/\mathrm{d}t$ can be represented in the vector space L over R by $\vec{v} = \dot{x}^T e_x$, where e_x denotes the inertial basis of L.

Two groups of constraints are imposed on the system; the conditions of the end-effector contact with the environment represented by a rigid, frictionless surface are considered as "material" constraints [13], and the next group represents "program" constraints which occur when the desired system behaviour is specified in advance. The program constraints may concern the desired end-effector trajectory, desired contact force profile, optimality conditions for the effective use of the robot redundancy, etc.

The conditions of the end-effector contact with the environment can be defined as a set of m independent analytical functions represented by a vectorial function $G(x, t) = 0$. The corresponding velocity relation is:

$$\tilde{V}(x, \dot{x}, t) = G_x(x, t)\dot{x} + G_0(x, t) = 0\,, \tag{1}$$

with $G_x = \partial G/\partial x$, and $G_0 = \partial G/\partial t$. Eq.(1) represents a set of material constraints imposed on the system.

Let the $n-m$ remaining degrees of freedom of the end-effector along the constraint surface be described by a vectorial function $H(x,t)$. By differentiating $H(x,t)$, the following independent velocity parameters can be defined:

$$V(x,\dot{x},t) = H_x(x,t)\dot{x} + H_0(x,t)\,, \tag{2}$$

with $H_x = \partial H/\partial x$ and $H_0 = \partial H/\partial t$.

The operations which led to Eqs.(1)-(2) can be interpreted as the introduction of a new set of coordinates, or the set of kinematic parameters V and $\tilde{V}$, where $\tilde{V}$ equals zero by virtue of the material constraint equations (1). Theoretically, the matrix H_x can be arbitrary, except that the matrix $[H_x^T\ G_x^T]^T$ must be nonsingular. In practice, the choice of H_x follows the physical analysis of the given problem and depends on the motion parameters to be programmed. The comparative study [14] elucidates the computational aspects of particular choices of independent kinematic parameters. With the above specified parameters, the velocity vector $\vec{v}$ can be represented in a new basis e_μ of L as $\vec{v} = [(\mu - H_0)^T\ (\tilde{\mu} - G_0)^T]e_\mu$, where the coordinate vector $-[H_0^T\ G_0^T]^T$ designates the velocity of the origin of the basis e_μ relative to the basis e_x, expressed in e_μ. After inverting the velocity relation built on the top of (1)-(2), the following relation for the velocity vector components $\dot{x}$ in the basis e_x is obtained:

$$\dot{x} = \begin{bmatrix} H_{G1} & H_{G2} \end{bmatrix} \left\{ \begin{bmatrix} V \\ \tilde{V} \end{bmatrix} - \begin{bmatrix} H_0 \\ G_0 \end{bmatrix} \right\}. \tag{3}$$

The inverse has been partitioned to have $H_{G1} \in R^{n\times(n-m)}$ and $H_{G2} \in R^{n\times m}$, with $G_x H_{G1} = 0$, $G_x H_{G2} = I_m$, $H_x H_{G1} = I_{k-m}$, and $H_x H_{G2} = 0$. In analytical mechanics, the approach leading to the velocity relations (1)-(3) is known as a part of formulation of Maggi's equations [14]. A similar constraint formulation has been used in [6, 8, 10] for robot hybrid position/force control. The methods presented in [5, 7] can be considered as special cases for this approach.

The reaction $\vec{R}$ of the material constraints (1) is naturally expressed in the space L^* which is dual to L: $\vec{R} = w^T e^x$, where e^x denotes the dual basis of e_x. The expression for the constraint wrench w is:

$$w = \begin{bmatrix} H_x^T & G_x^T \end{bmatrix} \begin{bmatrix} \tilde{\lambda} \\ \lambda \end{bmatrix}, \tag{4}$$

where $\tilde{\lambda}$ and λ denote the $(n-m)$- and m-dimensional vectors, respectively, of Lagrange multipliers. In the case of ideal material constraints $\tilde{\lambda} = 0$, and the constraint reaction is orthogonal to the constraint surface: $w = G_x^T\lambda$.

3 Task Space Decomposition

A generalized inner product is introduced in the basis e_μ of L and in the dual basis e^μ of L^*. Since the velocity parameters μ and $\tilde{\mu}$ constitute the independent velocity vector components in e_μ, and the force parameters $\tilde{\lambda}$ and λ represent the components of the contact force vector along the axes of e^μ, it is meaningful to choose the unity matrix for the matrix representation of the generalized inner product. The unity matrix will be the inner product matrix in both bases e_μ of L and e^μ of L^*. Other aspects of the system are simplified by this particular choice of metrics. The meaning of this particular choice of metrics is that the vectors directed along the axes of e_μ (or e^μ) can be considered as pseudo-orthogonal.

In L, the transition matrix from e_μ to e_x is $[H_x^T \; G_x^T]$. The matrix representation of the generalized inner product in e_x is found to be:

$$\Psi = \begin{bmatrix} H_x \\ G_x \end{bmatrix}^T \begin{bmatrix} H_x \\ G_x \end{bmatrix} = H_x^T H_x + G_x^T G_x \,. \tag{5}$$

Similarly, the corresponding matrix in the basis e^x of L^* is obtained:

$$\psi = \begin{bmatrix} H_{G1} & H_{G2} \end{bmatrix} \begin{bmatrix} H_{G1}^T \\ H_{G2}^T \end{bmatrix} = H_{G1} H_{G1}^T + H_{G2} H_{G2}^T \,. \tag{6}$$

It is obvious that, as expected, $\psi = \Psi^{-1}$.

The instantenous work (i.e., power) done by the reaction $\vec{R}$ in producing the velocity $\vec{v}$ is defined by the invariant $\mathrm{d}W/\mathrm{d}t = \vec{R} \cdot \vec{v}$. The corresponding bilinear form which represents the instantenous work can be obtained by taking the components of both $\vec{R}$ and $\vec{v}$ in one vector space (L or L^*). Representing $\vec{R}$ in the twist space L, the following vanishing bilinear mapping can be defined when $\lambda = 0$ and $\tilde{\mu} = 0$:

$$\frac{\mathrm{d}W}{\mathrm{d}t} = \lambda^T H_{G2}^T \Psi H_{G1} \mu = 0 \quad \text{with} \quad H_{G2}^T \Psi H_{G1} = 0 \,. \tag{7}$$

By representing the reaction and velocity by covectors in L^*, a similar relation can be obtained for the wrench space:

$$\frac{\mathrm{d}W}{\mathrm{d}t} = \lambda^T G_x \psi H_x^T \mu = 0 \quad \text{with} \quad G_x \psi H_x^T = 0 \,. \tag{8}$$

Having (8) in mind, the inversion of (1)-(2) to get (3) can be interpreted as taking the following weighted generalized inverses:

$$H_{G1} = (H_x)_\psi^+ = \psi H_x^T (H_x \psi H_x^T)^{-1} \,, \tag{9}$$

$$H_{G2} = (G_x)_\psi^+ = \psi G_x^T (G_x \psi G_x^T)^{-1} \,. \tag{10}$$

The geometric interpretation of the above operations on vector spaces can be carried out following the terminology of [11]. The twist space L has been decomposed into two complementary subspaces $[H_{G1} \; H_{G2}]$ in e_x, which are related by the vanishing of the bilinear form (7). Similarly, the wrench space L^* has been decomposed into two complementary subspaces $[G_x^T \; H_x^T]$ in e^x, which are related by (8). The robot end-effector is free to move about any twist in H_{G1} while exerting any wrench in G_x^T. H_{G1} and G_x^T together are referred to as a kinestatic model of the environment. It may also be desirable to define the projection matrices onto the range spaces of H_{G1}^T and G_x, to filter the given (measured) twist and wrench such that they are, respectively, in H_{G1} and G_x^T. The corresponding filters can be constructed as

$$P_h = H_{G1} (H_{G1})_\Psi^+ \,, \quad \text{or} \quad P_h = H_{G1} H_x \,, \tag{11}$$

$$P_g = G_x^T \left(G_x^T\right)_\psi^+ \,, \quad \text{or} \quad P_g = G_x^T H_{G2}^T \,. \tag{12}$$

No explicit expressions for the weighting matrices ψ or Ψ are needed to find the filters if the matrix inversion has been applied to determine H_{G1} and H_{G2} given H_x and G_x.

By analysing the method in which the bilinear forms on the right hand sides of Eqs.(7)-(8) have been defined, it becomes clear that, although they relate the subspaces of the twist and wrench spaces, respectively, the physical meaning of the expressions from which they originate is that of the power. As the bilinear forms on the left hand sides of Eqs.(7)-(8) vanish, the consistency of units should be verified on the basis of the pseudo-power relations.

4 Inverse Dynamics Control

4.1 System Dynamic Equations

The equations of motion of a robot with rigid joints, subject to material constraints (1), can be presented in the following matrix form:

$$A(q)\ddot{q} + B(q,\dot{q}) = u + J^T G_x^T \lambda \,. \tag{13}$$

The dynamic model of a robot with flexible joints, introduced using several commonly adopted assumptions, can be described by the following set of matrix equations:

$$\begin{aligned} A(q)\ddot{q} + B(q,\dot{q}) + K(q-\phi) &= J^T G_x^T \lambda \,, & (14)\\ I_r\ddot{\phi} + B_\phi\dot{\phi} - K(q-\phi) &= u \,. & (15) \end{aligned}$$

The symbols used in Eqs.(13)-(15) are as follows: $A(q)$ is the symmetric inertia matrix, $B(q,\dot{q})$ is the vector of centrifugal, Coriolis and gravitational terms, $J^T G_x^T \lambda$ is the vector of generalized constraint forces exerted by the environment on the robot end-effector, $\phi = [\phi_1 \ldots \phi_n]^T$ is the vector of angles of rotation of actuator rotors, $I_r = \mathrm{diag}[I_{ri}]$, where I_{ri} is the moment of inertia of the i-th rotor, $B_\phi = \mathrm{diag}[B_{\phi i}]$, where $B_{\phi i}$ is the viscous friction coefficient, $K = \mathrm{diag}[K_i]$, where K_i is the elastic constant of the i-th joint, and u is the vector of torques applied to the actuator rotors.

4.2 Control Law Formulation

Control objectives are represented by a set of desired motion/force characteristics (program constraints), imposed on the system. The first group of program constraints specifies the end-effector motion along the constraint surface, which results in the velocity relation $V(x,\dot{x},t) = V_d(t)$. A complementary set of m program constraints is defined by specifying the desired values of contact forces (i.e., material constraint reactions): $\lambda = \lambda_d(t)$.

The inverse dynamics control law can be constructed on the basis of the system dynamic models which need to be presented in a form reflecting direct relationship between the inputs and outputs. Since the considered system is subject to a set of material constraints (1), the joint accelerations $\ddot{q}$ and the joint snaps $\overset{(4)}{q}$ can be evaluated in function of the time derivatives of V, with $\tilde{V}$ equal to zero.

For the rigid joint robots, after substituting the expression for $\ddot{q}$ one obtains:

$$u = \hat{A}J^{-1}\left[\hat{H}_{G1}\left(\underline{\dot{V}} - \hat{H}_1\right) - \hat{H}_{G2}\hat{G}_1 - J_1\right] + \hat{B} - J^T\hat{G}_x^T\underline{\lambda}\,, \tag{16}$$

where the hat above a vector or a matrix indicates its estimated value, and H_1, G_1, and J_1 contain the terms which appear when performing the differentiations. The underlined quantities in (16) result from the modified program constraints, and for the basic inverse dynamics controller equal:

$$\underline{\dot{V}} = \dot{V}_d - k_{p1}(V - V_d) - k_{p0}(H - H_d)\,, \tag{17}$$

$$\underline{\lambda} = \lambda_d, \tag{18}$$

where k_{p1} and k_{p0} designate diagonal feedback gain matrices chosen to guarantee the desired properties of the transient response of the system.

For flexible joint robots, the system dynamic model (14)-(15) can be presented in the required form by differentiating Eq.(14) twice with respect to time, and then solving it for ϕ and substituting the result into Eq.(15) [8, 10]. After substituting the expression for $\overset{(4)}{q}$ one obtains:

$$\begin{aligned} u &= \hat{I}_r \left\{ \ddot{q} + \hat{K}^{-1} \left\{ \hat{A} J^{-1} \left[\hat{H}_{G1} \left(\overset{(3)}{\underline{V}} - \hat{H}_3 \right) - \hat{H}_{G2} \hat{G}_3 - J_3 \right] \right. \right. \\ &\quad \left. \left. + \hat{B}^* - J^T \hat{G}_x^T \underline{\ddot{\lambda}} - \hat{J}_\lambda \right\} \right\} + \hat{B}_\phi \dot{\phi} - \hat{K}(q - \phi), \end{aligned} \tag{19}$$

where $\overset{(i)}{s}$ designates the ith order time derivative of s, and B^*, J_λ, H_3, G_3, and J_3 contain the terms which appear when performing the differentiations.

To guarantee the stable transient response of the system, the underlined quantities are calculated from the following equations:

$$\overset{(3)}{\underline{V}} = \overset{(3)}{V}_d - k_{p3} \overset{(3)}{\varepsilon}_p - k_{p2} \ddot{\varepsilon}_p - k_{p1} \dot{\varepsilon}_p - k_{p0} \varepsilon_p, \tag{20}$$

$$\underline{\ddot{\lambda}} = \ddot{\lambda}_d - k_{f1} \dot{\varepsilon}_f - k_{f0} \varepsilon_f, \tag{21}$$

with the appropriate diagonal feedback gain matrices k_{p3}, ..., k_{f0}.

If the dynamic model of the system corresponds exactly with its actual counterpart (i.e., if $\hat{A} = A$, $\hat{B} = B$, etc.), the substitution of control laws (16) and (19) into the corresponding system dynamic equations, results in obtaining a number of decoupled subsystems governed by linear error equations.

During numerical simulations and real-time experiments it has been observed that the inverse dynamics control law is robust to bounded modeling errors and external disturbances. The tracking errors are especially small if a straightforward approach to robustness is used [10]. This relies on adding servocompensators to the control inputs, consisting of strings of integrators driven by the error signals. This approach enhances avoidance of problems caused by the time delay in the algebraic loop in the contact force subsystem [15]. In order to consider a larger class of robust controllers, the error space equations for the systems with the control laws (16) or (19) and inexact parameters can be formulated. It can be shown that the resulting systems satisfy matching assumptions. To render the system practically stable, various classes of controllers can be constructed [16, 17].

5 Illustrative Example

To illustrate the methodology presented in this paper, an example of a planar three-link flexible joint robot operating a joystick is considered. This problem has been chosen as an example of a task in which the orientation of the robot end-effector is restricted. No results of simulation or experimental studies for a similar case have been presented to date.

Let the generalized coordinates of the robot be $q \in R^3$, and the position and orientation of the end-effector be described by the vector $x = [x_e \; y_e \; \phi_e]^T$. Assume that the joystick axis of rotation moves in a slot with the velocity v_r which cannot be altered by the motion of the considered system (Fig. 1).

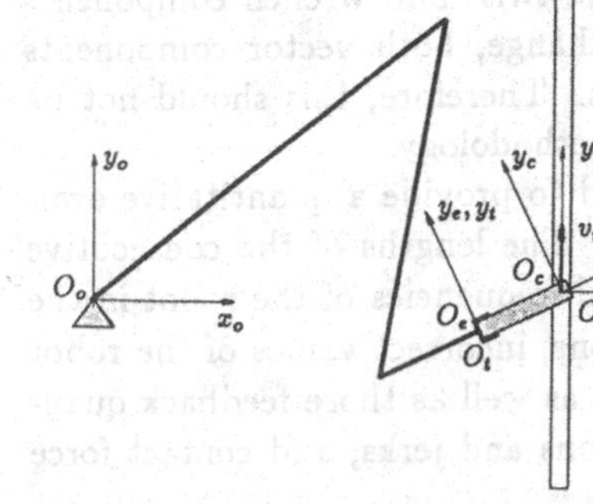

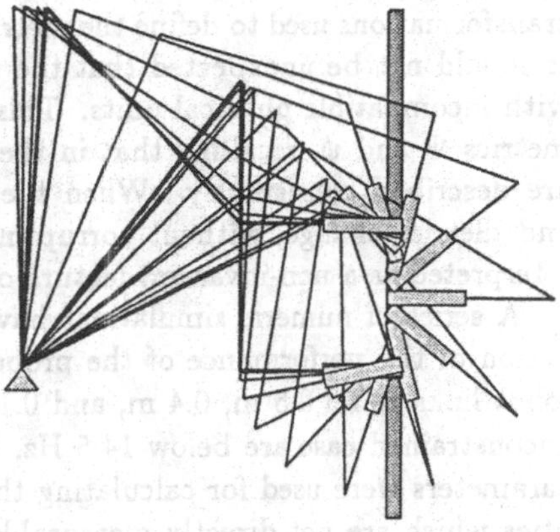

Fig. 2: System time evolution.

The material constraint equation (1) may be obtained by projecting the joystick axis velocity onto the axes of the moving frame $O_c x_c y_c$, which results in:

$$G_x = \begin{bmatrix} c\phi_e & s\phi_e & 0 \\ -s\phi_e & c\phi_e & r \end{bmatrix}, \quad G_0 = -\begin{bmatrix} v_{rx}c\phi_e + v_{ry}s\phi_e \\ v_{ry}c\phi_e - v_{rx}s\phi_e \end{bmatrix}. \tag{22}$$

A natural way to program the joystick motion is to specify the rotation angle ϕ_e as a function of time. The matrices in Eq.(2) become:

$$H_x = \begin{bmatrix} 0 & 0 & 1 \end{bmatrix}, \quad H_0 = 0. \tag{23}$$

The Lagrange multipliers λ_1 and λ_2 in the constraint reaction definition (4) correspond to the modulus of two forces acting on the joystick along $O_c x_c$ and $O_c y_c$ axes, respectively. The final program constraints to be defined are: $\lambda_1 = \lambda_{1d}(t)$ and $\lambda_2 = \lambda_{2d}(t)$.

The matrices H_{G1} and H_{G2} in (3) equal:

$$H_{G1} = \begin{bmatrix} rs\phi_e \\ -rc\phi_e \\ 1 \end{bmatrix}, \quad H_{G2} = \begin{bmatrix} c\phi_e & -s\phi_e \\ s\phi_e & c\phi_e \\ 0 & 0 \end{bmatrix}. \tag{24}$$

The twist space metric in e_x is:

$$\Psi = \begin{bmatrix} H_x \\ G_x \end{bmatrix}^T \begin{bmatrix} H_x \\ G_x \end{bmatrix} = \begin{bmatrix} 1 & 0 & -rs\phi_e \\ 0 & 1 & rc\phi_e \\ -rs\phi_e & rc\phi_e & r^2+1 \end{bmatrix}, \tag{25}$$

and the wrench space metric in e^x is given by

$$\psi = \begin{bmatrix} H_{G1} & H_{G2} \end{bmatrix} \begin{bmatrix} H_{G1}^T \\ H_{G2}^T \end{bmatrix} = \begin{bmatrix} r^2(s\phi_e)^2+1 & -r^2 s\phi_e c\phi_e & rs\phi_e \\ -r^2 s\phi_e c\phi_e & r^2(c\phi_e)^2+1 & -rc\phi_e \\ rs\phi_e & -rc\phi_e & 1 \end{bmatrix}. \tag{26}$$

The basis e_x is located at the contact point O_e, with its axes parallel to $O_o x_o y_o$ (see Fig. 1). The basis e_μ is located at the rotation axis of the joystick and corresponds to the axes of the frame $O_c x_c y_c$ enhanced by the rotation axis perpendicular to the plane

of motion. The dual bases e^x and e^μ have the axes parallel to e_x and e_μ. As the basis transformations used to define the metrics describe the combined rotation and translation, it should not be unexpected that the matrix products (25) and (26) contain the terms with incompatible physical units. This is a logical consequence of the definitions of the metrics Ψ and ψ, recalling that in the chosen bases, the twist and wrench components are described consistently. When the physical units change, both vector components and metrics change, without corrupting the final result. Therefore, this should not be interpreted as a non-invariant feature of the proposed methodology.

A series of numeric simulations have been performed to provide a quantitative evaluation of the performance of the proposed control law. The lengths of the consecutive robot links equal 0.5 m, 0.4 m, and 0.12 m. The natural frequencies of the robot in the unconstrained case are below 14.5 Hz. During simulations, incorrect values of the robot parameters were used for calculating the control inputs, as well as those feedback quantities which are not directly measurable: link accelerations and jerks, and contact force time derivatives.

The simulations have been conducted on a multitude of trajectories. The results presented here concern one particular trajectory observed to be instructive. As seen on the time evolution of the system motion (Fig. 2), the joystick is initially in the horizontal position, with rigidly attached end-effector and $\phi_e = 0$. The robot task is to turn the joystick around its axis of rotation which is moving vertically with the velocity v_r, powered by an independent mechanism. The additional task is to apply specified efforts to the joystick.

The analysis of the results of simulation for discrete-time control with sampling frequency $f = 500$ Hz reveals that very small tracking errors occur during the simulated motion. These findings are especiallly worthy because a realistic control scheme has been considered, together with modeling errors affecting control algorithm implementation. For the described tests, the maximum error has reached 2.7% for λ_1.

6 Conclusions

A nonlinear feedback control based on inverse dynamics has been proposed for robot constrained motion task execution. Based on the constrained system formalism, the control scheme presented achieves simultaneous, independent control of both position and contact force at the robot end-effector. The method is based on the introduction of a new set of metrics for the robot task space. This approach contributes to better understanding of the constrained task decomposition, and provides a consistent interpretation of the analytical procedures used for constraint formulation. It is expected that the remarks presented in this paper will clarify some misunderstandings among the robotics community related to the non-invariance issues in the robot hybrid control. The application of the methodology has been illustrated for the case of a three-link robot operating a moving joystick. The orientation of the end-effector is restricted when performing this task. The results of numerical simulation have been used to show the effectiveness of the proposed controller and its robustness to modeling errors. The analysis of results proves that the method can be useful for practical applications. The additional performance evaluation of the proposed controllers will be carried out during real-time experiments.

References

[1] Raibert, M. H., and Craig, J. J., 1981,"Hybrid Position/Force Control of Manipulators," *ASME J. Dyn. Syst. Meas. Control*, Vol. 102, pp. 126-133.

[2] Mason, M. T., 1981, "Compliance and Force Control for Computer Controlled Manipulators," *IEEE Tr. Syst. Man Cybern.*, Vol. SMC-11, pp. 418-432.

[3] West, H., and Asada, H., 1985, "A Method for the Design of Hybrid Position/Force Controllers for Manipulators Constrained by Contact with the Environment," *Proc. IEEE Conf. Rob. Autom.*, St. Louis, MO, pp. 251-259.

[4] Khatib, O., 1987, "A Unified Approach for Motion and Force Control of Robot Manipulators: The Operational space Formulation," *IEEE J. Rob. Autom.*, Vol. RA-3, pp. 43-53.

[5] Yoshikawa, T., 1987, "Dynamic Hybrid Position/Force Control of Robot Manipulators - Description of Hand Constraints and Calculation of Joint Driving Force," *IEEE J. Rob. Autom.*, Vol. RA-3, pp. 386-392.

[6] Kankaanranta, R. K., and Koivo, H. N., 1988, "Dynamics and Simulation of Compliant Motion of a Manipulator," *IEEE J. Rob. Autom.*, Vol. 4, pp. 163-173.

[7] McClamroch, N. H., and Wang, D., 1988, "Feedback Stabilization and Tracking of Constrained Robots," *IEEE Tr. Autom. Control*, Vol. 33, pp. 419-426.

[8] Jankowski, K. P., and ElMaraghy, H. A., 1991, "Dynamic control of flexible joint robots with constrained end-effector motion," *Prepr. IFAC Symp. Rob. Control SYROCO'91*, Vienna, Austria, pp. 345-350.

[9] Jankowski, K. P., and ElMaraghy, H. A., 1992, "Dynamic Decoupling for Hybrid Control of Rigid-/Flexible-Joint Robots Interacting with the Environment," *IEEE Tr. Rob. Autom.*, Vol. 8, pp. 519-534.

[10] Jankowski, K. P., and ElMaraghy, H. A., 1992, "Inverse Dynamics and Feedforward Controllers for High Precision Position/Force Tracking of Flexible Joint Robots," *IEEE Conf. Dec. Control*, Tucson, AZ, pp. 317-322.

[11] Lipkin, H., and Duffy, J., 1988, "Hybrid Twist and Wrench Control for a Robotic Manipulator," *Tr. ASME J. Mech. Transm. Autom. Design*, Vol. 110, pp. 138-144.

[12] Abbati-Marescotti, A., Bonivento, C., and Melchiorri, C., 1990, "On the invariance of the hybrid position/force control," *J. Intel. Rob. Syst.*, Vol. 3, pp. 233-250.

[13] Jankowski, K. P., 1989, "Dynamics of Controlled Mechanical Systems with Material and Program Constraints, Part I-III," *Mech. Mach. Theory*, Vol. 24, pp. 175-193.

[14] Kurdila, A., Papastavridis, J. G., and Kamat, M. P., 1990, "Role of Maggi's equations in computational methods for constrained multibody systems," *J. Guidance*, Vol. 13, pp. 113-120.

[15] Wen, J. T., and Kreutz-Delgado, K., 1992, "Motion and force control of multiple robotic manipulators," *Automatica*, Vol. 28, pp. 729-743.

[16] Leitmann, G., 1981, "On the efficacy of nonlinear control in uncertain linear systems," *ASME J. Dyn. Syst. Meas. Control*, Vol. 102, pp. 95-102.

[17] Chen, Y. H., 1988, "Design of robust controllers for uncertain dynamical systems," *IEEE Tr. Autom. Control*, Vol. AC-33, pp. 487-491.

SIMULATION EXPERIMENTS WITH FUZZY LOGIC-BASED ROBOT CONTROL

M. Vukobratović and B. Karan
M. Pupin Institute, Belgrade, Yugoslavia

Abstract

The paper describes trajectory tracking simulation experiments with a hybrid approach to robot control that combines traditional model-based and fuzzy logic-based control techniques. The combined method is developed by extending a model-based decentralized control scheme with fuzzy logic-based tuners for modifying parameters of joint servo controllers. The simulation experiments conducted on a real-scale six-degree-of-freedom industrial robot demonstrate suitability of fuzzy logic-based methods for improving the performance of the robot control system.

1 Introduction

Central issues in model-based robot control schemes are complexity and uncertainty of the internal robot model [1]. It may become a very complex system of nonlinear differential equations and it is always more or less an approximation of the real robot.

A viable possibility for attacking the problems of robot model complexity and uncertainty is offered by fuzzy logic-based controllers (FLC) and it has been explored by several researchers. The initial works that attempted to control manipulation robots directly with FLC [2, 3] have shown the applicability of the method, but they have also exposed considerable problems. The first problem is manifested by the lack of appropriate analytic tools for control design, i.e. selection of FLC parameters. A more tight connection between FLC and standard control methods was proposed by Tzafestas and Papanikolopoulos [4], who suggested to employ a two-level hierarchy in which FLC-based expert system is used for fine tuning of low-level PID control. The similar approach was applied to robot control by Popović and Shekhawat [5]. However, the two-level hierarchy does not actually solve the second problem, that is manifested in weak performance. Ordinary FLC schemes displayed performance characteristics that are similar or just slightly better than with simple constant-gain PID schemes. This fact shows that knowledge of the readily available mathematical model of robot dynamics should not be ignored. Most importantly, it may be employed to decrease the nonlinear dynamic coupling between its joint subsystems. Thus, a combined approach is preferred, and it may yield superior control schemes than both simple model-based or fuzzy logic-based approaches. The similar concept was formulated by de Silva and MacFarlane [6], who tested the concept by simulation of a two-link with an assumption of idealized effectiveness of the low-level global nonlinear feedback.

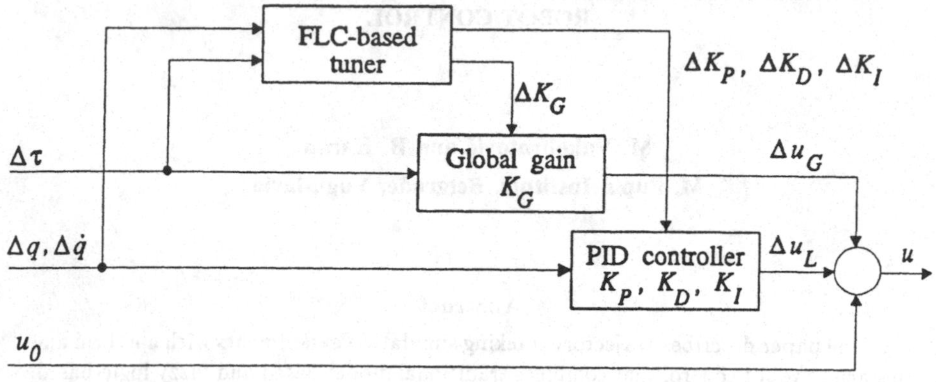

Figure 1: Hybrid control scheme

In this work, the performance of the hybrid approach is analyzed on a real-scale robot operating in free space. The purpose of this research is twofold. First, the expected improvements in robot control performance are demonstrated. An equally important aspect of this work is the question of how much it is possible to simplify the internal robot model by engaging the more sophisticated fuzzy logic-based servoing. Thus, the second goal is investigation of the extent of possible simplifications in traditional control, while maintaining the same level of control quality.

2 The hybrid approach

The hybrid design is an extension of decentralized control strategy and it consists of a set of subsystems closed around individual robot joints. Each of the subsystems comprises two components: the traditional model-based controller and optional fuzzy logic-based tuner (see Fig. 1). The tuner is designed as a fuzzy logic-based controller that monitors joint response characteristics and modifies the gains to provide better responses for large deviations of the monitored quantities.

The inputs to ith joint subsystem, $i = 1, \ldots, n$, where n is the number of actuated joints, are nominal (feedforward) control signal u_{i0}, joint position error Δq_i, joint velocity error $\Delta \dot{q}_i$, and optional computed or measured deviation of dynamic torque $\Delta \tau_i$ acting at the joint. Thus, the decentralized control of the general form

$$u_i = u_{i0} + \Delta u_{Li} + \Delta u_{Gi} \tag{1}$$

$$\Delta u_{Li} = K_{Pi} \Delta q_i + K_{Di} \Delta \dot{q}_i + K_{Ii} \int_0^t \Delta q_i dt \tag{2}$$

$$\Delta u_{Gi} = \frac{K_G}{C_{mi}/r_{mi}} \Delta \tau_i \tag{3}$$

is considered, where K_{Pi}, K_{Di}, and K_{Ii} are PID gains, K_{Gi} is the global control gain, and C_{mi}/r_{mi} denotes net static gain of the motor/reducer combination.

The nominal u_{i0} is calculated for prescribed trajectory on basis of the internal model of robot dynamics. Having in mind the numerical complexity of the actual expressions for computation of the nominal, it is profitable to explore the possibilities for using its

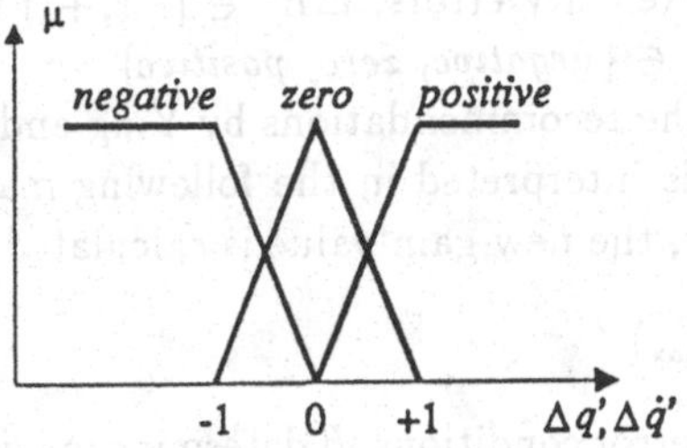

Figure 2: Primary fuzzy sets

Table 1: Gain tuning rules

		$\Delta q'$		
		negative	zero	positive
$\Delta \dot{q}'$	positive	zero	zero	positive
	zero	positive	negative	positive
	negative	positive	zero	zero

approximate forms. In this work, the following approximations of the robot arm dynamics were analyzed: (1) the model in which only linear models of actuators and gravity effects are taken into account, (2) the model which additionally incorporates self-inertia terms, (3) the model with added cross-inertia terms, and (4) the complete form of rigid-body model.

Gains of local PID controllers are synthesized independently for each joint to stabilize free (decoupled) subsystems. An important point is that the gain selection is actually made as a compromise between opposite requirements among the assumably constant PID gains. For example, K_{Pi} should be made large to provide fast response whereas it should be small enough to prevent resonance oscillations of the mechanical structure of the robot. Therefore, variable gains may yield more efficient control and thus the importance of fuzzy logic-based control as a effective mean for designing performance-driven variable-gain control schemes becomes apparent.

The general structure of the fuzzy controller permits to construct sophisticated control rules for tuning the gains of joint servo systems. However, in this work it is estimated that, for the purpose of comparison to traditional schemes, it is sufficient to consider a simple decentralized structure which consists of independent joint servo tuners, operating on the basis of current joint position error Δq and joint velocity error $\Delta \dot{q}$. Further, in order to examine the extent to which the hybrid approach may be utilized to eliminate the need for the full dynamic compensation, it is assumed that the global force feedback is not implemented in the schemes employing the fuzzy logic-based tuners.

The same set of fuzzy rules (although with different parameters) is used for tuning all PID gains in all joint subsystems. In order to express the rules mathematically, the normalized universe of discourse $[-1,+1]$ for each of the inputs Δq, $\Delta \dot{q}$ is partitioned into three primary fuzzy sets, labeled as ***negative, zero,*** and ***positive,*** and characterized by triangle-shaped grade of membership functions (see Fig. 2). Normalization of inputs is accomplished by a nonlinear transformation

$$x' = \text{sign}(x) \cdot \min\left(|x/x_{\max}|^{\alpha}, 1\right) \tag{4}$$

where the parameter $\alpha > 0$ is used for tuning the boundary between "zero" and "non-zero" fuzzy error regions. The rules are schematically displayed in the state-action diagram in Table 1, where each entry represents a consequent of a fuzzy implication

$$\textit{Rule } r\textit{: if } \Delta q' \textit{ is } \tilde{A}_r \textit{ and } \Delta \dot{q}' \textit{ is } \tilde{B}_r \textit{ then } \Delta K' \textit{ is } \tilde{C}_r \tag{5}$$

in which $\Delta q'$, $\Delta\dot{q}'$ are normalized joint position/velocity errors, $\Delta K' \in [-1, +1]$ is the normalized relative change in gain, and $\tilde{A}_r, \tilde{B}_r, \tilde{C}_r \in \{negative, zero, positive\}$.

The control rules (5) are evaluated following the recommendations by Ying and Siler [7]. The obtained crisp output $\Delta K' \in [-1, +1]$ is interpreted in the following manner. Given the value of the gain $K(t)$ at time instant t, the new gain value is calculated as

$$K(t + \Delta t) = \min(\max(K(t) \cdot e^{\beta \Delta K'}, K_0), K_{\max}) \tag{6}$$

where K_0 is the initial gain corresponding to small-error conditions, β determine maximum rate of change, and $K_{\max}$ determine the upper limit of the control coefficient. To reduce computational burden, the values of $e^{\beta \Delta K'}$ for proportional/derivative/integral gains are precalculated for equidistant values of $\Delta q'$, $\Delta\dot{q}'$ and stored in a lookup table. When the table has $2^m \cdot 2^m$ entries, $2m$ comparisons are needed to find an entry that corresponds to given errors Δq, $\Delta\dot{q}$. Since (6) is applied to all three types of gain, totally $2m + 6$ comparisons and 3 multiplications are required to implement each joint tuner.

Parameters of the proposed tuners are those appearing in input (4) and output (6) transformations. Input normalization is fixed by adopting maximum expected errors $\Delta q_{i\max}$, $\Delta\dot{q}_{i\max}$, and the exponents α_{q_i}, $\alpha_{\dot{q}_i}$ that determine the actual threshold levels between "small" and "large" errors. Parameters β_{Pi} and $K_{Pi\max}$, appearing in (6), were essentially determined by trial-and-error. Afterward, the corresponding values for derivative and integral control coefficients were determined to keep the stability level and damping ratio approximately constant throughout the error space.

3 Simulation results

In order to verify the effectiveness of the described hybrid approach, a simulation case study has been made with Manutec-R3 six-degree-of-freedom industrial robot [8]. The assumed trajectory was a straight-line motion with end points $[0.3, -0.35, -0.1]^T$ and $[1.0, +0.35, 0.6]^T$. The trajectory was generated by assuming the trapezoidal velocity profile with maximum velocity of 2m/s and the acceleration/deceleration percentage time of 10%.

Fig. 3 illustrates typical time histories of tracking errors obtained with constant-gain PID control (dashed curves) and fuzzy logic-based PID control (solid curves). A sistematic view to the results obtained with different control schemes provides Table 2, which shows that fuzzy PID control yields in all cases reduction of errors. The last row in Table 2 is obtained for the case of constant-gain PID control with feedforward calculated on basis of complete robot dynamics *and* with added global control term with the global control gain set to $K_{Gi} = 0.5$ for $i = 1, 2, \ldots, 6$. It is seen that such a control is still inferior compared to the fuzzy PID control with all effects included in feedforward.

In all trajectory tracking simulation experiments, it was found that the shape of control signals generated by fuzzy PID controllers closely follows the shape of control signals of corresponding constant-gain PID controllers. The differences occur primarily at the control peaks, where the fuzzy PID controllers displayed, by the rule, somewhat smaller peak values. Therefore, the experiments suggest that the PID gain adjustment introduced by the fuzzy logic control (a) does not influence much the shape of power and energy time

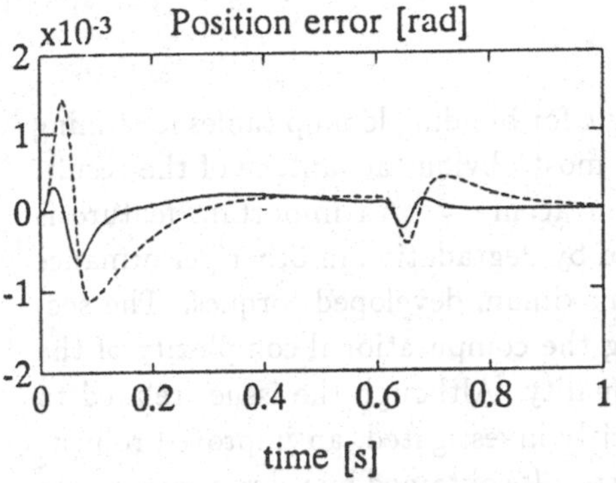

Figure 3: Tracking errors for the first joint

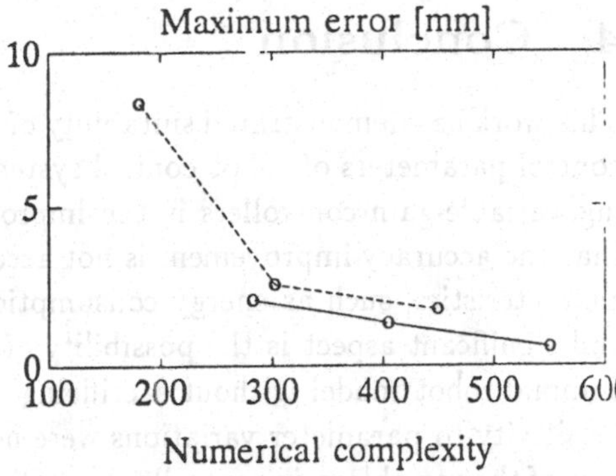

Figure 4: Accuracy/complexity diagrams

Table 2: Maximum tracking errors [mm]

Effects included in feedforward	Basic scheme PID	PID+FLC
Actuators+gravity	8.34	2.04
Same+self-inertia	4.03	1.90
Same+full inertia	2.52	1.31
Same+velocity terms	1.75	0.56
Same+global control	1.51	–

Table 3: Complexity of approximate models

Effects included in dynamic model	Operations sin/cos	*	+/–
Gravity terms	5	76	45
Same+self-inertia	–	–	–
Same+full inertia	5	141	102
Same+veloc. terms	5	219	170

histories and (b) it may reduce the maximum power and total energy demands. For the considered trajectory, the reduction of maximum power is considerable, ranging from 10 percents for the first two joints up to 30 percents for the third joint.

In order to examine the possibility of reducing the computational requirements by mean of the fuzzy logic-based tuners, the number of floating-point operations necessary to implement various approximations of Manutec-R3 dynamic model (see Table 3), were combined with accuracy characteristics from Table 2 to build the accuracy/complexity diagrams, Fig. 4. The "numerical complexity" in these diagrams represents the total normalized time

$$t/T_{+,-} = \Sigma N_{+,-} + R_{*}\Sigma N_{*} + R_{\sin,\cos}\Sigma N_{\sin,\cos}$$

necessary to compute mechanical torques and (optionally) implement PID tuners, where the values $R_{*} = T_{*}/T_{+,-} \approx 1$ and $R_{\sin,\cos} = T_{\sin,\cos}/T_{+,-} \approx 12$ were assumed.

From the accuracy/complexity diagrams, the benefits of the variable-gain PID control become more apparent. For the example trajectory, it is seen that the adoption of the fuzzy PID control and the simple model that incorporates only gravity effects would lead to a slight decrease in tracking accuracy, compared to the constant-gain PID control with complete feedforward. However, such a small degradation in quality is compensated by the reduction of numerical complexity of almost 40 percents.

4 Conclusion

This work has demonstrated suitability of fuzzy logic for building lookup tables for tuning control parameters of robot control systems. The most obvious advantage of the resulting variable-gain controllers is the improvement in accuracy. An important feature is that the accuracy improvement is not accompanied by degradation in other performance characteristics, such as energy consumption and maximum developed torques. The second significant aspect is the possibility of reducing the computational complexity of the nominal robot model without sacrificing control quality. Although the issues related to sensitivity to parameter variations were not explicitly investigated, an improved robustness of the variable-gain controller is implied by the results obtained by using approximate robot models.

From the perspective of the available knowledge of the controlled system, this case study may be regarded as a non-typical application of fuzzy logic-based control. A rigid robot operating in free space is a deterministic system for which a complete mathematical description is *a priori* available. Thus, this research has concentrated on the possibilities of high-quality tracking of fast trajectories using simplified models of robot dynamics. Further considerable improvements may be expected from the proposed hybrid control in adaptive robot control schemes where uncertainties in manipulation objects or in dynamic environment are taken into account.

References

[1] M. Vukobratović, D. Stokić, and N. Kirćanski, *Non-Adaptive and Adaptive Control of Manipulation Robots.* Berlin: Springer, 1985.

[2] N. J. Mandič, E. M. Scharf, and E. H. Mamdani, "Practical application of a heuristic fuzzy rule-based controller to the dynamic control of a robot arm," *IEE Proc. Control Theory and Applications*, vol. 132, pp. 190–203, July 1985.

[3] R. Tanscheit and E. M. Scharf, "Experiments with the use of a rule-based self-organizing controller for robotics applications," *Fuzzy Sets and Systems*, vol. 26, no. 1, pp. 195–214, 1988.

[4] S. Tzafestas and N. P. Papanikolopoulos, "Incremental fuzzy expert PID control," *IEEE Trans. Industrial Electronics*, vol. 37, pp. 365–371, Oct. 1990.

[5] D. Popović and R. S. Shekhawat, "A fuzzy expert tuner for robot controller," in *Proc. 1991 IFAC/IFIP/IMACS Symp. Robot Control*, (Vienna), pp. 229–233, 1991.

[6] C. W. de Silva and A. G. J. MacFarlane, *Knowledge-Based Control with Application to Robots.* Berlin: Springer, 1989.

[7] H. Ying, W. Siler, and J. J. Buckley, "Fuzzy control theory: A nonlinear case," *Automatica*, vol. 26, pp. 513–520, May 1990.

[8] B. Karan, *Application of Fuzzy Logic to Free-Space Robot Motion Control.* Study Report, Mihajlo Pupin Institute, Belgrade, 1993.

APPLICATION OF NEURAL NETS FOR CONTROL OF ROBOTIC MANIPULATORS

T. Uhl and M. Szymkat
University of Mining and Metallurgy, Krakow, Poland

1. Introduction

Fine motion control of robotic manipulators has become a desired goal in the last few years as a result of new robot morphology and the definition of new tasks involving high velocity motions and end effector tracking precision. In order to achieve better performance of robotic manipulators the artificial intelligence can be introduced into the control system. One way for accomplishing this is application of neural nets. Main advantage claimed for neural based controllers is their ability to learn and generalize from partial data and ability to perform parall processing. Many papers present possibilities of neural nets application for control [Hunt,92],[Sontag 93],[Anstaklis,92]. They have discussed the use in adaptive control, predictive control, optimal control, gain sheduling in PID with variable gain. A large area of neural nets application for control is control of nonlinear systems. Neural nets can be applied as a forward or inverse model of manipulator. Such neural net can learn a behaviour of the real system, without knowledge about robot's model structure.

The control of robotic manipulators is realized in many cases of industrial robots by hierarchic control system. There are two most important layers: trajectory planning and trajectory tracking. These two tasks are closely interconnected, for example for more smooth trajectory tracking, small effort of control system is required.

The first layer of hierarchic control system consists of finding the optimal trajectory for realization of given task, defined in Cartesian space. This task can be solved with application of optimization methods. Optimization criteria may be defined mathematically in quadratic form. The problem stated in most cases of robotic manipulators structure is strongly nonlinear, and solution of the problem is a time consuming task. The problem can be solved using local linearization or without linearization using simplified trajectory generation method and direct optimization. In the paper application of neural net for optimization problem solution is presented.

The control system consist of trajectory tracking system for each axis of robotic manipulators. The most commonly used controllers in robotic manipulators realize PD control law in stabilization loop, and inverse dynamics model for torque/force signal generation. The PD controllers are linear controller but robotic manipulators in many cases are nonlinear systems. This is a reason of non satisfatory performance of such controllers for robotic manipulator and justification for application of neural nets for control of this type of systems.

2. Application of neural nets for dynamic control of robotic manipulators

The robot motion control problem consists in obtaining dynamic model of manipulator and using this model to determine the control inputs. The main task of control system is here to track the required trajectory with given accuracy even in the presence of disturbances. This task is often referred to as the dynamic control [Craig,90]. Various methods are employed here, just to mention, conventional PD or PID controllers, linear state feedback, inverse dynamics modeling, resolved acceleration or computed torque approach. The block diagram of the typical conventional control system of the robot motion is shown in Figure 1. It represents an approach that will be referred to as the classical one.

The new trend in control of nonlinear systems tries to exploit neural networks as advanced nonlinear controllers. These attempts are justified by the approximation theorem [Funahashi,89] stating that virtually any continuous nonlinear mapping can be approximated with arbitrary accuracy by a neural net satisfying certain structural assumptions. Some neural network properties, such as their inherent computational parallelism, "fuzzification" of the knowledge represented by particular patterns, relative parameter insensitivity, together with the effective neural net training (weight setting) algorithms made this new tool attractive also for the control of robotic manipulators. We shall concentrate now on the problem of design of a neural-based controller for trajectory tracking of the manipulation robot. The performance of such controller will be compared with the one achieved using the classical approach. The integral quadratic error criterion will be used as the quantitative performance measure. The block diagram of the proposed control system is shown in Figure 2. In this scheme neural net acts both as the approximate inverse dynamics model and the feedback controller.

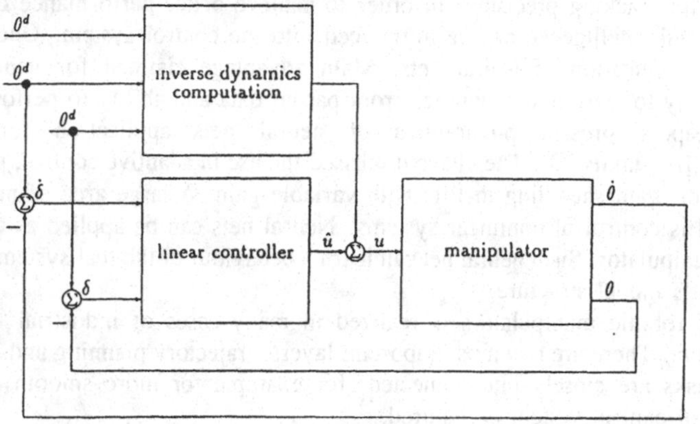

Fig.1. Classical control system

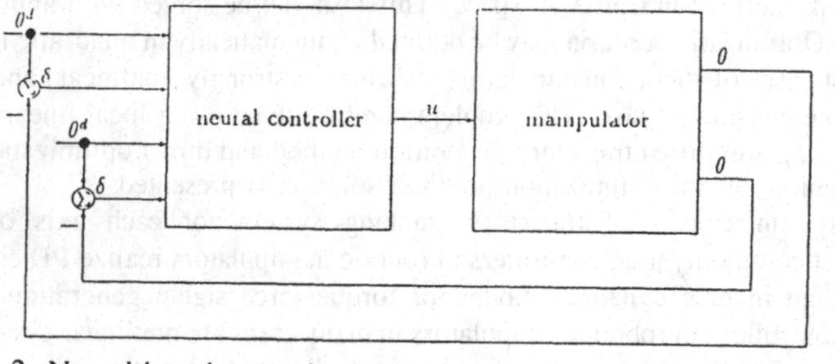

Fig.2. Neural-based control system

Simulation results obtained for the two-joint, two arms, rigid 2R manipulator including actuator models will be discussed in detail. The accuracy of desired trajectory following in the presence of clearance will be analyzed to illustrate the actual robustness of the proposed neural controller. In order to perform the analysis of the classical control system shown in Figure 1, particular desired joint trajectory was chosen. The desired trajectories, together with their derivatives up to the second order, were supplied to the inverse dynamics computation block and needed control signals were computed. The goal of the linear controller in the feedback loop (P,PD,PID) was to assure the disturbance rejection of the overall control system. Since robotic manipulators are strongly nonlinear, the problem of tuning the constants Kp, Kd is not simple. In the literature few methods suitable for the solution of this problem are proposed [Golla,81]. In the reported case the heuristic method has been applied. The representative error trajectory was chosen (representative trajectory means here the trajectory with all possible combinations of mutual movements) and the closed loop system with fixed gains was simulated. Next, the controller gains were changed in order to minimize the objective function.

The use of neural networks for control of motion of robotic manipulators was proposed in the number papers [Kawato,89],[Kawato,90]. Our neural-based control scheme, shown diagrammatically in Figure 2, represents a modified version of the one proposed by Kawato in [Kawato,89]. The major differences are the "incorporation" of the linear controller into the (trained) neural network and the lack of angular acceleration inputs. These modifications were motivated by the fact that we put more emphasis on the learning potential of the neural controller and its ability to improve system performance even when less information is supplied. Another differences pointed out later concern the weight initialization and the training strategy.

In our simulations we have used neural network of the multilayer perceptron type with sigmoid neuron activation function. The network had 8 inputs and 2 outputs, 8 neurons in the input layer, 8 neurons in the only hidden layer and 2 neurons in the output layer. All simulations were carried out on IBM RS/6000 workstation using NC class library described in the previous paper [Szymkat,91]. We have decided to use the method of initialization of networks weights in a way that the network would mimic the linear controller. The method called skeleton match, described in more detail in [Uhl,92], consists in creating of a subnetwork, within the neural controller, that is specialized in producing output control signal "almost linear" with respect to error inputs. All weights related to "non-skeleton" connections are initialized in a way that they compensate each other.

A lot of training methods was proposed in the context of application of neural networks in general control problems, see [Barto,90], some of them relate directly to the control of robot arms,[Kawato,90]. In the case of multi-layer perceptrons there exist two different main approaches to systematic network weight changing in order to achieve the better system performance. The first is based on gradient-type minimization of the network output using an iterative procedure called the error back-propagation. The second, called reinforcement method, originates from stochastic optimization, uses random weight perturbations and a scalar function to assess the network performance.

The results presented in the paper were obtained using the reinforcement procedure. In this method the need for model inversion (for training) has been avoided [Uhl,92]. We have used a variant of the Threshold Accepting algorithm. Few hundreds of iterations were sufficient to achieve a significant improvement. Anyway, the convergence could be even faster if some kind of non-scalar information were utilized. In the reported research we try to combine the minimization of an integral measure with the idea of the back- propagation through time, see also [Szymkat,91] for the exposition of some earlier results. The quality of neural-based control, even in presence of typical disturbances in the structure of the controlled system, is good enough for an industrial application.

The performance of the neural controller shows clearly that it was possible not only to "implant" into neural controller the internal representation of the inverse dynamics model, but also to improve the error rejection functions of the feedback controller, preserving in the same time a certain degree of parametric insensitivity what follows from the example with clearance in the first joint.

The results obtained for classical and neural based controller were compared quantitatively. The particular controlled variables errors are shown in Figures 3 and 4 for the system without and with clearance, respectively. The continuous lines correspond to the performance of the classical control system and the dashed lines to the results obtained using neural-based controller.

Summarizing the presented results we conclude the following:

- neural-based approach leads to the efficient control of motion of robot arms with avoidance of explicit formulation of the inverse dynamics model,
- the information supplied to the neural controller may be incomplete
- neural-based control system is less sensitive to structural changes in the
- dynamics of the controlled manipulator (e.g. the clearance appearing in joint due to the wear of elements)

However, when applying this novel control scheme one has take into account the

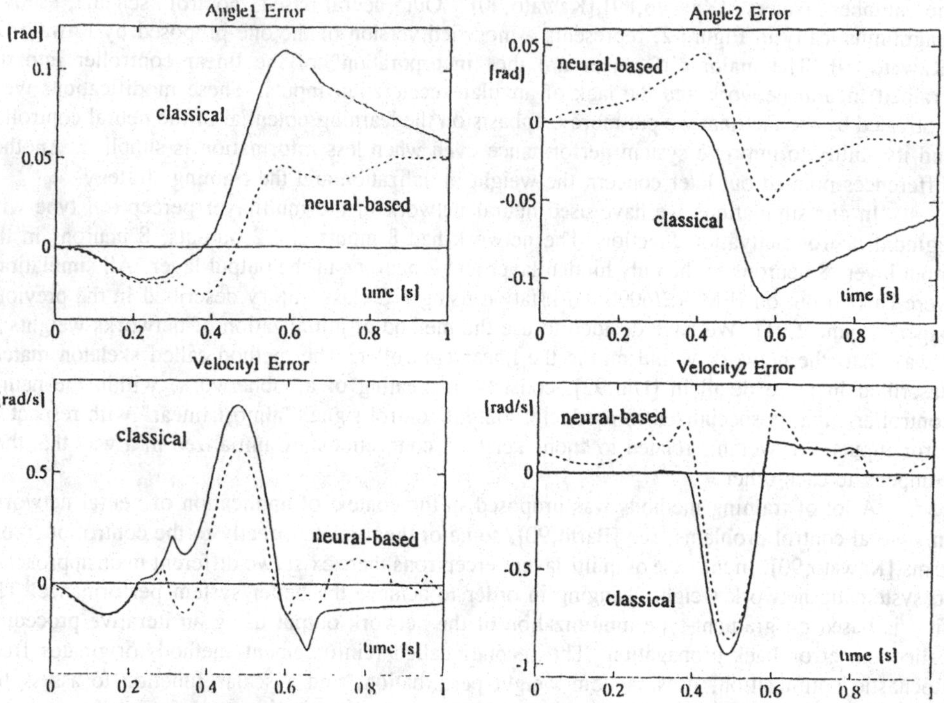

Fig.3. Comparison of control errors for the robotic manipulator

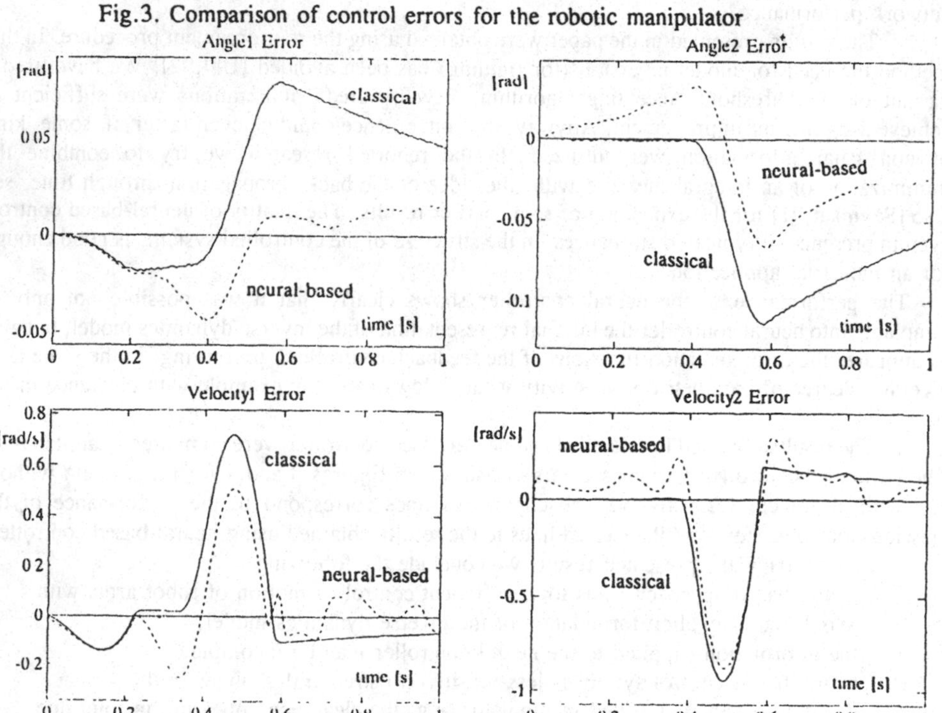

Fig.4. Comparision of control errors for the robotic manipulator with clearance

difficulties related to the implementation of the efficient neural-network training strategy. It is important to point out here that we have used the fixed trajectories, what could be sufficient in repetitive task control, but the generalization for arbitrary trajectories will need other training strategies, including also the trajectory planning level.

3.Application of neural nets for trajectory planning

The quality of the robot action depends strongly on the complexity of required trajectory that should be accomplished by the end-effector of the robot. When the trajectories are very smooth, their realization does not require much control effort and control system structure may be relatively simple. The requirements specified in trajectory planning concern the terminal conditions, path continuity and smoothness. The design objectives include singularity avoidance [Zvi,89], obstacle avoidance [Yoshikawa.85], torque minimization [Kawato,93], and energy consumption minimization [Abetisjan,87]. Various authors applied different optimization criteria including minimal task realization time, minimal driving torque or minimal energy, all of them with and without geometric constraints [Rieswijk,92]. There are also attempts to tackle combined or multi criteria formulation [Malladi,92]. Many papers present research dealing with neural nets application in upper levels of robotic control systems such as task planning, hierarhical control structures where there is still a lot of open problems, see [Kawato,90],[Kawato,93].

In the current paper the general methodology for the robot trajectory planning is presented, following [Szymkat.93]. We decompose it into following stages:

- specification of the desired path requirements (terminal positions, dynamical limitations,etc.) in terms of the manipulation coordinate frame.
- determination of the initial design parameters expressing the actual preferences associated with various kinematic and dynamic characteristics of the trajectories (integral of the sum squared torques, maximal absolute deviation from the nominal path).
- neural network supported generation of more specific requirements imposed on trajectories basing on previously chosen design parameters (the additional requirements will be exemplified later as intermediate point positions),
- generation of the joint trajectories using explicit computational scheme based on quadratic criteria. presented in detail in [Szymkat.94]. It is important to say here that the use of neural net was motivated by the necessity of exclusion of the time consuming calculations involving multi-objective nonlinear optimization from the on-line design loop. The whole burden of memorizing the particular shape of attainable set in decision parameter space has been transformed into neural network training phase.

In [Szymkat,94], an uniform approach to the on-line trajectory planning of robotic manipulators is presented. The general optimization problem formulation with linear constraints and quadratic criteria is used. The basic tool employed for the trajectory definition are under-determined spline functions. The trajectory planning procedure is decomposed into geometric planning in terms of Cartesian coordinates of the end-effector (manipulation frame) resulting in the formation of the so called reference trajectory. and kinematic/dynamic planning of the joint coordinates trajectories. The piecewise linearization of the manipulator's dynamics along the reference trajectory was used in the second stage enabling the application of the same general approach in torque optimization.

The methodology given in the paper is designed to optimize the joint trajectories while the end-effector follows possibly close the desired path and control system effort is small as possible. The desired (nominal) path is assumed to be a given curve connecting the initial and terminal points. The torque command which regulates both the speed and the acceleration of specific joints may be improved by the appropriate choice of intermediate positions (via-points) on the nominal path that there are passed to the spline generation module. That is why we turn our trajectory planning problem into a search problem of optimal position via-points along the nominal path. The structure of the new "inteligent" path planner module is presented on the diagram presented in Figure.5.

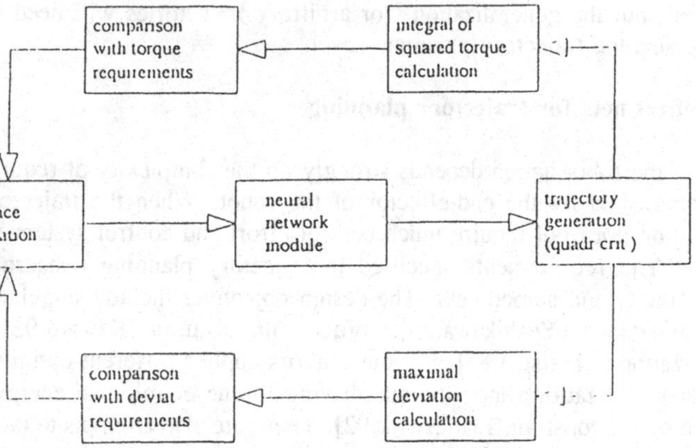

Fig.5. Neural-based trajectory planner

The general formulation of spline trajectories leaves a lot of freedom for chosing the right trajectory for a given task. In the very beginning, the interpolation nodes sequence should be specified. Then the type and order of interpolating functions for particular segments are to be selected. Finally, certain values of the weighting factors in the criterion function should be assumed. The type of required manipulators behaviour determines in practice the particular solution of the trade-off between the geometric accuracy of the end-effector's path and the performance of joint driving systems. This is why we propose a hierarchical procedure for the design of the optimal trajectory. The first phase is performed off-line In the first stage the combined measure of the overall quality of the trajectory is to be constructed. It should involved both certain preference variables as well as certain indicators of the design objectives. For simple example presented in the next section this measure has the particular form:

$\alpha \sum [\textit{integral of squared torques}](\lambda) + (1-\alpha)\max[\textit{absolute deviation from nominal trajectory}](\lambda)$

It is being parameterized with the preference variable α that is used to express the relative importance of the particular criterion. Generally, the preference variable may be a vector, and the selection of specific criteria is not limited. The second variable λ represents the optimization parameters that influence the form of the obtained solution. In the example presented later λ corresponds to relative via-points position on nominal trajectory.

The second stage of the off-line phase is the training of the neural network to memorize the solution of the "inverse optimization problem". The target network should simply give the values of λ that correspond to the optimal value of combined criterion for a given α. It is important to note that however the trajectories are generated using kinematic-based criteria the network is presented during the training phase the values obtained from optimization of the combined measure as the desired outputs. This requires some additional calculations of torques and deviations in each iteration of the training process. In order to economize the computational effort here it is desirable to tabularize these values in preprocessing phase. In this stage an auxiliary network may be used to store the shape of the mapping between via-point positions and corresponding values of criteria or alternatively this mapping is tabularized. These data are used later by standard optimizer to give the pattern values for the final network to be used by the trajectory planner module. The second phase may be performed on-line since there is no direct optimization here. The trajectory planner module manipulates the preference variables only in order to meet the predefined torque and deviation requirements.

We will illustrate the above described trajectory planning methodology using the simple example of three degree of freedom kinematically non-redundant 3R manipulator. The first joint connecting the manipulator's base and its first arm allows rotations of the first arm with respect to rotating horizontal axis and fixed vertical axis. The second joint connects the two arms allowing the rotations of the second arm around the axis parallel to the horizontal axis.

The inverse kinematic task has been solved with assumption $\theta_3>0$. The inverse transformation is needed to obtain the values of intermediate joint posiotions corresponding to via-points of robot's end-effector. Alternatively another neural network may be used for this task in manner similar to the one described in [Simon 93]. The joint trajectories were 5-th order algebraic splines, continuous with their derivatives up to the 3-rd order. The neural network used in the simulation experiment had three layers: the input layer with 4 neurons, the hidden layer with another 4 neurons (with hyperbolic tangent-like sigmoidal transfer functions), and the output layer with the number of linear neurons equal to the number of via-point positions considered in specific case. The number of network inputs dependended on the number of preference variables. The training of the network was performed with the function *trainbpx* from MATLAB's Neural Network Toolbox,implementing the variant of the backpropagation algorithm with adaptive learning rate and momnetum.The results for the two via-points case are shown in Figure 6.

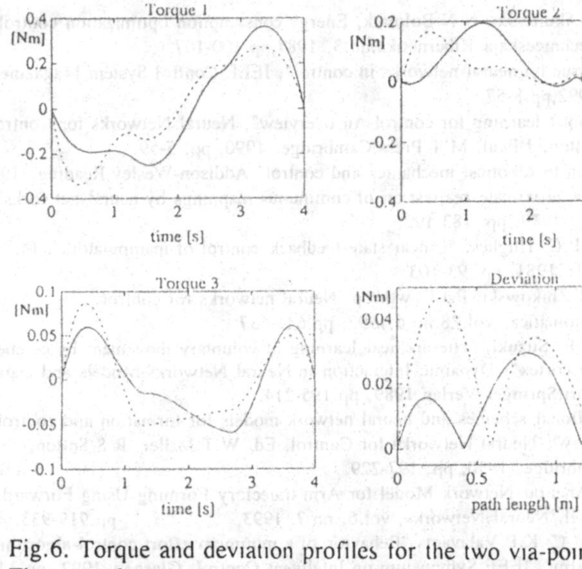

Fig.6. Torque and deviation profiles for the two via-points case

The sum of squared torques for $\lambda_1=0.419$, $\lambda_2=1.167$ was 0.2347 with maximal deviation of 4.63 cm while for $\lambda_1=0.588$, $\lambda_2=1.256$ it reached 0.2934 with maximal deviation of 2.93 cm only, that was less than 2% of the length of the planned trajectory (equal to approx 1.61 m). The obtained results show that increase of the via-points number can cause the significant decrease of the deviation with moderate influence on torque profiles. The neural network "advisor" can be used later as the relative importance of the specific criteria varies with the different tasks the repetitive actions in slightly changing environment, e.g. when terminal positions remain in certain neighbourhoods of some constant locations.

4.Conclusions

The research reported here brings an example of the architecture of the robot's trajectory planning and tracking combining standard analytical algorithm with neural-based approach. Neural network serves as the part that memorizes the difficult nonlinear dependencies , while the less computationally intensive tasks are accomplished with the conventional methods. The heteregenous approaches of this kind seem to be particularly promising for practical applications. The dynamic

properties are given so much attention since they constitute important factors influancing the cost and complexity of the required control and driving system of the robot. Neural-based approach leads to the efficient control of motion of robot arms with avoidance of explicit formulation of the inverse dynamics model.
The information supplied to the neural controller may be incomplete.Neural-based control system is less sensitive to the structural changes in the dynamics of the controlled manipulator. However, when applying this novel control scheme one has take into account the difficulties related to the implementation of the efficient neural-network training strategy.

4.Acknowledgment

This research has been supported by the Commision of the European Communities under the Copernicus project CP93:10119.

5.References

[Abetisjan,87] W.W.Abetisjan, L.D.,Akulienko, N.N.Bolotnik, Energy consumption optimization control of robotic manipulator, techniceskaja Kibernetika,no.3, 1987,pp 100-107.

[Antsaklis,92] P.Ansaklis, "Special issue on neural networks in control", IEEE Control System Magazine, vol.12, no.2, Aprill 1992,pp.8-57.

[Barto,90] A.G.Barto, "Connectionist learning for control-An overview", Neural Networks for Control, Ed.W.T.Miller,R.S.Sutton, J Paul, MIT Press,Cambridge, 1990, pp. 5-59.

[Craig,90] J.J. Craig, "Introduction to robotics, mechanics and control",Addison-Wesley,Reading, 1990.

[Funahashi,89] K.Funahashi, "On the approximate realization of continuous mappings by neural networks", Neural Networks, Vol.2,1989, pp 183-192.

[Golla,81] D.F. Golla,S.C. Garg,P.C Hughes, "Linear state-feedback control of manipulators", Mech. Machine Theory, Vol 16,1981, pp 93-103.

[Hunt,92] K.J.Hunt,D.Sabraro, R Zbikowski, P.J.Gawthrop, Neural networks for control systems - a survey. Automatica, vol.28.no.6,1992, pp.641-657

[Kawato,89] M. Kawato, M. Isobe, R. Suzuki, "Hierarchical learning of voluntary movement by cerebellum and sensory association cortex", Dynamic Interaction in Neural Networks-Models and Data, Ed.by A.Arbib, S.Amori,Springer Verlag,1989, pp.195-214.

[Kawato,90] M. Kawato, "Computational schemes and neural network models for formation and control of multi joint arm trajectory", Neural Networks for Control, Ed. W.T.Miller, R.S.Sutton, J.Paul,MIT Press, Cambridge, 1990, pp. 197-229.

[Kawato,93] M.Kawato, Y.Wada,"Aneural Network Model for Arm trajectory Forming Using Forward and Inverse Dynamics Model, Neural Networks, vol.6, no 7, 1993, pp.919-933.

[Malladi,92] S.R.Malladi, Mulder M C .K.P.Valavanis,"Behavior of a minimum effort control algorithm for a multi-joined robotic arm", IEEE Symposium on Inteligent Control, Glasgow,1992, pp.34-40.

[Rieswijk,92] T.A.Riejswijk, G.G.,Brown,"A robust and efficient approach for the time optimization of path constrained motion of robotic manipulators incorporating actuator torque and jerk constraints, IEEE Inter. Symposium on Inteligent Control, Glasgow, 1992,pp.507-513.

[Simon,93] D.Simon, C.Isik, "The generation and optimization of trigonometric joint trajectories for robotic manipulators",Int. J.of Control vol.42 no.2,1993.

[Sontag,93] E.D.Sontag,"Neural Networks for Control"in Essays on Control: Perspectives in Theory and it's Applications,Ed.Trentelman H,L.,Birkhauser, Boston, 1993.

[Szymkat,92] M.Szymkat, M.Brdyś, J.Sacha, "NC an object oriented programming tool for neural control design, Neuro-Nimes'91, pp. 741-744.

[Szymkat,93] M. Szymkat, T.Uhl, "Neural network approach to planning and tracking of robots trajectories",IEEE Inteligent Control Conference, Chicago, 1992, pp.244-268

[Szymkat,94] M. Szymkat, T. Uhl, "On-line robot arms trajectory planning using quadratic criteria" submitted to Int. J. of Mechatronics.

[Uhl,92] T.Uhl,Szymkat M.,"A comparision of the classical and neural-based approach to control of Manipulation robots,XVII IMSA, Leuven, September, 1992, pp.255-284.

[Zvi,89] S.Zvi, D.Steven, "Robot path planning with obstacles, actuator, gripper and payload, International Journal of Robotic Research, vol 8,no 6, 1989,pp.3-18.

A STRUCTURE AND PRINCIPLE OF OPERATION OF THE ADAPTIVE CONTROL SYSTEM OF THE UNDERWATER VEHICLE MOTION

Z. Kitowski and J. Garus
Naval Academy, Gdynia, Poland

Abstract

There are two characteristic states of an underwater vehicle. The first one is a state of manoeuvring during searching and approaching the target and the second one is a state of keeping at constant distance above or to the target. In the paper some aspects of an optimal control of the vehicle are considered. The main assumption is that the non-linear mathematical model describing its motion can be replaced by the linear model. The identification of parameters of the linear model enables calculation of quasi optimal control signals, that taking into account the influence of external perturbations, drive the vehicle along the given trajectory. The structure description, principle of operation of the adaptive control system and results of computer simulation are presented.

1. Introduction

Underwater vehicles (or robots) differ themselves in constructions, shapes, overall dimensions, kinds of drives and control systems. Taking into account all these differences it can be noticed that character of their motion has got many common properties. One of them is that a mathematical description of control process of motion of one type of the underwater

vehicle can be used (with changes of little importance) to motion analysis of the another type [4].

For computer analysis of the vehicle motion, the mathematical description of the process control of the autonomous vehicle that can freely move in horizontal and vertical plane, may be used as a basic model. The full model of the three-dimensional motion of the solid body in water environment is required only when a displacement in any direction and violent changes of a trajectory are considered. In the case when stationary operating conditions are considered, such as motion and stabilisation of the vehicle on demanded depth or direction, draught or emerge, etc. a simplified model can be applied.

2. Dynamics and mathematical description of underwater vehicles

To analysis of the underwater vehicle dynamics it is necessary to work out the mathematical model of its motion. There are some difficulties in the description of the underwater vehicle as a control object. The first one is the description of forces operating on it regarding to characteristics of the power transmission system and control system. These systems characteristics depend on a speed vector of the water jet and changes of the vehicle resistance. It is often assumed that forces generated by the power transmission system and control system do not depend on kinematics of the vehicle's motion. The second one is evaluation of coefficients of state equations (both by theoretical and experimental methods) because of variations in hulls' form. It is a very complicated process using in theoretical research experimental data for this task. A lack of analytical dependencies cause that it is necessary to make use of approximation methods.

For these reasons a purposeful is using of an adaptive control system for steering of the underwater vehicle. This system has to contain an identification system to determination of parameters of the mathematical model having a definite structure. This kind of solution is applied below and the structure of equations is defined basing on experimental research described in works [3,6], whereas the identification system evaluates in real time, changing in a stochastic way, coefficients of these equations.

Dynamics equations of the three-dimensional motion of the underwater vehicle can be formulated as follows [5,6]:

$$\left.\begin{aligned}
& m_x\frac{d\upsilon_x}{dt}+m_z\omega_y\upsilon_z-m_y\omega_z\upsilon_y+\lambda_{35}\omega_y^2-\lambda_{26}\omega_z^2=F_x\\
& m_y\frac{d\upsilon_y}{dt}+\lambda_{26}\frac{d\omega_z}{dt}+m_x\omega_z\upsilon_x-m_z\omega_x\upsilon_z-\lambda_{35}\omega_x\omega_y=F_y\\
& m_z\frac{d\upsilon_z}{dt}+\lambda_{35}\frac{d\omega_y}{dt}+m_y\omega_x\upsilon_y-m_x\omega_y\upsilon_x+\lambda_{26}\omega_x\omega_z=F_z\\
& I_x\frac{d\omega_x}{dt}+(\lambda_{26}+\lambda_{35})(\omega_y\upsilon_y-\omega_z\upsilon_z)=M_x\\
& I_y\frac{d\omega_y}{dt}+\lambda_{35}\frac{d\upsilon_z}{dt}+\omega_x\omega_y(I_x-I_z)+\upsilon_x\upsilon_z(m_x-m_z)-\lambda_{26}\omega_x\upsilon_y-\lambda_{35}\omega_y\omega_x=M_y\\
& I_z\frac{d\omega_z}{dt}+\lambda_{26}\frac{d\upsilon_y}{dt}+\omega_x\omega_y(I_y-I_x)+\upsilon_x\upsilon_y(m_y-m_x)+\lambda_{35}\omega_x\upsilon_z+\lambda_{26}\omega_z\upsilon_x=M_z
\end{aligned}\right\}\quad (1)$$

where:

$\upsilon_x, \upsilon_y, \upsilon_z$ - **linear velocities in the co-ordinate system connected with the underwater vehicle,**

$\omega_x, \omega_y, \omega_z$ - **angular velocities,**

m - **vehicle's weight**

$m_x = (1+k_{11})\,m,\ m_y = (1+k_{22})\,m,\ m_z = (1+k_{33})\,m,$

k_{ij} - **nondimensional coeficients being a function of principal dimensions,**

I_x, I_y, I_z - **moments of inertia of the vehicle,**

F_x, F_y, F_z - **components of forces reacting on the vehicle,**

M_x, M_y, M_z - **components of moments reacting on the vehicle.**

Right side of the equation (1) is determined by hydrodynamic forces and moments reacting on the vehicle and values of its weight, displacement and thrust of propellers.

3. Structure and principle of operation of adaptive control system

A control system of the underwater vehicle motion realises the following tasks:

- **take measurements and filtering of results of measurements of state variables describing dynamics and kinematics of its motion,**
- **identify of the vehicle parameters i.e. determine values of parameters characterising its dynamic features,**

- calculate optimal values of command signals for each of propellers installed on its board.

The structure of adaptive control system of the underwater vehicle is presented in Figure 1.

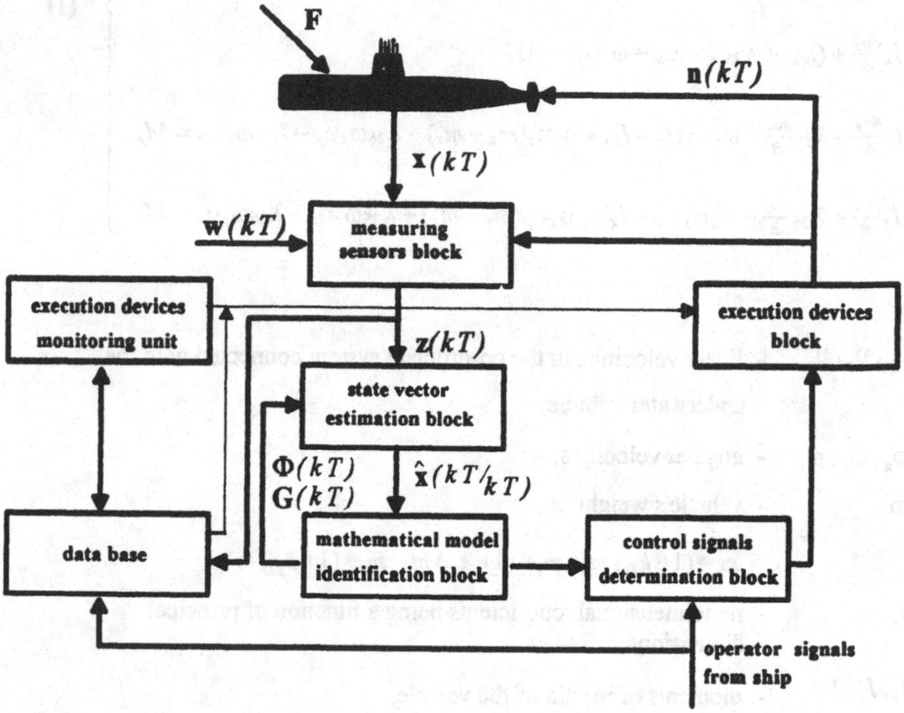

Fig. 1. Block diagram of adaptive control system of the underwater vehicle motion
F - external perturbations vector, **x** - state vector, **w** - measuring noises vector,
z - output vector, $\hat{\mathbf{x}}$ - state vector estimate, **n** - command signals vector.

A principle of operation of this system is based on using for the goal of control the discrete linear mathematical model forming in consequence of the approximation of equations system (1):

$$\mathbf{x}[(k+1)T] = \mathbf{\Phi}(T)\mathbf{x}(kT) + \mathbf{G}(T)\mathbf{n}(kT) \tag{2}$$

where:

$\mathbf{x}(kT) = [x_0(kT), x_1(kT), ..., x_n(kT)]^T$ – state vector $(n \times n)$,

$\mathbf{u}(kT) = [u_0(kT), u_1(kT), ..., u_r(kT)]^T$ – control vector $(r \times l)$,

$\Phi(T)$ - process matrix $(n \times n)$,

$\mathbf{G}(T)$ - control matrix $(n \times r)$,

l - quantity of control signals.

External forces reacting on the vehicle, noise of meter circuits, changes of its hydrodynamic characteristics, etc. influence directly on measured values of state vector and control vector. Hence, as it is mentioned above, a synthesis of the control system requires taking the following problems into account:

- estimation of the state vector,
- identification of the vehicle dynamic parameters i.e. coefficients of motion equations,
- evaluation of quasi optimal command signals.

A task of the estimation of the state vector is based on an assumption that measured output vector $\mathbf{z}(kT)$ and state vector $\mathbf{x}(kT)$ are connected by means of the linear function:

$$\mathbf{z}(kT) = \mathbf{M}\mathbf{x}(kT) + \mathbf{w}(kT) \tag{3}$$

where:

M - measuring matrix transforming the state vector $\mathbf{x}(kT)$ into the output vector $\mathbf{z}(kT)$ in case of lack of noise,

$\mathbf{w}(kT)$ - white noise having average zero.

Availing of the procedure describing in [7], which is based on properties of orthogonal projection, the problem of evaluation of the estimate of the state vector that minimise the meansquare error can be formulated as below:

$$\bar{\varepsilon}^2 = \mathrm{E}\left\{[\mathbf{x}(kT) - \hat{\mathbf{x}}(kT/jT)]^T[\mathbf{x}(kT) - \hat{\mathbf{x}}(kT/jT)]\right\} \tag{4}$$

where:

$\hat{\mathbf{x}}(kT/jT)$ - estimate of the state vector $\mathbf{x}(kT)$ calculated using measured values of the output vector $\mathbf{z}(jT)$ for $0 \le j \le k$.

The Walsh functions technique is used to determine the most probable values of coefficients of matrices $\Phi(T)$ and $G(T)$ by means of calculated estimates of vectors $x(kT)$ and $n(kT)$. Current values of the unknown coefficients of matrices $\Phi(T)$ and $G(T)$ can be easily computed by using of formula [1,2]:

$$\Phi G = R(BQ)^T\left[BQ(BQ)^T\right]^{-1} \qquad (5)$$

where:

$$\Phi G = [\Phi(T), G(T)],$$

$$BQ = \begin{bmatrix} B \\ Q \end{bmatrix},$$

$$B = FSU^T,$$

$$Q = HSU^T,$$

$$R = TU^T,$$

$$U^T = [u(1), u(2), ..., u(n+r)],$$

$u(k)$ - the kth vector of discrete values of Walsh functions,

F, H - matrices of coefficients of orthogonal expansion of state vector and control vector respectively,

S - matrix for summation operation of discrete values of Walsh functions,

Z - matrix for shift operation of discrete values of Walsh functions.

The next step in optimisation of adaptive control system is synthesis of the controller driving a vehicle along the given trajectory. There is described the algorithm of solving this problem in the work [5]. In this algorithm equation (2) is repeatedly used leading to the following expression:

$$M(T)n^T(iT) = x(k) - \Phi(nT)x(0) \qquad i = \overline{0, n-1} \qquad (6)$$

where:

$x(0)$ - vector of initial state for the succeeding control step ($x(0) = x(iT)$ for i=0,1,2,...),

$x(k)$ - assignment vector of final state,

$M(T) = [\Phi(n\text{-}1)TG(T);\Phi(n\text{-}2)TG(T);\Phi(n\text{-}3)TG(T);...;\Phi(T)G(T);G(T)]$.

A solution of equation (6) for each control step gives *l* values of command signals. Thesesignals are sent to propellers located on the vehicle and in this way it is kept at given trajectory.

There is also a decisive system in the adaptive control system which task is checked on currently a correctness of the controller acting. The decisive system acts basing on the set of information referring to measured data, former experiences, etc. and it is assigned to a quick reacting in the case of appearance of extreme operating conditions of the control system.

4. Computer simulation of underwater vehicle motion

The analysis of motion in vertical and horizontal plane investigated for three types of underwater vehicles. Table 1 shows their constructional parameters.

No	Name	Symbol	Unit of measure	Values of parameters		
				I	II	III
1.	hull length	L	m^2	7.20	5.40	3.60
2.	diameter	d	m	0.533	0.533	0.533
3.	volume of displacement	v	m^3	1.465	1.065	0.665
4.	mass	m	kg	1490	1083	676
5.	moment of inertia towards Z-axis	I_z	kgm^2	4573	2129	729
6.	nondimensional coefficient of virtual mass along X-axis	k_{11}	-	0.025	0.03	0.05
7.	nondimensional coefficient of virtual mass along Y-axis	k_{22}	-	0.975	0.97	0.92
8.	nondimensional coefficient of virtual moment of inertia towards Z-axis	k_{66}	-	0.91	0.88	0.80
9.	stabilisers area	S_{st}	m^2	0.12	0.12	0.12
10.	diving planes area	S_{sg}	m^2	0.05	0.05	0.05
11.	rudders area	S_{sk}	m^2	0.05	0.05	0.05

Table 1. Constructional parameters of underwater vehicles

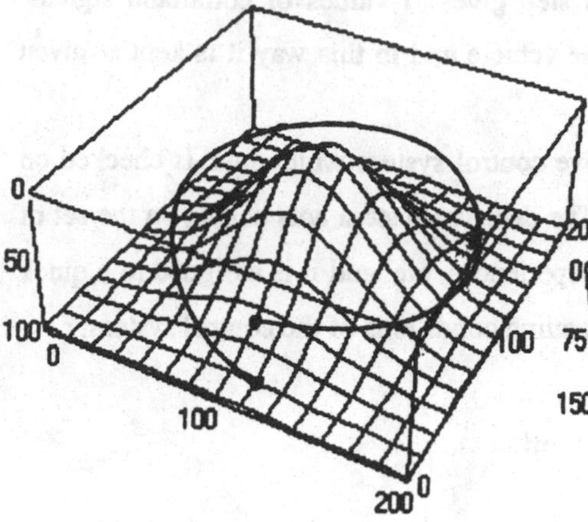

Fig. 2. Example of vehicle's motion in vertical plane

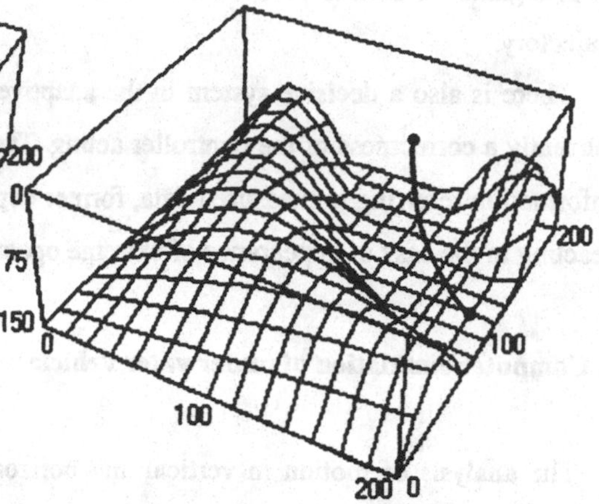

Fig. 3. Example of vehicle's motion in horizontal plane

Some experiments were accomplished. It was assumed that disturbing forces acting on the vehicle were a residual floating **A** and moment of a static unbalance $\mathbf{M_A}$. The examples of motion trajectories of the underwater vehicles are presented in Fig. 2 and Fig. 3

5. Conclusion

The computer simulation of underwater vehicle motion carried out for variety of operating conditions confirms correctness of worked out algorithms of identification of its dynamic parameters, digital processing of measured signals and determination of optimal control signals driving it along desired trajectory independently of motion type and realised tasks. Farther design works aim in direction of unification and optimisation of control algorithms irrespective of type and quantity of propellers installed on the vehicle's board and used steering gears.

References

[1] Ahmed N., Rao K.R.: Orthogonal transform for digital signal processing. Springer-Verlag, Berlin, 1975

[2] Ing-Rong Horng, Shinn-Jang Ho: Synthesis of digital control systems using discrete Walsh polynomials. Int. J. of Systems Science, 1986, vol. 17, No 10, pp.1399-1408

[3] Jastrebow W.S., Armiszew S.W.: Algoritmy adaptiwnowo dwiżenija podwodnowo robota. Nauka, Moscow, 1988 (in Russian)

[4] Kitowski Z., Morecki A., Ostachowicz W.: Underwater Robotics in Poland. Proceedings of the 24th International Symposium on Industrial Robots, Tokyo, 1993

[5] Kitowski Z.: Synteza układu utrzymania okrętu na zadanej trajektorii z dużą dokładnością. Zeszyty Naukowe WSMW, No 87A, Gdynia, 1985 (in Polish)

[6] Łukomskij J.A., Czugunow W.S.: Sistemy uprawlenija morskimi podwiżnymi obiektami. Sudostrojenije, Sankt Petersburg, 1988 (in Russian)

[7] Tou T.J.: Optimum design of digital control systems. Academic Press, New York, 1963

PARALLEL COMPUTATION AND CONTROL METHOD FOR DYNAMIC MOTION CONTROL OF ROBOT BY UTILIZING DIGITAL SIGNAL PROCESSOR (DSP)

Y. Sankai

University of Tsukuba, Tsukuba, Japan

ABSTRACT

Our aim is to have a high speed and effective control strategy for nonlinear systems such as an inverted pendulum or a robot link system or a legged locomotion robot. The control method by using inverse dynamics compensation is used for controlling of nonlinear systems. But the increase of degrees of freedom would be too much time consuming for calculations of the inverse dynamics. In this paper, we proposed a parallel computing method and control strategies for dynamic systems ,especially, nonlinear systems. We tried to develop a DSP based parallel computation and control method and a small DSP unit for installation. The DSP parallel computation enables to realize a high speed calculation of the inverse / forward dynamics. Every module has same calculation routine and can adopt parallel LQ controller proposed in this research. The size of a small DSP unit constructed in our laboratory is (5cm x 6cm).

1. INTRODUCTION

Nonlinear systems such as inverted pendulums or link mechanisms have a nonlinearity relating to displacements and velocities of the pendulums or links. The linear quadratic [LQ] control method has been sometimes used in the linear or linearized system, but it is not capable of fully controlling the nonlinear system. The nonlinearity of controlled systems should not be ignored. We can adopt a control method using the inverse dynamics compensation. But the increase of degrees of freedom would be too much time consuming for calculations of the inverse dynamics.

We proposed and developed a DSP based parallel computation and a small DSP unit (5cm x 6cm). We also propose a parallel control method. Each link has a LQ controller influenced by only both sides of the link. We aim in our research to have a high speed and effective control strategy for the nonlinear system such as an inverted pendulum or a robot link system or a legged locomotion robot. In this paper, the following methods are referred in this order:

1) Parallel Computation and Control 2) DSP system

2. METHODS and RESULTS

2.1 Parallel Computation and a parallel Control

Generally, Lagrange method and Newton-Euler method are used to calculate motion equations. Luh's algorithm[1] is one of famous parallel computation methods. These methods need to decompose the calculation of dynamics into sub-tasks and some schedulings before actual parallel computation.

In this research, the parallel computation is performed every link unit. The calculation process and routine of each link unit are uniform and we can treat the computation unit as a common module. The dynamics and control operations are distributed into each link object. These objects correspond to operation units such as DSP respectively. The link system and the link object constructed by data and method are shown in Fig.1. This link system can be regarded as object oriented system.

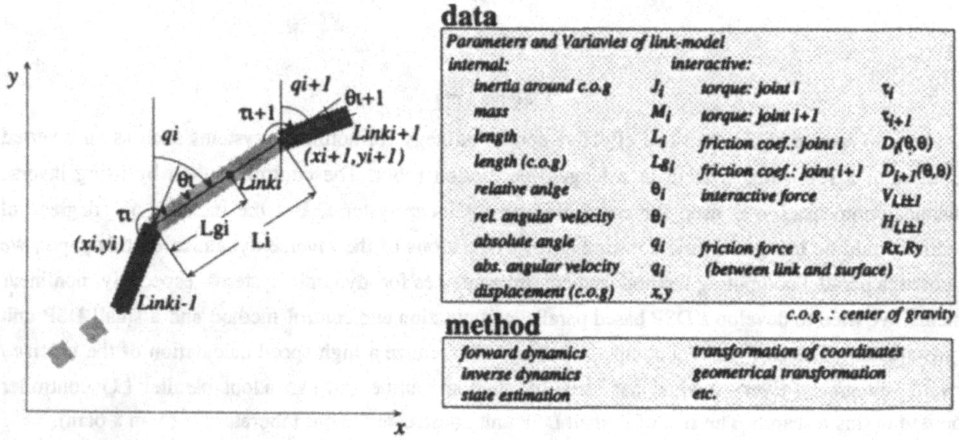

Fig.1 Schema of link system

The interactive forces among these links can be separated into the internal force and the external force. And then, the motion equations of this system are represented as follows,

$$M_i\ddot{x}_i = H_{i-1} - H_{i+1}\ , \qquad M_i\ddot{y}_i + M_i g = V_{i-1} - V_{i+1} \tag{1}$$

$$J_i\ddot{q}_i = Lg_i[V_{i-1}\sin q_i - H_{i-1}\cos q_i] + (L_i - Lg_i)[V_{i+1}\sin q_i - H_{i+1}\cos q_i] - D_i(\dot{q}_i - \dot{q}_{i-1}) - D_{i+1}(\dot{q}_i - \dot{q}_{i+1}) + \tau_i - \tau_{i+1} \tag{2}$$

By rearranging as follows, the equations of parallel computing module are represented as simple forms.

$$\begin{pmatrix} x_{i+1} \\ y_{i+1} \end{pmatrix} = \begin{pmatrix} x_i \\ y_i \end{pmatrix} + L_i \begin{pmatrix} \sin q_i \\ \cos q_i \end{pmatrix} \tag{3}$$

$$\begin{pmatrix}\ddot{x}_1\\ \ddot{y}_1\end{pmatrix} = \begin{pmatrix}\ddot{x}_0\\ \ddot{y}_0\end{pmatrix} + L_{g1}J_1 \tag{4}$$

$$\begin{pmatrix}\ddot{x}_j\\ \ddot{y}_j\end{pmatrix} = \begin{pmatrix}\ddot{x}_{j-1}\\ \ddot{y}_{j-1}\end{pmatrix} + (L_{j-1} - L_{gj-1})J_{j-1} + L_{gj}J_j \quad (2 \le j \le n) \tag{5}$$

,where $J_j = \begin{pmatrix} \ddot{q}_j cosq_j - \dot{q}^2 sinq_j \\ -\ddot{q}_j sinq_j - \dot{q}^2 cosq_j \end{pmatrix}$.

The interactive forces at both ends of a link are represented as follows,

$$H_{i,i-1} = -\sum_{j=1}^{i-1}(M_j\ddot{x}_j) \quad , \qquad H_{i,i+1} = \sum_{j=i+1}^{n}(M_j\ddot{x}_j) \tag{6}$$

$$V_{i,i-1} = -\sum_{j=1}^{i-1}(M_j\ddot{y}_j + M_j g) \quad , \qquad V_{i,i+1} = \sum_{j=i+1}^{n}(M_j\ddot{y}_j + M_j g) \tag{7}$$

Therefore, Eq.(2) can be calculated by Eq.(4),(5),(6),(7) in turn. The inverse dynamics compensation is given by Eq(8).

$$\tau_i = \tau_{i+1} + J_i u_i + D_i(\dot{q}_i - \dot{q}_{i-1}) + D_{i+1}(\dot{q}_i - \dot{q}_{i+1}) - T_i \tag{8}$$

,where $T_i = L_{gi}(V_{i-1}\sin q_i - H_{i-1}\cos q_i) + (L_i - L_{gi})(V_{i+1}\sin q_i - H_{i+1}\cos\theta_i)$.

The number of calculation steps in the inverted dynamics calculation is as follows,

	add	multiply
calc. of acceleration by Eq(5)	7n	9n-5
calc. of trque by Eq(8)	3n+6	3n+5
Total	10n+6	12n

Table 1. The number of calculation for n-link system

Attention should be paid to the calculation of the forward dynamics. A small error will be accumulated by use of the implicit solving method which uses one digit delay data. In this case, following a duplicating calculation is effective to calculate more precisely. In actual use, it is not necessary to use.

$$\begin{pmatrix} J_{i-1} & 0 & 0 \\ 0 & J_i & 0 \\ 0 & 0 & J_{i+1}\end{pmatrix}\begin{pmatrix}\ddot{q}_{i-1}\\ \ddot{q}_i \\ \ddot{q}_{i+1}\end{pmatrix} = \begin{pmatrix} T_{i-1} - D_{i-1}(\dot{q}_{i-1} - \dot{q}_{i-2}) - D_i(\dot{q}_{i-1} - \dot{q}_i) + \tau_{i-1} - \tau_i \\ T_i - D_i(\dot{q}_i - \dot{q}_{i-1}) - D_{i+1}(\dot{q}_i - \dot{q}_{i+1}) + \tau_i - \tau_{i+1} \\ T_{i+1} - D_{i+1}(\dot{q}_{i+1} - \dot{q}_i) - D_{i+2}(\dot{q}_{i+1} - \dot{q}_{i+2}) + \tau_{i+1} - \tau_{i+2}\end{pmatrix} \tag{9}$$

We can find other parallel computation method such as Luh's algorithm, which need to decompose the calculation of dynamics into sub-tasks and some schedulings before actual parallel computation. The merits of the parallel computation method in this research are summarized as follows,

1) Content and volume of calculation are same in every link module.

2) Calculation of inverse dynamics / forward dynamics can be calculated easily by using state variables.

3) Parameters / variables for control are represented explicitly.

4) Actual installation into the DSP(Digital Signal Processor) system can be realized easily.

A simulation of the forward dynamics calculation for 4-link system is shown in Fig.2. This system is 4-link pendulum influenced by only gravity.

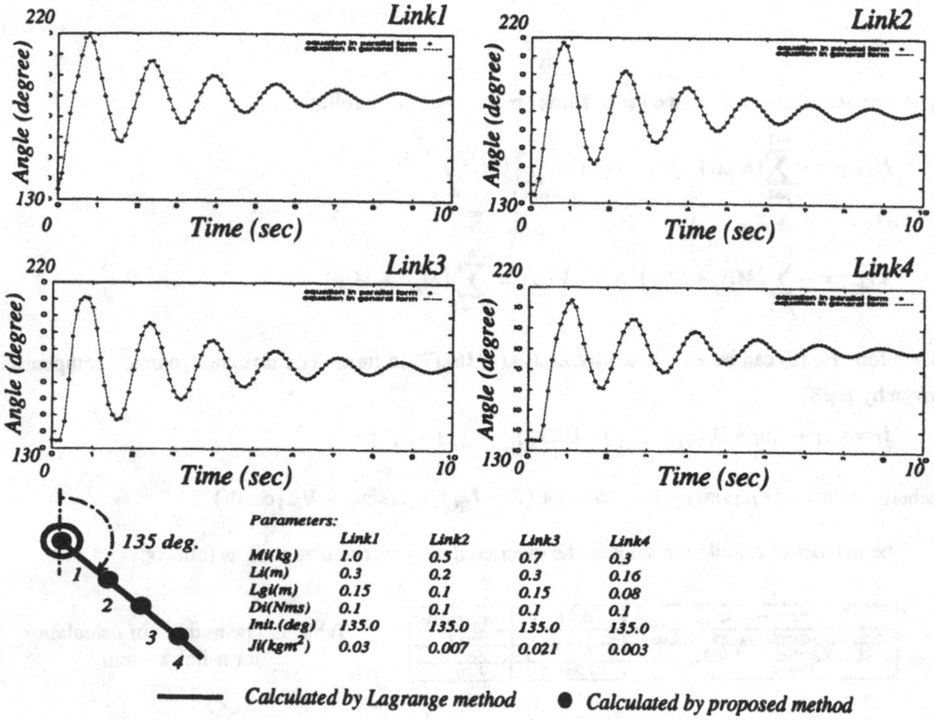

Fig.2 Simulation of 4-link pendulum system

2.2 A parallel control method

To control a complex and large scale system has some difficulties for the total control design. It would be difficult to design or decide many parameters of complex systems. We propose a distributed parallel control method to realize a simple control design and a cooperative control. This method is a localized design method which considers both neighborhood links. The nonlinear system such as Eq(2) is linearized as Eq(8) by the compensation of inverse dynamics. Considering the state feedback including information of neighborhood links, a control input vector and a state feedback gain matrix are obtained as follows,

$$\textbf{State equation:}\quad \frac{d}{dt}\begin{pmatrix} q_i \\ \dot{q}_i \end{pmatrix} = \begin{pmatrix} 0 & \mathrm{I} \\ 0 & 0 \end{pmatrix}\begin{pmatrix} q_i \\ \dot{q}_i \end{pmatrix} + \begin{pmatrix} 0 \\ \mathrm{I} \end{pmatrix}\mathbf{u}_i \qquad \mathbf{u}_i = \begin{pmatrix} u_{i-1} \\ u_i \\ u_{i+1} \end{pmatrix} \quad \text{:state variables: (6 x 1)} \tag{10}$$

$$\textbf{Control input::}\quad u_i = -\mathbf{G}_i\begin{pmatrix} q_i \\ \dot{q}_i \end{pmatrix} \tag{11}$$

$$\textbf{LQ gain:}\quad \mathbf{G}_i = \begin{pmatrix} g_{11} & g_{12} & g_{13} & g_{14} & g_{15} & g_{16} \\ g_{21} & g_{22} & g_{23} & g_{24} & g_{25} & g_{26} \\ g_{31} & g_{32} & g_{33} & g_{34} & g_{35} & g_{36} \end{pmatrix} \tag{12}$$

In this case, a feedback gain is calculated by LQ control theory from Eq.(10). And Gains relating to $\mathbf{u}_i$ surrounded by a quadrilateral should be selected as actual $\mathbf{G}_i$. The both ends of n-link system are treated by only single side information. An example of simulation of this control method is shown in Fig.3. The control performance is similar to the traditional control.

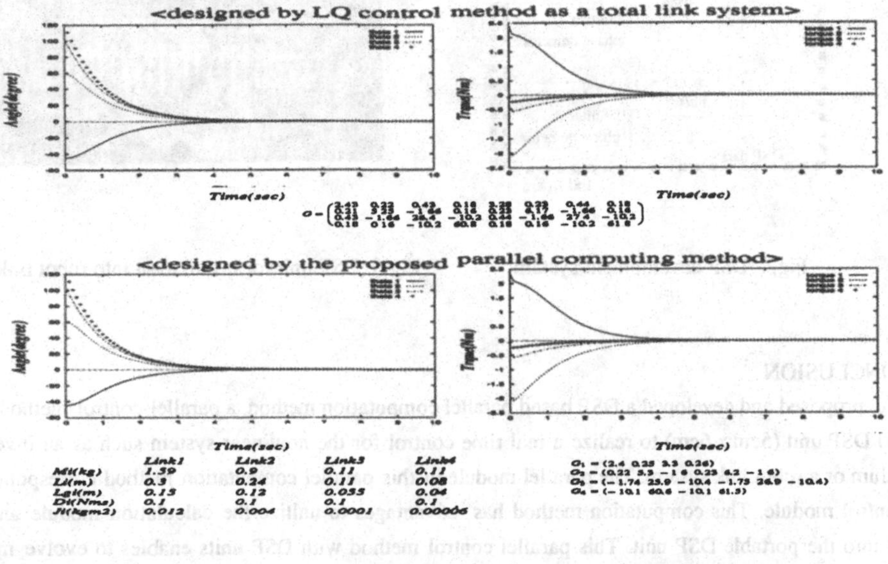

Fig.3 Comparison of traditional LQ control and proposed parallel LQ control

2.3 DSP system

Specifications of the parallel computation and control for the robot link system are 1)high speed and real time operation ,2)easy installation of proposed computation method mentioned above ,3)compact and simple architecture, 4)low cost. The remarkable feature of DSP(Digital Signal Processor) is superior treatments of array data and high speed operations of matrices. DSP 56002(Motorola) is selected as an actual DSP unit according to these specifications. [Specifications of DSP 56002: 1)24 bit digital data paths., 2)20 MIPS (40MHz), 3)simple structure, 4)low cost, etc.]

A schema of the DSP development system is shown in Fig.4. This DSP board is put on a PC bus

(ISA-bus) and it provides hardware and software test environments. Further more, we constructed small DSP units for mounting on a link. This DSP unit is shown in Fig.5. One chip IC including a counter circuit and a phase detector can be put on the DSP unit. Communication is performed by a serial port of DSP.

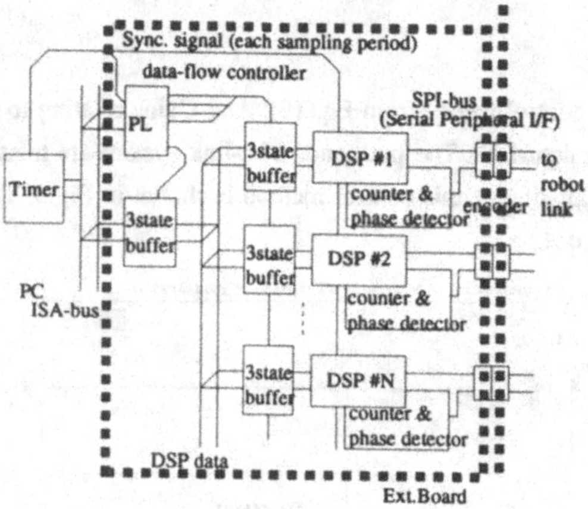

Fig.4 DSP development system

Fig.5 DSP unit for installation into robot links

3. CONCLUSION

We proposed and developed a DSP based parallel computation method, a parallel control method and a small DSP unit (5cm x 6cm) to realize a real time control for the nonlinear system such as an inverted pendulum or a robot link system. The parallel module in this parallel computation method corresponds to the control module. This computation method has advantages to unitize the calculation module and to install into the portable DSP unit. This parallel control method with DSP units enables to evolve multi-agent systems.

ACKNOWLEDGEMENT

Acknowledgements are due to members in our laboratory for their technical assistance.

REFERENCES

[1] J.Y.S.Luh, C.S.Lin, "Scheduling of Parallel Computation for a Computer-Controlled Mechanical Manipulator", IEEE Trans. SMC-12,2,pp214-234, 1982.
[2] "DSP56000 Digital Signal Processor Family Manual",.Motrola Inc., 1992.

Part VI
Synthesis and Design I

ANALYSIS AND DESIGN OF POSITION/ORIENTATION DECOUPLED PARALLEL MANIPULATORS

S.P. Patarinski† and M. Uchiyama
Tohoku University, Sendai, Japan

Abstract—Parallel manipulators (PM) are known for their inherent high mechanical stability, accuracy, load capacity, and stiffness, but their kinematics, statics and dynamics are quite involved and rich of singularities. In the paper a new kinematic structure of PM, the decoupled parallel manipulator (DPM) is proposed, decoupling the end-effector's position and orientation. Geometric and instantaneous kinematics as well as statics of DPM are studied, and a number of mechanical designs based thereof are derived. The new structure substantially simplifies PM modeling and real-time control, and allows for achieving very fast and accurate motions.

I. INTRODUCTION

Parallel manipulators (PM) are basically defined by their mechanical structure, comprising two links, *a base* and *an end-effector*, connected through a number of open loop linkages (*legs*) arranged in parallel. As a result, PM are spatial, huge closed loop mechanisms with both active, single degree of freedom (d.o.f.), directly actuated joints, and passive, multiple-d.o.f. ones. They possess an inherent mechanical stability and provide better positioning accuracy and higher stiffness relative to the more conventional serial manipulators (SM). Reviews on PM have been reported recently by Patarinski [9].

There are many specifities in the design of PM due to the existing huge choice among possible kinematic structures and types of active/passive joints, as well as to the personal preference and experience of the designers. It is noteworthy that no design is known yet, for which end-effector's position and orientation are decoupled. Although a number of studies [1], [3], [13], [15], [16], [17] have been reported during the last decade on *the serial/parallel duality*, the question *"What is the dual of the common SM with three wrists axes intersecting at one and the same point?"* has not found any answer yet.

In this paper a new kinematic structure, *the decoupled parallel manipulator* (DPM), with six legs is proposed, providing position/orientation decoupling of the end-effector. Geometric and instantaneous kinematics, as well as statics, both inverse and forward, are evaluated explicitly, and their singularities are analysed. Other related structures are also derived, and possible mechanical designs of DPM are discussed. The new structure substantially simplifies PM modeling and real-time control and allows for achieving very fast and accurate motions.

II. NOMENCLATURE AND BASIC RELATIONS

The basic kinematic structure of a DPM is shown in Fig. 1. It is symmetric, both the base and the end-effector are shaped as equilateral triangles with vertices A_i, B_i ($i \in \overline{1,3}$) and centres O and O', at which coordinate frames, O_{xyz} and $O'_{x'y'z'}$ are originated, the $\mathbf{x}$ and $\mathbf{x}'$ axes passing through A_1 and B_1, respectively. The one ends of the legs are arranged in couples, q_i and q_{3+i}, attached to the base at A_i, while their other ends are connected to the end-effector at O' or B_i respectively, $i \in \overline{1,3}$. All passive joints are spherical (S-joints), while the active ones are prismatic (T-joints).

With a regard to Fig. 1, the following nomenclature is used: $\mathbf{r}_i = \vec{OA_i}$, $\mathbf{r}'_i = \vec{O'B_i}$; $\mathbf{q}_i = \vec{A_iO'}$, $\mathbf{q}_{3+i} = \vec{A_iB_i}$; $q_i = (\mathbf{q}_i^T\mathbf{q}_i)^{\frac{1}{2}} > 0$ (for physical reasons), where ^{T} means transposition; q^{min} and q^{max} denote the minimum and the maximum lengths of the legs, respectively; $\mathbf{p} = [p_x \mid p_y \mid p_z]^T = \vec{OO'}$.

†The first author acknowledges the support by the Bulgarian National Science Research Foundation under grant NIMM 44/91 "Control of Intelligent Machines." Unfortunately, he died of a heart attack on April 11, 1993 in Sendai, Japan. This work is based on his work before his death.

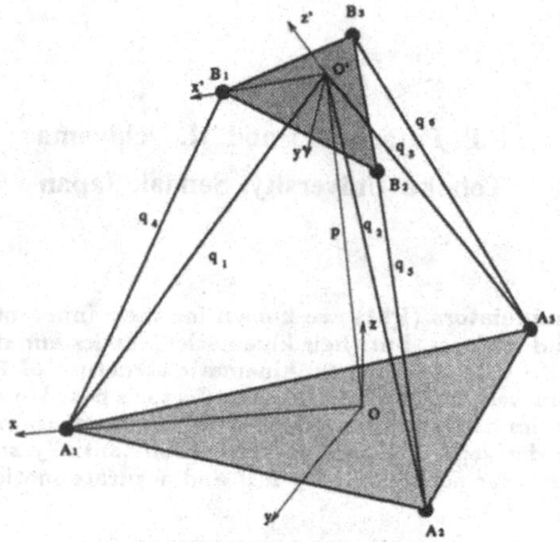

Fig. 1. The basic kinematic structure.

The pose[1] of the end-effector is uniquely defined by the Cartesian coordinates p_x, p_y, p_z of its centre O', and either the cosine directions matrix:

$$\mathbf{R} = [\mathbf{k} \mid \mathbf{l} \mid \mathbf{m}] = \begin{bmatrix} k_1 & l_1 & m_1 \\ k_2 & l_2 & m_2 \\ k_3 & l_3 & m_3 \end{bmatrix} \tag{1}$$

or any proper set of three angles. In the case of Bryant angles $\boldsymbol{\varphi} = [\varphi_1 \mid \varphi_2 \mid \varphi_3]^T$ [14] (this choice is discussed below) the entries of the $\mathbf{R}$ matrix in (1) are:

$$\mathbf{R} = \begin{bmatrix} c_1 c_2 & c_1 s_2 s_3 - s_1 c_3 & c_1 s_2 c_3 + s_1 s_3 \\ s_1 c_2 & s_1 s_2 s_3 + c_1 c_3 & s_1 s_2 c_3 - c_1 s_3 \\ -s_2 & c_2 s_3 & c_2 c_3 \end{bmatrix} \tag{2}$$

where $s_i = \sin\varphi_i$ and $c_i = \cos\varphi_i$, $i \in \overline{1,3}$.

The mechanism of position/orientation decoupling is readily seen from Fig. 1: three of the legs, viz. $\mathbf{q}_1$, $\mathbf{q}_2$ and $\mathbf{q}_3$ define the position, while the rest, $\mathbf{q}_4$, $\mathbf{q}_5$ and $\mathbf{q}_6$, define the orientation of the end-effector.

The following basic relations stream directly from Fig. 1 and the choice of coordinate frames:

$$\mathbf{r}_1 = [r \mid 0 \mid 0]^T; \quad \mathbf{r}_2 = [-\frac{1}{2}r \mid \frac{1}{2}\bar{r} \mid 0]^T; \quad \mathbf{r}_3 = [-\frac{1}{2}r \mid -\frac{1}{2}\bar{r} \mid 0]^T$$

$$\mathbf{r}'_1 = [r' \mid 0 \mid 0]^T; \quad \mathbf{r}'_2 = [-\frac{1}{2}r' \mid \frac{1}{2}\bar{r}' \mid 0]^T; \quad \mathbf{r}'_3 = [-\frac{1}{2}r' \mid -\frac{1}{2}\bar{r}' \mid 0]^T$$

$$\bar{\mathbf{r}}'_1 = r'\mathbf{k}; \quad \bar{\mathbf{r}}'_2 = -\frac{r'}{2}(\mathbf{k} - \bar{\mathbf{l}}); \quad \bar{\mathbf{r}}'_3 = -\frac{r'}{2}(\mathbf{k} + \bar{\mathbf{l}}) \tag{3}$$

$$\mathbf{p} = \mathbf{r}_i + \mathbf{q}_i; \quad \bar{\mathbf{r}}'_i = \mathbf{q}_{3+i} - \mathbf{q}_i, \quad i \in \overline{1,3}$$

where: $\bar{r} = r\sqrt{3}$, $\bar{r}' = r'\sqrt{3}$, $\bar{\mathbf{l}} = \mathbf{l}\sqrt{3}$; $\mathbf{r}'_i$ is presented in the $O'_{x'y'z'}$ coordinate frame, and $\bar{\mathbf{r}}'_i = \mathbf{R}\mathbf{r}'_i$, $i \in \overline{1,3}$.

These relations are quite elementary, of course, and they are recalled here just to illustrate the concept of decoupling. Equation (3) give the position of the end-effector as a function of only three joint variables, q_1, q_2, q_3, and for any fixed position $\mathbf{p}$ its orientation is defined through the rest three joint variables, q_4, q_5, q_6.

[1] In this paper *pose* means *position and orientation* of the end-effector.

III. POSE KINEMATICS

Inverse kinematics can be derived separately for position/orientation and respective expressions follow directly from (3):

$$q_i = (r_i^2 + p^2 - 2\mathbf{p}^T\mathbf{r}_i)^{\frac{1}{2}}, \quad i \in \overline{1,3} \tag{4}$$

and

$$q_i = (r'^2_{i-3} + q^2_{i-3} + 2\mathbf{q}^T_{i-3}\bar{\mathbf{r}}'_{i-3})^{\frac{1}{2}}, \quad i \in \overline{4,6} \tag{5}$$

If the orientation of the end-effector is prescribed in terms of Bryant angles, the respective expressions from (2) and (3) should be substituted in (5).

Forward kinematics can be readily resolved from (4) with regard to the end-effector's position, yielding:

$$p = (r_i^2 + q_i^2 + 2\mathbf{q}_i^T\mathbf{r}_i)^{\frac{1}{2}} \tag{6}$$

for any $i \in \overline{1,3}$. It may be shown that all values under the square roots in (6) will be positive if

$$q_{min} > r \tag{7}$$

Therefore, as far as only the position of the end-effector is concerned, (7) gives a necessary condition for avoiding *architecture singularities* [6] in DPM.

Although PM forward kinematics evaluation with respect to the orientation of the end-effector is generally quite involved indeed, it admits an explicit solution here, due to the decoupled structure of DPM—(3) yields:

$$\mathbf{R}\mathbf{r}'_i = \mathbf{q}_{3+i} - \mathbf{q}_i, \quad i \in \overline{1,3} \tag{8}$$

from where the **k**, **l** and **m** cosine directions vectors can be easily calculated.

Equation (8) are nonlinear, and generally have more than one solution, defining physically admissible orientation of the end-effector. Configurations for which two solutions coincide, can be expected to be singular.

Equations (4), (5), (6) and (8) completely describe the geometric kinematics of DPM in an explicit form, demonstrating that no algorithmic singularities are involved therein.

If Bryant angles are used instead, just **k**, l_3 and m_3 are required—the components of φ are uniquely defined (using the *arctangent of two arguments* function) by [14]:

$$s_2 = k_3; \quad c_2 = \pm(1 - s_2^2)^{\frac{1}{2}}; \quad s_1 = -l_3/c_2; \quad c_1 = m_3/c_2; \quad s_3 = -k_2/c_2; \quad c_3 = k_1/c_2 \tag{9}$$

Relations (9) are singular for $\varphi_2 = \pm\frac{\pi}{2}$, that is seldom admissible due to the limited dexterity of PM. This is one of the advantages when using Bryant angles (cf. the common θ, ψ, ϕ Euler angles [14], introducing a unavoidable algorithmic singularity at $\theta = 0$).

IV. INSTANTANEOUS KINEMATICS

Let $\dot{\mathbf{p}}$ and $\boldsymbol{\omega}$ be the linear and the angular velocities of the end-effector, respectively. Then the joint velocities $\dot{\mathbf{q}}$ are related to the end-effector's twist $\mathbf{w} = [\dot{\mathbf{p}}^T \mid \boldsymbol{\omega}^T]^T$, through the Jacobian matrix **J** of DPM:

$$\dot{\mathbf{q}} = \mathbf{J}\mathbf{w} \tag{10}$$

Although the **J** matrix can be derived differentiating the inverse kinematics relations (4) and (5) directly, this is quite tedious and impractical. Instead, it can be presented as [5], [7], [8]:

$$\mathbf{J} = \left[\begin{array}{c|c} \mathbf{J}_{11} & \mathbf{0}_{3,3} \\ \hline \mathbf{J}_{21} & \mathbf{J}_{22} \end{array}\right] = \left[\begin{array}{c|c} \mathbf{q}_1^T/q_1 & \\ \mathbf{q}_2^T/q_2 & \mathbf{0}_{3,3} \\ \mathbf{q}_3^T/q_3 & \\ \hline \mathbf{q}_4^T/q_4 & (\bar{\mathbf{r}}'_1 \times \mathbf{q}_4)^T/q_4 \\ \mathbf{q}_5^T/q_5 & (\bar{\mathbf{r}}'_2 \times \mathbf{q}_5)^T/q_5 \\ \mathbf{q}_6^T/q_6 & (\bar{\mathbf{r}}'_3 \times \mathbf{q}_6)^T/q_6 \end{array}\right] \tag{11}$$

where $\mathbf{O}_{3,3}$ is the 3×3 null matrix. Explicit expressions for its entries can be easily found substituting (1) or (2) and (3) into (11). Equations (10) and (11) describe the inverse instantaneous kinematics of DPM—as usual, it has a unique solution and no singularities.

The use of Bryant angles does not introduce *algorithmic singularities* in DPM inverse kinematics, because [14]:

$$\omega = \mathbf{T}\dot{\varphi} \tag{12}$$

where:

$$\mathbf{T} = \begin{bmatrix} c_2c_3 & s_3 & 0 \\ -c_2s_3 & c_3 & 0 \\ s_2 & 0 & 1 \end{bmatrix}$$

Forward kinematics evaluation is subject to the inversion of the Jacobian matrix, and more particularly of its $\mathbf{J}_{11}$ and $\mathbf{J}_{22}$ blocks.

It is straightforward seeing, either from (11) directly, or from (4) and (6), or for geometric reasons referring to Fig. 1, that $\mathbf{J}_{11}$ is always non-singular—two of its rows will be linearly dependent, if their respective elements are equal, that is never possible. Therefore (10) uniquely defines the linear velocity of the end-effector for any given set of joint velocities $\dot{q}_1$, $\dot{q}_2$, $\dot{q}_3$.

On contrary, rank $\mathbf{J}_{22}$ degrades whenever two or all the three of its rows become either zero vectors, or linearly dependent, i.e. when one or more of the following conditions hold:

$$\begin{gathered} \bar{\mathbf{r}}'_i \times \mathbf{q}_{i+3} = \mathbf{0}_3, \quad i \in \overline{1,3} \\ \bar{\mathbf{r}}'_1 \times \mathbf{q}_4 = \alpha(\bar{\mathbf{r}}'_2 \times \mathbf{q}_5); \quad \bar{\mathbf{r}}'_2 \times \mathbf{q}_5 = \beta(\bar{\mathbf{r}}'_3 \times \mathbf{q}_6); \quad \bar{\mathbf{r}}'_3 \times \mathbf{q}_6 = \gamma(\bar{\mathbf{r}}'_1 \times \mathbf{q}_4) \end{gathered} \tag{13}$$

where $\mathbf{0}_3$ is the 3 zero vector and α, β, γ are scalar constants. In fact, (13) are *the necessary and sufficient conditions* for *configuration singularity* [6] in the DPM. The detailed analysis and the physical meaning of these conditions are both important and interesting tasks, but due to the lack of space here the respective results will be reported later. Anyway, it is straightforward seeing that (13) provides for an explicit singularity resolution.

If the $\mathbf{J}_{22}$ matrix has a full rank, the forward instantaneous kinematics of DPM are described by:

$$\mathbf{w} = \mathbf{J}^{-1}\dot{\mathbf{q}} \tag{14}$$

where:

$$\mathbf{J}^{-1} = \left[\begin{array}{c|c} \mathbf{J}_{11}^{-1} & \mathbf{0}_{33} \\ \hline -\mathbf{J}_{22}^{-1}\mathbf{J}_{21}\mathbf{J}_{11}^{-1} & \mathbf{J}_{22}^{-1} \end{array} \right]$$

If Bryant angles are used, the substitution of $\omega = \mathbf{T}^{-1}\dot{\varphi}$ from (12) into (14) with

$$\mathbf{T}^{-1} = \begin{bmatrix} c_3/c_2 & -s_3/c_2 & 0 \\ s_3 & c_3 & 0 \\ -c_3t_2 & s_3t_2 & 1 \end{bmatrix}$$

where $t_2 = \tan\varphi_2$, will introduce *algorithmic singularities* into the forward kinematics of DPM whenever $\varphi_2 = \pm\frac{\pi}{2}$ (if this is an admissible orientation). Alternatively, DPM motion can be presented as seen from the coordinate frame associated with the desired trajectory (see [4] for details)—then $\mathbf{T}^{-1} \approx \mathbf{I}_3$, the 3×3 unit matrix, and $\dot{\varphi} \approx \omega$ [14]. This is another reason for giving preference to the Bryant angles when describing the end-effector's orientation in PM.

It is noteworthy, that the block-triangular structure of the Jacobian matrix provides for a *substantial reduction of the computational burden* required for DPM instantaneous kinematics, both inverse (10) and forward (14), evaluation.

V. STATICS

The static forces in DPM are related by:

$$\mathbf{f} = \mathbf{J}^T\mathbf{t} \tag{15}$$

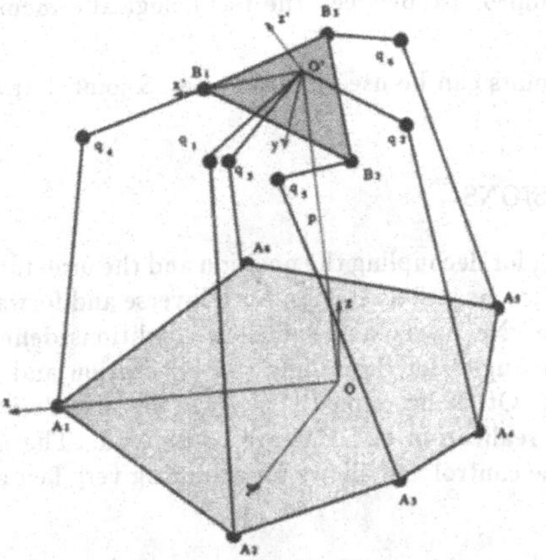

Fig. 2. DPM with R-joints.

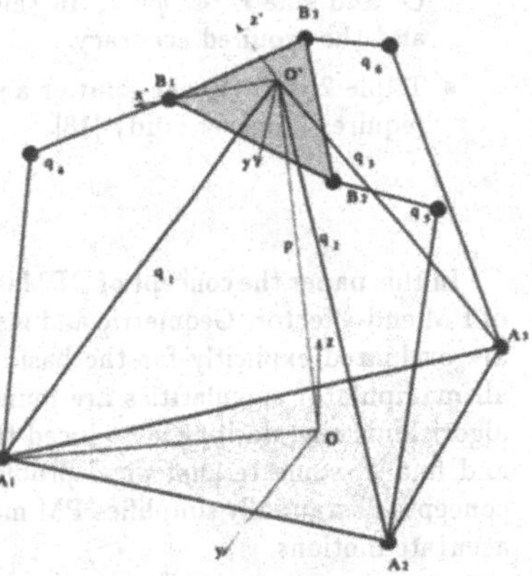

Fig. 3. DPM with T and R-joints.

where: f is the external wrench applied by/to the end-effector, and t are the legs' forces. Equation (15) describes the forward statics, and its inversion, if rank $\mathbf{J} = 6$, i.e. if rank $\mathbf{J}_{22} = 3$—the inverse statics of DPM. Both the analysis carried out and the conclusions made in the previous section, apply here, too.

Denoting $\bar{\mathbf{K}} = \text{diag}\{\kappa_1, ..., \kappa_6\}$, where κ_i, $i \in \overline{1,6}$ are the respective legs stiffnesses due to legs elasticity, to backlash in the gears/transmissions, to the control (desired stiffness), etc., allows the end-effector's stiffness matrix, K, to be defined [3] from the relation:

$$\delta \mathbf{f} = \mathbf{K} \delta \mathbf{w}$$

where δ means the respective infinitesimal increments, as:

$$\mathbf{K} = \mathbf{J}^T \bar{\mathbf{K}} \mathbf{J} \tag{16}$$

Partitioning the matrices in (16) into 3 × 3 blocks yields:

$$\mathbf{K} = \left[\begin{array}{c|c} \mathbf{J}_{11}^T \bar{\mathbf{K}}_{11} \mathbf{J}_{11} + \mathbf{J}_{21}^T \bar{\mathbf{K}}_{22} \mathbf{J}_{21} & \mathbf{J}_{21}^T \bar{\mathbf{K}}_{22} \mathbf{J}_{22} \\ \hline \mathbf{J}_{22}^T \bar{\mathbf{K}}_{22} \mathbf{J}_{21} & \mathbf{J}_{22}^T \bar{\mathbf{K}}_{22} \mathbf{J}_{22} \end{array} \right]$$

that demonstrates once again the decoupled nature of DPM.

VI. MECHANICAL DESIGNS

For mechanical design reasons it is often preferable [2], [12] to use R instead of T-joints. Then the basic arrangement of Fig. 1 is not very suitable any more (because of the possible collisions between the legs)—the kinematic scheme shown in Fig. 2 with the six legs arranged at the vertices of a regular hexaon on the base may appear to be a better choice.

Another interesting structure, providing higher dexterity, is shown in Fig. 3—three of the legs (defining the position) comprise T-joints, while the rest three (defining the orientation of the end-effector) R-joints. Of course, other schemes are also possible and may appear practically useful.

Although developing particular mechanical designs of DPM is not an aim of this paper, a common feature of all kinematic schemes proposed is the use of a triple S-joint, that creates serious technological problems indeed. Possible solutions include, but are not limited to, the following:

- **The ideas presented in [10] can be used for relatively simple mechanical designs closely approximating a triple S-joint. One of the ideas is a flexible ball joint or a set of three such joints. The application of this type of design is reported in [11] for a micro-PM.**

- Three S-joints can be used instead of one, located at the vertices of a equilateral triangle with centre O' and side $r'' \ll q^{min}$. In this case r'' should compromise between the technological difficulties and the required accuracy.
- Triple 2 d.o.f. Hooke joint or a set of three Hooke joints can be used instead of an S-joint, but this requires a deeper study [18].

VII. CONCLUSIONS

In this paper the concept of DPM is proposed, providing for decoupling the position and the orientation of PM end-effector. Geometric and instantaneous kinematics, as well as statics, both inverse and forward, are evaluated explicitly for the basic kinematic structure. Necessary and sufficient conditions defining all manipulator singularities are found. The use of three angles for describing the orientation and the algorithmic singularities introduced thereby are analysed. Other decoupled structures are also derived, and finally, some technological problems concerning the realization of DPM are considered. The new concept substantially simplifies PM modeling and real-time control and allows for achieving very fast and accurate motions.

REFERENCES

[1] K. H. Hunt. Structural kinematics of in-parallel-actuated robot arms. *Trans. ASME. J. Mech.. Transmiss.. Automat. Des.*, Vol. 105, 1983, pp. 705–712.

[2] H. Inoue, Y. Tsusaka, and T. Fukuizumi. Parallel manipulator. *Proc. 3rd ISRR*. Gouvieux, October 7–11, 1985, pp. 321–327.

[3] D. R. Kerr, M. Griffis, D. J. Sangar, and J. Duffy. Redundant grasps, redundant manipulators, and their dual relationships. *J. Robotic Systems*, Vol. 9, No. 7, 1992, pp. 973–1000.

[4] M. S. Konstantinov, S. P. Patarinski, V. B. Zamanov, and D. N. Nenchev. A new approach to the solution of the inverse kinematic problem for industrial robots. *Proc. 6th IFToMM World Congress*, Vol. 2, New Delhi, December 15–20, 1983, pp. 970–973.

[5] M. S. Konstantinov, Z. M. Sotirov, V. B. Zamanov, and D. N. Nenchev. Force feedback control of parallel topology manipulating systems. *Proc. 15th ISIR*, Vol. 1, Tokyo, September 11–13, 1985, pp. 181–188.

[6] O. Ma, and J. Angeles. Optimum architecture design of platform manipulators. *Proc. ICAR'91*, Vol. 2, Pisa, June 19–22, 1991, pp. 1130–1135.

[7] O. Ma, and J. Angeles. Architecture singularities of platform manipulators. *Proc. IEEE ICRA '91*, Vol. 2, Sacramento, April 9–11, 1991, pp. 1542–1547.

[8] M. G. Mohamed, and J. Duffy. A direct determination of the instantaneous kinematics of fully parallel robot manipulators. *Trans. ASME, J. Mech., Transmiss., Automat. Des.*, Vol. 107, June 1985, pp. 226–229.

[9] S. P. Patarinski. Parallel robots: A review. *Technical Report*. Dept. Mechatronics and Precision Eng., Tohoku University, 1993, pp. 1–30.

[10] S. P. Patarinski, and M. Uchiyama. Position/orientation decoupled parallel manipulators. *Proc. ICAR'93*, Tokyo, November 1–2, 1993, pp. 153–158.

[11] R. Stoughton, and T. Arai. Kinematic optimization of a chopsticks-like micromanipulator. *Proc. Japan/USA Symp. Flexible Automation*, Vol. 1, San Francisco, July 13–15, 1992, pp. 151–157.

[12] M. Uchiyama, K.-I. Iimura, F. Pierrot, P. Dauchez, K. Unno, and O. Toyama. A new design of a very fast 6-dof parallel robot. *Proc. 23rd ISIR*, Barcelona, October 6–9, 1992, pp. 771–776.

[13] K. J. Waldron, and K. H. Hunt. Series-parallel dualities in actively coordinated mechanisms. *Proc. 4th ISRR*, Santa Cruz, August 9–14, 1987, pp. 176–181.

[14] J. Wittenburg. *Dynamics of Systems of Rigid Bodies*. B. G. Teubner, Stuttgart, 1977.

[15] V. B. Zamanov, S. M. Sotirov, and S. P. Patarinski. Structures and kinematics of parallel topology manipulating systems. *Proc. Int. Symp. Design and Synthesis*, Tokyo, July 11–13, 1984, pp. 453–458.

[16] V. B. Zamanov, and Z. M. Sotirov. Duality in mechanical properties of sequential and parallel manipulators. *Proc. 20th ISIR*, Tokyo, October 4–6, 1989, pp. 1041–1050.

[17] V. B. Zamanov, and Z. M. Sotirov. A contribution to the serial and parallel manipulator duality. *Proc. 8th IFToMM World Congress*, Prague, September 26–31, 1992, pp. 517–520.

[18] K. E. Zanganeh, and J. Angeles. Instantaneous kinematics and design of a novel redundant parallel manipulator. *Proc. IEEE ICRA '94*, Vol. 4, San Diego, May 8–13, 1994, pp. 3043–3048.

PARALLEL MANIPULATORS WITH CONTROLLED COMPLIANACE

D. Chakarov

Bulgarian Academy of Sciences, Sofia, Bulgaria

Abstract. The present paper studies parallel manipulators and the possibilities for synthesis of their impedance parameters – damping and compliance. A mechanical model is presented, accounting for the closeness of the mechanical structure. Synthesis of impedance parameters of the end effector is performed as a function of the impedance parameters in the manipulator joints, at different manipulator structures. The influence of the manipulator position and geometry on the impedance parameters is considered. An optimization procedure for stiffness synthesis is proposed, by choosing the optimization parameters to be stiffnesses in joints as well as manipulator geometric parameters

1. INTRODUCTION

One of the possible approaches to building robots and manipulators is using parallel structures. The parallel manipulators are suitable for performing a number of technological operations related to force interaction between a manipulator and the environment. These operations are performed by principles of control based on the impedance parameters inertia, damping and stiffness [1]

The present paper aims at analyzing the factors having effect on the impedance parameters and to propose a procedure of synthesis of these parameters

2. MECHANICAL MODEL

Parallel mechanisms are characteristic by the presence of closed loops which encompass the kinematic joints and links in the mechanism. If denote by m the number of the loops, denote by p the number of kinematic joints and denote by n the number of links in the mechanism, then the following relation between them holds [2], [3]:

$$m = p - n + 1. \tag{1}$$

The parameters of relative motion in all joints of the mechanism are considered as generalized parameters:

$$\theta = [\theta_1, \theta_2, ..., \theta_\lambda]^T. \tag{2}$$

The number of these parameters is related to the number of loops m and the general number of degrees of mobility of the mechanism h [4]:

$$\lambda = h + bm \tag{3}$$

where $b = \begin{cases} 3 - \text{at planar mechanisms} \\ 6 - \text{at spatial mechanisms.} \end{cases}$

For the study of parallel manipulators, the generalized parameters can be represented by the vector:

$$\theta = [q; w]^T \tag{4}$$

where q is an $(h \times 1)$ vector of parameters in the shortest branch between the end effector and the immobile base, and w is a $(bm \times 1)$ vector of parameters in the remaining joints. Each closed loop in the parallel structure of the manipulator implies the appearance of connections between the generalized parameters These connections are expressed by m vector-matrix equations or bm scalar functions:

$$F_i(\theta) = 0, i = 1, ..., bm \tag{5}$$

These equations allow to determine the components of the vector w, whose number is bm, as function of the parameters q. The relations between the generalized speeds $\dot{q}$ and $\dot{w}$ are linear and are obtained after differentiation of equation (5):

$$H_q\dot{q} + H_w\dot{w} = 0 \tag{6}$$

The matrix H_q is of $(bm \times h)$ dimension and the matrix H_w is of $(bm \times bm)$ dimension.

The position X and the speed $\dot{X}$ of the end effector are function of the generalized parameters q and speeds $\dot{q}$ in the branch between the end effector and the base, which implements the degrees of mobility of the manipulator.

$$X = \Phi(q) \tag{7}$$

$$\dot{X} = J\dot{q} \tag{8}$$

In the above, J denotes a $(\nu \times h)$ matrix of the partial derivatives, $\nu \leq 6$.

The relation between the generalized speeds in the joints of the manipulator $\dot{\theta}$ and the end effector speed $\dot{X}$ is obtained using (6) and (8):

$$\dot{\theta} = I\dot{X} \tag{9}$$

Here, $$I = \begin{bmatrix} J^{-1} \\ .... \\ -H_w^{-1}H_qJ^{-1} \end{bmatrix} \tag{10}$$

denotes a block $(h+bm) \times \nu$ matrix. Equation (9) has a solution, if rank $J = h = \nu$ and rank $H_w = bm$.

The desired dynamics of a manipulator with impedance control is provided by appropriate values of the coefficients in the linearized differential equations [1], [5]:

$$B_x\dot{X} + K_x[X - X_r] = P_{ext} \tag{11}$$

The coefficients B_x and K_x are a symmetrical $(\nu \times \nu)$ matrices of end effector damping and stiffness. In the above, P_{ext} is a $(\nu \times 1)$ vector of the external forces applied, and X_r is a $(\nu \times 1)$ reference position vector of the end effector. By assuming that the manipulator links are ideally rigid, equation (11) can be represented [4] by the matrix (10) in the form:

$$I^T B_\theta I\dot{X} + I^T K_\theta I[X - X_r] = P_{ext} \tag{12}$$

Here, B_θ and K_θ are $(h+bm) \times (h+bm)$ diagonal matrices of damping and stiffness of the manipulator joints.

3. SYNTHESIS OF IMPEDANCE PARAMETERS

The impedance parameters damping B_x and stiffness K_x of the end effector, according to (11) and (12), are represented by the following matrix equations

$$B_x = I^T B_\theta I \tag{13}$$

$$K_x = I^T K_\theta I \tag{14}$$

and include

$$\mu = \nu(\nu + 1)/2 \tag{15}$$

independent components.

By assuming the manipulator links to be ideally rigid, they are function of: a) the impedance parameters B_θ and K_θ in the manipulator joints [5]; b) the manipulator position q [6]; c) the manipulator geometry R.

The number π of the components of impedance parameters in the manipulator joints is equal to or less than the number λ of the generalized parameters of the relative motions in these joints (3):

$$\pi \leq \lambda = h + bm \tag{16}$$

For a given manipulator position q^* and given geometry R^* the synthesis of impedance parameters B_x and K_x can be performed by solving systems of linear equations (13) and (14) [5], by finding $\pi = \mu$ compliant joints. According to (16) the possible number of compliant joints depends on the manipulator structure defined by the number of degrees of mobility h and the number of closed loops m.

For planar mechanisms with $h = 2$ and 3 and spatial mechanisms with $h = 2,..., 6$ degrees of mobility and $m = 1, 2,$ and 3 closed loops, the possible number of compliant joints λ is shown in Table 1 [7].

Table 1

	h	μ	m = 1		m = 2		m = 3	
			λ	α	λ	α	λ	α
Planar mechanism	2	3	5	2	8	5	11	8
	3	6	6	0	9	3	12	6
Spatial mechanism	2	3	8	5	14	11	20	17
	3	6	9	3	15	9	21	15
	4	10	10	0	16	6	22	12
	5	15	11	−4	17	2	23	8
	6	21	12	−9	18	−3	24	3

The table also shows the difference

$$\alpha = \lambda - \mu \tag{17}$$

which indicates that the possibilities of synthesis by introducing compliant joints are greater at a greater number of loops m and lower number of the degrees of mobility h.

For the cases where $\alpha > 0$, the performance of the synthesis depends on the correct choice of compliant joints $\pi = \mu$ out of all possible joints λ. For example, for a planar manipulator with $h = 2$, $m = 3$ and a structural diagram shown in Fig. 1, the incorrect choice of 3 compliant joints out of 11 possible, is shown in Fig. 1a).

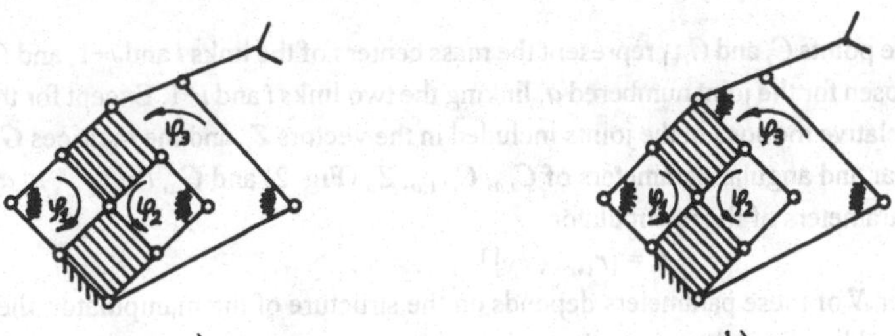

a) b)

Fig. 1. Choice of compliant joints: a) incorrect; b) correct.

In this case, the constructed systems of equations (13) or (14) are incompatible and they do not yield a univalent solution. This is not the case for the solution shown ion Fig. 1b). Along with this solution, there exist other variations of a correct choice of the compliant joints. In these variants, the loops including joints of the main branch jointly have:

$$\pi' \leq h'(h'+1)/2 \tag{18}$$

compliant joints. Here, h' is the total number of degrees of mobility in the joints of the main branch included in the loops So, the loop φ_3 can include at most 3 compliant joints chosen from the five joints in the loops, by 6 variants Also, the loop φ_3 can include 2 compliant joints, and the loops φ_1 and φ_2 can jointly include 1 compliant joint, as shown in Fig. 1b). In this case the two compliant joints can be chosen out of 4 joints by 6 variants, and the one compliant joint can be chosen out of 7 joints. There exist a total of $6 + 6.7 = 48$ variants for choosing three compliant joints in the manipulator in Fig 1.

4. OPTIMIZATION PROCEDURE FOR STIFFNESS SYNTHESIS

At a greater number of degrees of mobility for the manipulator, e. g. $h = 4, 5, 6...$, the number of the independent components of matrices (13) and (14) is great, e. g. $\mu = 10, 15, 21...$ and not for all structure there exists a possibility to choose the required number of compliant joints $\pi = \lambda$, $\lambda < \mu$. Also, the it is not always justified to include a great number of motors. At many problems, it is required to ensure the preset values of the impedance parameters in several positions of the manipulator q_l, $l = 1,..., \sigma$. Then, the number of the synthesized parameters is also great

$$\mu.\sigma > \lambda \tag{19}$$

In these cases, the synthesis of the impedance parameters can be accomplished at a less number of compliant joints and choosing an appropriate manipulator geometry.

The manipulator geometry includes linear and angular parameters that uniquely define the position of the manipulator and remain unchanged at its motion. The relative position of a link numbered i+1 towards its neighbouring link numbered i can be described [8] by means of the (3 x 1) vectors $C_{i,a}$, $C_{i+1,a}$, Z_a (Fig 2) and the (3 x 3) transformation matrix G_a between the coordinate systems of the objects $C_{i,}e^i$ and $C_{i+1,}e^{i+1}$.

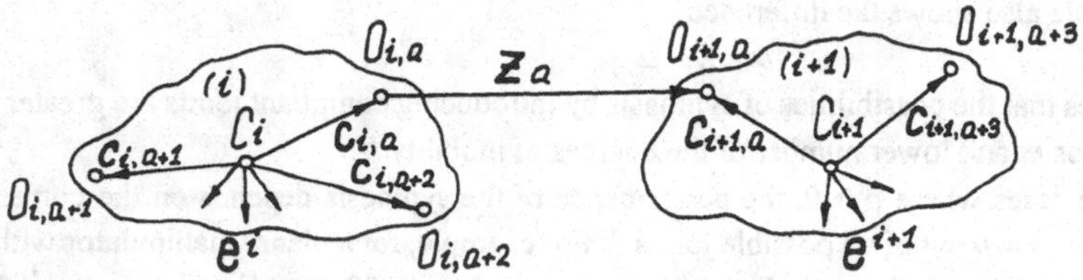

Fig. 2 Values describing the position of link i+1 towards link i.

In Fig. 2 the points C_i and C_{i+1} represent the mass centers of the links i and i+1, and $O_{i,a}$ and $O_{i+1,a}$ are joint points chosen for the joint numbered a, linking the two links i and i+1. Except for the parameters that describe the relative motions in the joints included in the vectors Z_a and the matrices G_a, $a = 1, .., p$, the remaining linear and angular parameters of $C_{i,a}$, $C_{i+1,a}$, Z_a (Fig. 2) and G_a, $i = 1, .. , n$, $a = 1, .., p$ are the geometrical parameters of the manipulator·

$$\boldsymbol{R} = [r_1, .., r_N]^{\mathrm{T}} \tag{20}$$

The number N of these parameters depends on the structure of the manipulator, the number and type of the joints and links and their mutual position.

If the impedance parameters of the end effector are considered not only as function of the impedance parameters in the joints, but as function of the geometrical parameters of the manipulator as well, the synthesis problem can be solved as an optimization one.

For each position of the manipulator q_l the equation (14) can be represented by μ scalar functions of the stiffness $\boldsymbol{K}_\theta$ in the joints and the manipulator geometry $\boldsymbol{R}$.

$$k_{x_{i,l}} = f_{i,l}(\boldsymbol{K}_\theta, \boldsymbol{R}), i = 1, ..., \mu \tag{21}$$

Above, $k_{x_{i,l}}$, $i = 1, ..., \mu$ denotes the independent components of the stiffness matrix $\boldsymbol{K}_x$. If $k^0_{x_{i,l}}$

are the desired values of the stiffness of the end effector in the chosen configurations q_l, $l = 1, \ldots, \sigma$, then the following objective is constructed to perform the synthesis

$$Q=\sum_{i=1}^{\mu}\sum_{l=1}^{\sigma}[k_{x_i,l}^{0}-f_{i,l}(K_{\theta},R)]^2 \tag{22}$$

The synthesis is performed in the following sequence

a) For a chosen number π and position of the compliant joints and a constant manipulator geometry R^*, minimizing (22) as function of the stiffness in joints

$$\min Q(K_{\theta})=Q(K_{\theta}^{0}),\ K_{\theta}\in A \tag{23}$$

b) For the obtained values of the stiffness in the joints K_{θ}^{0}, finding a zero minimum of (22) as function of the manipulator geometry

$$\min Q(R)=Q(R^{0})\rightarrow 0,\ R\in B \tag{24}$$

The synthesis is performed by taking into account the design restriction on the values of stiffness in the joints and on the values of the geometric parameters represented by the sets

$$A=\{K_{\theta}: k_{\theta_j}^{\min}\le k_{\theta_j}\le k_{\theta_j}^{\max};\ j=1,\ldots,\pi\} \tag{25}$$

$$B=\{R: r_j^{\min}\le r_j\le r_j^{\max};\ j=1,\ldots,N\} \tag{26}$$

The optimization procedure can be repeated at another choice of compliant joints and another choice of the geometrical parameters for synthesis

5. NUMERICAL EXAMPLE

A parallel manipulator with a structure according to Fig 1a has been created in order to carry out experiments The manipulator is a spatial one and it includes rotating joints of class 4 and 5 in the main chain The number of the degrees of freedom of the end effector is $h = \nu = 3$, for which the number of the independent components of its stiffness is $\mu = 6$ The manipulator has three compliant joints, which are located according to Fig 1a), but are shaped as linear joints The values of the stiffness in these joints are $k_{\theta_j} = 30000$ [N/m], $j = 1, 2$ and 3 For the chosen initial geometry, the general view of the manipulator is shown in Fig 3a)

For the position shown, the end effector stiffness is represented by means of a stiffness ellipsoid [9] shown in Fig 4a) with main stiffness values $k_{x1} = 1266$ [N/m], $k_{x2} = 20408$ [N/m] and $k_{x3} = 11236$ [N/m]

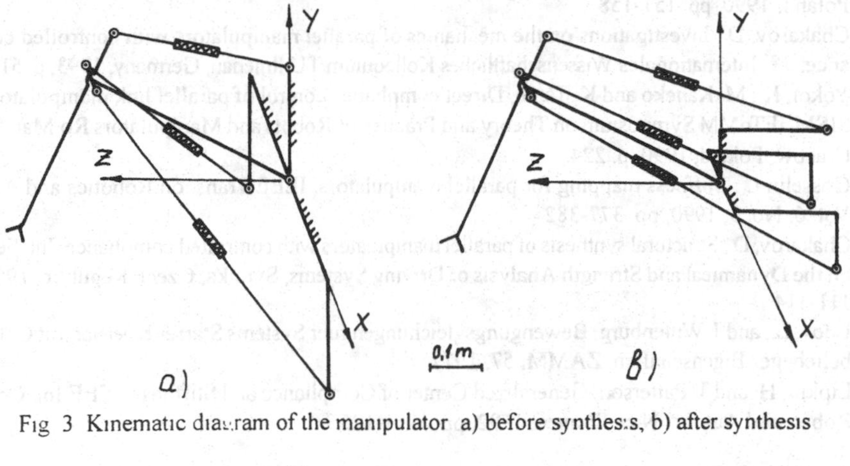

Fig 3 Kinematic diagram of the manipulator a) before synthesis, b) after synthesis

The problem is to synthesize another end effector stiffness represented by the ellipsoid in Fig. 4b). After optimization by nine geometrical parameters the general view of the manipulator is shown in Fig 3b) The main values of the end effector stiffness in this case are k_{x1} = 1449 [N/m], k_{x2} = 1961 [N/m] and k_{x3} = 1724 [N/m]. The numerical method of random search is used to perform the optimization

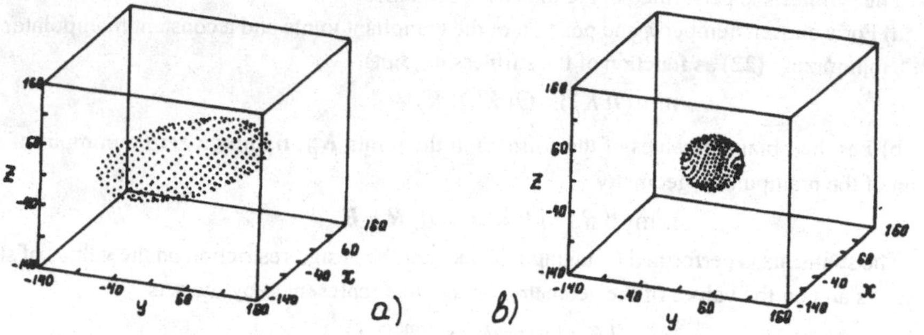

Fig. 4. Stiffness ellipsoid: a) before synthesis, b) after synthesis

6. CONCLUSION

The synthesis procedure shown can be used both for implementing compliant control and for design of parallel manipulators Also, end effector damping can be synthesized instead of stiffness Such a procedure can be used for complex synthesis of the impedance parameters – damping and stiffness

ACKNOWLEDGMENTS. The work is supported by the Ministry of Science and Education through Grant 25/91

REFERENCES

[1] Hogan, N., Impedance control an approach to manipulation, ASME Journal of Dynamic Systems, Measurement and Control, 1985, pp 1-24

[2] Konstantinov, M., Methodologic der Rechnerunterstuetzen Struktursynthese und Kinematic Analyse von Mechanismen und Robotern, Int Symp in Education of TMM and Robotics, SEMEMATRO'82, Blagoevgrad, Bulgaria, 1982, Vol 1, p 179

[3] Parushev, P., D Chakarov, Structural investigation of manipulators with linear drives, 8th CISM-IFToMM Symposium on Theory and Practice of Robots and Manipulators Ro Man Sy'90, Cracow, Poland, 1990, pp 151-158

[4] Chakarov, D., Investigations on the mechanics of parallel manipulators with controlled compliance, 38 Internationales Wissenschaftliches Kolloquium TU Ilmenau, Germany, 1993, p 516-520

[5] Yokoi, K , M Kaneko and K Tanie, Direct compliance control of parallel link manipulators, 8th SISM-IFToMM Symposium on Theory and Practice of Robots and Manipulators Ro Man Sy'90, Cracow, Poland, 1990, p. 224

[6] Gosselin C , Stiffness mapping for parallel manipulators, IEEE Trans. on Robotics and Autom , Vol 6, No 3, 1990, pp 377-382

[7]. Chakarov, D , Structural synthesis of parallel manipulators with controlled compliance, Int Seminar on the Dynamical and Strength Analysis of Driving Systems, Svratka, Czech Republic, 1993, pp 111-114

[8] Lilov, L., and J Wittenburg, Bewengungsgleichungen fuer Systems Starrer Koerper mit Gelenken beliebeger Eigenschaften, ZAMM, 57, 1977.

[9] Lipkin, H and T. Patterson, Generalized Center of Compliance and Stiffness, IEEE Int Conf on Robot and Autom , Nice, France, 1992, pp 1251-1256

A MULTIOBJECTIVE OPTIMISATION APPROACH TO ROBOT KINEMATIC DESIGN

A.P. Pashkevich and D.E. Khmel
Minsk Radioengineering Institute, Minsk, Republic of Belarussia

Introduction

Arising requirements to robot productivity and quality of work challenge to robot designers abilities. The proper performance of industrial robots can be achieved through the careful design of manipulators, drive systems and controllers. A number of performance measures have been proposed which provide a representation of selected aspects of manipulator performance. Since these each represent alternative facets of performance it is desirable to include several of them as objectives in optimisation problem and treat it as multiobjective one. So as the objectives will be competing with one another during the optimisation sequence, there is no unique solution to this problem. Rather, there is a set of solutions termed Pareto-optimal solutions that may be regarded as a trade-off surface from which designer selects a compromise solution.

Design Objectives

There have been many investigations which have attempted to meet increased performance requirements and there have been proposed a number of symbolic and numerical techniques to enable designers to evaluate robot features. Sophisticated dynamic models that contain many configuration dependent parameters and variables may be too time-consuming and expensive to use during the initial stages of design procedure.

As a scalar performance measures can be used kinematic and dynamic manipulability, inertia ellipsoids, speed and acceleration radii, condition number, measure of isotropy and others /1-4,6/. Most of the measures are based on Jacobian matrix analysis from differential motion equation

$$\dot{x} = J(q)\,\dot{q}, \qquad (1)$$

where $\dot{x}$, $\dot{q}$ are vectors of end-effector and joint velocities respectively, $J(q)$ - Jacobian matrix that is derived from primary robot kinematic equation $x = F(q)$.

There are two main approaches to quantifying manipulator dexterity using the linear mapping measures. They are an approach based on Jacobian matrix singular values analysis and an approach based on speed (acceleration) hull geometrical investigation.

The first one is widely used due to availability of efficient matrix analysis software tools. The second approach deals with more complicated performance measures that require sophisticated computational algorithms. But the latter operates with more conventional measures which have clear physical meaning (velocities, accelerations, etc.).

The paper is devoted to robot kinematic design using the second approach. To estimate manipulator velocity properties let us consider the mapping from joint velocity space $\{\dot{q}\}$ to end-effector velocity space $\{\dot{x}\}$. According to linear equation (1) parallelepiped of feasible joint velocities $\{\dot{q} : |\dot{q}_i| \le \dot{q}_i^{max}\}$ is mapped into polyhedron of achievable end-effector velocities in $\dot{x}$ -space. The polyhedron geometry completely determines the manipulator velocity properties. But for design purpose we will use only two scalar measures

$$\hat{V} = max\ Hull\left\{\dot{x} : \dot{x} = J(q)\dot{q}, |\dot{q}_j| \le \dot{q}_i^{max}, i = 1,2,\ldots,n\right\}$$
$$\check{V} = min\ Hull\left\{\dot{x} : \dot{x} = J(q)\dot{q}, |\dot{q}_i| \le \dot{q}_i^{max}, i = 1,2,\ldots,n\right\}$$

where $\check{V}$ and $\hat{V}$ are the longest and the shortest distance from the origin of the hull respectively. These performance indices were proposed in /3,4,7/.

Since the values of all the all above mentioned indices are configuration dependent, an optimisation approach that seeks to include all working points must integrate the local index over the working space or calculate its upper or lower bound in a given region. In this paper we will use the upper bound of local indices in whole workspace as a global measure.

Problem Statement

Due to complexity of general manipulator design problem it is, as a rule, divided into several repeatable stages (kinematic, dynamic and actuator design). The paper deals with kinematic design that results an appropriate selection of joint velocities and link lengths. It is usually desirable to achieve a certain performance or, in our case, a velocity in a certain workspace. This objective can be expressed in terms of isotropy and velocity measures described in the previous section. As mentioned above it is preferable to use several performance indices simultaneously and generate a set of noninferior solutions to a designer with useful information for making the final decision.

A variety of techniques exist for obtaining the set of solutions, the one favoured by us is the goal attainment by Gembicki /5/. This method is based on function scalarisation and requires a standard nonlinear programming algorithm to perform optimisation.
Using this method the optimisation problem

$$\min_{p} \left[f_1(p), f_2(p), \ldots, f_k(p)\right] \tag{3}$$

may be recast as

$$\min_{p,\lambda} \lambda \tag{4}$$

where $p=[p_1,p_2,...,p_n]$ is a set of minimising parameters, λ is an unrestricted scalar variable. The set of objectives are transformed to the set of goals

$$f_i = f_i^* - w_i\lambda \quad (5)$$

where $w_i \geq 0$ are weighting coefficients.

Applying this approach to manipulator kinematic design we can define the multiobjective optimisation problem

$$\min\left[\dot{q}_n^{max},...,\dot{q}_n^{max}\right]$$

subject to

$$r_1 \geq r_1^*,..., r_k \geq r_k^*$$

where r_i is manipulator performance measure, r_i^* is a lower bound of its desired value, $p=\left[\dot{q}_1^{max},...,\dot{q}_n^{max},l_1,...,l_n\right]$ is a set of minimising parameters, $\dot{q}_i^{max}$ is i-th joint velocity, l_i is i-th link length.

Objective function scalarisation based on goal-attainment approach gives

$$\min_{p,\lambda} \lambda \quad (6)$$

subject to

$$q_i^{max} - w_i\lambda \leq 0; \quad i=1,...,n$$
$$-r_j \leq -r_j^*; \quad j=1,...,k$$

where $p=\left[\dot{q}_1^{max},...,\dot{q}_n^{max},l_1,...,l_n\right]$ is a set of minimising parameters, λ is an auxiliary scalar value. Weights $w_{n+1},...,w_{n+k}$ are set to zero to ensure the corresponding objectives are realised. The remaining weights $w_1,...,w_n$ are at the disposal of the designer who will choose suitable combination in order to identify the trade-off available within design.

To ensure correctness of the optimisation problem from engineering point of view additional constraint should be imposed on design parameters

$$\sum_{i=1}^{n} l_i = const \quad (7)$$

otherwise the desired values of performance indices may be achieved by increasing link lengths. This constraint can be treated as workspace dimension one and also can be converted into a performance index (structural length index) /7/.

Design Examples

Consider the kinematic design problem for a two-link planar manipulator

$$x = l_1c_1 + l_2c_{12}, \quad y = l_1s_1 + l_2s_{12} \quad (8)$$

where l_1, l_2 are link lengths, $c_1 = cos(q_1)$, $s_1 = sin(q_1)$, $c_{12} = cos(q_1+q_2)$, $s_{12} = sin(q_1+q_2)$. This manipulator is a touchstone of any design technique.

The task is to determine manipulator kinematic parameters $\dot q_1^{max}, \dot q_2^{max}, l_1, l_2$ provided that angular velocities are minimum for given values of global measures of $\tilde V$ and $\hat V$ (maximum and guaranteed mobilities).

$$\dot q_1^{max} \rightarrow min, \; \dot q_2^{max} \rightarrow min \tag{9}$$

subject to

$$\tilde V \geq \tilde V^{*}, \; \hat V \geq \hat V^{*}, \; l_1 + l_2 = l.$$

Let us get analytical expressions for mobility measures $\tilde V$ and $\hat V$ that are functions of design parameters $p = \left[\dot q_1^{max}, \dot q_2^{max}, l_1, l_2\right]$ and position in the workspace ρ (the distance from the origin to the workpoint). Jacobian matrix of the manipulator that used for calculating of performance indices is

$$J(q) = \begin{bmatrix} -l_1 s_1 - l_2 s_{12} & -l_2 s_{12} \\ l_1 c_1 + l_2 c_{12} & l_2 c_{12} \end{bmatrix} \tag{10}$$

The longest and the shortest distance from the origin of the speed hull are

$$\begin{aligned} &\tilde V(\alpha,\rho) = \sqrt{\left(l_1 \dot q_1^{max}\right)^2 + \left(l_2 \dot q_2^{max}\right)^2 + \left|\rho^2 + l_1^2 + l_1^2\right| q_1^{max} q_2^{max}} \\ &\hat V(a,r) = min \left\langle \hat V_1, \hat V_2 \right\rangle \end{aligned} \tag{11}$$

where $\hat V_1 = l_1 s_2 \dot q_1^{max}$, $\hat V_2 = \dfrac{l_1 l_2 s_2}{\rho} \dot q_2^{max}$, $\rho^2 = \left|l_1^2 + l_2^2 + 2 l_1 l_2 c_2\right|$, $\alpha = \dfrac{l_2}{l_1 + l_2}$

To derive global measures (the extreme values of the performance indices in workspace) let us analyse the functions $\tilde V(\alpha,\rho)$ and $\hat V(a,\rho)$. The maximum value of the $\tilde V(\alpha,\rho)$ may be obtained from the geometrical consideration and is

$$\tilde V^{max}(\alpha) = (l_1 + l_2)\dot q_1^{max} + l_2 \dot q_2^{max}$$

In accordance with the expression (11) maximum of $\hat V(a,\rho)$ can be found as minimum of three values

$$\hat V^{max}(\alpha) = min\left\{ \hat V_1^{max}, \hat V_2^{max}, roots(V_1 = V_2) > 0 \right\} \tag{12}$$

that correspond to maximum of $\hat V_1^{max}$, $\hat V_2^{max}$ and their intersection point (Fig. 1).

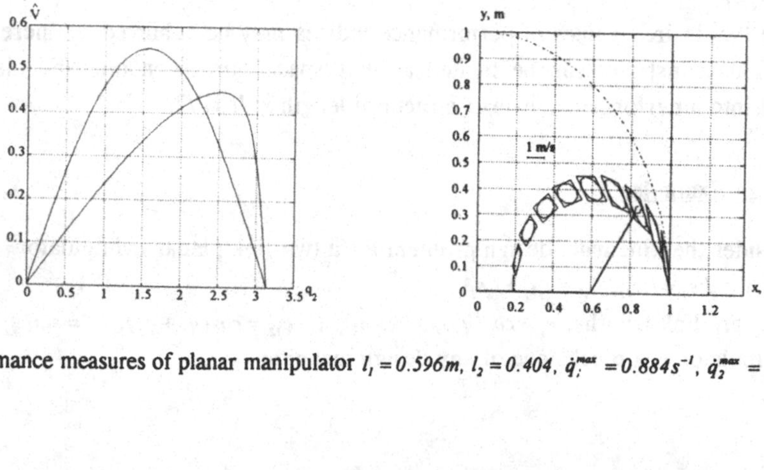

Fig.1 Performance measures of planar manipulator $l_1 = 0.596m$, $l_2 = 0.404$, $\dot q_1^{max} = 0.884 s^{-1}$, $\dot q_2^{max} = 1.325 s^{-1}$

Applying standard technique we can find maximum values of $\hat{V}_1(\alpha,\rho)$ and $\hat{V}_2(\alpha,\rho)$. They are

$$\hat{V}_1^{max}(\alpha)=\dot{q}_1^{max} l_1;\ \ \hat{V}_2^{max}=\dot{q}_2^{max}\min\langle l_1,l_2\rangle.$$

The mobility value in the intersection point is equal to

$$\hat{V}_{int}=l_1 s_2^{*}\dot{q}_1^{max},$$

where $s_2^{*}=\sqrt{1-\left(c_2^{*}\right)^2}$, $c_2^{*}=\dfrac{\left(l_2\,\dot{q}_2^{max}/\dot{q}_1^{max}\right)^2-l_1^2-l_2^2}{2l_1l_2}$

The final expression for $\hat{V}^{max}(\alpha)$ is

$$\hat{V}^{max}(\alpha)=\min\left\{\dot{q}_1^{max} l_1;\ \dot{q}_2^{max}\min\langle l_1,l_2\rangle,\ l_1 s_2^{*}\dot{q}_1^{max}\right\} \tag{13}$$

Using the goal-attainment approach the optimisation problem for the two-link manipulator may be stated as

$$\lambda\rightarrow\min_{p,\lambda} \tag{14}$$

subject to

$$\dot{q}_1^{max}-w_1\lambda\le 0;\quad \tilde{V}_1^{max}\le\tilde{V}_1^{*};\quad \dot{q}_2^{max}-w_2\lambda\le 0;\quad \hat{V}_2^{max}\le\hat{V}_1^{*};\ l_1+l_2=l$$

where $p=\left[\dot{q}_1^{max},\dot{q}_2^{max},l_1,l_2\right]$ is the vector of design variables, λ is an unrestricted scalar variable, w_1, w_2 are weights that are at the designer disposal. Weights w_3 and w_4 are equal to zero to ensure that corresponding objectives are realised.

For the particular case of a two-link planar manipulator, the problem may be normalised using length unit l, velocity unit $\hat{V}_2^{max}$ and angular velocity unit $\hat{V}_2^{max}/l$. Normalisation is performed to obtain a set of scalable solutions (nomogram).

The optimisation results presented on Fig.2 correspond to a general case and can be applied to any particular case by scaling. Analysing these results we see that goal $\hat{V}^{max}$ is over-attained for all points of curve 1. Curve 2,3,4 correspond to case of competing goals $\tilde{V}^{max},\hat{V}^{max}$. As explained in /5/ it is prudent to select $\dot{q}_1^{max},\dot{q}_2^{max}$ in the region indicated in the Figure.

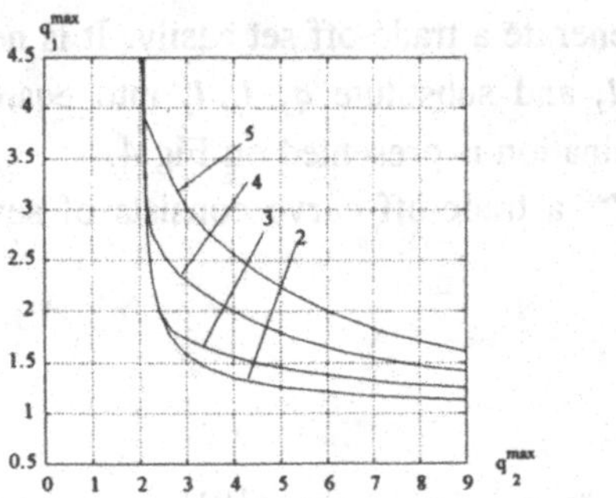

ig.2 Pareto-optimal solutions for different $\tilde{V}^{*}$, $\hat{V}^{*}$

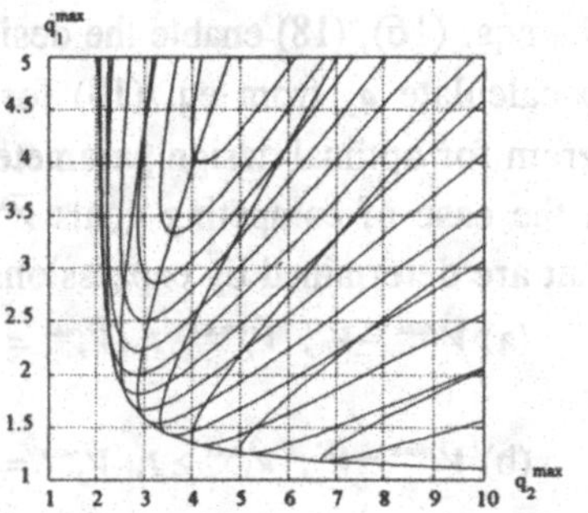

Fig.3 Set of guaranteed mobilty curves and their envelo

Using results obtained for normalised manipulator we can design any manipulator by scaling design objectives and variables. For example, if it is necessary to design a planar manipulator with the specifications $l=0.8m$, $\tilde{V}^{max}\geq 2\,m/s$, $\hat{V}^{max}\geq 0.5\,m/s$, we can use solution indicated on the curve 3 (Fig. 2). Scaling of the solution gives ($\dot{q}_1^{max}=1.025s^{-1}$, $\dot{q}_2^{max}=1.904s^{-1}$, $l_1=0.39m$, $l_2=0.41m$); ($\dot{q}_1^{max}=1.175s^{-1}$, $\dot{q}_2^{max}=1.437s^{-1}$, $l_1=0.381m$, $l_2=0.419m$); ($\dot{q}_1^{max}=1.387s^{-1}$, $\dot{q}_2^{max}=1.135s^{-1}$, $l_1=0.367m$, $l_2=0.433m$).

In the particular case of a two-link planar manipulator the set of noninferior solutions can be obtained analytically. Let us consider firstly the case of active goal $\hat{V}^{max}$. Taking into account that $\hat{V}_1^{max}$ and $\hat{V}_2^{max}$ are linear functions of $\dot{q}_1^{max}$ and $\dot{q}_2^{max}$ respectively and all variables are normalised, the optimal solution should satisfy the equations

$$\hat{V}_1^{max}=1,\quad \hat{V}_2^{max}=1 \tag{15}$$

Otherwise, it would be possible to improve solution by simultaneous reduction of $\dot{q}_1^{max}$ and $\dot{q}_2^{max}$.

Using equations (11) and (15) we can define a set of planar curves (Fig. 3)

$$\dot{q}_1^{max}=\frac{1}{l_1 s_2},\quad \dot{q}_2^{max}=\frac{\rho}{l_1 l_2 s_2} \tag{16}$$

that depends on two parameters $\dot{q}_2^{max}$ and l_2 (design parameter l_1 can derived from the constraint $l_1+l_2=l$). Analysis of Fig.3 shows that Pareto-optimal solutions form an envelope for the curves (16). Hence, a desired trade-off set can be found from equations (16) and additional expression

$$\det\begin{bmatrix}\dfrac{\partial\dot{q}_1^{max}}{\partial q_2} & \dfrac{\partial\dot{q}_2^{max}}{\partial q_2}\\[2mm] \dfrac{\partial\dot{q}_1^{max}}{\partial l_2} & \dfrac{\partial\dot{q}_2^{max}}{\partial l_2}\end{bmatrix}=0 \tag{17}$$

that can be transformed to an equation for q_2

$$a_2\cos^2(q_2)+a_1\cos(q_2)+a_0=0 \tag{18}$$

where $a_2=\alpha(1-\alpha)$, $a_1=(1-\alpha)$, $a_0=\alpha^2, \alpha={l_2}/{l_1+l_2}$

The eqs. (16), (18) enable the designer to generate a trade-off set easily. It is necessary to calculate q_2 from eq. (18) for various l_2 and substitute q_2, l_1, l_2 into eqs.(16). Nomogram for optimal design parameters determination is presented on Fig.4.

In the case of competing goals $\tilde{V}^{max}$ and $\hat{V}^{max}$ a trade-off curve consists of several parts that are determined by expressions

(a) $\tilde{V}^{max}=\tilde{V}^*,\ \hat{V}_1^{max}=1,\ \hat{V}_2^{max}=1$

or

(b) $\tilde{V}^{max}=\tilde{V}^*,\ \hat{V}_1^{max}\geq 1,\ \hat{V}_2^{max}=1$

or (19)

(c) $\tilde{V}^{max}=\tilde{V}^*,\ \hat{V}_1^{max}=1,\ \hat{V}_2^{max}\geq 1$

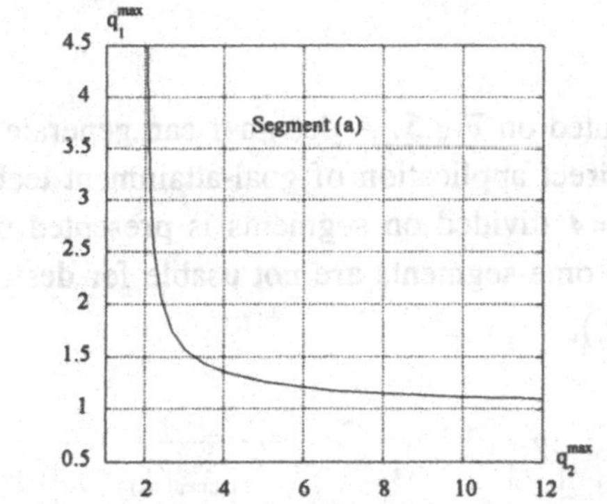

Fig.4 Nomogram for optimal design parameters detemination (active goal $\hat{V}^*$).

$\dot{q}_1^{max}$	$\dot{q}_2^{max}$	α	$\tilde{V}^{max}$
1.1117	10.0055	0.1004	2.1167
1.2564	5.0258	0.2030	2.2768
1.4582	3.4025	0.3071	2.5034
1.5945	2.9612	0.3577	2.6537
1.9909	2.4334	0.4423	3.0674
2.2845	2.2845	0.4693	3.3568
2.6707	2.1851	0.4853	3.7312
3.1793	2.1195	0.4934	4.2252
3.8558	2.0762	0.4972	4.8882
4.7773	2.0474	0.4989	5.7988
12.7550	2.1688	0.4831	3.8030

or

$$\text{(d) } \tilde{V}^{max} = \tilde{V}^*, \hat{V}_1^{max} = 1,\ \hat{V}_2^{max} = 1$$

The number of segments and their sequence depends on value of $\tilde{V}^{max}$Segment (a) of the curve corresponds to the case of noncompeting goals that has been considered above (see eqs.(16)-(18)).

For segment (b) a similiar set of planar curves is described by equations

$\dot{q}_1^{max} = \tilde{V}^* - l_2\dot{q}_2^{max};\ \dot{q}_2^{max} = \dfrac{\rho}{l_1 l_2 s_2}$Applying the same approach we can obtain an an equation for q_2

$$\left(c_2 + {}^{l_2}\!/_{l_1}\right)\left(c_2 + {}^{l_1}\!/_{l_2}\right) = 0,$$

that leads to the following expressions for an envelope

$$\dot{q}_1^{max} = \tilde{V}^* - 1;\ \dot{q}_2^{max} = 1/\alpha;\quad 0 < \alpha \le 0.5$$

and

$$\dot{q}_1^{max} = \tilde{V}^* - \alpha/(1-\alpha);\ \dot{q}_2^{max} = 1/(1-\alpha);\quad 0.5 \le \alpha < 1$$

Segment (c) of the curve is an envelope for the curves

$$\dot{q}_1^{max} = 1/l_1 s_2;\ \dot{q}_2^{max} = \tilde{V}^*/l_2 - \dot{q}_1^{max}/l_2$$

that is determined by equations

$$\dot{q}_1^{max} = 1/(1-\alpha);\ \dot{q}_2^{max} = \tilde{V}^*/\alpha - 1/\alpha(1-\alpha)$$

Segment (d) is determined by a set of equations

$$\dot{q}_1^{max} = 1/(1-\alpha)s_2;\ \dot{q}_2^{max} = \dot{q}_1^{max}\,\rho/\alpha;\ \dot{q}_1^{max} + \alpha\dot{q}_2^{max} = r$$

After elimination of $\dot{q}_1^{max}, \dot{q}_2^{max}$ we get an equation

$$1+\rho = r(1-\alpha)s_2$$

that can be solved for α.

Nomograms for segments (a)-(d) are presented on Fig.5. A designer can generate a trade-off curve using both the nomograms or direct application of goal-attainment technique. An example of trade-off curve for $\tilde{V}^{max} = 4$ divided on segments is presented on Fig.6. Careful analysis of the curve yields that some segments are not usable for design for design because the value α is too high $(l_2 > l_1)$.

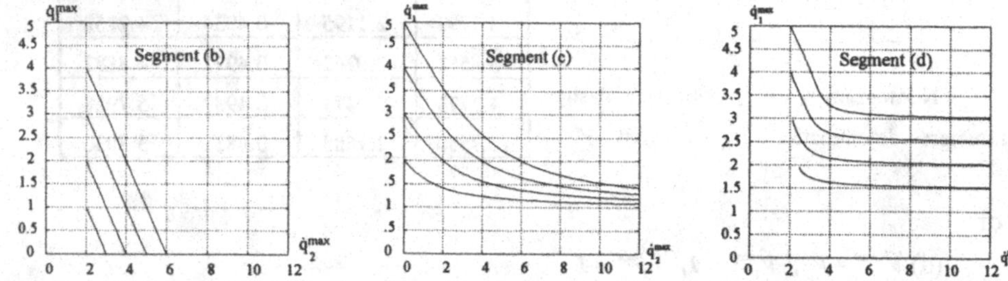

Fig.5 Nomogram for optimal design parameters detemination (competing goals $\tilde{V}^*$ and $\hat{V}^*$).

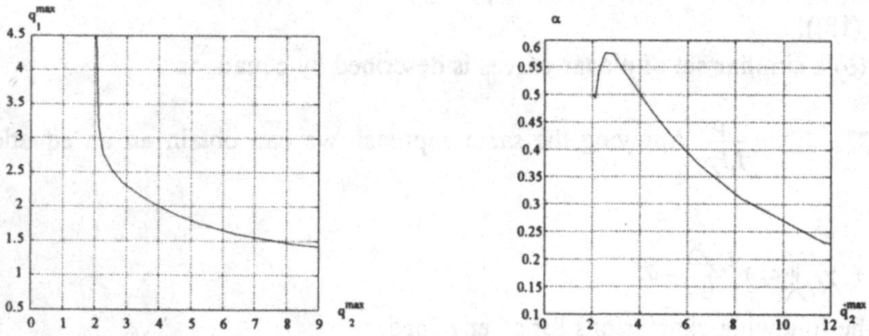

Fig.6 Analytically obtained trade-off curve for $\tilde{V}^* = 4$

The latter is not desirable from engineering point of view and leads to very strong requirements to actuators. In this case is preferable to change design specification $\hat{V}$ in order to keep ratio $\tilde{V}^{max}/\hat{V}^{max}$ about 2.

Conclusion

In the paper a novel approach to robot kinematic design has been proposed that based on multiobjective optimisation of mobility performance measures using goal-attainment technique of Gembicki. The proposed approach has been applied to SCARA manipulator design and yielded tables and nomograms that can be directly used by mechanical engineers.

This work is a part of main project on robotic cells design that currently is carried out in Robotics Laboratory of MRTI. The project also includes manipulator dynamics optimisation and robotic cell layout design.

References

1. Graettinger R.J., Krough B.M. "The Acceleration Radius: A Global Performance Measure for Robotic Manipulator" / IEEE Journal of Robotics and Automation, vol.4, no.1, February, 1988, pp.60-69.

2. Kim J.O., Khosla P.K. "Dexterity Measures for Design and Control of Manipulators" / IEEE/IRS International Workshop Robots and Systems IROS'91, Nov., Osaka, Japan.

3. Kobrinskij A.A., Kobrinskij A.E. "Manipulation system of Robots", Nauka, 1985 (in Russian).

4. Pashkevich A.P. "Estimation of Kinematic Properties of Manipulators", Automation and Computers, no.18, 1989 (in Russian).

5. Paskevich A.P., Fleming P.J. " A Multyobjective Optimisation Approach to Robotic Manipulator Design" / Proc. Int. Symp. on Computer-Aided Design of Control Systems, CADCS'91, UK, Swansea, 1991, 6 pp.

6. Yoshikawa T. "Mobility and Redundancy Control of Robotic Mechanisms" / Proc. IEEE Int. Conf. Robotics and Automation, 1985, pp.1004-1009.

7.Waldron K. "Design of Arms" in "International Encyclopedia of Robotics",R.Dorf and S.Nof, Editors, John Wiley and Sons, 1988.

8.Wang C.C. "The Optimal Design of Robot Drive System-Actuator Gains"/ Proc. IEEE Int. Conf. Robotics and Automation, 1987, pp.816-821.

DEVELOPMENT OF ROBOTIC SURGICAL INSTRUMENTS

Y. Louhisalmi and T. Leinonen
University of Oulu, Oulu, Finland

ABSTRACT

Robotic theory, components and constructions have been studied and applied for neurosurgical instrumentation to improve the treatment accuracy and to cut the operation length. A passive six-jointed sensored arm was developed for localization of neurosurgical targets. The arm was moved by the human hand. This experience promotes now the development of an active arm that moves also itself. In this paper three topics of this research are presented with preliminary results; tasks for the active arm, configuration of its linkage and the actuator-sensor construction.

1 INTRODUCTION

The robotic surgical treatment is based on digital imaging modalities, such as computed tomography (CT), magnetic resonance imaging (MRI) and ultrasound (US). They give out digital data transferable for any visualization, computer-aided or robotic treatments. Physical laws and data distortions cause the only limitations for accuracy. Distortions may occur from object artefacts or movements, as well as from image misinterpretation or scanning inaccuracy. However, image processing and three-dimensional visualization methods produce adequate and useful representations of the target volume on computer display. Computer-aided treatment methods may help to confirm and clarify the exact trajectory and target coordinates before operation and will show them during operation. Robotic treatment methods may control or guide surgical laser or other instruments.

Kwoh published the use of a commercial PUMA 200 robot for neurosurgery in 1988 [Kwoh]. Other studies have also been presented later [Benabid, Drake, Frankhauser, Koyama]. Lavallee has done large studies of the methodology of computer assisted medical interventions [Lavallee]. We also recognized during the development of a passive localization system [Louhisalmi 1992A] that the system could include also active properties. It might move and stop in any desired positions by actuators and brakes, for example. Further, some routine neurosurgical tasks might be done by a such system. However, an intelligent robotic assistant should be developed for neurosurgical procedures instead of the use of the commercial robot [Louhisalmi 1992B].

2 THEORETICAL CONSIDERATIONS OF THE ACTIVE ROBOTIC SURGERY

We have developed the passive localization system that consists of two parts; a passive six-jointed sensored arm and a visualization software in a UNIX workstation, fig. 1. The arm is moved by the human hand.

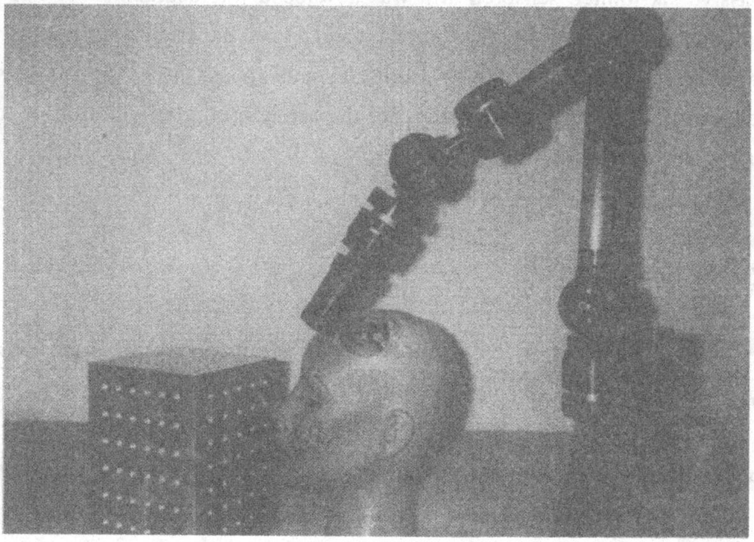

Fig. 1. The passive localization arm.

Movements are registered by joint encoders and microcontrollers, and are used to visualize image data three-dimensionally on the computer display in respect to the tip of the arm.
The next work is now the development of an active system. The arm includes actuators that can be controlled and programmed by computer. The new system should include both the active robotic use and the passive human hand use in the same system. Three topics of this work are written in following sections; tasks for the new arm, configuration of its linkage and the actuator-sensor construction.

2.1 TASKS FOR THE ACTIVE ARM

Following tasks are formed by the experience of the passive arm by which more than 25 operations have been done [Louhisalmi 1992B]. The active drive is proposed there, where it might be argued. Results will guide the development of the active arm.

The calibration procedure is the first task. It fixes the image coordinates and the patient coordinates together. Calibration marks are recognized and coordinated in the image data before operation. The same marks must also be found from patient at the beginning of operation. Then, the transformation matrix from the image coordinates to the patient coordinates can be computed. When the patient is fixed and non-movable, the robotic system may be used to coordinate these marks. If it is done by the active drive, there are possible hazards, such bumping the patient by the robotic arm. The passive human hand use is more recommended or the arm must be equipped by touch sensors. Also rough recognizing can be done on the human hand guidance and only highly precise centering of marks would be done by the active drive.

The next task is the sterile draping. The active arm can be first draped by driving it into the sterile bag. Then, it might help to sterilize and drape the patient. Before sterilizing only non-sterile instruments are used and they are changed to sterile instruments during sterilizing. Instrument change might be done without human help.

At the beginning of the image-based orientation procedure the robotic arm is used to locate the previously, manually and experimentally, chosen route to the target. When the active arm is used, changes to this route can be done by moving the arm by the hand and its response is either new orientation to the same target point, or new target point.

Task that can be done only by the active arm is the assistant procedure. It includes instrument holding tasks and tasks for picking knives, needles or whatever instruments are needed during normal brain operations. Also the arm could be programmed to do routine treatment procedures semi-automatically. It is possible to move laser, US-aspirator or suction tube inside the programmed volume as well as take 3-D US scans by the active arm. US scans are used for dynamic visualization of the remaining target and movements of the tissue during treatment.

Concluding presented tasks include the robotic and the human hand use alternately, that enter additional claims to the arm configuration and the joint construction compared to commercial robots. Secondly tasks must always be done under surgeon's supervision and with good interaction. Thirdly the robotic arm must be specially designed for sterile procedures.

2.2 CONFIGURATION OF THE ACTIVE ARM

Our passive mechanical arm is one device for accurate image guided surgery. The other might be armless systems, that are based on magnetic field, light wave or sonic wave applications. Benefits for armless systems are free rotations and movements of the hand and instruments. Certify of faultless detection and avoid of dead angles are the biggest problems. If the human hand is connected to a mechanical linkage, then those two arms may conflict because of their different or reduced degrees of freedoms. Also external loads to the human hand need balancing. Because the mechanical linkage is the spine of any such active robotic systems, this work is stressed to find such configuration that provides routine surgical working.

Configuration of our passive arm was formed from following directions. When the hand piece lift, fallen, pushed forward, or pulled backward the linkage had to follow smoothly. There were six degrees of freedom. Rough movements were done by first three joints, fine movements mainly by last three joints. Distance from the center of patient's head to the base of the arm was chosen to be 300 mm. The work volume included calibration points and route to the operation target. The arm was used to locate target points, lines or volumes by short term use.

We recognized if long term use is needed then more degrees of freedom might be arranged for the instrument to ensure convenient and accurate surgery. Ergonomic studies of the human hand are also very important. Fluent free-hand and sub-millimetric surgical work needs well-designed instruments to guarantee accurate long term use without fatigue. Ulnar deviation, radial deviation, palmar flexion or dorsiflexion of the human hand should be avoided [Nunez]. The wrist should be straight. All fingers and the thumb might be used for grip. Control of the instrument, such as on-off switch or control of power, might be done without loss of accuracy. If it is done by the thumb it might be done by moving it along the instrument. If it is done by the palm it might be done by pressing smoothly. Shoulders, elbows and forearms should lie normally and in rest. The hand may not be too straight, the forearm not too extended or in flexed position. The most natural use of the human hand might be done in the level of the elbow. Devices must be developed for the right and left hand use. Mass center of instruments must be laid optimally.

2.3 ACTUATOR-SENSOR CONSTRUCTION

The third topic is the actuator-sensor construction. The human hand use may alternate with the robotic use. Such construction is needed that releases gear and motor inertias during the human hand use, but keeps encoders connected to ensure accuracy. Conventional encoder-motor-gear-clutch-brake combinations used for robots, are too big and weight too much. Available and useful combinations are under study. They will be developed or replaced by better construction.

3 RESULTS AND CONCLUSION

Experience from the development of the passive arm for neurosurgical use was encouraged the development of the active arm. In this paper three topics of this research were presented with preliminary results; tasks for the active arm, its configuration and construction of joints. Theoretical results are under testing by prototypes.

ACKNOWLEDGMENT

This work was financially supported by grants from the Academy of Finland, Tekniikan edistamissäätiö and Tauno Tönning foundation.

REFERENCES

Benabid, A.L., Lavallee, S., Hoffmann, D., Cinquin, P., Demongeot, J., Danel, F., Potential use of robots in endoscopic neurosurgery. Acta Neurochir., Suppl. 54, pp. 93-97, 1992.

Drake, J.M., Joy, M., Goldenberg, A., Kreindler, D., Computer- and robot-assisted resection of thalamic astrocytomas in children. Neurosurgery, Vol. 29, No. 1, pp. 27-33, 1991.

Frankhauser, H., Glauser, D., Flury, P., Villotte, N., Meuli, R.A., Prototype of a robot for stereotactic neurosurgery. Acta Neurochir., Vol. 117, Fasc. 1-2, 1992

Koyama, H., Uchida, T., Funakubo, H., Takakura, K., Frankhauser, H., Development of a new microsurgical robot for stereotactic neurosurgery. Stereotact Funct Neurosurg., Vol. 54+55, pp. 462-467, 1990.

Kwoh, Y.S., Hou, J., Jonckheere ,E., Hayati, S., A robot with improved absolute positioning accuracy for CT guided stereotactic brain surgery. IEEE Trans Biomed Eng., Vol. 35, No. 2, pp. 153-160, 1988.

Lavallee, S., Cinquin, P., Computer assisted medical interventions. 3D Imaging in Medicine. NATO ASI series, F60 (Ed. by K.H. Höhne et al.) Springer-Verlag, pp. 301-311, 1990.

Louhisalmi, Y., Alakuijala, J., Oikarinen, J., Ying, X., Koivukangas J., Development of a localization arm for neurosurgery. Proc. of the IEEE Eng Med Biol Conf. Vol. 14 , pp. 1085-1086, 1992A.

Louhisalmi, Y., The idea of a robotic assistant for neurosurgery. Laboratory of Machine Design, Research papers 1992. University of Oulu, Report No. 89, pp. 48-51, 1992B.

Nunez, G., Kaufman, H., Ergonomic considerations in the design of neurosurgery instruments. J Neurosurg., Vol. 69, pp. 436-441,1988.

OPTIMIZATION OF KINEMATICS PERFORMANCES OF MANIPULATORS UNDER SPECIFIED TASK CONDITIONS

S. Zeghloul, B. Blanchard and J.A. Pamanes
U.R.A. - C.N.R.S., Poitiers, France

ABSTRACT

This paper presents a method for dealing with the problem of designing a robotized cell. The problem dealt with here, concerns determining the robot's placement, for any given task. The approach presented is based on an optimization technic, which consists of determining a placement which minimizes an objective function describing the task. In the proposed formulation, during the placement search, this method considers the constraints due to obstacles as well as the robot's joint limits. Through the different examples presented we will show the effectiveness of this method.

INTRODUCTION

Robot choice and placement together, constitute an important problem in the domain of robot cell design. In fact the ability of the robot to reach and perform any given task depends almost entirely on the robot's structure and placement. By choosing an adequate placement, for any given robot, it is possible to ensure task accessibility and to improve the robot's performance (geometric, kinematic).

Solutions to similar problems have been presented by authors proposing the optimization of some performance criterion. For example, Nelson [1] proposes a method which locates assembly tasks in the manipulator workspace in order to optimize manipulability. The compatibility index [2], can be applied to find the relative manipulator/task position which optimizes the accuracy and/or the magnitude of force and velocity. A criterion which allows keeping all the manipulator joint angles away from its limits as far as it is possible, has been applied by Pamanes [3]. Another approach [4] deals with the problem by minimizing the difference between the volumes of the workspace and the taskspace, or the distance of their gravity centers.

Most of the above works apply a single criterion in order to determine the relative manipulator/task placement. However, the use of multiple criteria would be advisable for this problem for industrial tasks. For instance, a task may have assembly zones and transfer zones. Clearly, at an assembly zone a good manipulability is desirable, whereas at a transfer zone the increase of the velocity in a certain direction is preferable. For a given task, the compatibility index proposed by Chiu allows combining four criteria. However, adding up other important criteria with that index does not seem feasible.

We have developed a method, close to the one presented in this paper, for dealing with multiple criteria, but its application was presented for one manipulator with six d.o.f [5].

2

In this paper, our formulation is extended in order to treat the most commonly used robots, to consider either the robot's placement or task placement. We have also included a local path planner, in the formulation, which considers, during the optimization process, only the placements for which a free path exists for the whole trajectory.

OPTIMAL PLACEMENT FORMULATION

The placement problem can be stated as follows: *Find a design vector (orientation and position) which minimizes an objective function under a task constraint*

The design vector is defined by the position and the orientation of the reference Σc (attached to robot base) related to the absolute coordinate system Σo if the problem consists of finding the robot placement in a workstation. Otherwise, the design vector describes the position and the orientation of the reference Σt (attached to task) related to the absolute coordinate system Σo if the problem consists of finding the task placement, in this case the robot base is fixed. In our approach the design vector is defined by the position in the Cartesian coordinates, it is written as r_{ox}, r_{oy} and r_{oz} and the orientation given by the pitch, roll and yaw angles written as λ, μ and ν.

The task of the manipulator is defined in the task coordinate system Σt by a set of end-effector locations, described by the transformation matrix $A^i te$., at the points $S_i\ (i = 1,2,\cdots,m)$ of the task (Figure 1). The joint coordinate vector q_i of the manipulator, at any point of the path is the solution of the inverse kinematics problem. This solution is obtained with end-effector location $A^i ce$ related to the coordinate system Σc attached to the robot base, this location is computed from that related to Σt by:

$$A^i ce = Aco\ Aot\ A^i te$$

where the matrix Aot represents the (4x4) transformation matrix describing the location (position and orientation) of Σt related to Σo, and Aco: the transformation matrix describing the location of Σo related to Σc. The matrix Aot is a function of the design vector. If the problem consists of finding the task placement, then the matrix Aco is equal to the identity matrix. Otherwise, the matrix Aot is equal to the identity matrix and Aco depends on the design vector.

PROBLEM CONSTRAINTS

In general, the constraints limit the evolution of the Design Vector. In our formulation we consider two types of constraints: the explicit constraints and the implicit constraints. The explicit constraints are directly applied to the components of the design vector and are expressed by (Eq. 1), while the implicit constraints are applied to variables that depend implicitly on the design vector.

$$\begin{array}{ll} \lambda_l \le \lambda \le \lambda_u & r_{ox_l} \le r_{ox} \le r_{ox_u} \\ \mu_l \le \mu \le \mu_u & r_{oy_l} \le r_{oy} \le r_{oy_u} \\ \nu_l \le \nu \le \nu_u & r_{oz_l} \le r_{oz} \le r_{oz_u} \end{array} \qquad (1)$$

3

where the subscripts l and u denote respectively lower and upper bounds, which can be determined from the available space at the workstation. In addition we consider the two sets of implicit constraints in order to avoid interference and the joint limits by :

$$(d_{hk})_i \geq d_{hkl} \quad h=1,2,\dots,n_e \quad i=1,2,\dots,m \quad k=1,2,\dots,n$$

where $(d_{hk})_i$ is the shortest distance between the h-th object of the environment and the k-th link of the manipulator for the i-th task point, with the number of objects in the environment being defined as n_e. The distance $(d_{hk})_i$ for any task point is bounded by a minimum admissible value d_{hkl}. The computation of the distance $(d_{hk})_i$ is carried out using an efficient formulation of the gradient projection method [7]. The parameters m and n are respectively the number of task points and the number of joint variables of the manipulator. The manipulator joint constraints can be expressed as :

$$q_{k_l} \leq q_{ki} \leq q_{k_u} \qquad k=1,2,\dots,n \quad i=1,2,\dots,m$$

where the q_{ki} is the k-th component of the joint vector q_i of the manipulator's configuration corresponding to the i-th path point.

OBJECTIVE FUNCTIONS

In our approach many functions can be used in the placement problem. Some of these functions consider only one criterion to be optimized and others integrate multiple criterion.

- Travel Time

As an objective function, one can use the travel time of the trajectory defined by the points S_i. In this case the optimization process calculates the placement that leads to the minimal trajectory travel time, respecting the problem constraints. In order to compute the travel time we consider the two popular velocity profiles used in the control of industrial robots without considering the robot's dynamic.

- Joint Limit Avoidance

This function, defined by (Eq. 2) determines the placement, allowing the robot to execute the task with configurations q_i that keep the joint of robots as far as possible from their limits. The objective of the optimization process is to minimize the following function, proposed by Pamanes [3].

$$\phi = \bar{k} + k_{\sigma} \tag{2}$$

where $\bar{k}$ is the mean, and k_{σ} the standard deviation of the $m \times n$ values of the ratio k_{ij} defined by:

$$k_{ij} = \left[\frac{\Delta q_{ij}}{\Delta q_{i\,\max}}\right]^2 \qquad j=1,\dots,m \quad i=1,\dots,n$$

where Δq_{ij}: represents the deviation of the i-th joint variable, with respect to the center of its permissible range, at the j-th task point, and $\Delta q_{i\,\max}$ represents the maximum deviation permissible for the i-th joint variable. This criterion, here termed joint limits avoidance, can be used to measure the ease of a manipulator to reach to a certain task.

4

- *Multi-Criterion Function*

It is also possible to determine the placement that minimizes many performance criterion at the same time. In this situation we assign a criterion to each task point S_i. In this approach we have used certain performance criterion available in literature such as:

*The manipulability measure introduced by Yoshikawa [6] is quantified by the index $w=\sqrt{\det(JJ^T)}$, where J is Jacobian matrix of the manipulator. The manipulability is a measure of ease when arbitrarily changing the position and orientation of the end effector. Thus, its maximization would be appreciated in task zones where relatively large deviations in the prescribed motion of the end effector are likely.

*The compatibility index introduced by Chiu [2] allows one to optimize the magnitude and accuracy of force and velocity of the manipulator on related displacement direction u. The compatibility index is based on the transmission ratios of force α and of velocity β in the considered direction. These ratios are given by :

$$\alpha=\left[u^T(JJ^T)u\right]^{-1/2} \quad , \quad \beta=\left[u^T(JJ^T)^{-1}u\right]^{-1/2}$$

where u is a unit vector in the direction of interest. If the magnitude of force or velocity are considered, α or β must be maximized. If the accuracy of force or velocity are considered, then α^{-1} or β^{-1} must be maximized.

In the multi-criterion approach, we choose p points ($p \le m$), from the m points defining the task, to which kinematic criteria are assigned, and we express as kj the index of the kinematic criterion assigned to some point S_i. The objective of the optimization is to locate the manipulator in such a way that the p indices kj are kept as great as possible. However, since in general the orders of values of these indices are different, they can't be compared effectively in an optimization process. We avoid this difficulty by introducing the following normalized index $Cj = kj / kj^*$, where the normalization factor kj^* is the maximum value that can be reached by kj^*. Thus, Cj is bounded as follows : $(0 \le Cj \le 1)$

Next, in order to complete the formulation, we define a small typical element of the set of p indices Cj as :

$$C=\overline{C}-C\sigma \tag{3}$$

where $\overline{C}$ and $C\sigma$ are respectively the mean and the standard deviation of the set. It is evident that the optimal set, the one with all its elements as great as possible, must have a C maximum.

The minimization of the objective functions (Eq. 2, Eq. 3) subject to problem constraints can be carried out using some classical methods of non-linear programming. We apply the Box method [8], which does not require the derivatives of the objective function in order to solve the placement problem.

The same approach is applied when computing the normalization factors kj^*. In fact, the normalization factor kj^* is the maximum value which can be obtained by kj. This maximum is determined by solving an optimization problem.

5

EXAMPLE

The method presented was used in order to determine the relative robot/task placement for the following example: it consists of an arc welding task, which is composed of four solder joints modelized by straight segments. In dealing with this example we considered two types of robots, a 6R robot and a robot consisting of a spherical type arm (RRP) and a 3R wrist. For both cases we calculated the placement while considering either the robot's base as fixed, or the task as fixed. In all cases, in order to calculate the placement we first of all look for a placement that will ensure the accessibility to all task points while respecting all of the constraints. Next, by assigning the performance criterion to particular points in the task, we look for the placement that will optimize all of the criterion. In this example we have assigned, the manipulability criterion to the intersection points of the four segments, and accuracy of velocity to all of the segment's middle points. In all cases, this method has allowed us to find the placement as well as improving the objective function, however we limit ourselves, in this paper, to presenting the results for the 6R robot where the design vector is defined by the position and orientation of the robot base. It should be mentioned that, as expected, the results obtained with the 6R were better than those produced by the spheric robot.

The placement problem have been solved using four independent variables: the component of r_o , and the angle ν (rotation about z axis). The figure 2 shows the important improvement on the C_j factors after optimization. The sequences of the figure 3 allows us to observe the configurations obtained for some path points, in both initial and optimal placement. Note: the number of task points is 20.

CONCLUSION

The problem of relative robot-task placement calculation is dealt with in this paper. Our approach is based on the optimization of an objective function describing the task. This method makes it possible to determine a placement which ensures that the robot can reach the task and improve robot performances during task execution. This method can be applied to different robot structures, for any task given, permitting evaluation of their capacity to accomplish the task. Therefore, the method can be used as a tool for selecting robots.

Our approach consists of finding a placement that optimizes an objective function describing the task. We propose optimizing different objective functions in order to determine the placement. These functions incorporate different performance criteria. In our formulation we consider the constraints that make it possible to determine a placement which ensures non-collision for the robot as well as respecting the joint limits.

The simulation tests that were done show the effectiveness of our method for determining robot placement in an environment containing obstacles.

6

REFERENCES

1. B. Nelson, K. Pedersen and M. Donath, "Locating assembly tasks in a manipulator's workspace". IEEE Proc. of the Int. Conf. on Robotics and Automation, pp. 1367-1372. (1987).
2. S.L. Chiu, "Task compatibility of manipultors postures". The Int. J of Robotics Research, Vol. 7, nø 5, pp. 13-21. (1988).
3. J.A. Pamanes-Garcia, "A criterion for optimal placement of robotic manipulators". IFAC Proc. of the 6th Sym. on Information Control Problems in Manufacturing Technology, pp. 183-187. (1989).
4. P. Chedmail, "Synthèse de Robots et de Sites Robotisés. Modélisation de Robots Souples". Thèse de Doctorat d'Etat, Univ de Nantes, E.N.S.M, France. (1990).
5. S. Zeghloul and J.A. Pamanes-Garcia, "Multi-criteria optimal placement of robots in constrained environments". Robotica Int J Vol 11, pp. 105-110. (1993).
6. J. Yoshikawa, "Manipulability of robotic mechanisms". The Int. J. of Robotic Research, Vol. 4, nø 2, pp. 3-9. (1985).
7. S. Zeghloul, "Developpement d'un Système de C.A.O-Robotique Integrant la Planification de Tâches et la Synthèse de Sites Robotisés". Thèse de Doctorat d'Etat, Univ de Poitiers, France (1991).
8. M. J. Box, "A new method of constrained optimization and a comparison with other methods". Computer J., Vol. 8, pp. 42-52. (1965).

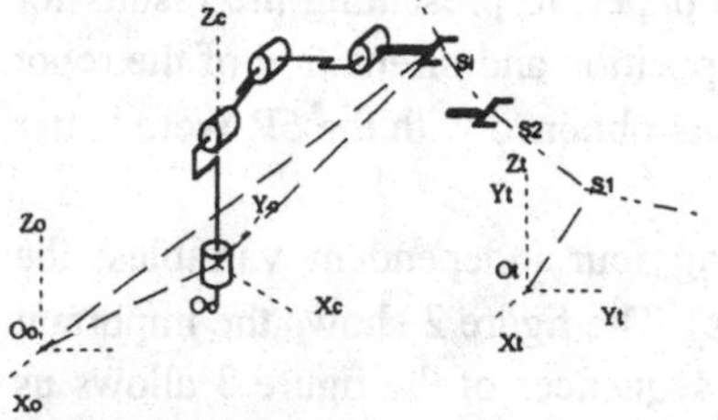

FIGURE 1: Task and coordinate systems Σo, Σc and Σt

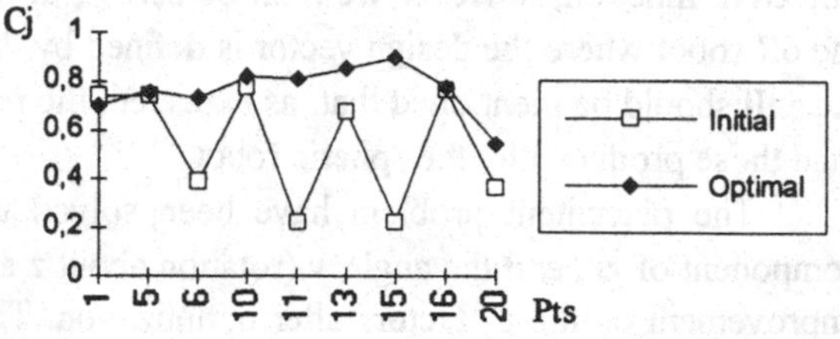

FIGURE 2 : Values of the indices Cj for the initial and the optimal placements

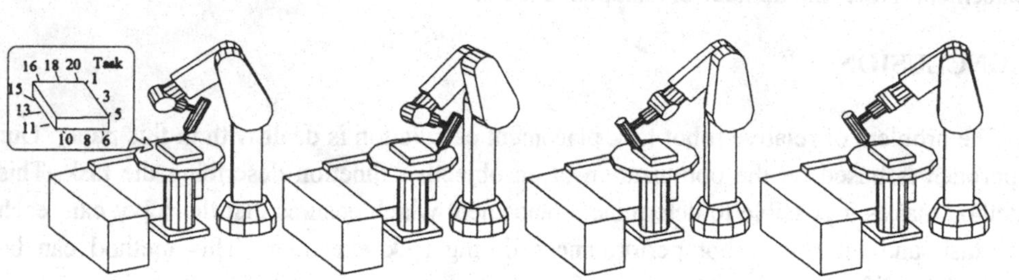

Initial Placement

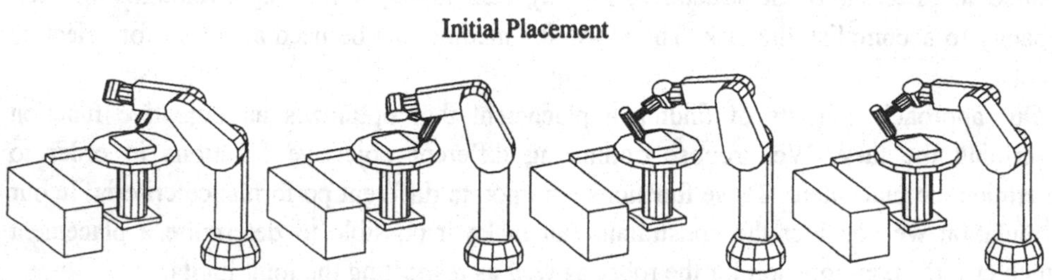

Optimal Placement

FIGURE 3: Simulation of the task for the initial and optimal placements

Part VII
Synthesis and Design II

DYNAMIC TESTING OF SPACE MANIPULATORS

M.R. Kujath and U.O. Akpan
Technical University of Nova Scotia, Halifax, N.S., Canada

Abstract
The paper presents several aspects of the orbital environment that affect the dynamic testing of space manipulators and that are beyond the earthly engineering experience. Many of the dynamic parameters of space manipulators cannot be verified or identified on ground because of the effect the gravity has on the manipulator structure. The in-orbit testing, on the other hand, has to be limited to a very short time compared to ground based experiments. The paper outlines the scope of ground preparations that are necessary to assure that the in-orbit tests are successful. Then, it discusses an optimization procedure of the space experiments using modal parameter sensitivity.

1. Introduction
One of Canada's contributions to the Space Station Program is the Space Station Remote Manipulator System (SSRMS), see Figure 1. The SSRMS is a large, flexible, mechanical arm, measuring approximately 17.6 m when fully extended and it is designed to manipulate a wide range of payload masses up to and including the Shuttle orbiter. It has seven revolute joints, two Latching End Effectors (LEEs) and two boom assemblies.

Verification tests to be carried out on the ground and in-orbit are currently being developed. The objectives of these tests include [1] characterization of robotics and structural dynamics parameters, and assessment of the fidelity of the SSRMS mathematical models.

2. Orbital Conditions and Technical Constraints
The Canadian built space manipulators are designed for the low orbits of the Shuttle orbiter or the planned space station. The orbits of these spacecrafts are relatively close to the Earth's surface because their altitudes measure a small fraction of the Earth diameter (less than 0.04; for comparison, the altitude of a geostationary satellite measures approximately 3 Earth diameters). A spacecraft circles the Earth on such a low orbit in about 90 minutes. Half the time it experiences the orbital night when it is shielded by Earth from the Sun and half the time it experiences the orbital day when it is fully exposed to the Sun's radiation. During the night the surface temperature drops to -150° C and during the day it rises to + 150° C. The ever changing orbital illumination is very critical if cameras are used for spatial measurements and the varying temperatures affect the electrical and mechanical properties of the

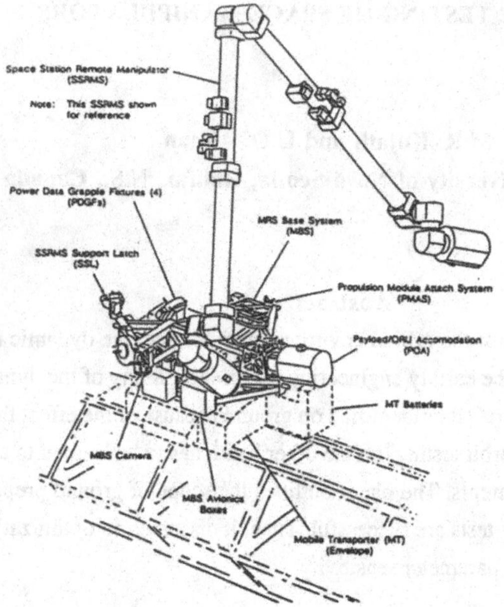

Figure 1. Space Station Remote Manipulator System (SSRMS).

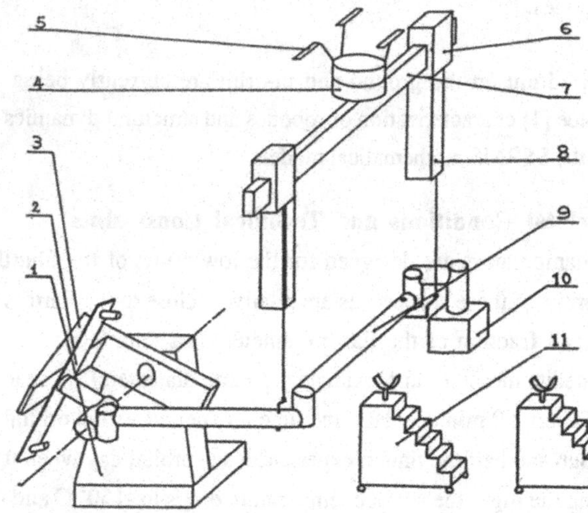

Figure 2. Suspended Manipulator of the Measurement Systems Test-bed (MST). (1) Stand, (2) Camera Carrousel, (3) Cradle, (4) Rotating Head, (5) Brace, (6) Trolley, (7) Spreader, (8) Counterbalance, (9) MST Arm, (10) Target, (11) Rolling Support.

manipulator components. These conditions are very different from typical ground-laboratory conditions where the illumination and the temperature can be stable for as long as it is needed.

A test on a space manipulator has to rely heavily on the flight instrumentation used for its normal operation because any additional instruments will add dramatically to the cost of the experiment. In the case of the SSRMS the flight instrumentation includes video cameras, joint-angle resolvers, force/moment sensors (if available), and motor current sensors. Additional instruments to measure the acceleration of the base of the SSRMS are considered as well. Again, this is a limitation when compared with ground tests where there is always an option of adding instrumentation.

The in-orbit testing of manipulators has to be limited to a very short time compared to ground based experiments. This is because of the high costs involved in any space operation. Therefore, the length of the space testing has to be measured in minutes instead of being measured in days as it is on ground. In addition, the maximum speeds of the space manipulators, when compared with their industrial counterparts, are also very low because of the power limitations. Further, space manipulators are designed to be light because of the high costs involved in launching and maneuvering masses in space. Therefore, they are flexible and have very slow dynamics.

In order to address the orbital conditions and the orbital technical constraints of the SSRMS, a special purpose Measurement Systems Test-bed and a set of optimization procedures are being developed.

3. Measurement Systems Test-bed

In support of the space testing of the SSRMS, a Measurement Systems Test-bed (MST) is proposed [1]. The objectives for the use of this facility include validation of the following: (1) The system concept of the experiments, (2) Measurement systems capabilities relative to the project's needs, (3) Calibration methods for the measurement systems, (4) Data processing and parameter identification algorithms, (5) In-orbit tests, (6) In-orbit operational configurations and procedures.

The MST will comprise a full-scale mock-up of the SSRMS (see Figure 2), with a model of the complete measurement system in place, along with an independent, higher accuracy measurement system to test against. The principal physical requirement of the MST is to emulate the SSRMS characteristic motion response in a three-dimensional fashion, while the principal testing requirement is to provide a means of generating realistic measurement data to permit the development of a set of tests which can be performed in-orbit with the highest degree of confidence and success, within available resources.

4. Optimization Example

In order to optimize the space experiments a modal parameter sensitivity study is being used. If the motion of the manipulator tip is selected as the most critical, then the measurements should be carried out at configurations where the tip oscillations are very sensitive to small perturbations of the manipulator model parameters. In the following study the motion of the manipulator is described in two coordinates: the joint global coordinate $\theta = [\theta_1, \theta_2, \theta_3, .., \theta_n]^T$ representing changes of the kinematic configurations, and the joint local coordinate $\mathbf{q} = [q_1, q_2, q_3, .., q_n]^T$ representing small joint oscillations around a given kinematic configuration. The local motion is studied when the manipulator does not move globally. Damping is assumed to be very small and negligible.

5. Definition of Observability Indices

The equation of local motion can be derived from the Lagrange Energy Principle [2]. Since it has been assumed that the oscillations are small and that they are investigated when the manipulator does not move globally, all the terms which are functions of $\dot{\theta}$ and $\ddot{\theta}$ vanish and the nonlinear terms can be neglected. Therefore, the homogenous form of the equation of motion has the form [3]

$$\ddot{\mathbf{q}} + \mathbf{M}(\theta)^{-1}\,\mathbf{K}\,\mathbf{q} = 0\,, \qquad \mathbf{M} \text{ - inertia matrix,} \quad \mathbf{K} \text{ - stiffness matrix} \tag{1}$$

The joint local motion is defined as follows

$$\mathbf{q}(t) = \mathbf{Q}e^{j\omega t}, \qquad \mathbf{Q} \text{ - modal amplitude vector,} \quad \omega \text{ - modal frequency} \tag{2}$$

Substitution of equation (2) into (1) results in the eigenvalue problem

$$(\mathbf{A} - \omega^2\mathbf{I})\,\mathbf{Q} = 0, \qquad \mathbf{A} = \mathbf{M}^{-1}\mathbf{K} \tag{3}$$

Since the modal amplitude of the local motion $\mathbf{Q}$ is small, it can be related to the amplitude of the manipulator tip oscillation $\mathbf{x}$ by the manipulator Jacobian $\mathbf{J}$ as follows

$$\mathbf{x} = \mathbf{J}\mathbf{Q}\,, \qquad \mathbf{x} = [\,x, y, z, \alpha_x, \alpha_y, \alpha_z\,]^T \tag{4}$$

If a small perturbation $d\varepsilon$ of the parameter vector ε takes place then the corresponding change in the tip oscillation amplitude vector $d\mathbf{x}$ can be defined as follows [4]

$$d\mathbf{x} = \frac{\partial \mathbf{x}}{\partial \varepsilon}\,d\varepsilon = \frac{\partial(\mathbf{J}\mathbf{Q})}{\partial \varepsilon}\,d\varepsilon = \mathbf{J}\,\frac{\partial(\mathbf{Q})}{\partial \varepsilon}\,d\varepsilon \tag{5}$$

The vector $d\mathbf{x}$ can be interpreted as a vector of the deviation between the expected nominal and the actually observed amplitude of the tip oscillations. The following norm can be used as an observability index of a single parameter ε_k

$$\mathbf{OBS1}(k) = \left|\frac{d\mathbf{x}}{d\varepsilon_k}\right| = \sqrt{\left(\sum_{i=1}^{n}\frac{dx_i}{d\varepsilon_k}\right)^2}, \qquad \frac{dx_i}{d\varepsilon_k} = \sum_{j=1}^{n} J_{ij}\,\frac{\partial Q_j}{\partial \varepsilon_k} \tag{6}$$

An overall observability index for all the identifiable parameters can be defined as

$$\mathbf{OBS\Sigma} = \prod_{k=1}^{n} \mathbf{OBS1}(k)_n\,, \qquad \mathbf{OBS1}(k)_n = \frac{\mathbf{OBS1}(k)}{\mathbf{OBS1}(k)_{max}} \tag{7}$$

6. Estimation of Structural Parameters

To estimate the structural parameter deviations $d\varepsilon$, in the first step a set of m optimal measurement configurations is selected. The next step requires measurements of the amplitude of the tip oscillations. The difference between the measured amplitudes and the amplitudes computed from the nominal parameters defines the vector dx. The information from the measurements can be gathered into the following equation, which can be solved for the parameter perturbation vector $d\varepsilon$

$$\mathbf{Y} = \mathbf{E}\, d\varepsilon \tag{8}$$

The vector **Y** contains the deviations dx and the matrix **E** contains the individual matrices $\mathbf{J}\frac{\partial(\mathbf{Q})}{\partial\varepsilon}$.

7. Numerical Example

Consider the 3 DOF planar manipulator shown in Figure 3 for which the manipulator Jacobian **J** can be easily derived [2]. The numerical simulation was performed for $\theta_1 = 0$ deg, $\theta_2 = 90$ deg, and a range of $\theta_3 = [-180,180]$ deg at 1 deg intervals. It was assumed that the links of the manipulator have equal lengths l_i; See Table 1 for mass m, stiffness k, and inertia I parameters. The logarithm of **OBSΣ** index is shown in Figure 4. Figure 5 shows an example of the **OBS1** indices.

8. Conclusions

This paper outlines the rationale and the initial concept for a measurement systems testbed to support space experiments. It discusses an optimization of the space experiments using modal parameter sensitivity. To optimize the manipulator configurations two observability indices have been defined. The following general conclusions can be made from the simulations. It is practical to use the introduced observability indices to evaluate different manipulator configurations for parameter observability. The observability indices show very sharp variations of the sensitivity for different configurations and different modes.

9. References

1. M. R. Kujath, W. B. Graham, Measurement System Testbed for the Robotic Evaluation and Characterization of the Space Station Remote Manipulator System, Cooperative intelligent robotics in space III, Boston, 16-18 Nov. 1992, SPIE vol. 1829, pp. 91-101.
2. E. I. Rivin, Mechanical Design of Robots ,McGraw-Hill Book Company ,1987.
3. M. R. Kujath, U. O. Akpan, Optimization of Dynamic Testing of Space Manipulators, 12th International Model Analysis Conference (IMAC), Honolulu, Jan. 31- Feb. 3, 1994.
4. R.Sharp, P. C. Brooks, Sensitivities of frequency response functions of linear dynamic systems to variations in design parameter values, J. of Sound and Vibr., 1988 vol 126(1) pp. 167-172.

Parameter	m_1	m_2	m_3	I_1	I_2	I_3	k_1	k_2	k_3
ε^{nom}	1.500	1.500	1.500	0.125	0.188	0.125	1.000	1.000	1.000
$d\varepsilon^{true}/\varepsilon^{nom}$ (%)	5.0	3.0	4.3	5.0	3.5	4.6	5.5	4.3	5.5

Table 1. Nominal values of structural parameters and the magnitude of parameter deviation used for simulation (units: mass kg, stiffness N/m, and inertia kgm^2).

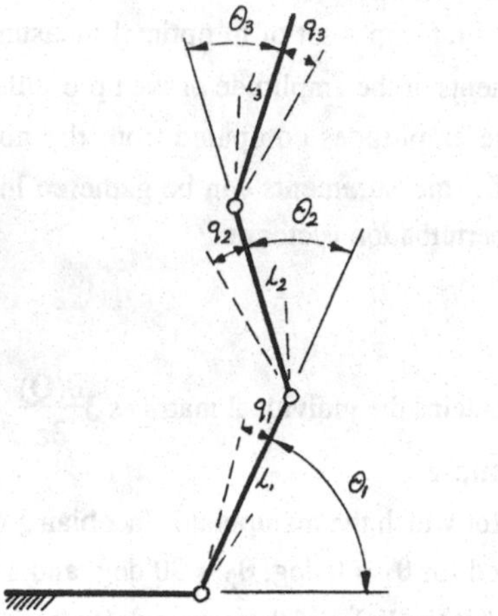

Figure 3. 3 Degrees-Of-Freedom (DOF) manipulator.

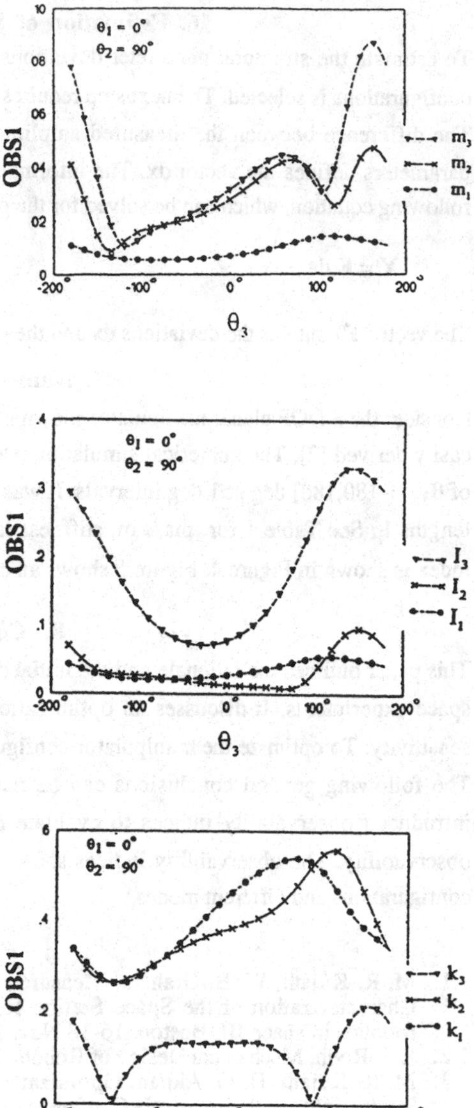

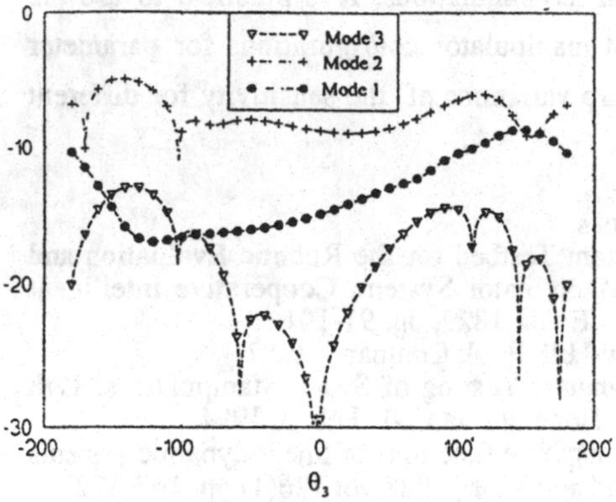

Figure 4. Overall observability index for dynamic parameters of a 3 DOF manipulator (log **OBSΣ**).

Figure 5. Parameter observability indices **OBS1** for the first vibration mode of a 3 DOF manipulator.

DESIGN OF A MANIPULATOR FOR PLANETARY ROVERS

O. Gaudry, F. Pierrot, E. Dombre and A. Liégeois
University of Montpellier, Montpellier, France

ABSTRACT

This paper addresses the stages for selecting the concept of an arm suited to carry out the various scientific operations of planetary robotic explorations: sample recovery, supply of the on-board analysers, soil drilling and testing capabilities, rock inspection, probe/beacon deposit, tool manipulation, etc. After a study of these requirements from a geometrical point of view, the comparison of the various candidate designs leads us to retain a 3 dof pantographic system. It should cooperate with a redundant parallel wrist. The location of the base, the lengths of the links and the joint ranges are then optimised with respect to the operational and folding constraints, given a sketch of the vehicle. A stiffness analysis allows us to select the link rigidities. Finally, the actuators and the drive mechanisms characteristics are derived. The result is a trade-off for taking into account the maximum allowed bending error, the mass and the energy consumption. The preliminary design described in this paper is devised to exert a 20 N force and to allow a 0.5 cm/s tip velocity for 30 W peak power. The arm can fold within a volume smaller than 20 liters. Its estimated mass is about 5 kg, and the control is simple, thanks to the pantographic arrangement.

INTRODUCTION

In 1989, the French National Center for Space Studies (CNES) has launched the V.A.P. (Automatic Planetary Rover) Program [1,2]. The Scientists proposed unmanned, Mars Rover missions as priorities, but the robotic aspects studied on this basis had to be generic for other applications on other planets of the solar system. Among the research teams involved in the preliminary designs and feasibility studies of the mission subsystems, the LIRMM laboratory was in charge of the manipulator(s) studies.

The most probable version of the rover had to be designed in the framework of a small vehicle having an overall mass of about 100 kg. Our first task has been to look after existing manipulator arms suitable for the corresponding requirements of low mass, low energy consumption, and expected performances shown in table 1, together with the space constraints from launching to landing, as well as during the operational phases on the planet surface.

The survey of the existing arms, prototypes and designs for operations by autonomous and teleoperated vehicles led us to discard the ones designed for nuclear, civil, underwater and agricultural applications because of the high masses and energy consumptions. Regarding manipulators for space, the J.P.L. has conducted experiments, some years ago, using the "CURV arm" [3], the pantographic structure of which posseses many advantages with respect to ease of control and "manipulability" index, while the Viking arm seemed to be not enough dexterous and flexible for our applications. On the other hand, the E.S.A. proposal [4] of using HERA type arms has the advantadge of simplifying the space qualifications of components but assumes a fairly large and heavy vehicle.

The main tasks of the arm for the V.A.P. are related to release of surface stations (beacon and probe deposit), and to supply various manipulative functions on the ground and on the mobile platform (soil analysis, sample manipulations). However, the corresponding arm motions required for that have demonstrated to be composed of relatively simple moves:

- lifts and deposits along axes approximately parallel to the local vertical axis (or to the yaw axis),
- large motions consisting in transfers in a vertical plane combined with rotations about a vertical axis.

For these reasons, planar arrangements of arm and forearm(s) have been considered, associated with a single shoulder yaw rotation. Serial and pantographic kinematics are compared in the next section.

Requirements	Values
Masses	Arm : 10 kg End effector+wrist : 5 kg.
Powers . idle mode . braked . in slave mode . peak	 1 watt 10 watts 20 watts 30 watts
Velocities . translational . rotational	 1 cm/s unloaded, 0.5 cm/s with full load 0.5 deg/s
Accuracies .absolute .resolution	 0.5 cm and 1deg 1mm and .0001 rd

Table 1: technological requirements

KINEMATICS ANALYSIS

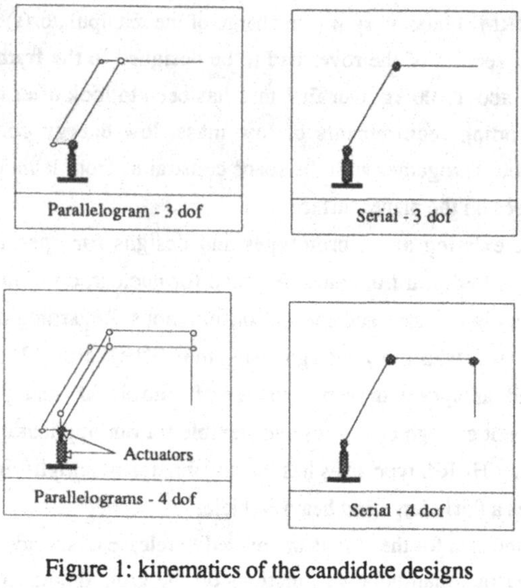

Figure 1: kinematics of the candidate designs

Four candidate concepts have been analysed (Figure 1): 3 degrees of freedom (dof) and 4 dof, both in a serial and a parallel version. It is assumed that the arm carries a redundant parallel wrist [5,6]. Since the orientation capabilities of the parallell wrist are intrinsically limited, a constraint is to move the wrist with a constant orientation. Another constraint is that when the rover is moving, the arm has to be folded within a restricted volume. The advantages and drawbacks of each candidate concept are shown in table 2.

KINEMATICS	ADVANTAGES	DRAWBACKS
SERIAL 3 AXES	- Simple - Easy folding - Large workspace - Fairly good force capabilities	- Dynamically non decoupled - Poor stiffness capabilities
SERIAL 4 AXES	- Simple - Natural folding - Large workspace - Fairly good force capabilities - Redundancy - Orientation of the wrist	- Dynamically non decoupled - Poor rigidity performances - Higher mass
SINGLE PARALLELOGRAM	- Good rigidity - Good accuracy - Good force capabilities - Dynamically decoupled - Low mass	- Limited workspace - Higher number of joints (Passive and active joints)
TRIPLE PARALLELOGRAM	- Good rigidity - Good accuracy - Good force capabilities - Dynamically decoupled - Low mass - Redundancy - Orientation of the wrist	- Limited workspace - Higher number of joints - Folding limitations

Table 2: evaluation of the candidate concepts

In order to satisfy the first constraint, the 3 dof serial arm is discarded. To progress toward the selection of a design solution, we present in the next section the stiffness analysis.

STIFFNESS ANALYSIS

The *Bresse* relationships give, for homogeneous and isotropic materials, the deformation of a beam (figure 2) considered at point B with respect to the one considered at point A, both in translation (denoted δ) and in rotation (denoted Ω) as follows:

$$\Omega_B = \Omega_A + \int_A^B \frac{M(G)}{EI} ds$$

$$\delta_B = \delta_A + \Omega_A \times AB + \int_A^B \frac{M(G)}{EI} GB\ ds$$

where: $M(G)$ is the torque applied to the beam, considered at point G (a point between A and B), I is the beam inertia momentum, and E is the Young's Modulus.

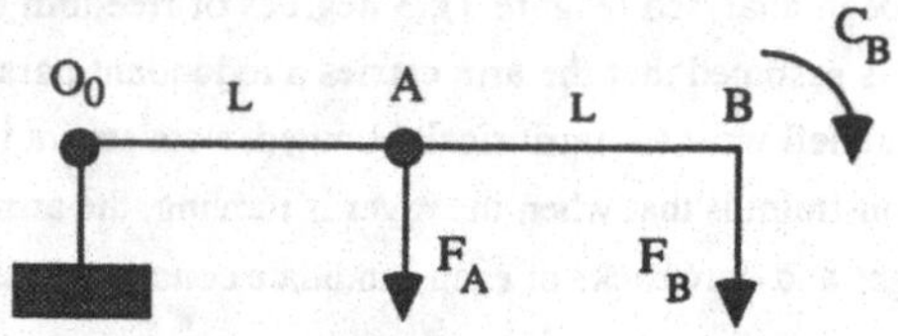

Figure 2: model of a serial robot for statics study

For a serial 2-dof robot the *Bresse* relationships are (Figure 2):

$$\delta_B = \delta_A + L\Omega_A + \frac{F_B L^3}{3EI} + \frac{C_B L^2}{2EI}$$

with:

$$\Omega_A = \frac{F_A L^2}{2EI} + \frac{F_B L^2}{2EI} + \frac{CL}{EI} \qquad \delta_A = \frac{F_A L^3}{3EI} + \frac{F_B L^3}{3EI} + \frac{CL^2}{2EI}$$

$$C = P_B \frac{L}{2} + C_B + F_B L \qquad F_A = P_S + P_A$$

This leads to: $\delta_B = \dfrac{5(P_S + P_A)L^3}{6EI} + \dfrac{8F_B L^3}{3EI} + \dfrac{2C_B L^2}{EI}$

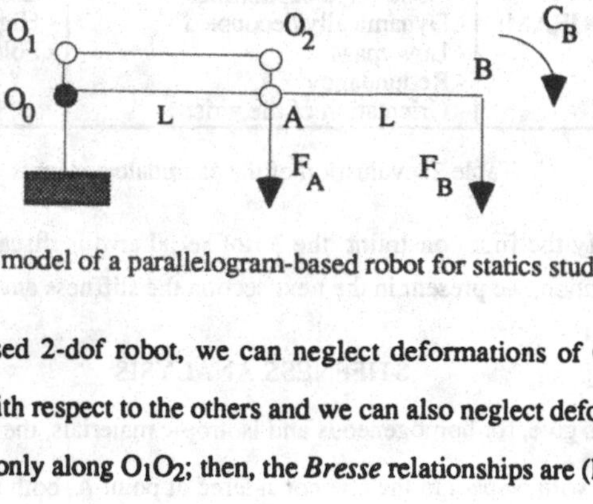

Figure 3: model of a parallelogram-based robot for statics study

For a parallelogram-based 2-dof robot, we can neglect deformations of O_0O_1 and AO_2 links because their lenght are small with respect to the others and we can also neglect deformation of O_1O_2 link because the mechanical stress is only along O_1O_2; then, the *Bresse* relationships are (Figure 3):

$$\delta_B = \delta_A + L\Omega_A + \frac{F_B L^3}{3EI} + \frac{C_B L^2}{2EI}$$

with: $\Omega_A = \dfrac{F_A L^2}{2EI} + \dfrac{F_B L^2}{2EI} \qquad \delta_A = \dfrac{F_A L^3}{3EI} + \dfrac{F_B L^3}{3EI} \qquad F_A = P_S$

This leads to: $\delta_B = \dfrac{5P_S L^3}{6EI} + \dfrac{7F_B L^3}{6EI} + \dfrac{C_B L^2}{2EI}$

The additionnal deformation encountered with a serial robot with respect to a parallelogram-based robot is then:

$$\Delta(\delta_B) = \frac{L^2}{6EI}(5P_A L + 9F_B L + 9C_B)$$

This result shows the main advantage of parallelograms which increase the mechanical stiffness. Then we *a priori* need a triple parallelogram system.

GEOMETRICAL VALIDATION

The parallelogram-based design has better stiffness performances but its workspace is drastically reduced. Moreover such a structure presents too many singular configurations to provide a good folding capability. Consequently we finally choose a pantographic structure (Figure 4). A CAD study has been performed in order to derive the optimal link lengths, joint ranges and base location on the rover with respect to operational and folding constraints. After several iterations, taking into account the dimensions of the rover (Length=1250 mm, Width=400 mm, Height=700 mm), the following dimensions have been chosen:

- arm length: L_2=900 mm,
- forearm length: L_3=700 mm,
- distance between the second axis and the top of the rover: R_1=1100 mm,
- shoulder yaw joint: $-180° \leq \theta_1 \leq +180°$,
- shoulder elevation: $-40° \leq \theta_2 \leq +100°$,
- elbow: $-170° \leq \theta_3 \leq -40°$.

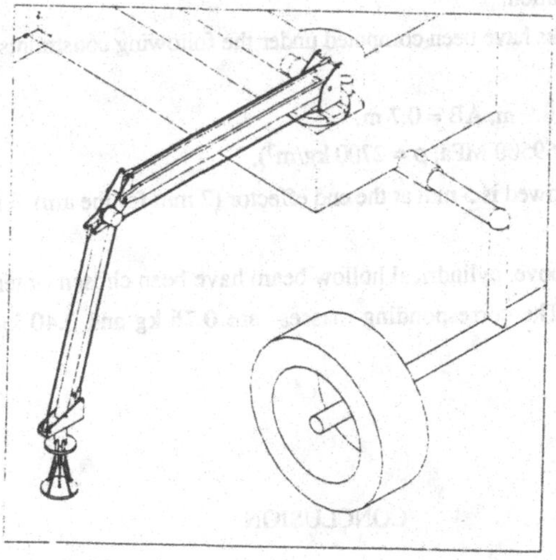

Figure 4: the arm on the rover and the parallel wrist.

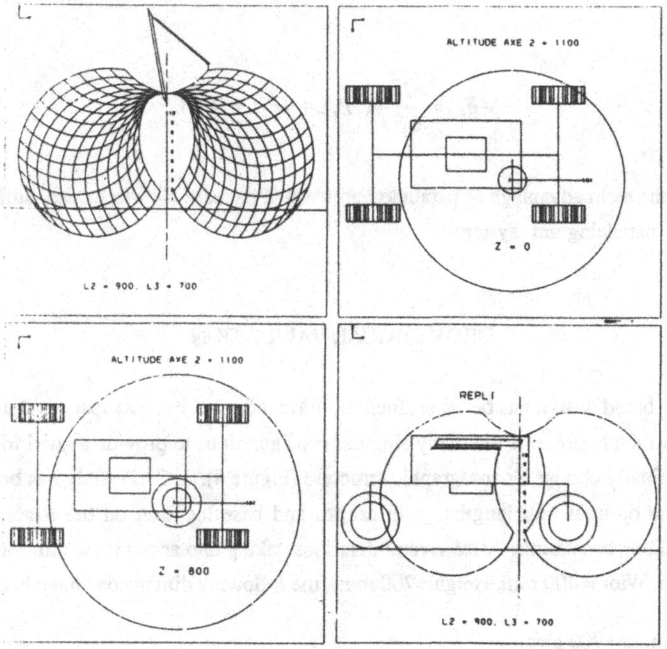

Figure 5: CAD simulations showing the workspace and the folded configuration of the pantographic arm.

The drawings of figure 5 show respectively: 1) the projection of the workspace in a vertical plane, 2) the projection of the workspace in a horizontal plane at the ground level (z=0), 3) and at z=800 mm, 4) the arm in the folded configuration.

The sections of the links have been computed under the following constraints:

- beam length: $O_0A = 0.9$ m, AB = 0.7 m,
- Al/Si/Mg alloy (E = 69500 MPa, $\rho = 2700$ kg/m³),
- maximum flexion allowed is 5 mm at the end effector (2 mm for the arm, 3 mm for the forearm).

Using the equations above, cylindrical hollow beam have been chosen (internal diameter: 36 mm, external diameter: 32 mm). The corresponding masses are 0.76 kg and 0.40 kg for the arm and the forearm respectively.

CONCLUSION

Due to the drastic limitations on power consumption, low power actuator are required. Since the external efforts are rather large (20 N and 5 Nm for external forces to which one must add 15 N and 2 Nm for the wrist), high reduction ratios are necessary to provide counteracting. Taking into account the load constraints due to the manipulator structure and to the actuators and drives, a possible solution is to make use of actuators such as SAGEM 10 MCM90 (25 w, 0.085 Nm, 0.3 Kg) and Harmonic drives HDUC IH-20 (torque: 40-80 Nm, reduction ratio 1/100, 0.4Kg) in addition to a conventional gear.

The study summarised in this paper has confirmed the advantages of a double parallelogram mechanism for performing a variety of operations in planetary explorations. The corresponding arm is stiffer than the equivalent serial one and it allows for positionning the parallel wrist and the end effector without any disturbance in orientation. The concept and the stages in the study allows for a straightforward adaptation of the methods to design of a variety of arms for use on autonomous vehicles.

[1] MOURA D.J.P., *et al*, " The French Preparatory Program for Future Mars Rover Missions". 41st Congress of the I.A.F., October 6-12 1990, Dresden, paper IAF 90-415.

[2] GIRALT G., BOISSIER L., "The French Planetary Rover VAP: Concept and Current Development". IEEE Int. Conf. on Intelligent Robots and Systems, Raleigh, 7-10 July 1992, pp. 1391-1398.

[3] BEJCZY A.K., "Allocation of Control Between Man and Computer in Remote Manipulation". 2nd RoManSy, Warsaw, September 14-17 1976, pp. 417-430.

[4] E.S.A. Mars Exploration Study Team, "Mission to Mars". ESA Report SP-1117, July 1989.

[5] REBOULET C., *et al.*, "Etude sur les robots à structures parallèles", Rapport final Contrat MRT 88.P.0715, 1990.

[6] GAILLET A., *et al*, "Projet VAP-Charge utile : concepts matériel et logiciel", Bulletin de liaison de la recherche en informatique et en automatique, N° 139, 1992, pp. 35-39.

THE DESIGN AND CONTROL OF A NOVEL ROBOT STRUCTURE WITH DIFFERENTIAL DRIVE UNITS

P.M. Taylor
The University of Hull, Hull, UK
and
J.C. Kieffer
Australian National University, Canberra, Australia
and
A.J. Wilkinson, J.M. Gilbert, Q. Guo, J. Grindley and R. Oldaker
The University of Hull, Hull, UK

Abstract

A novel robot has been designed and constructed as part of a project to develop water hydraulic systems. A planar five-bar mechanism is driven through two co-axial shafts to position the end effector anywhere within a circular working area in the horizontal plane. No end stops are required. Each shaft is driven by two Fenner 06 water motors or two rotary oil motors via a differential drive unit (DDU). This allows great flexibility over motor speeds and control strategies, in particular allowing low and zero speed output shaft movements whilst keeping the motors running at well-controllable speeds. The robot design is described with practical results from different control strategies for velocity and position control.

Introduction

This work arose from a project carried out at Hull on 'Water Hydraulics' funded by the DTI and J. H. Fenner plc. The Department of Electronic Engineering was responsible for the electronics and control systems for the motors and valves being developed by J. H. Fenner. It was decided that a robot be used to demonstrate the technology and to focus the research effort. The principal two axes were constructed as described below. A major design issue was the predicted high stalling speed (~200rpm) of the water motors [1]. Given a then predicted maximum speed of ~3000rpm it becomes impossible to achieve well controllable zero to high speed motion with a traditional gearbox and so the concept of using two motors on each axis driving a Differential Drive Unit (DDU) was introduced. The robot and the DDU are now described.

An overview of the robot

Figures 1 and 2 depict the five bar linkage which forms the primary two axes of the robot. The novel feature of this linkage is its 3-D nature which allows infinite rotations of θ_1 and θ_2 and positioning of the

2

end effector directly in the centre of the circular working area. The kinematic equations are simple. If the position of the end effector is expressed in polar (R, ∅) form then

$$R = 2L\cos\left\{\frac{\theta_1-\theta_2}{2}\right\}$$

and
$$\varnothing = (\theta_1+\theta_2)/2$$

with its inverse form
$$\begin{bmatrix}\theta_1\\ \theta_2\end{bmatrix}=\begin{bmatrix}1 & 1\\ 1 & -1\end{bmatrix}\begin{bmatrix}\varnothing\\ \cos^{-1}(R/2L)\end{bmatrix}.$$

When $\theta_1=\theta_2$, the arm is fully stretched and, when perturbed, OABC can potentially take on a non-parallelogram form. A toothed belt between O and A is used to keep links OC and AB parallel and thereby prevents this occurrence.

Figure 3 illustrates the robot inertia model. The links are numbered from one to four and have respective masses M_i $(i=1,2,3,4)$. Each link has a moment of inertia I_i about its mass centre which is a distance d_i from the joint axis. J_1 and J_2 are the moments of inertia of the coaxial shafts which rotate about the illustrated origin with angles θ_1 and θ_2 respectively.

If a lumped mass load, M_e, is placed at point P and a counterbalance mass, M_ι, is placed at point B, the dynamic equations take the form [3]:

$$\begin{bmatrix}\tau_1\\ \tau_2\end{bmatrix}=\begin{bmatrix}I_{11} & I_{12}\mathrm{Cos}(\theta_1-\theta_2)\\ I_{12}\mathrm{Cos}(\theta_1-\theta_2) & I_{22}\end{bmatrix}\begin{bmatrix}\ddot{\theta}_1\\ \ddot{\theta}_2\end{bmatrix}+I_{12}\mathrm{Sin}(\theta_1-\theta_2)\begin{bmatrix}\dot{\theta}_2^2\\ -\dot{\theta}_2\end{bmatrix}$$

where
$$I_{11}=M_2d_2^2+I_2+M_3L^2+M_4d_4^2+I_4+M_eL^2+M_\iota L^2+J_1$$
$$I_{22}=M_1d_1^2+I_1+M_2K^2+M_3d_3^2+I_3+M_eL^2+M_\iota K^2+J_2$$
$$I_{12}=-Kd_2M_2+Ld_3M_3+L^2M_e-Lk\,M_\iota$$

and so the inertia matrix may be made diagonal $(I_{12}=0)$ for a given load at P by a suitable choice of counterbalance mass M_ι.

The robot structure thus has the following features:

- High structural stiffness from the five-bar linkage
- Relatively low mass and inertias: both drive units are mounted in the base
- Drive inertias decoupled and configuration independent given an appropriate balancing load
- The end effector may be positioned anywhere within the working circle, including the origin
- No end stops required, increasing the robot's applicability.

3

The Differential Drive Unit and Control Strategies

A harmonic drive gearbox shown in figure 4 forms the heart of each DDU, being driven such that

$$w_5 = \frac{1}{n_{15}} w_1 + \frac{1}{n_{25}} w_2$$

where w_5 is the velocity of the output shaft connected to axis θ_1 or θ_2, and w_1 and w_2 are the two motor velocities. Two design options were considered. The first would have $n_{15} = n_{25}$ to maximise the range of output velocities. The second would have $n_{15} >> n_{25}$, motor 1 being used for fine positioning and motor 2 for coarse positioning. A compromise was adopted with $n_{15} = 200$ and $n_{25} = 24$. These figures include the final spur gear reduction [2].

In the first control strategy, velocity control of w_5 is achieved by having one motor controlled to run at an appropriate speed in its operating range and using closed loop control around the other motor to provide an offset and to compensate for perturbations in the velocity of the first motor and the effects of external disturbances. If w_1 and w_2 are of opposite sign then, in theory, the output w_5 can be completely at rest. The second strategy proposed, for position control, combines the action of the two motors as follows:

$$u_{motor1} = \begin{cases} k_{p2}e + k_{d2}\dot{e} + k_{i2}\int edt & \text{if } \left|e + k\dot{\theta}_d\right| \geq e_{max} \\ k_{p3}e + k_{d3}\dot{e} + k_{i3}\int edt & \text{if } \left|e + k\dot{\theta}_d\right| < e_{max} \end{cases}$$

$$u_{motor2} = \begin{cases} k_{p1}e + k_{d1}\dot{e} + k_{i1}\int edt & \text{if } \left|e + k\dot{\theta}_d\right| \geq e_{max} \\ 0 & \text{if } \left|e + k\dot{\theta}_d\right| < e_{max} \end{cases}$$

where u_{motor1} is the input to the fine motor, u_{motor2} is the input to the coarse motor, e is the position error and $\dot{\theta}_d$ is the rate of change of demand signal, k_{pj}, k_{dj} and k_{ij} (j=1,2,3), k and e_{max} are constants.

Using this strategy, when either the position error or the rate of change of demand is large both motors will operate under PID control, giving a fast response and, in the case of a rapidly changing demand, a small steady state error. For slowly varying demands, once a small position error is achieved, the coarse motor is halted and the fine motor used to obtain precise positioning. The transition between the two modes of operation is controlled by e_{max} which is chosen so that the coarse motor has sufficient time to stop before causing overshoot. k is then chosen so that the coarse motor becomes active if the rate of change of demand is such that it could not be followed using the fine motor alone. k_{pj}, k_{dj} and k_{ij} (j=1,2,3) may be chosen as in a traditional PID controller.

Experimental Validation of Dynamic Decoupling

It is apparent from the dynamic equations for the mechanism that if the load is correctly counterbalanced the coupling matrix becomes diagonal and so any torque applied to one axis will not affect the second. If however the counterbalance effect is incomplete, and a constant torque is applied to one joint the velocity of the second will vary in synchronisation with $\cos(\theta_1-\theta_2)$ and, provided the variation is small and the second joint is operating at a velocity where the friction is locally linear, the amplitude will be proportional to M_c - M_e. By measuring the amplitude of the velocity variation in the second joint any coupling effect

4

may be determined. Repeating this procedure for a number of counterbalance masses it is possible to see the variation in the coupling effect. Figure 5 shows the amplitude of the interaction dependant velocity variation for a range of experiments in which a load mass of 8.2kg was used. It is clear that for a value of M_c = 3.9kg the interaction effect drops to zero. It should be noted that below this value the velocity variation is in phase with $\cos(\theta_1-\theta_2)$ while above it the variation is 180° out of phase.

Experimental Results for Proposed Control Strategies

Both of the control strategies proposed were implemented using oil hydraulic motors with manual tuning of controller parameters. Results for PID control using only the coarse motor were also obtained. Results using only the fine motor showed that the joint velocity could not reach that demanded in either strategy.

Strategy 1 (Velocity Control): To test the perfomance using this strategy a demand profile involving high, low and zero velocity motion was used. Although the joint velocity is reversed, the DDU means that neither of the motors need stop or reverse. Figure 6 shows the velocity of the robot joint using this strategy. Figure 7 shows the velocity error along with the error obtained when the same trajectory was attempted using only the coarse motor. It is clear that the dual motor strategy gives significantly reduced velocity errors and that it is possible to obtain zero joint velocity without halting either motor.

Strategy 2 (Position Control): Figure 8 shows the position response for a trapeziodal demand using only the coarse motor while Figure 9 shows the response using the dual motor strategy. The position errors for both cases are shown in Figure 10. The velocities of the two motors are shown in figure 11, illustrating how their combined effort is used to generate the required joint motion. It is clear that the new strategy gives a significantly reduced error during both the transient and the steady state. The large position error obtained using the coarse motor is a result of the high level of motor stiction.

Conclusions

The novel structure and differential drive unit of a two degree of freedom robot have been described and it has been shown that, for a suitable counterbalance mass, it is possible to decouple the dynamics of the two links. Strategies for velocity and position control have been proposed making use of the flexibility provided by the differential drive unit and it has been shown that in both cases the performance obtained is superior to that which would be obtained if only a single motor were available.

References

1. P M Taylor, J Kieffer, A J Wilkinson, R Oldaker & J Gilbert, 'A Water Hydraulic Robot', CDC Dec '93.

2. Q Guo, 'Mechanism Design and Analysis for a Differentially Driven, Five Bar, Hydraulic Robot', August 1992, University of Hull MSc Dissertation.

3. J Grindley, 'Friction and Inertia Modelling for a Differentially Driven, Five Bar, Hydraulic Robot', September 1993, University of Hull MSc Dissertation.

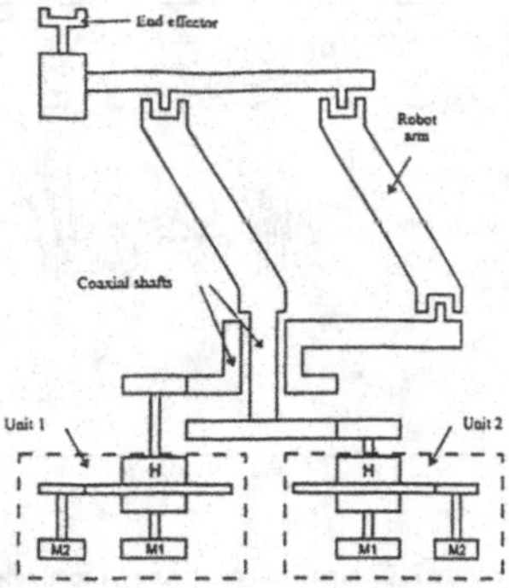

Figure 1. Robot structure in elevation.

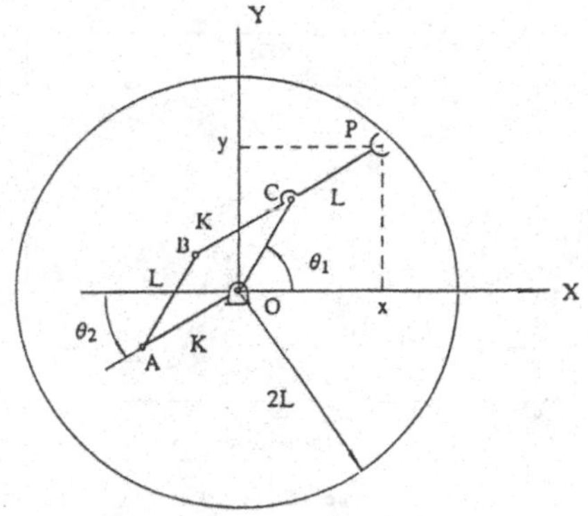

Figure 2. Schematic plan view.

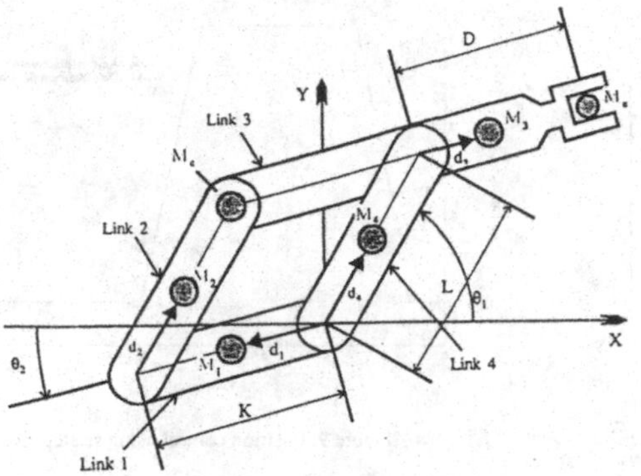

Figure 3. Robot inertia model.

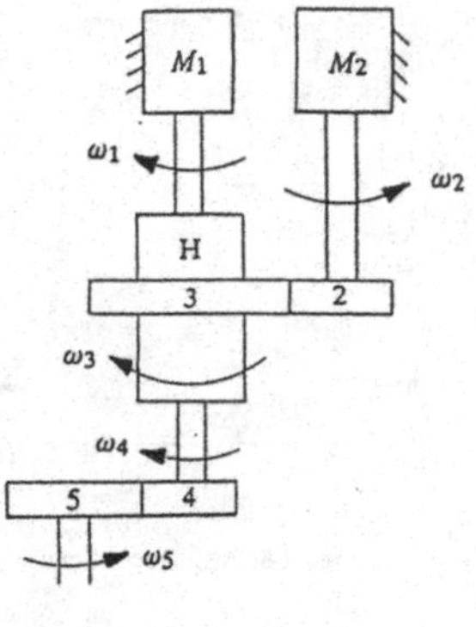

Figure 4. Differential Drive Unit.

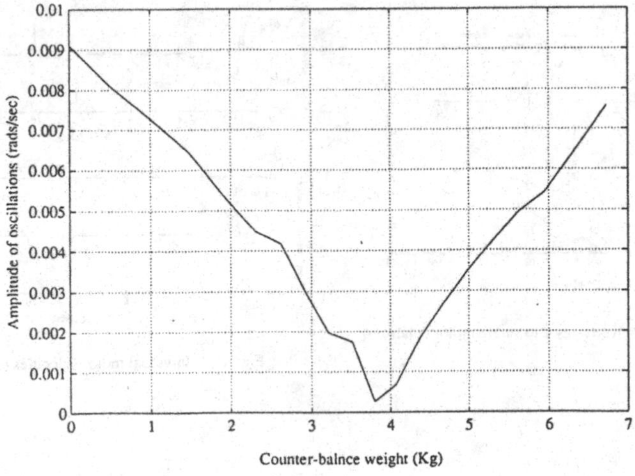

Figure 5. Axes interaction.

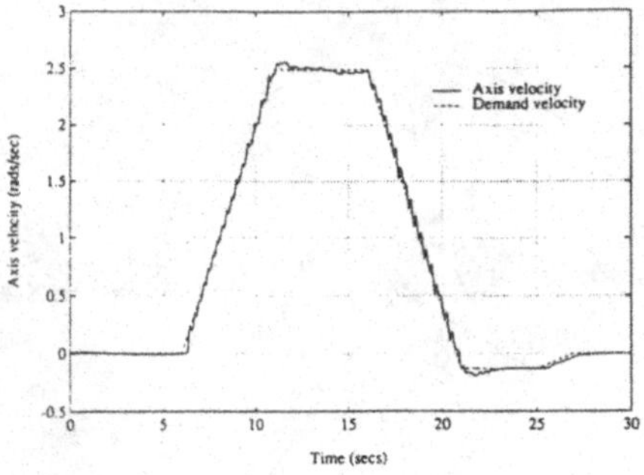

Figure 6. Strategy 1.

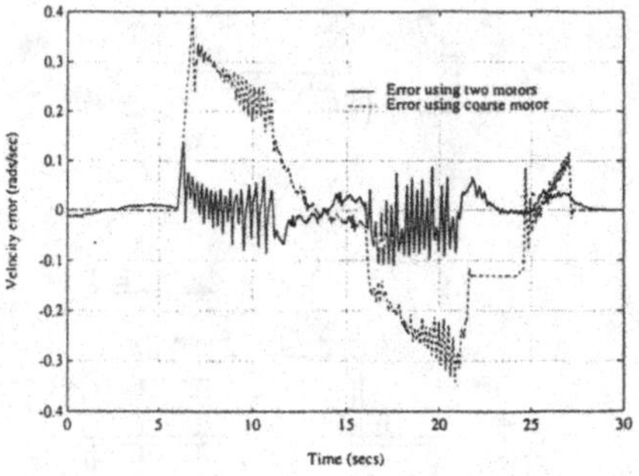

Figure 7. Comparison of strategy 1 to single motor control.

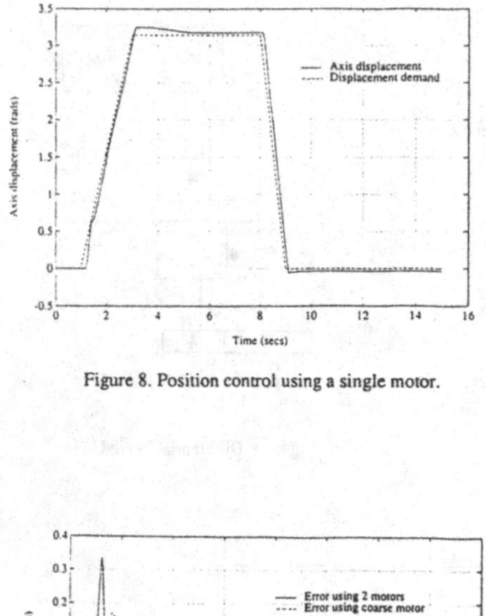

Figure 8. Position control using a single motor.

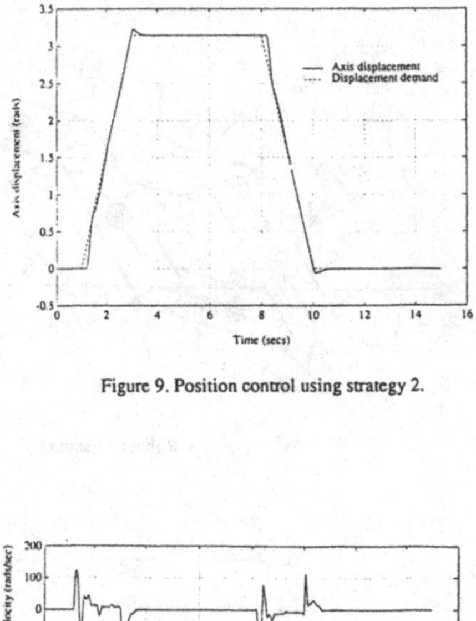

Figure 9. Position control using strategy 2.

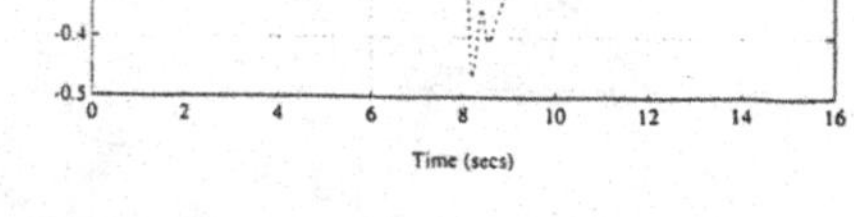

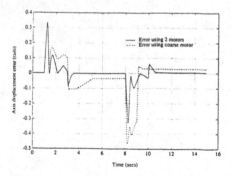

Figure 10. Comparison of strategy 2 to single motor control.

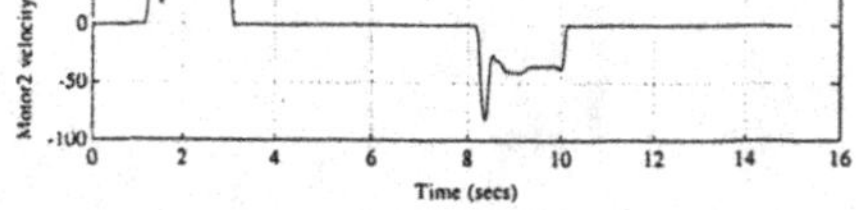

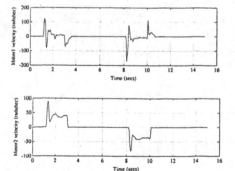

Figure 11. Inividual motor velocities using strategy 2.

DEVELOPMENT OF A SYNCHRONOUS MOTOR WITH THREE DEGREES OF FREEDOM

T. Yano
Mechanical Engineering Laboratory, Tsukuba, Japan
and
M. Kaneko
University of Hiroshima, Higashi-Hiroshima, Japan
and
M. Sonoda
Kumamoto Industrial Research Institute, Kumamoto, Japan

ABSTRACT

The second prototype synchronous motor with three degrees of freedom is developed and tested. First, the principle of the synchronous motor with three degrees of freedom is shown briefly and the structure of the developed synchronous motor is presented. Next, the control system for the synchronous motor is presented. Finally, experimental results of changing the orientation of the output shaft and the output torque measured are shown.

1. INTRODUCTION

Generally, a system with multi degrees of freedom such as a robot manipulator uses the same number of actuators as the number of degrees of freedom. In the case of the serial link manipulator, the actuator nearer the hand becomes the load of the actuator nearer the shoulder and the actuator nearer the shoulder is liable to become larger and larger like a snow ball. The total weight of the manipulator often becomes very heavy in comparison to the payload. In addition to the above problem, a manipulator has a problem that it needs a large number of calculations for the inverse kinematics. If we can use actuators with three degrees of freedom for the wrist and the shoulder joints and a conventional one for the elbow joint, a manipulator with seven degrees of freedom needs only three actuators and the weight of the manipulator is expected to be much lighter than a conventional one. And if the rotation center of the actuator with three degrees of freedom is identical for any direction of rotation, the inverse kinematics of the manipulator above can be solved geometrically. So, it is also expected to reduce the calculation time and to control the manipulator in real time.

Recentry, several types of actuators with multi degrees of freedom have been considered and/or developed. They are the extension of DC servo motor [1], stepping motor [2], magnetic bearing [3], and magnetic levitation servo [4]. But they needs many calculations in real time and/or have small working area for the manipulator joints.

We have extended the principle of a conventional induction motor and a synchronous motor to the three-dimensional space, and developed a prototype induction motor and a prototype synchronous motor, each having three degrees of freedom. The experimental results showed that each motor rotated according to the principle but produced only a small torque and could not change the orientation of the output shaft by itself [5].

The second prototype synchronous motor reported in this paper generates maximum torque of more than 0.2 Nm and can hold the output shaft in any orientation.

2. STRUCTURE OF A SYNCHRONOUS MOTOR

The principle of a conventional synchronous motor is shown in Fig.1(a). The inner rotor and the outer rotor are permanent magnets. If we turn the outer rotor, the inner rotor rotates in the same direction by the magnetism. If we rotate the magnetic field electrically, we get a synchronous motor. Extend the principle to the three dimensional space as in Fig.1(b). The inner rotor and the outer rotor are permanent magnets. They are supported so as to be able to rotate in any direction. If we rotate the outer rotor in an arbitrary direction, the inner rotor rotates in the same direction by the magnetism. If we rotate the magnetic field electrically, we get a synchronous motor with three degrees of freedom.

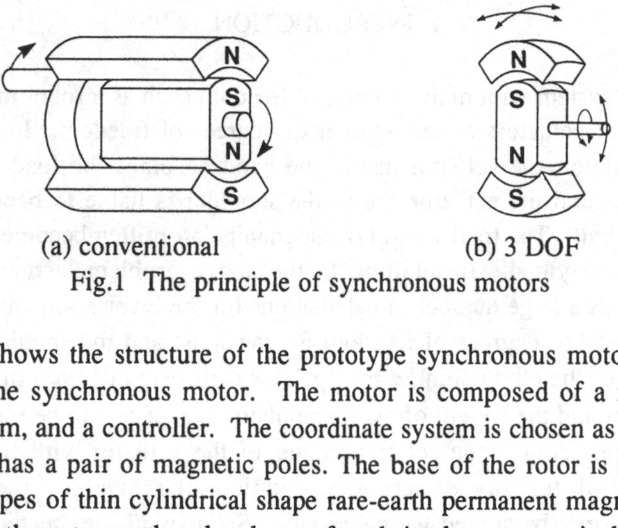

Fig.1 The principle of synchronous motors

Figure 2 shows the structure of the prototype synchronous motor and Fig.3 is a photograph of the synchronous motor. The motor is composed of a rotor, a stator, a gimbal mechanism, and a controller. The coordinate system is chosen as shown in Fig.2.

The rotor has a pair of magnetic poles. The base of the rotor is made of carbon-ferrite. Three types of thin cylindrical shape rare-earth permanent magnets are mounted on both sides of the rotor so that the shape of each is almost a semisphere and the flux density decreases gradually along the distance from the center of the semisphere. The diameter of the rotor is 4 cm.

The stator has three pairs of armature poles. The turns of each armature winding is 300. The armature poles are arranged 18.0 degrees tilted from the horizontal plane and 120 degrees apart from each other. As the orientation vectors of the three pairs of armature poles span the three dimensional space, we can generate a magnetic field rotating about an arbitrary axis in the three dimensional space by supplying the armature windings with sinusoidal waveform currents, instead of the complicated waveform currents if the assumption is satisfied that the flux distribution of each pair of armature pole is sinusoidal.

The gimbal mechanism supports the rotor so that the rotor can rotate in the three dimensional space freely.

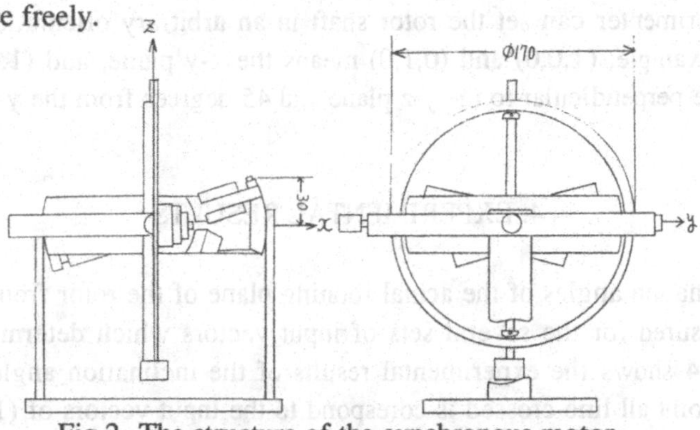

Fig.2 The structure of the synchronous motor

Fig.3 A photograph of the synchronous motor

3. CONTROL SYSTEM

The controller is composed of a microprocessor and a current generator. The input data is the magnitude and frequency of the rotating magnetic field, and two input vectors which determine the rotation plane of the magnetic field, that is considered to be the same as the rotation plane of the rotor. The microprocessor receives the data from the experimenter and decomposes the rotating magnetic field into three sinusoidally oscillating magnetic fields, whose orientations are the same as those of armature poles. Then, it calculates the magnitude and the phase delay of each sinusoidal armature currents to generate the respective oscillating magnetic field. Those calculations are done prior to the control. During the control, the microprocessor calculates the phase of each sinusoidal current in real time by adding the constant $\omega\Delta t$, where ω is the angle velocity of the rotating magnetic field and Δt is the time of one control cycle, to the phase and send it with the magnitude of the current to the current generator. The current generator uses a sine-ROM and a multiplier DA converter to generate the instant magnitudes of the currents. Then three DC amplifiers deliver the sinusoidal currents to the armature windings.

The experimenter can set the rotor shaft in an arbitrary orientation by two input vectors. For example, (1,0,0) and (0,1,0) means the x-y plane, and (1,0,0) and (0,1,1) means the plane perpendicular to the y-z plane and 45 degrees from the y-axis.

4. EXPERIMENTAL RESULTS

The inclination angles of the actual rotating plane of the rotor from the x-axis and y-axis are measured for the several sets of input vectors which determine the rotating plane. Figure 4 shows the experimental results of the inclination angles for the input vectors. The point all line crossed is corespond to the input vectors of (1,0,0) to (0,1,0), and each line is marked for every 0.1 interval of a. The real inclination angle was smaller than the input. We consider that a large open space around the z-axis makes the real flux distribution differ from the assumption and causes it. For a small a, the orientation of inclination are approximately the same as the input. But as a increases, the orientation differs from the input. There is also some offset in the inclination angle. The reason of these are considered to come from the fact that the air gap between the rotor and the stators is not uniform because of the difficulty in assembly. But the positioning is repeatable within the error angle of 1 degree, and if we use an input-output angle table and/or angle feedback, we will be able to position the output axis in an arbitrary orientation fairly accurately within the circle shown in Fig.4.

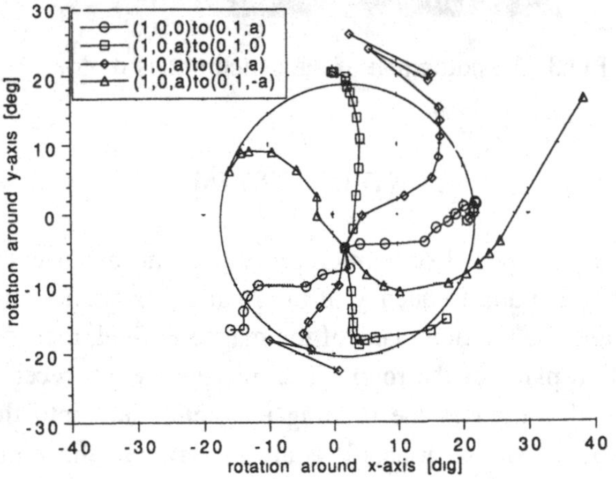

Fig.4 Measured inclination angle for the input

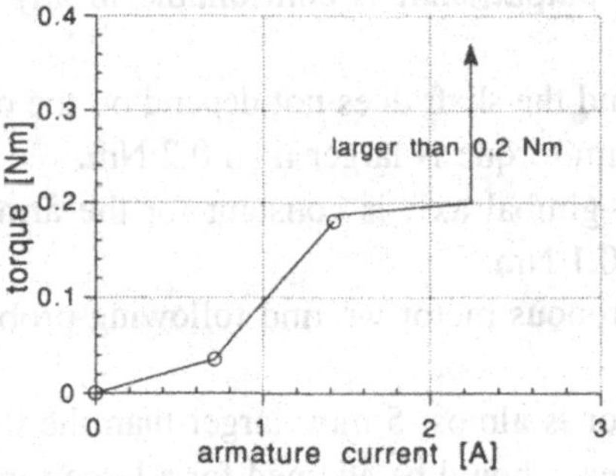

Fig.5 Output torque around output shaft

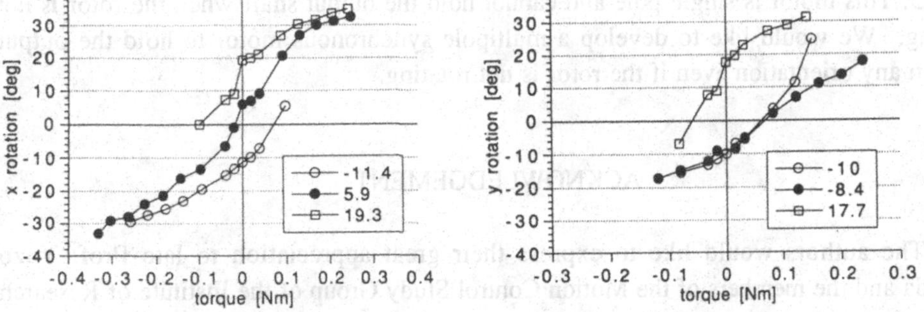

(a) x-axis (b) y-axis

Fig.6 Torque around x-axis and y-axis of the gimbal

The torque generated by the motor is also measured. Figure 5 shows the torques around the output shaft for the armature currents when the rotor is rotating in the x-y plane. When the armature current is 2.12 A, the output torque is larger than 0.2 Nm, the maximum range of the torque sensor. The experimental results also show that when the input magnitude of the rotating magnetic field is the same, the output torque is the same within the range of 16.7 degrees of inclination angle.

Figure 6(a) and (b) show the torques for the angle displacement around the x-axis and the y-axis of the gimbal mechanism each other. From Fig.6, we find that the gimbal mechanism is like a spring and the maximum torque is more than 0.1 Nm. The torque is constant for the armature current ranging from 0.71 A to 2.12 A.

5. CONCLUSIONS

The second prototype synchronous motor is developed and tested. From the experimental results, we get following features.

1. The orientation of the output shaft is controllable in any direction within the circle of the x-y plane.

2. The output torque around the shaft does not depend on the orientation but on the armature currents. The maximum torque is larger than 0.2 Nm.

3. The torque around the gimbal axis is constant for the armature currents. The maximum torque is larger than 0.1 Nm.

With the prototype synchronous motor we find following problems to be solved in our future work.

1. The air gap of the motor is almost 5 mm, larger than the 0.1 mm or less of the conventional motor. A smaller gap should be attained for a large torque.

2. More detailed designing and more precise modelling and assembly should be carried out for a higher performance of the motor.

3. This motor is single pole and cannot hold the output shaft when the rotor is not rotating. We would like to develop a multipole synchronous motor to hold the output shaft in any orientation even if the rotor is not rotating.

ACKNOWLEDGEMENT

The authors would like to express their great appreciation to late Prof. Yuzo Yoshida and the members of the Motion Control Study Group of the Institute of Research for Intelligent System Technology in Kumamoto for their cooperation in designing the experimental synchronous motor.

REFERENCES

[1] K. Kaneko, I. Yamada, and K. Itao, "A Spherical DC Servo Motor with Three Degrees of Freedom", Trans. ASME, Journal of Dynamic Systems, Measurement, and Control, Vol.111, p398, 1989

[2] K. Lee and C. Kwan, "Design Concept Development of a Spherical Stepper for Robotic Applications", IEEE Trans. Robotics and Automation, Vol.7, No.1, p175, 1991

[3] M. Tsuda, H. Higuchi, and S. Fujiwara, "Magnetic Levitation Servo for Flexible Assembly Automation", The International Journal of Robotic Research, Vol.11, No.4, p329, 1992

[4] R. L. Hollis, S. E. Salcudean, and A. P. Allen, "A Six-Degrees-of-Freedom Magnetically Levitate Variable Compliance Fine-Motion Wrist: Design, Modeling, and Control", IEEE Trans. Robotics and Automation, Vol.7, No.3, p320, 1991

[5] T. Yano and M. Kaneko, "Development of an Actuator with Multi Degrees of Freedom", Proc. 2nd International Conference on Automation, Robotics, and Computer Vision (ICARCV'92), p.RO-2.2.1, 1992

DESIGN OF THE FAST MANIPULATOR WITH ELIMINATED JOINT LIMITS AND REDUCED DYNAMIC INTERACTIONS

K. Nazarczuk
Warsaw University of Technology, Warsaw, Poland

Abstract

In this paper a new concept of the manipulator design is proposed. This design due to application of a direct-drive motors with special transmission mechanisms and proper mass distribution, ensure the elimination of the dynamic interaction as well as joint limits in the manipulator arm. The first link of the arm is really driven directly, while the second and the third links are driven remotely through mechanisms, the transmission ratio of which equals 1. Primary structural assumptions for 6R manipulator with such a direct-drive arm and a spherical wrist driven through gears have been formulated. According to these assumptions the 1:5 model has been built. The first DOF of the arm together with all three DOFs of the wrist exhibit the motion range without any limits. In the two others joints the ranges of motion are limited to about 8π due to electrical cables twisting. Basing on the presented dynamic analysis, one can state, that the design being proposed ensure a considerable simplification of the control system and improves the properties of a fast direct-drive robot.

1. Introduction

Improvement of a robot performance by means of versatility, dexterity, positioning accuracy, high speed motion and tracking precision, requires a permanent design study.

Recently however, the possibility of introducing the direct-drive systems into industrial robots has created both a new perspective and problems in manipulator design. In a direct-drive manipulator arm high torque motors are directly coupled to each joint. This results in high mechanical stiffness, no backlash and low friction observed in the structure, but also implies significant interactions between the individual degrees of freedom (DOF), further intensified due to very high speed that the arm can reach.

Up till now, direct drive manipulators become increasingly massive and bulky with the increase in the DOFs number. Asada and Youcef-Tuomi (1984), have shown that dynamic interactions can be eliminated by the mass redistribution and modification of the arm structure. The design guidelines for this procedure, based on the concept of decoupled and configuration-invariant inertia matrix, formulated by Asada and Youcef-Tuomi (1987) concern, however, a simple manipulator arm with two DOFs. The prototype of the manipulator arm with three DOFs presented by them has a diagonal (decoupled) but not invariant inertia matrix.

Basing on the suggestions given by Asada and Youcef-Tuomi, Nazarczuk (1993) proposed two different designs of a manipulator arm with three revolute DOFs, for which it is possible to eliminate both dynamic interactions and joint limits.

One of the aforementioned designs, possibly of greater practical interest, is shown in details in the present contribution. It is the manipulator with the 6 DOFs, in which direct-drive motors realize the arm drive, while the wrist driven through gears.

2. Scheme and primary structural assumptions of the manipulator design

A simplified manipulator scheme is presented in Fig.1. Like in the SCARA type robots, the first and the second joints, respectively have vertical axes, while the third joint axis goes horizontally. The first two DOFs are driven by MD_1 and MD_2 direct-drive motors mounted in a collinear way on the fixed frame L_0. The first link L_1 in form of a big hollow crank is driven directly by MD_1 motor. The second link L_2 is driven remotely by MD_2 motor through the so-called skew parallelogram SP_2, mounted inside the L_1 link.

The skew parallelogram consists of three parallel oblique coupling bars joining up two disks, one of which is mounted on the MD_2 rotor, while the second one is mounted on the output shift. The skew parallelogram displays a variety of advantages in comparison with the parallel-drive mechanism which have been used in manipulators up to now, e.q. five- bar mechanism, steel band or chain drives etc.

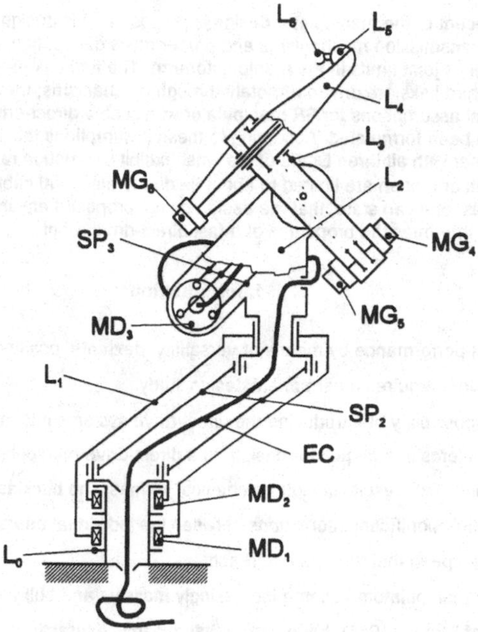

Fig. 1

Main advantages of the skew parallelogram should be mentioned: practically unlimited range of the transmitted rotation, no singular positions appearing and high stiffness independent of the rotation angle value. Besides, inside the mechanism a group of elastic electric or pneumatic conductors can be mounted, which not twisting under unlimited rotation of the link L_1 are subject only a slight bending. The only disadvantage displayed by the mechanism is the necessity for the high-accuracy manufacturing of each element due to the redundant constrains existence and elimination of the backlash. MD_3 direct-drive motor of horizontal axis driving the third DOF through the skew parallelogram SP_3 is fixed on the link L_2, being in fact also a counterbalance ensuring the adequate position of the L_2 link center of mass. The link L_3 has a special shape ensuring the proper location of motors driving the spherical wrist. These motors, denoted as MG_4, MG_5 and MG_6 respectively, are supplied with reducers. Electric cables EC coming to them through the special duct going e.g. inside direct-drive motors and parallelogram SP_2 and SP_3. The kinematic scheme of manipulator together with the structural assumptions made ensure that the first DOF of the arm together with all three DOFs of the wrist exhibit the motion range without any limits. In two other joints of the arm the ranges of motion are limited to $\pm 4\pi$ due to the electrical cables twisting.

3. Reduction of dynamic interactions

The remote drive of the link L_2 through a parallelogram enables relatively simple dynamical decoupling of the two first manipulator DOFs in the case when within links $L_2 \div L_6$ no relative motions exist, though they can be assumed to form one rigid system. We need only to ensure the center of mass of the system (composed of links $L_2 \div L_6$) to be located on the second joint axis.

Dynamic decoupling of the first three DOFs seems to be a much more difficult problem, since the system of links $L_3 \div L_6$ representing a rigid body should additionally satisfy the two conditions given by Korendyasev, Salamandra and Tyves (1988). The first of them requires that the third joint axis overlaps one of the central principal axes of inertia of the $L_3 \div L_6$ system. The second one requires the inertial moments of the $L_3 \div L_6$ system with respect to the two other principal axes of inertia to be equal. There is no difficulty in satisfying the first condition. The second one however, demands for unique structural designs and the third DOF inertia to be increased. It is noteworthy to consider the usefulness of the aforementioned procedure. In the case when the all the aforementioned conditions are satisfied, except the last one, the equations of the motion of manipulator arm scheme of which is shown in Fig. 1, at a given wrist configuration can be written as follows

$$h_1\ddot{q}_1 = \tau_1 \tag{1}$$

$$(h_2 + I_{x3}\sin^2 q_3 + I_{y3}\cos^2 q_3)\ddot{q}_2 + (I_{x3} - I_{y3})\dot{q}_2\dot{q}_3 \sin(2q_3) = \tau_2 \tag{2}$$

$$h_3\ddot{q}_3 - 0{,}5(I_{x3} - I_{y3})\dot{q}_2^{\,2}\sin(2q_3) = \tau_3 \tag{3}$$

where (for i = 1, 2, 3)

h_i - constant (invariant of the configuration) component of the diagonal element of the inertia matrix,

q_i - actuator generalised co-ordinate, when $q_1 = \Theta_1$, $q_2 = \Theta_1 + \Theta_2$, $q_3 = \Theta_3$,
where Θ_i - joint co-ordinate defined according to the Denavit - Hartenberg convention,

τ_i - i-th driving torque,

I_{x3}, I_{y3} - inertia moments of the links $L_3 \div L_6$ system with respect to two of the central principal axes of inertia perpendicular to the third joint axis.

From equations (1 ÷ 3) it can be shown that if $I_{x3} \neq I_{y3}$ the inertia matrix of the manipulator arm under consideration is a diagonal one, depending however on the configuration. The dynamic interactions magnitudes depend in this case on the co-ordinate q_3 and the angular velocity $\dot{q}_2$ and $\dot{q}_3$ values. It should be noted that in typical constructions the link L_3 has a form of beam, on the end of which the other links $L_4 \div L_6$ are mounted, forming the wrist.

If the beam is fixed along the axis x_3 we can assume that $I_{x3} = 0$ and $I_{y3} = h_3$. Let us consider a simple way of the dynamic interactions reduction by means of the I_{x3} value increasing in terms additional masses mounted along the axis y_3. It can be written then $h_3 = I_{x3} + I_{y3}$ and from equations (2), (3) it results that for $I_{x3} = I_{y3}$ the dynamic interactions vanish, limiting on the other hand, ranges of the accelerations $\ddot{q}_2 = \ddot{q}_{02}$ and $\ddot{q}_3 = \ddot{q}_{03}$ reached at zero values of $\dot{q}_2$ and $\dot{q}_3$ velocities. It should be noted that the range of $\ddot{q}_3$ becomes of two times smaller than that for a typical case. The dynamic interactions reduction is necessary for high velocity motions, however. In typical structures in which $I_{x3} = 0$ and $I_{y3} = h_3$, when the angular velocity $\dot{q}_2$ satisfied condition

$$0.5\,\dot{q}_2^2 > \left|\ddot{q}_{03}\right|_{max} \tag{4}$$

for $q_3 = \frac{\pi}{4}$ the value of angular acceleration $\ddot{q}_3$ is negative for any permissible value of the torque τ_3 (see equation (3))

In this case it is impossible for the manipulator to realise some typical motions .

To ensure manipulator performance to the comparable with the SCARA type direct-drive robots (maximal speed value aprox. 10 m s^{-1} and maximal acceleration aprox. 50 m s^{-2}) the condition (4) should be satisfied.

In the presented design the condition that $I_{x3} = I_{y3}$ together with the other conditions of the dynamic interactions reduction are satisfied mainly due to the proper arrangement of driving motors MD_3 and $MG_4 \div MG_5$.

This arrangement is shown in Fig. 2. presenting two configurations of the 1:5 manipulator model, which has been built It should be noted, that joint limits elimination enables the manipulator mass distribution to be properly tested and corrected by means of dynamic balance.

The wrist structure being designed should ensure the condition $I_{x3} = I_{y3}$ to be fulfilled at required accuracy despite the configuration.

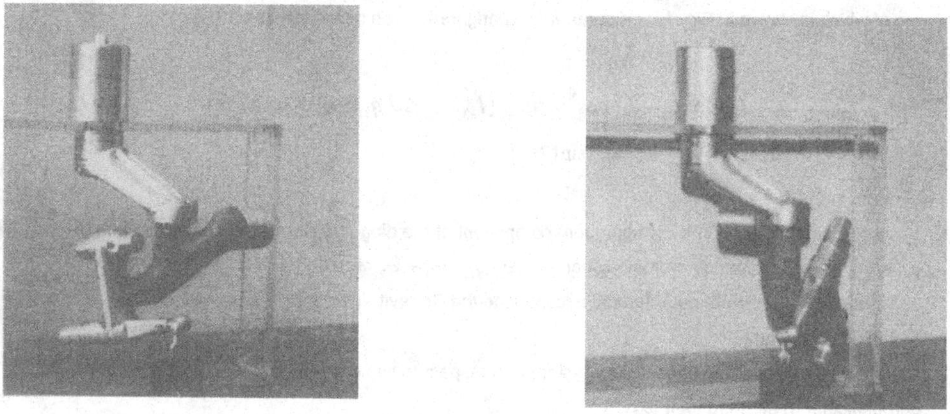

Fig. 2.

The reduction of dynamic interactions due to wrist motions, which can not be eliminated completely emerges as a special problem. These dynamic interactions are limited however and independent of configuration since motors driving the wrist are supplied with speed reducers.

4. Some remarks concerning the time-optimal control.

The advantage of the present simplified model of the manipulator are obvious. For the decoupled and invariant inertia matrix, the arm with n DOFs can be treated as a system of n independent linear subsystems with constant parameters. As a result, the control system of the manipulator can be simplified and the control performance can be improved due to reduced dynamic complexity. For the manipulator design under investigation this assumption concerns the arm with three DOFs (n=3) realising free motions or subject to a properly balanced payload. It is worth mentioning that the arm joint limits elimination results also in essential profits, primarily due to the fact that the change in the co-ordinate q can be made in two different ways. This is of crucial importance for the time-optimal control over the joint velocity profile shown in Fig. 3b, since it enables to find the shorter time trajectory presented on the phase-plane (see Fig. 3a). Co-ordinates q_b and $q_b - 2\pi$ correspond to the same joint configuration. The time required for passing from q_a to q_b is much shorter along the trajectory 2 then along the trajectory 1. The second considerable advantage consist in the fact that there is no such region in the workspace, within which the manipulator reveals poorer performance due to the limited number of configurations or the necessity of braking in front of joint limit. The region of a phase-plane hatched in Fig. 3a consists of all permissible states of the considered DOF in the case when a zero-dimensional limit appears in the joint. After elimination of this limit, the area of permissible states is restricted only by the maximum speed condition

$$|\dot{q}| \le \dot{q}_{max}$$

It can be easily shown that under the time-optimal; control in the joint space the maximum time interval required for any change of state of the decoupled dynamic manipulator with zero-dimensional joint limits is approximately twice as long as after the elimination of these limits.

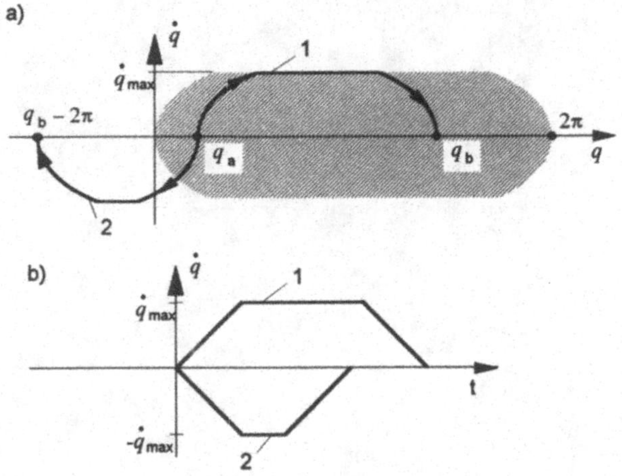

Fig. 3

5. Conclusions.

The manipulator arm design, described above, in which the dynamic interactions together with the joint limits are eliminated may considerably improve the properties of electric direct-drive robots.

Profits from such a design consist in substantial simplification of control system, extension of the workspace, kinematic flexibility increase and considerable shortening of the task realisation time. The latter concerns mainly extent motions in which trajectory profile from the initial to the final state is of secondary importance. In the case

of the motion performed along a specified path the easy choice of the proper configuration may also ensure reaching a higher value of speed.

References.

[1]. Asada H. Youcef-Tuomi K., 1984, *Analysis and Design of a Direct - Drive Arm with a Five-Bar-Link Parallel Drive Mechanism*, Trans of ASME J. of Dynamic Systems, Measurments and Control, 106, 225-230

[2] Asada H. Youcef-Tuomi K., 1987, *Dirtect - Drive Robots. Theory and Practice.* The MIT Press,

[3] Korendyasev A.I., Salamandra B.L., Tyves L.I. 1988, *Mechanics of Robots with Dynamically Decoupled Motions*, Proceedings of 7th CISM-IFToMM Symposium on Theory and Practice of Robots and Manipulators, Romancy 86, Udine, Italy, Ed. Hermes 1990 448-455,

[4] Nazarczuk K., 1993, *Improvement of Robot Performance Due to Elimination of Both Dynamic Interactions and Joint Limits in the Manipulator Arm.* Journal of Theoretical and Applied Mechanics, No 3. Vol. 31, Warsaw 621-636

[5] Voukobratovic M., Filaterow V., 1992, *Static Balancing and Dynamic Decoupling of the Motion Robots*, Proc. of the 23 ISIR, Barcelona, Spain, 207-212

Part VIII
Synthesis and Design III

MICRO-ROBOT WITH INTEGRATED PARALLEL LINKS

K. Matsushima and K. Nishi
Toin University of Yokohama, Yokohama, Japan

ABSTRACT

This paper presents on a micro-robot constructed with integrated parallel links which were made by means of wire electro-discharge machining. The structure of the robot is of rectangular type and the mechanisms of its X and Y axes are constructed by the parallel links of which joints are formed with circular shaped notches. These mechanism are actuated by laminated piezo-actuators. The mechanism and movement of the Z axis are realized by a bimorph piezo-actuator. The damping of this robot body is very small, so that it is controlled by a state feedback controller with the state observer for suppressing the strong resonance. The results obtained experimentally are as follows, (1) dynamic range : 40 μm, (2) bandwidth : about 200 Hz , (3) resolution : submicron order

1. INTRODUCTION

Recently, many kinds of micromechanisms or micro-robots have been studied under requirement from biological or medical fields. We have developed a micro-robot for micro-manipulation under a microscope[1]. The structure of this robot is of rectangular type and the mechanisms of its X and Y axes are constructed by parallel links and lever-mechanisms of which joints are formed with circular shaped notches. These mechanisms are actuated by laminated piezo-actuators. The movement of Z axis is realized by a bimorph piezo-actuator. The damping of this robot body is very small, so it is controlled by a state feedback controller with the state observer for suppressing the strong resonance. In section 2, we design the robot body and present its static and dynamic characteristic. In section 3, the overall characteristic of the robot is presented. Section 4 concludes the paper.

2. STRUCTURE AND CHARACTERISTICS OF ROBOT BODY

The robot body is of rectangular type and is constructed by a parallel link mechanismof which joints are formed with circular shaped notches. And the movements of X and Y axes are realized by its mechanism. The structure of the robot body is shown in Fig. 1. This mechanism was made by wire cut electro-discharge machining. Its circular notched joint is a elastic one, so it is friction free. As a design specification of the robot, we give the maximum range 50 μm of X and Y axis . This mechanism is actuated by the laminated piezo-actuators of which maximum displacement is 16 μm /150 v. Therefore, its displacement is magnified by a lever mechanism which lever ratio is 3.4:1 (the lever length is 12 mm). In this case, the maximum bending angle of the notch is 0.24 deg.

The spring constant of the circular notched joint ,as shown in Fig.2, is calculated as follows2),

$$K_b \cong \frac{2Ebt^{2.5}}{9\pi\rho^{0.5}} \tag{2-1}$$

where E: Young's modulus, b: width of link, t: minimum thickness of notch,
ρ: radius of circular notch.

On the other hand, the relation between the maximum torque and maximum stress is

$$\sigma_m = \frac{6}{t^2 b}\tau_m \tag{2-2}$$

From eq.(2-1) and eq.(2-2), maximum allowable bending angle of notch is represented by eq.(2-3).

$$\theta_m = \frac{\tau_m}{K_b} = \frac{3\pi t^{-0.5}\rho^{0.5}}{4E}\sigma_m \tag{2-3}$$

In our case, we use soft steel SKD-11 as the material of the robot body. Its Young's modulus is $E = 2.1\times10^{4}$ Kgf/mm2 and its maximum stress is 85~125 Kgf/mm2.

The dimension of the link is designed as b=10 mm, t= 0.5 mm and ρ=1.0 mm.

So we get the maximum allowable bending angle θ_m = 0.37~0.54 deg. From the above calculation, the bending angle of the notch is within the θ_m, the robot body works in the elastic limit of the material.

The piezo-actuator has hysteresis between the supply voltage and the displacement.

Fig. 3 shows the hysteresis characteristics of our robot body. The measurement shows that X axis and Y axis have about 10 μm as maximum hysteresis width, respectively, and Z axis about 20 μm .

As an example, the normalized frequency response of X axis is shown in Fig. 4 .

From the results of the static and frequency response tests, the dynamics of each axis are approximated respectively as follows,

The approximated transfer function of X (Y) axis :

$$G_{p,x,y}(s) = \frac{3.2476\times10^7}{s^2 + 7.6843\times10^1 s + 5.9048\times10^7} \quad [\mu m/V] \qquad (2\text{-}4)$$

where $\omega_n = 7.6843 \times 10^3$ [rad/s] , $\zeta = 0.005$, gain constant $b_p = 0.55$ [μm/V].

The approximated transfer function of Z axis:

$$G_{p,z}(s) = \frac{9.7370\times10^6}{s^2 + 6.1196\times10^1 s + 3.7450\times10^6} \quad [\mu m/V] \qquad (2\text{-}5)$$

where $\omega_n = 1.9352 \times 10^3$ [rad/s] , $\zeta = 0.016$, gain constant $b_p = 2.6$ [μm/V].

3. DESIGN OF CONTROLLERS AND CHARACTERISTICS OF ROBOT

The robot body has very small damping as shown in the previous section. So in order to suppress such strong resonance, we basically designed the controller of each axis as optimal robust servo controllers[3)] with the minimum order state observer which estimates the velocity of each axis. Figs. 5 (a) and (b) are the block diagrams of X (Y) and Z axes, respectively. Hence a parameter k in each block diagram is the adjusting parameter explained below. The case k=1 is the system designed as the linear optimal robust servo. But in the minor state feedback loop of X (Y) axis having k=1, the limit cycle of about 6 KHz which is considered to be caused by the hysteresis of the piezo-actuator were observed. So we adjusted k=1 to k=0.056 , at the sacrifice of optimality, for eliminating the limit cycle. The limit cycle in the minor state feedback loop of Z axis was also observed and k was adjusted to 0.05.

The frequency characteristics of each axis after the adjustment of k are shown in Figs. 6 (a) and (b). Figs. 7 (a) and (b) show the step responses of X and Z axes. The static characteristics of X and Y axes are shown in Figs. 8 (a) and (b).

Fig. 9 shows an example of the movement of the robot when it executed the command ' write RO-MAN-SY IFToMM , one letter size 4 × 4 μm ' which was programmed on a computer. In this example, the executed time was about 9 sec.

links. At first, we showed the structure of the robot body and its dynamic characteristics. Then, we designed the controllers of each axis on the basis of optimal robust servo. But in each axis of the real system, the limit cycle caused by the hysteresis of the piezo actuator occurred, so we, at the sacrifice of optimality and response time, eliminated its limit cycle by reducing the loop gain. As the results, we obtained the overall characteristics of our robot as follows, the dynamic ranges of X and Y axis are 40 μm, Z axis 100 μm, the bandwidth of X and Y axis are about 200 Hz, Z axis about 80 Hz and the resolutions of each axis are submicron order, respectively. Finally, as an example of the movement of the robot , we made it write the letters " RO-MAN-SY IFToMM" of which each size has 4 × 4 μm/one letter and showed it has good response and high resolution.

We would like to thank Mr. Yosito Sakai and Mr. Yasuhiro Tanaka , undergraduate students of Toin University of Yokohama, for their help on this study.

REFERENCE

1) H. Fukasawa : Design of Micro-Robot with Integrated Parallel Links; MS Thesis of Univ. of Tsukuba, 1990 (in Japanese)

2) M. Tanaka and K. Nakayama : The Bending Elasticity of a Circular Notched Bar as the Monolithic Flexure Hinges in a Precision Mechanism ; Vol.50, No. 3, Appl. Phys. 1981 (in Japanese)

3) K. Furuta et.al. : Mechanical Systems Control ; Ohmu Sha , (in Japanese)

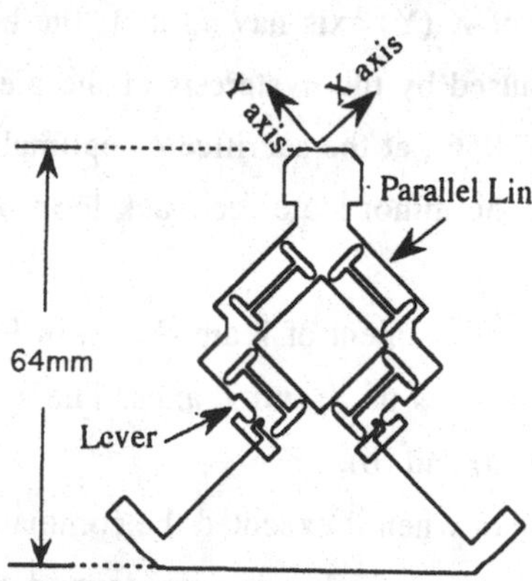

Fig. 1 Structure of Robot Body

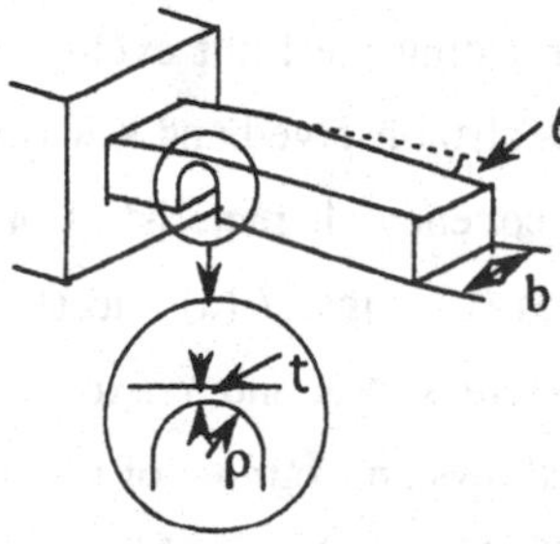

Fig. 2 Circular Notch

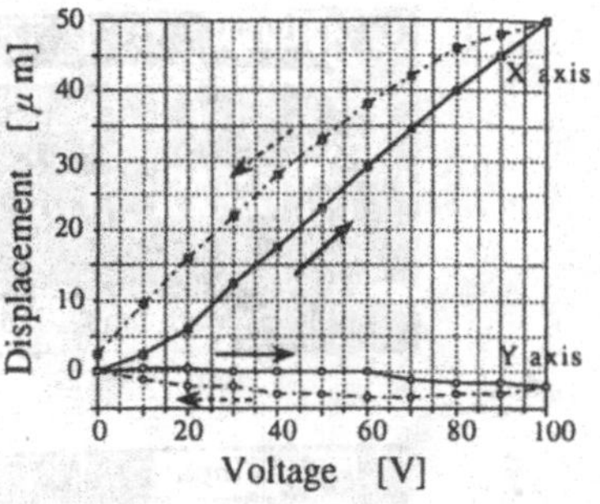

Fig. 3 Hysteresis of Robot Body (X axis)

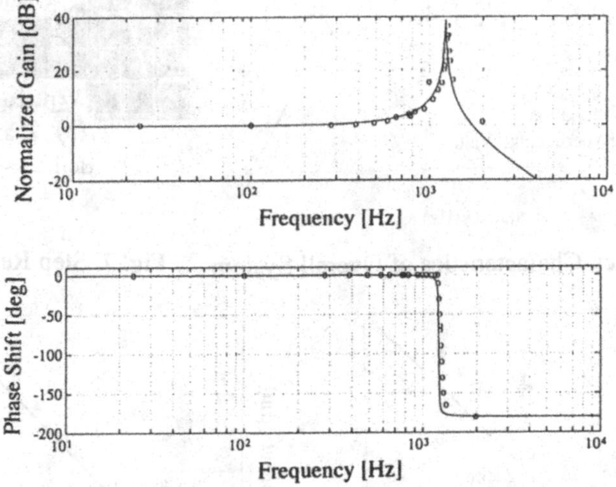

Fig. 4 Frequency Characteristics of Robot Body

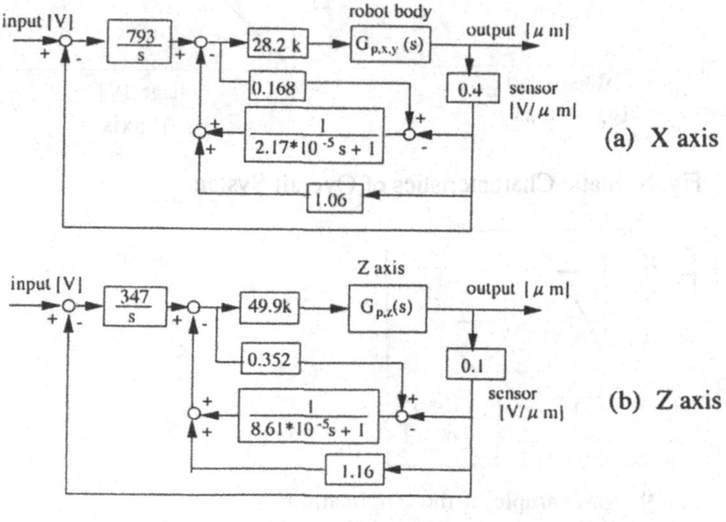

Fig. 5 Block Diagram of Robot Control System

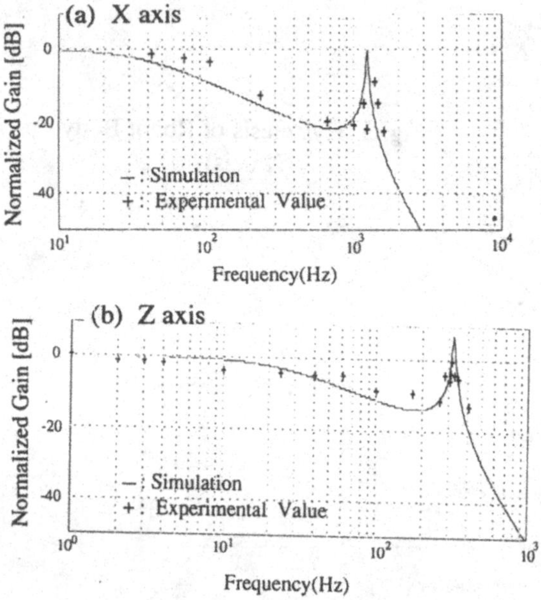

Fig. 6 Frequency Characteristics of Overall System

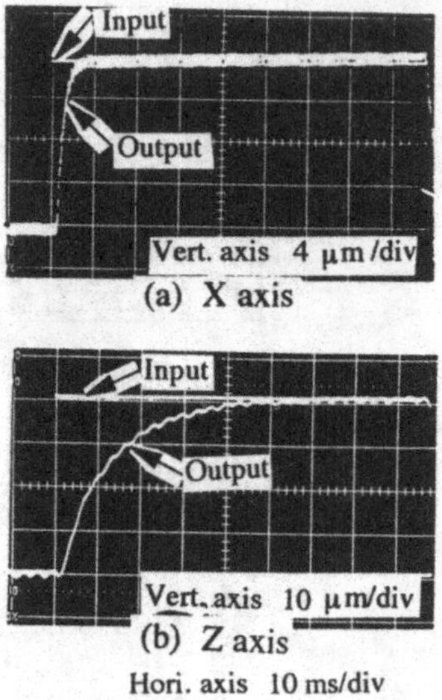

Fig. 7 Step Response

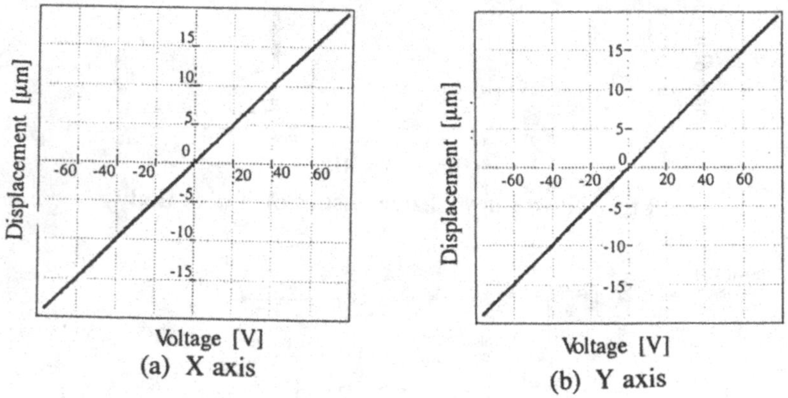

Fig. 8 Static Characteristics of Overall System

Fig. 9 An example of the Application

THE MODELLING OF A HAND WITH 16 DEGREES OF FREEDOM INTEGRATING SURFACIC ELASTIC CONTACTS

J-N. Gaborieau, M. Arsicault and J.P. Lallemand
U.R.A. - C.N.R.S., Poitiers France

ABSTRACT

This paper presents a comprehensive study of the power grasp which takes into account the elasticity of human fingers and as far as possible, real life conditions of contact. Therefore we modelize the hand in order to have a distribution of forces as well as a mechanical behaviour similar to that of the human hand. First, an experimental study has allowed us to know the characteristics and the nature of an "object - human finger" contact. Then, this contact is modelized in the form of relations between the resultants of applied forces and the relative displacements of the finger. Then, with the aim of modelizing the hand so that it can reproduce the greatest possible number of movements similar to the human hand, the hand consists of a 3D polyarticulated mechanical system with sixteen degrees of freedom that presents a kinematics similar to that of the human hand. Finally, we present the formalism which allows us to determine the forces generated at contacts in order to ensure a stable grasp. The resolution of the former is presented in the form of a non-linear programming problem.

1. INTRODUCTION

For about twenty years, a great deal of research has been developed in the area of grasping objects with a multi-fingered polyarticulated mechanical hand [1,2,4]. The development of such systems necessitates considering an essential condition : that of stability. The study of stability has been widely developed for rigid contacts and elastic contacts [5,6,8]. However these works have especially focused on grasping objects with fingertips. Thus, studies have recently concentrated on power grasp [7,9] which enables a better stability. Since models are considered as having 2D mechanical architecture and rigid contacts between the grasped object and the phalanxes, most studies about power grasp are somewhat limited in scope. Therefore this paper deals with the study of the power grasp, taking into account as far as possible, real life conditions of contact.

2

2. EXPERIMENTAL STUDY

The aim of this study is to evaluate the properties and the nature of an "object - human finger" contact. An experimental site [3] determines the elasticity of human fingers as well as the mechanical properties of human skin. This site provides precise information on forces and positions in the three directions X, Y and Z (fig. 1a). The forces are exerted by a circular section feeler on a human finger. The different measurements have been taken to the middle of one subject's forefinger distal phalanx. These measurements give representative results of the elastic and mechanical properties of fingers. The different figures (1b, 1c, 1d) show the forces exerted on the finger according to the displacements in Z, X and Y directions respectively.

Figure 1b shows the loading and unloading curves obtained for feelers whose contact surface is 4 mm^2 or 100 mm^2. Other experiments have been done with intermediate section feelers. These experiments also allow us to associate "force - displacement" evolution laws with the contact surface.

Figures 1c, 1d show the "tangential force - displacement" curves along the X and Y axes. These curves have been made with a 4 mm^2 section feeler. They are obtained for a normal force equal to 0.35 and 1.0 Newton respectively. For the same normal force given, the slopes of curves obtained along the X and Y axes are different. The longitudinal elasticity of the human finger is therefore distinct from the transversal elasticity. These results bring to the fore a sliding phenomenum at points A_i and B_i (figures 1c, 1d). Moreover, the "tangential force - displacement" relations are different according to the considered normal force. Therefore, we need to consider a different relation for each given normal force. So, we can express these different ratios K_{Xi} and K_{Yi} (fig. 1c and 1d) according to the normal force N. As a consequence, a normal force and tangential forces correspond to each displacement generated along the different axes.

3. MECHANICAL SYSTEM

With the aim of modelizing the hand so that it can reproduce as many movements as possible, we have chosen a kinematics similar to that of the human hand. The 3D mechanical system (fig. 2) is made up of four three-phalanx fingers. One finger, representing the thumb, is placed opposite to the other three fingers. The hand will be fitted with not only the movements of finger opening and closing but also with "abduction-adduction" movements which allow finger movement, outside the plane. The "abduction-adduction" joint axe always remains normal to the proximal phalanx (fig. 2). Each joint is represented by a revolute joint type link assumed to be perfect. As a consequence, the hand possesses sixteen degrees of freedom in all. Contrary to a three-fingered hand, four

3

fingers are necessary and sufficient to obtain an easier manipulation while ensuring a good stability. The links have different lengths representative of the sizes of human hand. Fingers possess a variable thickness which is due to their elastic nature, which will be taken into account during the formalism implementation. We have considered the points of contact with friction but without sliding. Since their mechanical behaviours are the result of the experimental study which brings in surfacic contacts, we will consider that the efforts distributed on the contact surface can be reduced to a single resultant force, applied to a point. The object to be gripped is assumed to be rigid.

4. ANALYSIS

The stability of the grasped object mainly depends on a good distribution of contact forces on the object. Thus, we suggest solving this issue using an optimisation problem. This analysis takes into account the deformability of fingers and the small displacement of the object contact points, which results from the torques applied at joints. The equilibrium of the mechanical system is therefore sought whatever the grasp orientation, which is characterized by both position parameters θ and ϕ of the object (fig. 3). The problem consists in maximizing the load P which represents the weight of the object, while imposing limits on the applied joint torques τ_{ik}. The non-linear programming formulation of this problem is given as follows :

Maximize : P

Subject to :

$$\delta A_{ij} = J_{ij}\delta q_i \quad (1)$$

$$\delta R_{ij} = \delta A_{ij} - H_{G_{ij}}\delta G \quad (2)$$

i = 1, ..., 4 : n° finger

$$J^T F = \tau \quad (3)$$

j = 1, ..., 3 : n° phalanx

$$NF' = -T_E \quad (4)$$

k = 1, ...,4 : n° joint

$$\left|T_{X_{ij}}\right| \le \mu_1 N_{ij} , \left|T_{Y_{ij}}\right| \le \mu_2 N_{ij} \quad (5)$$

(for the palm : ij = p)

$$\delta R_{ij_N} \ge 0 \quad (6)$$

$$N_{ij} = g_1\left(\delta R_{ij_N}\right) \quad (7)$$

$$T_{X_{ij}} = K_X\left(N_{ij}\right)\delta R_{ij_{Tx}} , T_{Y_{ij}} = K_Y\left(N_{ij}\right)\delta R_{ij_{Ty}} \quad (8)$$

The equation (1) is given by the kinematic equations, where δA_{ij} is the small displacement vector of the contact points A_{ij} belonging to the links assumed to be rigid, J_{ij} is the Jacobian matrix relative to the link j belonging to the finger i, and δq_i is the small variation vector of joint coordinates q_{ik} belonging to the finger i.

The equation (2) gives the relative displacements at contact points A_{ij}, where δR_{ij} is the relative displacement due to the deformation of the link or the palm, and $H_{G_{ij}}\delta G$ is the

4

displacement of the object contact points A_{ij}. δG is the vector of the small generalized displacements of the object defined at point G.

The equation (3) is given by the static equilibrium equations for the mechanical system, where F is the vector of the forces exerted by the fingers, τ is the vector of the torques applied at the joints and J^T is the transpose of J which is Jacobian matrix of the mechanical system.

The equation (4) is given by the equilibrium equations for the object, where N is the matrix of Plucker coordinates of the lines of action of the forces exerted by fingers F and by the palm F_P, $F' = \left(F^T, F_P^T\right)^T$ and $T_E = \left(R_E^T, M_E^T\right)_G^T$ is the external wrench exerted on the object at centre of gravity G.

The relationship (5) gives the non-sliding contraints at contact points, where $T_{X_{ij}}$ and $T_{Y_{ij}}$ are the vectors of the tangential forces, N_{ij} is the vector of the normal forces, and μ_1, μ_2 are the coefficients of friction along the transversal and longitudinal axis of finger.

The relationship (6) requires that all phalanxes and the palm be in contact with the object, where δR_{ij_N} is the relative displacement along the normal axis at contact.

Equations (7) and (8) are the "force - displacement" relations obtained by the experimental study.

The results of the non-linear programming problem are plotted in figure 4, which shows the maximum load obtained for any given grasp orientation, on the plane determined by $\theta = 90°$. These results are obtained for a mechanical system (fig. 3) which is made up of two three-knuckle fingers opposed to one two-knuckle finger (the thumb). The contact points are situated in the middle of both phalanxes and palm. In the beginning of the optimisation process, the joint parameters q_{ij} are set at 60°. The "normal force - displacement" law corresponds to a 100 mm² section feeler. The experimental results give coefficients of friction which more or less approximate to 0.55 along the longitudinal and transversal axis. We also give (table fig. 4) the configuration for which the weight is the best ($\phi = 164°$). To these results, we add the valeur of joint parameters q_{ij}, as well as the forces exerted by the fingers and the palm. This table also presents results obtained on the plane determined by $\theta = 180°$. We precise the various parameters above mentionned for $\phi = 180°$ (configuration for which the weight is the best) and for $\phi = 90°$, that is the say when the weight P is directed along the axe Z.

5. CONCLUSION

This paper has presented a comprehensive study of the power grasp. Taking into account the real elasticity of the human finger. The approach of the power grasp stability is studied by a non-linear programming method. It enables one to solve different problems such as the optimisation of the weight, while imposing limits on the applied joint torques.

5

Contrary to the load which is obtained when the object is supported by the distal phalanxes, the results have shown that non-sliding and contact constraints on each phalanx and the palm greatly limit the maximum load on the palm. The resolution of the non-lincar programming problem will be extended to other problems, such as the achievement of the best distribution of the forces on both phalanxes and palm, for any given weight of an object.

6. REFERENCES

[1] M. S. ALI, K. J. KYRIAKOPOULOS and H. E. STEPHANOU. *"The kinematics of the anthrobot-2 dextrous hand"*. Proceedings of the 1993 IEEE International Conference on Robotics and Automation, Atlanta, Georgia, volume 3, pp. 705-710, May 1993.

[2] C. BONIVENTO, E. FALDELLA and G. VASSURA. *"The university of Bologna robotic hand project : Current state and future developments"*. Proceedings of the Fifth International Conference on Advanced Robotics (1991 I.C.A.R.), pp. 349-356, Pisa, Italy, June 1991.

[3] J. N. GABORIEAU, M. ARSICAULT, J. P. LALLEMAND and S. ZEGHLOUL. *"A new approach to the modelling of power grasp"*. Proceedings of the Ninth CISM-IFToMM Symposium on Theory and Practice of Robots and Manipulators (RoManSy 92), pp. 1-10, Udine, Italy, September 1992.

[4] S. C. JACOBSEN, E. K. IVERSEN, D. F. KNUTTI, R. T. JOHNSON and K. B. BIGGERS. *"Design of the UTAH/M.I.T. dextrous hand"*. Proceedings of the 1986 IEEE International Conference on Robotics and Automation, pp. 1520-1532, San Francisco, California, April 1986.

[5] J. W. JAMESON and L: J. LEIFER. *"Quasi-static analysis : a method for predicting grasp stability"*. Proceedings of the 1986 IEEE International Conference on Robotics and Automation, pp. 876-883, San Francisco, California, April 1986.

[6] L. LIANGYI and D. KOHLI. *"Analysis of conditions of stable prehension of a robot hand with elastic fingers"*. Proceedings of the Robotic Intelligence and Productivity Conference, pp. 99-104, Detroit, Michigan, 1983.

[7] K. MIRZA and D. E. ORIN. *"Force distribution for power grasp in the digits system"*. Proceedings of the Eighth CISM-IFToMM Symposium on Theory and Practice of Robots and Manipulators (RoManSy 90), Cracow, Poland, July 1990.

[8] D. J. MONTANA. *"The condition for contact grasp stability"*. Proceedings of the 1991 IEEE International Conference on Robotics and Automation, pp. 412-417, Sacramento, California, April 1991.

[9] N. ULRICH, V. KUMAR, R. PAUL and R. BAJCSY. *"Grasping with mechanical intelligence"*. Proceedings of the Eighth CISM-IFToMM Symposium on Theory and Practice of Robots and Manipulators (RoManSy 90), pp. 1-8, Cracow, Poland, July 1990.

6

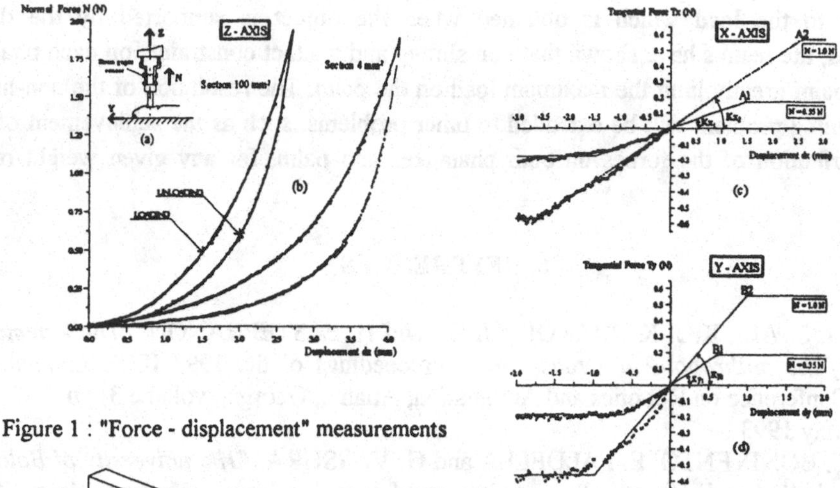

Figure 1 : "Force - displacement" measurements

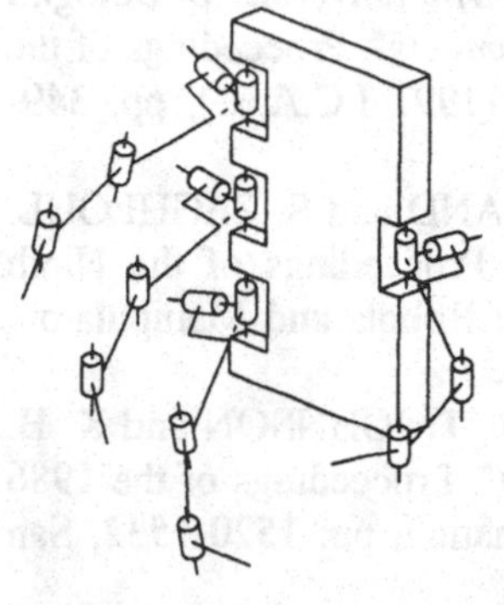

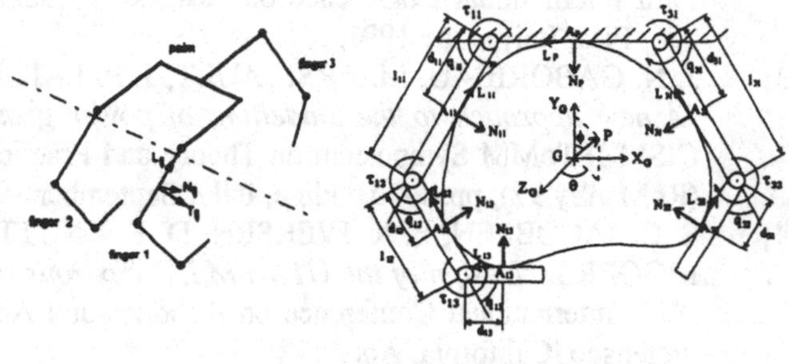

Figure 2 : Four-fingered hand

Figure 3 : Three-fingered hand

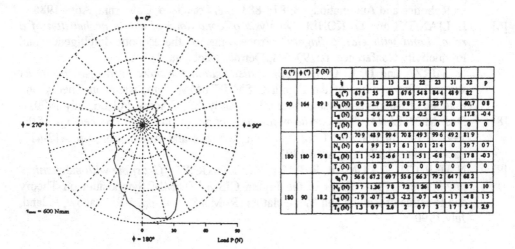

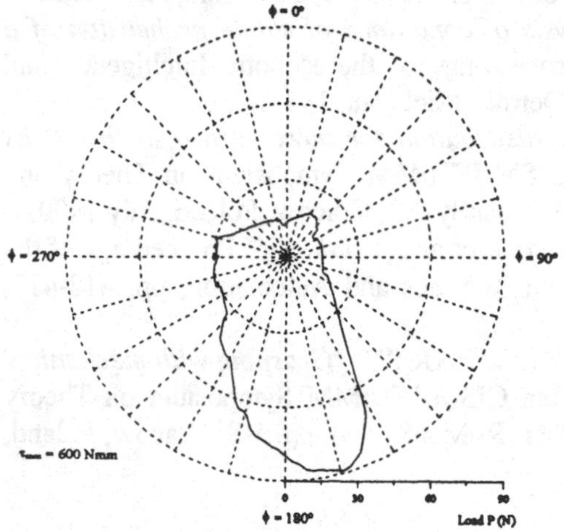

θ (°)	ϕ (°)	P (N)										
			ij	11	12	13	21	22	23	31	32	p
90	164	89 1	q_{ij} (°)	67 6	55	83	67 6	54 8	84 4	48 9	82	
			N_{ij} (N)	0 9	2.9	22.8	0 8	2 5	22 7	0	40.7	0 8
			L_{ij} (N)	0.3	-0 6	-3.7	0.3	-0.5	-4.5	0	17.8	-0 4
			T_{ij} (N)	0	0	0	0	0	0	0	0	0
180	180	79 8	q_{ij} (°)	70 9	48 9	99 4	70 8	49 3	99 6	49 2	81 9	
			N_{ij} (N)	6 4	9 9	21 7	6 1	10 1	21 4	0	39 7	0 7
			L_{ij} (N)	1 1	-5 2	-6 6	1 1	-5 1	-6 8	0	17 8	-0 3
			T_{ij} (N)	0	0	0	0	0	0	0	0	0
180	90	18.2	q_{ij} (°)	56 6	67.2	69 7	55 6	66 3	79 2	64 7	68 2	
			N_{ij} (N)	3 7	1.26	7 8	7.2	1 26	10	3	8 7	10
			L_{ij} (N)	-1 9	-0 7	-4.3	-2.2	-0 7	-4 9	-1 7	-4 8	1 5
			T_{ij} (N)	1.3	0 7	2.6	2	0 7	3	1 7	3 4	2.9

Figure : 4 Results of the non-linear programming problem

METHODICAL DESIGN OF NEW PARALLEL ROBOTS VIA THE LIE GROUP OF DISPLACEMENTS

J.M. Hervé

Ecole Centrale de Paris, Chatenay Malabry, France

and

F. Sparacino

Polytechnical University of Milan, Milan, Italy

ABSTRACT

Our aim is to give a complete presentation of the application of Lie Group Theory to the structural design of manipulator robots. We focused our attention on parallel manipulator robots and in particular those capable of spatial translation. This is justified by many industrial applications which do not need the orientation of the end-effector in the space. The advantage of this method is that we can derive systematically all kinematics chains which produce the desired displacement subgroup. Hence, an entire family of robots results from our investigation. The Y-STAR manipulator is now a working device. H -ROBOT is also being constructed. Both manipulators respond to the increasing demand of fast working rhythms in modern production at a low cost and are suited for any kind of pick and place jobs like sorting, arranging on palettes, packaging and assembly.

INTRODUCTION

The mathematical theory of groups can be applied to the set of displacements. If we call (D) the set of all possible displacements, it is proved, according to this theory, that (D) have a group structure. The most remarkable movements of a rigid body are then represented by subgroups of (D) . This method leads to a classification of mechanisms (1). The main step for establishing such a classification is the derivation of an exhaustive inventory of the subgroups of the displacement group. This can be done by a direct reasoning by examining al the kinds of products of rotations and translations (2).

However, a much more effective method consists in using Lie Group Theory (3), (4). Lie Groups are defined by analytical transformations depending on a finite numbei of real parameters. The displacement group (D) is a special case of a Lie Group of dimension six.

Our interest in Lie group theory is also confirmed by a general gain of interest in this field of mathematics by other researchers. These people utilize Lie Groups in various different fields of application all related to robotics as for example control theory (5) and vision (6), (7).

LIE'S THEORY

Within the framework of Lie's theory, we associate infinitesimal transformations making up a Lie algebra with finite operations which are obtained from the previous ones by exponentiation. Continuous analytical groups are described by the exponential of differential operators which correspond to the infinitesimal transformations of the group. Furthermore, group properties are interpreted by the algebraic structure of Lie algebra of the differential operators and conversely. We recall the main definition axiom of a Lie algebra : a Lie algebra is a vector space endowed with a bilinear skew symmetric closed product. It is well know (8), that the set of screw velocity fields is a vector space of dimension six for the natural operations at a given point N.

By following the steps indicated in (3) we can produce the exhaustive list of the Lie subgroups of the group of Euclidean displacements (D) (see synoptical list 1). This is done by first defining a differential operator associated with the velocity field.

Then, by exponentiation, we derive the formal Lie expression of finite displacements which are shown to be equivalent to affine direct orthonormal transformations. Lie sub-algebras of screw velocity fields lead to the description of the displacement subgroups.

THE (X (w)) SUBGROUP

In order to generate spatial translation with parallel mechanisms, we are led to look for displacement subgroups the intersection of which is the spatial translation subgroup (T). We will consider only the cases for which the intersection subgroup is strictly included in the two "parallel" subgroups. The most important case of this sort is the parallel association of two (X (w)) subgroups with two distinct vector directions **w** and **w'**. It is easy to prove :

$$(X(\mathbf{w})) \cap (X(\mathbf{w}')) = (T) \qquad \mathbf{w} \neq \mathbf{w}'$$

The subgroup (X(w)) plays a prominent role in mechanism design. This subgroup combines spatial translation with rotation about a movable axis which remains parallel to a given direction **w**, well defined by the unit vector **w**. Physical implementations of (X (w)) mechanical liaisons can be obtained by ordering in series kinematic pairs represented by subgroups of (X (w)). Practically only prismatic, revolute and screw pairs P, R, H are used to build robots (the cylindric pair C combines in a compact way a prismatic pair and a revolute pair). A complete list of all possible combinations of these kinematic pairs generating the (X (w)) subgroup is given in (9).

Two geometrical conditions have to be satisfied in the series : the rotation axes and the screw axes are parallel to the given vector **w** ; there is no passive mobility.

The displacement operator for the (X (w)) subgroup, acting on point M is :

$$M \rightarrow N + a\mathbf{u} + b\mathbf{v} + c\mathbf{w} + \exp(h\mathbf{w}\wedge)\,\mathbf{NM}$$

$\wedge$ is the symbol of the vector product.

Point N and the vectors **u, v, w** make up an orthogonal frame of reference in the space and a, b, c, h are the four parameters of the subgroup which has the dimension 4.

The advantage of using this method to find mathematical expression is that it is independent of the choice of a particular frame of reference of the Euclidian affine space (the set of points).

PARALLEL ROBOTS FOR SPATIAL TRANSLATION

To produce spatial translation it is sufficient to place two mechanical generators of the subgroups (X(w)) and (X (w')), **w** ≠ **w'**, in parallel, between a mobile platform and a fixed platform. If we want to built a robot which has only fixed motors then three generators of the three subgroups (X(w)), (X(w')), (X(w")), **w** ≠ **w'**, is needed. Any series of P, R or H pairs which constitute a mechanical generator of the (X(w)) subgroup can be implemented. Morever, these three mechanical generators may be different or the same depending on the desired kinematic results. This wide range of combinations gives rise to an entire family of robots capable of spatial translation. Simulation of the most interesting architectures can easily be achieved and the choice of the robot to be constructed can therefore meet the needs of the commissioner.

Clavel's Delta robot belongs to this family as it is based on the same kinematic principles (10).

THE PARALLEL MANIPULATOR Y-STAR

STAR (19) is made up by three cooperating arms which generate the subgroups (X(u)), (X(u')), (X(u")) (fig. 1 and 2). The three arms are identical and each one generates a subgroup (X(u)) by the series RHPaR where Pa represents the circular translation liaison determined by the two opposite bars of a planar hinged parallelogram. The axes of the two revolute pairs and of the screw pair must be parallel in order to generate a (X(u)) subgroup. Hence we write R1uHuPaR2u. For each arm, the first two pairs, i.e. the coaxial revolute pair and the screw pair, constitute the fixed part of the robot and form at the same time the mechanical structure of an electric jack which can be fixed in the frame. All the three axes lie on the same plane and divide it into three identical parts thus forming a Y shape. Hence the angle between any two axes is always $2\pi/3$. The mobile part of the robot is made up by three PaR series that all converge to a common point below which the mobile platform is located. The platform stays parallel to the reference plane and cannot rotate about the axis perpendicular to this plane. Any kind of appropiate end effector can be placed on this mobile platform.

The three deformable parallelograms ensure the stability and increase the rigidity of the whole. As we have already mentioned above, the axis of the last revolute pair R2u has to be parallel to the one of the first revolute pair R1u. This theoretical condition needs to be carefully applied in practice.

The derivation of the (T) subgroup, which proves the mobile platform can only translate in the space, is given in (11).

THE H- ROBOT

For a great majority of parallel robots including the Delta Robot and the Y Star, the working volume of the end effector is small relative to the bulkiness of the whole device. It is the essential drawback of such a kind of manipulator. In order to avoid this native narrowness of the working volume, it can be imagine to implement three input electric jacks with three parallel axes instead of converging axes. Three arms similar to those of the Y Star cannot be employed : the intersection set of three equal set (X (v)) will be equal to (X (v)) instead of (T).Hence, designing the new H-Robot (19), we have chosen two arms of the Y Star type and a third pattern which may be compared with the Delta arms. This third mechanism begin from the fixed frame with a motorized prismatic pair parallel to the first two electric jacks. It is followed by a hinged planar parallelogram which is free to rotate around an axis perpendicular to the P pair thank to a revolute pair R linking the slider of the P pair and a bar of the parallelogram. The opposite bar is connected to the mobile platform via a revolute pair R of parallel axis. This property is maintained when the parallelogram changes of shape (with one degree of freedom).

In a first prototype built at "IUT de Ville d'Avray" (France) by a team a students directed by the professor Pastoré, a H-Robot implements 3 systems screws (1) / nut (2) with a large pitch, which allow rapid movements. It is hold by bearings (6) and animated by the actuators M. Three planar hinged parallelograms, on both sides (4) and at the center (5) make the connection from the nuts to the horizontal platform (3). The stand (7) supports the whole structure (fig. 3, 4).

The side screws permit rotation and translation along their axes. The central nut does not allow the rotation of the parallelogram plane about the screw axis.

The mobile platform can only translate with 3 degrees of freedom inside the working space which may be assimilated to a half-cylinder.

The main advantage of this device is that the working volume is directly proportional to the length of the parallel axes and it can be made considerably large.

The prototype have a 1 meter long rotation axis. All the actuators are in a fixed position.

The three arms made of composite materials (carbon fibre) are stiff and lightweighted. The resulting movements of the mobile platform made of honeycomb material are then very fast.

H-Robot seems to be a better solution than Y-Star especially in those cases where the size of the working volume becomes important.

CONCLUSIONS

The importance of Lie group theory, expecially for kinematics is recognized from various sources (13), (14), (15), (16), (17), (18). Investigation of new parallel robots generating pure translation led us to the construction of two prototypes : Y-Star and

H-Robot. Increasing performances and the low cost of fabrication make these robots attractive for modern industry. They are presented as an alternative to the DELTA robot and have the classical parallel robot advantages for positioning, precision, rapidity and fixed motor location.

REFERENCES

(1) HERVE J. M., "Analyse structurelle des mécanismes par groupe des déplacements", Mech. Mach. Theory 13, pp. 437-450 (1978).

(2) FANGHELLA P., "Kinematics of Spatial Linkage by Group Algebra : a structure based approach", Mech. Mach. Theory 23, n° 3 pp. 171-183 (1988).

(3) HERVE J. M., "The mathematical group structure of the set of displacements" Mech. Mach. Theory 29, n°1 pp. 73-81 (1994).

(4) HERVE J. M., "Intrinsic formulation of problems of geometry and kinematics of mechanism", Mech. Mach. Theory 17,pp. 179-184 (1982).

(5) ISIDORI ALBERTO,"Nonlinear Control Systems", Springer-Verlag, (1985).

(6) TANAKA M., KURITA T., UMEYAMA S., "Image Understanding via Representation of the Projected Motion Group", IEEE Conference on Computer Vision and Pattern Recognition ICCV, June 1993.

(7) HOFFMAN W.C., "The Lie algebra of visual perception", J. Math. Psych. 3, 1966, pp. 65-98.

(8) SUGIMOTO K., DUFFY J.,"Application of linear Algebra to Screw Systems", Mech. Mach. Theory , Vol. 17 pp. 73 -83 (1982).

(9) HERVE J. M., SPARACINO F.,"Structural Synthesis of Parallel Robots Generating Spatial Translation" 5th Int. Conf. on Adv. Robotics, IEEE n° 91TH0367-4, Vol 1, pp. 808-813, 1991.

(10) CLAVEL R.,"Delta, a fast robot with parallel geometry", Proc. Int. Symp. on Industrial Robots, April 1988, pp 91 -100.

(11) HERVE J. M., SPARACINO F.,"Star, a New Concept in Robotics", 3rd Intern. Workshop on Advances in Robot Kinematics, Sept. 7-9, 1992, Ferrara, Italy pp. 176-183.

(12) MERLET J.P.,"Les robots parallèles", Hermès, Paris 1990.

(13) KARGER A., NOVACK J., Space Kinematics and Lie Groups, Gordon and Breach Science Publishers, 1985.

(14) CHEVALLIER D.P.,"Lie Algebras, Modules, Dual Quaternions and Algebraic Methods in Kinematics", Mech. Mach. Theory , Vol. 26, n°6, pp. 613-627 (1991).

(15) POPPLESTONE R.J.,"Group Theory and Robotics", in Robotics Research. The First Int. Symp., M. Brady and R. Paul Eds., Cambridge, MA.MIT Press 1984.

(16) ANGELES J.,"Spatial kinematics chains",Spring Verlag, Berlin, 1982.

(17) HILLER M.,WOERNIE C.,"A Unified Representation of Spatial Displacements", Mech. Mach. Theory , Vol. 19 pp. 477-486 (1984).

(18) SAMUEL A.E., Mc AREE P.R.,HUNT K.H.,"Unifying Screw Geometry and Matrix Transformations", The International Journal of Robotics Research, Vol. 10, n°5, October 1991.

(19) HERVE J.M., "Dispositif pour le déplacement en translation spatiale d'un élément dans l'espace, en particulier pour robot mécanique", French patent n° 9100286 of january 11, 1991. European patent n° 91403521.7 of december 23, 1991.

SYNOPTICAL LIST 1

SUBGROUPS OF THE DISPLACEMENT GROUP

(E)	identity
(T(D))	translations parallel to the straighline D
(R(N,**u**))	rotations around the axis determined by the pair N,**u** (or and equivalent pair N',**u** with NN'$\wedge$**u** = O)
(H(N,**u**,p))	screw motions with the axis N,**u** and the pitch p = 2πk
(T(P))	translations parallel to the plane P
(C(N,**u**))	combined rotations and translations along an axis (N,**u**)
(T)	spatial translation
(G(P))	planar movements parallel to the plane P
(Y(**w**,p))	screw translations allowing plane translations perpendicular to **w** and screw motions of pitch p along any axis parallel to **w**
(S(N))	spheric rotations around the point N
(X(**w**))	translating hinge motions allowing spatial translations and rotations around any axis parallel to **w**
(D)	general rigid body motions

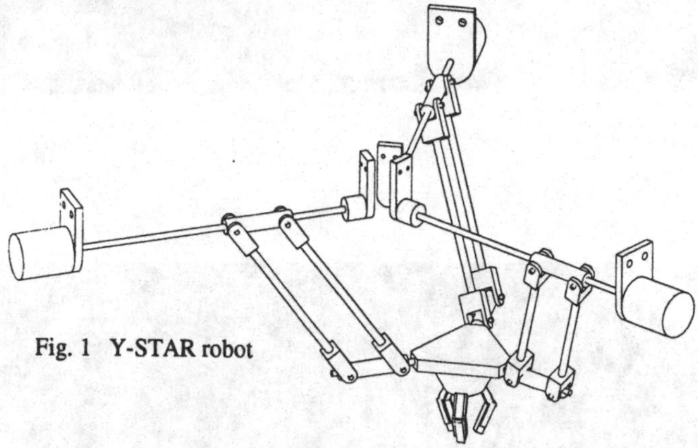

Fig. 1 Y-STAR robot

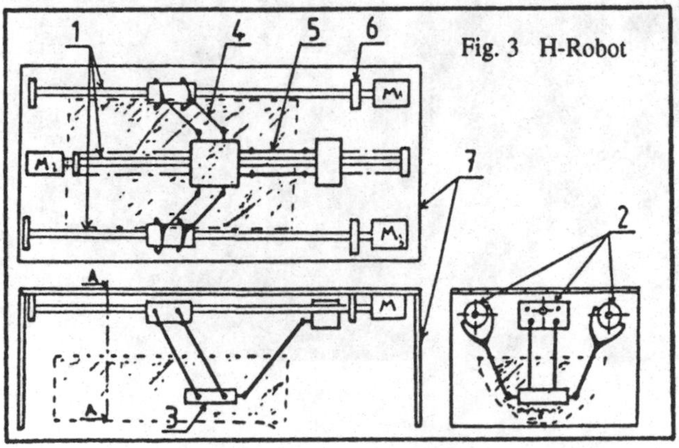

Fig. 3 H-Robot

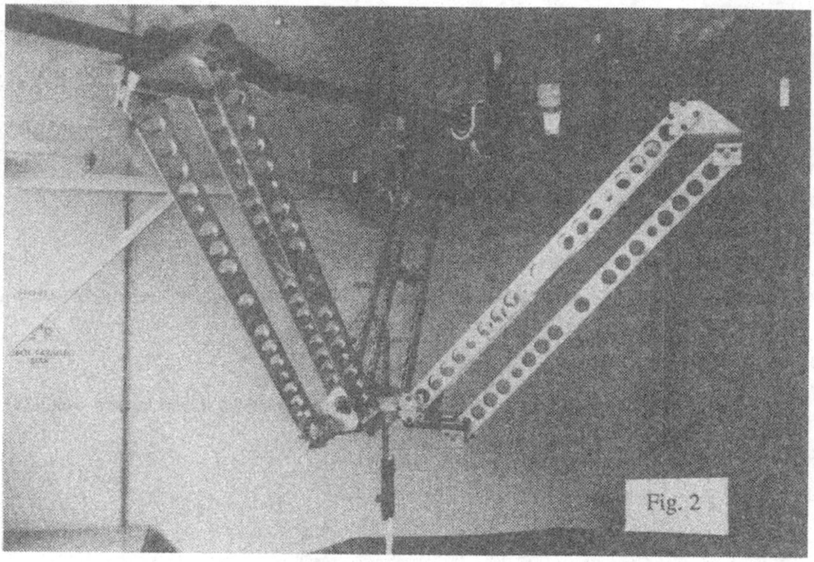

Fig. 2

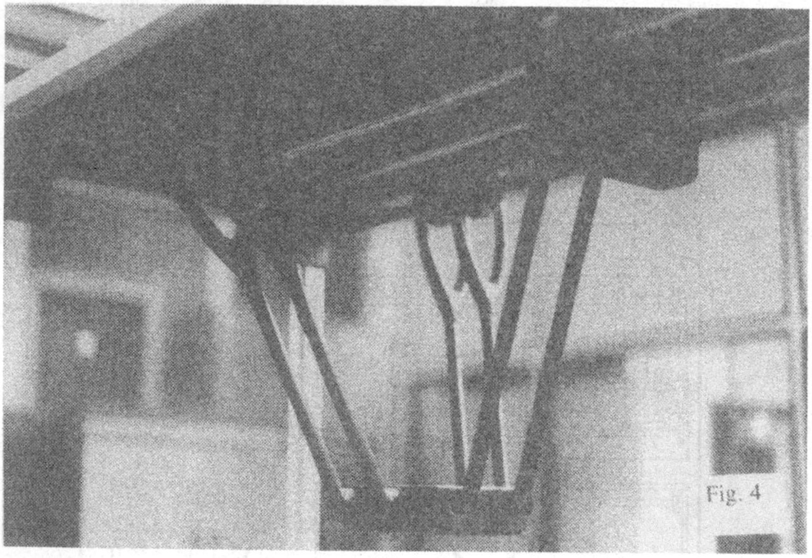

Fig. 4

VIBRATION CONTROL OF A ONE LINK FLEXIBLE ARM: EXPERIMENTAL RESULTS

M.M. Gola and A. Somà
Polytechnical University of Turin, Turin, Italy

Abstract:

In the present paper an open-loop technique is experimentally investigated in conjunction with closed-loop controller in order to reduce vibration after the move of a one link flexible manipulator. The vibration compensator scheme consists of an impulse sequence filter which time delay is given by the structural frequencies to be controlled. Two and three impulse sequence filters are used to compare compensator with different robustness under system parameter uncertainty. Satisfactory performances have been achieved with point-to point trajectory in the case of one link flexible arm operated through brushless motor and controlled by a DSP real time board. Experiments are performed in order to compare the efficiency of the method of control increasing the joint speed and acceleration. Besides, the other tests are implemented to eliminate the residual vibration due to the second bending mode.

1. Introduction

The present work is developed in the frame of the "Progetto Finalizzato Robotica" of the C.N.R. of Italy. The final result is a full scale industrial robot in line with light mechanism that can manipulate heavier payloads with top speed a precise end effector tracking. Such performance requires to investigate the practical limitation of the control of the robot as an elastic multibody system and enhance the knowledge in control/structure interaction. The experimental testbed described in this paper is a one link very flexible arm useful to improve the performance of the system of control that will be applied to the full scale machine.

An examination of the recent literature concerning robot control shows that flexible manipulators are currently under very active investigation. Many papers have been presented on these topics and experimentally tests are normally included (1,2,3,4); however the major drawback is the lack of robustness of the vibration compensator in different operating conditions.

The present work will focus on the point to point control problem that is of significance because many practical engineering systems require to move from an initial point to a desired final point in a desired time. In this case several open loop schemes based on input preshaping techniques have been recently developed in order to modify the input command so that the endpoint vibration can be eliminated. Among these methods often filters in frequency domain are proposed for input signal conditioning. This approach gives poor robustness in case uncertainty of system modal parameters. Other approaches include the construction of input command from versine or ramped sinusoid functions. Also optimal control approaches have been used to generate input profiles. In both cases each motion requires the recomputation of the control law.

In the present work the command input is preshaped by an impulse sequence filter whose time delay is given by structural frequencies to be controlled. Each impulse is scaled taking into account the modal damping. The controlled trajectory is so obtained through convolution of the impulse sequence filter together with the desired input command.

The theoretical development of the preshaping theory will not be reported here, but is addressed to the original work of Singer & Seering (5). In the case of real time control the simplicity of the impulse sequence filter allows the direct use as closed loop when is not possible to recalculate the trajectory (6). The result of numerical simulation applied to the robot model have been presented by the authors in previous works (7,8). In this paper the preshaping technique is operated in conjunction with closed-loop PI controller in order to experimentally investigate the vibration reduction.

The experimental modal analysis is first of all performed to update the dynamic parameters (frequency and damping) including the effect of the brushless motors and the gear train. Results are briefly described in order to outline the dynamic parameters whose knowledge is of relevance to apply the method of control. The so validated model allows to preshape input velocity command. The trajectory is fed to a DSP controller and the endpoint vibrations with and without compensator are measured by means of accelerometers.

Applications to different joint speeds are investigated in order to improve the performance of the preshaping technique with two and three impulses.

2. The Experimental System

2.1 Real time controller

The control system is based on a TMS 320 DSP on a Dspace control board. Together with the DS1001 processor board a digital/analog and dedicate incremental encoder boards are connected via PHS-BUS. The open loop preshaping technique investigated in the present work does not allow to use the control system at its full capacity. Future applications to reduce vibration of the robot prototype will enhance the performance of the DSP as real time controller.

2.2 Brushless Motor and PI controller

The manipulator joint is operated though a MOOG brushless servo motor. The electrical connection is direct from the motor connector to stator windings; in particular the absence of brushes and commutator means there is no mechanical limit or peak motor current to provide high peak torque. Among the possible choices the goal of high dynamic performance suggests to choose the motor that provides the best torque/rotor inertia ratio. The controller unit operates the commutation of the servomotor, close the motor speed loop and give out a sinusoidal 3-phase motor current. The velocity card contains the PI-velocity controller that has to adjust respect the system dynamic. The tuning of the proportional and integral gains allows to obtain the system stability and increase the joint stiffness. The set-up procedure was performed by monitoring experimentally the tachometer output with PI gain adjusting with low current reference input.

2.3 The aluminium arm

Fig. 1) shows the schematic representation of the very flexible aluminium arm chosen as testbed to investigate the practical limitation of the preshaping control. The geometrical and material parameters of the arm are listed in table I).

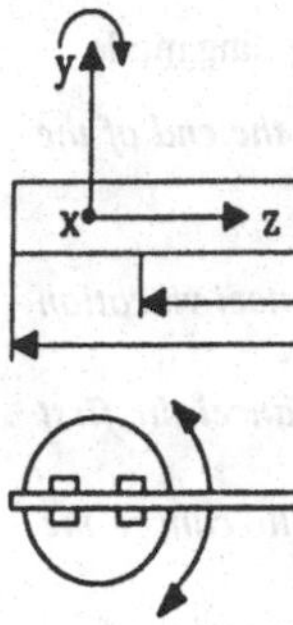

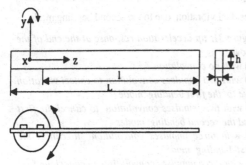

figure 1: aluminium arm scheme

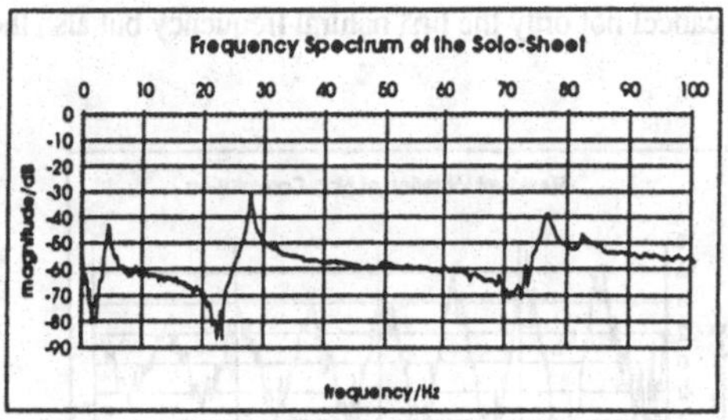

figure 2: natural frequencies of the system

3. System identification

In order to update the dynamic properties of the system taking into account the effect of the control/structure interaction a frequency domain system identification is performed. Extended treatment of the model updating methods were presented in previous work (8). In the present work the simpler identification procedure is obtained through the following experimental step:

- a reference zero input signal is fed to the Moog servo system to introduce stiffening effect of the motor and PI controller;
- an impact excitation test with an instrumented hammer is performed and the tip response is measured with accelerometers
- at the end the transfer function (fig. 2) between the excitation source and the accelerometer sensor is obtained trough a FFT analyser board.

In table II) the devices used in the identification procedure are listed. Fig. 2) shows the graphical representation of the first three structural bending frequencies as well to see by the high peaks. Of interest are the first and the second and the relative damping used in the experimental implementation of the preshaping method. The modal parameters, listed in table III), are extracted by using a curve fitting algorithm (Star Modal 4.0). The first bending mode occurs at 4.45 Hz with 1% of damping, while the second bending frequency is 27.6 Hz with 0.5% of damping.

Table I: parameters of the arm

l=1 m	L = 1.2 m
b= 0.006 m	h = 0.08 m
E = 70 GPa	m = 1.66 kg
I = 1.44E-9 m4	J = 3.5E-6 kg*m2

Table II: measurements devices

analyser	Data Physics FFT analyser board
accelerometer	B&K type 4395 sens. 5.5 mV/g
instrumented hammer	PCB type GK291 sens. 9.3 mV/N
modal analysis software	Star Modal 4.0

4 Experiments and Results

Experiments were made to test the theory of convolution. Their purpose is to get a comparison of the residual vibration without convolution, with two and with three impulses. Besides, other experiments were implements, to cancel not only the first natural frequency but also the residual vibration due to the second bending mode.

a)

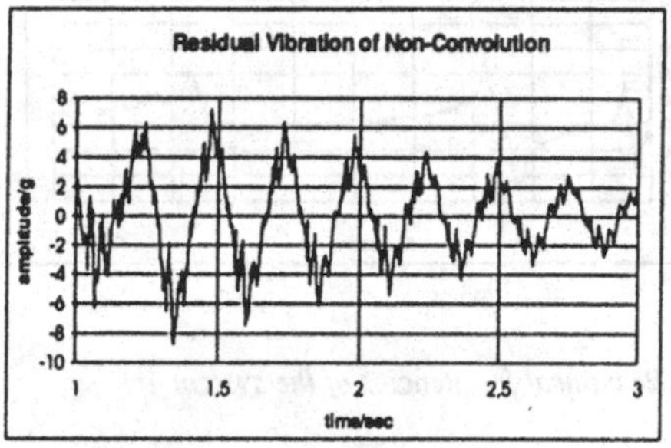

figure 3): tip acceleration response at the end of the move
a) with no convolution
b) with two impulses convolution to cancel vibration due to the first bending mode
c) with two impulses convolution to cancel the first and the second bending modes
d) with three impulses convolution to cancel the first bending mode
e) with three impulses convolution to cancel the first and the second bending modes

b)

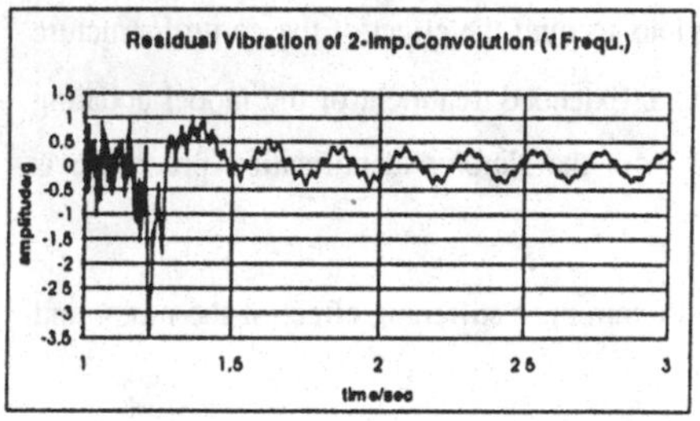

c)

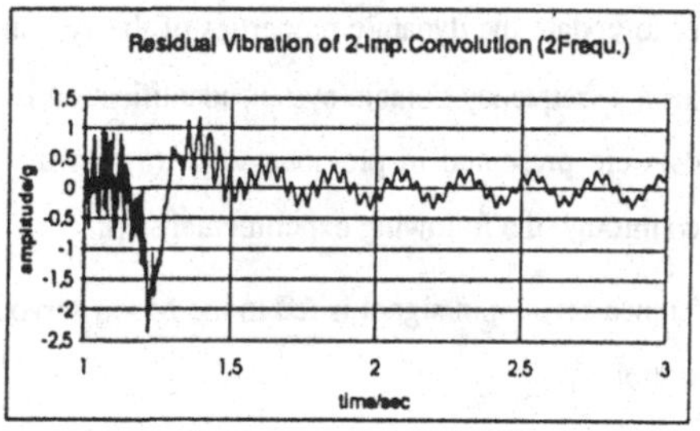

d)

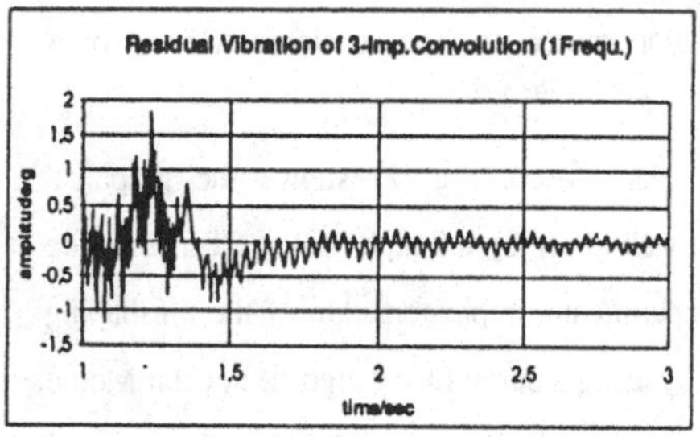

e)

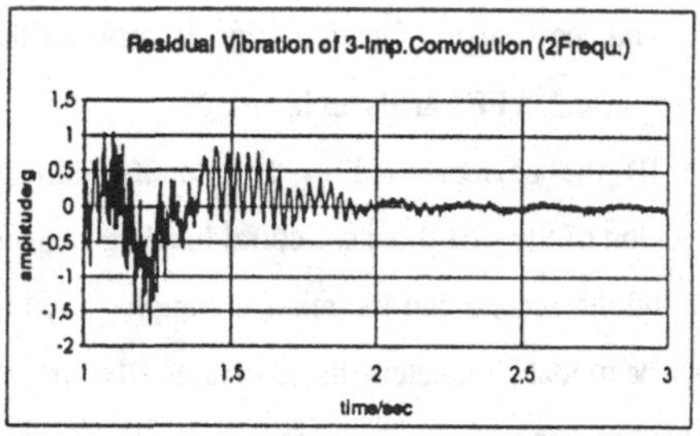

The point-to-point control trajectory is performed using as command input a trapezoidal velocity ramp. Limiting the total displacement the main parameters of the velocity ramp is obviously represented by the joint acceleration and the maximum speed. The behaviour of the flexible arm is investigated at two different velocity levels (2-6 rad/s) increasing the joint acceleration from 2g to 8g with steps of 1g.

At each pair of velocity and acceleration parameters corresponds 5 measurements. In fact the velocity ramp is first of all calculated choosing the two dynamic parameters (joint acceleration and maximum velocity). The so obtained input signal is then convoluted with four different impulse filters:

- two impulses to cancel the vibration due to the first bending mode
- two impulses to cancel the vibration due to the first and the second bending modes
- three impulses to cancel the vibration due to the first bending mode
- three impulses to cancel the vibration due to the first and the second bending modes

In a classical open loop scheme the velocity ramp is then sampled and sent to the motor. The sampling rate of the DSP controller is set for all the measurements at 1 msec.

At the end of each velocity command the data of the bending response at the tip of the arm are stored. The graphs of fig. 3) show the tip response at the end of the move for the five velocity ramps obtained in the case with 6 rad/s of maximum speed and 4g of joint acceleration. Fig. 3a) shows the response without preshaping. The acceleration amplitude is scaled by the gravity acceleration. The diagrams of fig. 3) show clearly that the 3 impulse sequence is more able to reduce residual vibrations increasing also the robustness under system uncertainty. Altogether is possible to underline that the residual vibrations present a threshold that cannot be avoided. This effect due to the disturbances of the motor-controller system and signal noise needs further investigation in order to obtain very precise tracking.

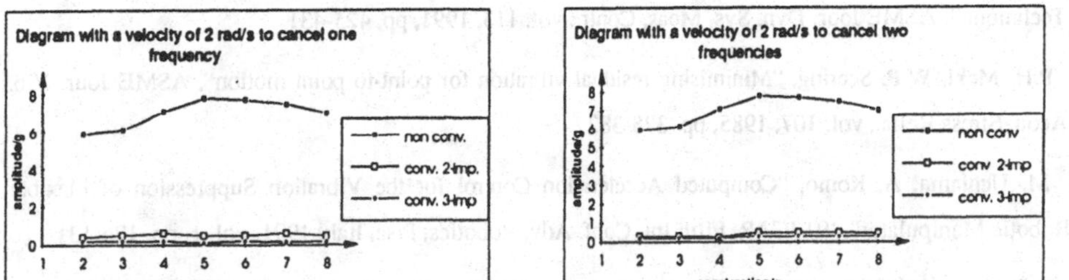

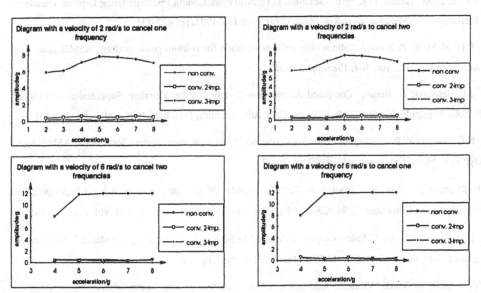

figure 4: maximum acceleration amplitude at the end of the move; comparison with non convoluted and convoluted with two and three impulse sequence:

a) maximum velocity 2 rad/s; preshaping on the first bending mode

b) maximum velocity 2 rad/s; preshaping on the first and the second bending modes

c) maximum velocity 6 rad/s; preshaping on the first bending mode

d) maximum velocity 6 rad/s; preshaping on the first and the second bending modes

The maximum acceleration amplitude at the end of the move is stored for all the measurements and graphically reported in fig. 4). In each diagram the non convoluted residual vibration is compared with the results obtained through two or three impulses convolution. Fig. 4a.b) refers the cases of 2 rad/s maximum joint velocity, while fig. 4c,d) those with 6 rad/s. The diagram of fig. 4) let see that increasing the joint speed the preshaping technique present a better performance. In fact even if at a higher speed correspond an increasing of non convoluted residual vibration, the suppression of the vibration gives successfully result increasing the ratio between the non convoluted / convoluted residual amplitude.

Conclusion

A technique for preshaping command input has been experimentally implemented to a one flexible manipulator. The joint is operated through a Moog brushless servosystem and the vibration controller is programmed on a DSP board. Experiments are performed in order to compare the efficiency of the method with different speed and acceleration condition. Satisfactory performance have been achieved showing the superiority of the more robust three impulse sequence filter.

References

(1) D.M. Aspinwall, "Acceleration profiles for minimising residual response", ASME Jour. Dyn. Meas. Contr., vol. 102, 19080, pp. 3-7

(2) S.P. Bhat, M. Tanaka, D.K. Miu, "Solutions to Point-to Point Control Problems Using Laplace Transform Technique", ASME Jour. Dyn. Sys. Meas. Contr., vol. 113, 1991, pp. 425-431

(3) P.H. Mekl, W.P. Seering, "Minimising residual vibration for point-to point motion", ASME Jour. Vib. Acou. Stress Relia., vol. 107, 1985, pp. 378-382

(4) M. Uchiama, A. Komo, "Computed Acceleration Control for the Vibration Suppression of Flexible Robotic Manipulator", '91 ICAR, Fifth Int. Conf. Adv. Robotics, Pisa, Italy 1991, vol. 1, pp. 126-131

(5) N.C. Singer, W.P. Seering, "Preshaping Command Inputs to Reduce System Vibration", ASME Jour. Dyn Syst. Meas. Contr., vol. 112, 1990, pp. 76-82

(6) F. Khorrami, S. Jain, W. Grosman, A. Tzes, W. Blesser, "Non-linear control with Input Preshaping for Flexible-Link Manipulators", '91 ICAR, V Int. Conf. Adv. Robotic, Pisa, Italy 1991, vol. 1, pp. 96-101

(7) M.M. Gola, A. Somà. "Mode superposition method in flexible robot trajectory simulation", Struceng & Femcad Conf. Grenoble Oct. 17-18 1990, Proc. I.I.T.T. Paris, pp. 65-70

(8) M.M. Gola, A. Somà, "Vibration reduction in a four mode flexible robot: identification and simulation", Proc. of the II Iasted Int. Conf., Alexandria, May, 5-7, 1992, pp. 79-84

Acknowledgement

This paper is developed within the frame of "Progetto finalizzato Robotica" of C.N.R. Italy grant n. 92.01069.67

Part IX
Sensing and Machine Intelligence

SENSOR-BASED REACTIVE ROBOT CONTROL

C. Zieliński and T. Zielińska
Warsaw University of Technology, Warsaw, Poland

Abstract

The paper presents a formalized approach to sensor-based reactive robot control. This type of control consists in actively using sensor readings to modify robot actions. First the sensor reading space is partitioned into subspaces. With each of these subspaces a reaction is associated. If during the realisation of the goal, the sensor readings "enter" a subspace associated with a certain action, then the realisation of the previous goal is interrupted and an appropriate reaction is triggered. As the sensor reading space, robot reactions and the global goal can be described formally, a formal specification of the robotic controller realising a given task is obtained. This specification is used as the basis for coding the software of a controller tailored to the needs of a specific task.

1 Introduction

Utilization of sensor information in robot control has been the subject of research for several years [1]. This research concentrates on: development of new types of robot sensors (e.g. [2]) and incorporation of many types of sensors into a robotic system (e.g. [3]). The problems of data aggregation (fusion) are considered too (e.g. [4, 5]). The researchers of robot control systems relying on sensor data take different approaches to the problems of: integration of multiple sensors into a robotic system (e.g. [6, 7, 8, 9]), formal task description for robots equipped with sensors (e.g. [10, 11]), sensor data aggregation and interpretation (e.g. [4, 5]). A comprehensive discussion of some of the above topics can be found in [5].

Artificial intelligence approach to robot control strongly relies on world models to execute a task. Sensors, in this case, are mainly used to update the world model, which in turn is used in the generation of the plan of actions. On the other hand, behavioural control concept does not need a world model to execute a task [12, 13, 14]. In this case the controller is built of several finite state automatons functioning in parallel, each achieving a single objective by a certain behaviour. The controller is constructed incrementally by adding ever more complex layers of behaviours on top of the more elementary ones. Upper layers examine data from lower levels and can suppress or inhibit their behaviours.

In this paper the goal that is to be achieved and a single layer of actions (that can also be called behaviours or reactions) that are executed when sensors detect appropriate conditions, are distinguished. Unlike the pure behavioural approach, where the partitioning of the system is intuitive, a formal path was followed.

2 Sensor-based reactive robot control

2.1 Theoretical considerations

A robotic system is decomposed into three subsystems: **effectors** (manipulator arm or arms, tool and the cooperating devices), **receptors (real sensors)**, and the **control subsystem** (e.g. memory). The state $s \in S$ of such a system is denoted in the following way:

$$s = < e, r, c >, \quad s \in S, \quad e \in E, \quad r \in R, \quad c \in C, \tag{1}$$

where

e	is the state of the effectors,	E	is the effector state space,
r	is the state of the real sensors,	R	real sensor reading space,
c	is the control subsystem state,	C	is its state space.

The raw data obtained from real sensors usually cannot be utilized directly to control the system. It has to be transformed into a useful form. This transformation is called **data aggregation**. As a result of this a **virtual sensor reading** v is obtained.

$$v = f(r), \quad v \in V \tag{2}$$

Vector function f is called an **aggregating function**. V is the virtual sensor reading space.

The majority of robotic systems is computer controlled, so the execution of a task is subdivided into steps. The initial state is labeled 0, and the consecutive intermediate states $i = 1, \ldots, i_T$, where i_T is the label of the terminal state. In the terminal state either the task is accomplished or an execution error is detected.

While the robot is executing the task, the sensors monitor the state of the system and the environment constantly. The task is executed according to some **general plan** (associated with the **main reaction**), but it can be hindered by certain external events. These events can be detected by sensors. Once this happens the system reacts to them by a **reaction**. The system exhibits a number of reactions $B_j, j = 0, \ldots, j_R$, where $j_R + 1$ is the number of reactions. Each of the reactions is executed by realising an instance of this reaction. A **reaction instance** b_j of reaction B_j is executed as a sequence of steps:

$$b_j = (e_j^0, c_j^0)\ (e_j^1, c_j^1) \ldots (e_j^T, c_j^T), \quad b_j \in B_j, \tag{3}$$

where (e_j^0, c_j^0) is the initial state and (e_j^T, c_j^T) is the terminal state of reaction instance $b_j \in B_j$

A reaction B_j is triggered by a virtual sensor reading belonging to a virtual sensor reading subspace $V_j \subset V$. There exists a virtual reading subspace $V_0 \subset V$, which does not trigger any specific reaction, i.e. while the virtual sensors are supplying readings from this subspace the previously executed reaction is continued. The currently executed reaction terminates either because virtual sensors trigger another reaction or because the job of this reaction is done (the final state of this reaction is reached). It is assumed that there exists a reaction B_0 (called the **main reaction**) which does not have to be triggered by any virtual sensor reading. The system initially executes reaction B_0 and whenever the execution of other reactions terminates, and no other reaction is triggered by adequate virtual sensor readings, execution of reaction B_0 is resumed.

The virtual sensor reading space is subdivided into subspaces $V_j, \ j = 0, \ldots, j_R$, where $j_R + 1$ is the number of these subspaces, in such a way that:

$$V = V_0 \cup \bigcup_{j=1}^{j_R} V_j, \quad \text{and} \quad \forall_{j \neq q} V_j \bigcap V_q = \emptyset, \quad q = 0, \ldots, j_R, \tag{4}$$

Each virtual sensor reading $v_j \in V_j$, $j = 1, \ldots, j_R$, triggers a reaction instance $b_j \in B_j$. If reaction B_j is triggered basing only on information contained in the virtual sensor reading v_j ($v_j \Rightarrow B_j$) then B_j is called **state independent reaction**, but if reaction B_j is triggered basing both on information contained in virtual sensor reading v_j and the current effector state e_i or control state c_i ($(v_j, e_i, c_i) \Rightarrow B_j$) then B_j is called **state dependent reaction**.

2.2 Example: simple maze running task

The task was to move a 4-bit touch-probe from the entrance of the maze to the exit area avoiding obstacles (maze walls) – fig.1. No map of the maze was available to the robot. It had to rely on local information obtained through the touch-probe. The touch probe can be pictured as a cylinder with 4 touch sensors pointing in the north (N), south (S), east (E) and west (W) directions perpendicular to the cylinder axis. The maze was positioned in such a way that a 5 d.o.f. robot could approximately align the N-S and E-W sensor pairs with the N-S and E-W directions of the maze respectively. The entrance to the maze was in its S-W corner and the exit area was defined as the northern bounding wall of the maze. A very simple maze with no cul-de-sacs was assumed. This assumption simplifies the path-finding algorithm (no backtracking in the N-S direction is necessary). The N direction in the maze coincides with the maze $Y+$ and the E direction with $X+$.

State of the manipulator probe in relation to the maze is expressed as three Cartesian coordinates and two Euler angles:

$$e = [e_x, e_y, e_z, e_\phi, e_\psi], \quad e \in E, \tag{5}$$

The goal G_* is formulated as:

$$G_* \ : \ e^{i_T} \in E_*^{i_T} \qquad E_*^{i_T} = E_{*x}^{i_T} \times E_{*y}^{i_T} \times E_{*z}^{i_T} \times E_{*\phi}^{i_T} \times E_{*\psi}^{i_T} \tag{6}$$

where $E_*^{i_T}$ describes the exit from the maze and i_T is (an initially unknown) number of executed steps.

Virtual sensor reading is expressed as:

$$v = [\ {}_Nv,\ {}_Sv,\ {}_Wv,\ {}_Ev\ ], \qquad {}_Nv,\ {}_Sv,\ {}_Wv,\ {}_Ev \in \{contact,\ no\_contact\} \tag{7}$$

In this simple case the virtual sensors are identical with real sensors.

The division of virtual sensor reading space V and the assignment of reactions B is:

$$\begin{cases} V_0 : & {}_Nv,\ {}_Sv,\ {}_Wv,\ {}_Ev\ =\ no\_contact & \Rightarrow B_j, \quad j = 0, \ldots, 4 \\ V_1 : & {}_Nv\ =\ contact,\quad {}_Sv,\ {}_Wv,\ {}_Ev\ =\ no\_contact & \Rightarrow B_1\ =\ B_N \\ V_2 : & {}_Wv\ =\ contact,\quad {}_Nv,\ {}_Sv,\ {}_Ev\ =\ no\_contact & \Rightarrow B_2\ =\ B_W \\ V_3 : & {}_Ev\ =\ contact,\quad {}_Nv,\ {}_Sv,\ {}_Wv\ =\ no\_contact & \Rightarrow B_3\ =\ B_E \\ V_4 : & \text{all other cases} & \Rightarrow B_4\ =\ B_{ERR} \end{cases} \tag{8}$$

If sensor readings enter subspace V_0 while reaction B_j, $j = 1, \ldots, 4$ is being executed, then the reaction B_j is continued, if it has not been terminated at that instant due to all of its steps being completed. If termination of reaction B_j occurred upon sensor readings entering subspace V_0, then reaction B_0 is continued.

The definition of a reaction is formulated by describing 'snapshots' of the state of the effectors and the control subsystem. This notation does not show how to obtain the next state, it only shows how the next state should look like. The method of attaining that state is up to the implementer of the system.

Reaction B_0 is defined as follows:

$$b_0 \in B_0, \quad b_0 = (e_0^0, c_0^0)\ (e_0^1, c_0^1), \ldots, (e_0^{i_T}, c_0^{i_T})$$

$$\begin{cases} e_0^0 & : \ e_x^0 = entry_x, \quad e_y^0 = entry_y, \quad e_z^0 = entry_z, \quad e_\phi^0 = entry_\phi, \quad e_\psi^0 = entry_\psi \\ c_0^0 & : \ \delta^0 = 1, \ i = 0 \\ e_0^i & : \ e_y^i = e_y^{i-1} + \Delta e_y, \quad \text{for } i = 1, \ldots, i_T \\ c_0^i & : \ \delta^i = \delta^{i-1}, \quad \text{for } i = 1, \ldots, i_T \\ e_0^{i_T} & : \ e_x^{i_T} = exit_x, \quad e_y^{i_T} = exit_y, \quad e_z^{i_T} = exit_z, \quad e_\phi^{i_T} = exit_\phi, \quad e_\psi^{i_T} = exit_\psi; \\ & \quad \text{where } exit_x \in E_{*x}^{i_T} \quad exit_y \in E_{*y}^{i_T} \\ c_0^{i_T} & : \ i = i_T \end{cases} \tag{9}$$

where:

i is the global task step number (associated with a counter), and

i_T is the number of steps needed to accomplish the task – it is initially unknown,

$entry$, $exit$ are the entry and exit positions to and from the maze respectively ($exit_z = entry_z$, $exit_\phi = entry_\phi$, $exit_\psi = entry_\psi$).

The reaction B_0 starts its execution with the probe positioned at the entrance ($entry$) to the maze, and the search-direction marker δ^0 (in step 0) set to 1 (search to the right). The global step counter is 0. In each following step i, the value of the marker (δ^i) does not change and the probe is transferred by Δe_y in the north direction (maze $Y+$ direction). Reaction B_0 is executed until the probe reaches the exit area.

The reaction B_N is defined as follows:

$$b_N \in B_N, \quad b_N = (e_N^0, c_N^0)\ (e_N^1, c_N^1),\ (e_N^2, c_N^2)$$

$$\begin{cases} e_N^0 & : \ \text{current effector state} \\ c_N^0 & : \ i_N = 0, \ \delta_N = \delta^i \\ e_N^1 & : \ e_y^1 = e_y^0 - \Delta e'_y \\ c_N^1 & : \ i_N = 1 \\ e_N^2 & : \ e_x^2 = e_x^1 + \delta_N * \Delta e_x \\ c_N^2 & : \ i_N = 2, \ \delta^i = \delta_N \end{cases} \tag{10}$$

where i_N is the reaction B_N execution step number.

The 0-th state of the effectors in reaction B_N is equivalent to the last state of the effectors in the previously executed reaction. In the 0-th step the reaction B_N execution step number i_N is 0 and the search-direction marker δ_N for reaction B_N assumes the same value as the value of the current search-direction marker δ^i. As reaction B_N is activated by detecting contact by $_N v$ (i.e. northern sensor), in the first step of reaction B_N the probe slightly (by Δe_y) backs off to the south. In the next step the probe either moves east or west depending on the value of the search-direction marker δ_N. With this motion the probe tries to avoid an obstacle (a wall).

Reaction B_E is defined as follows:

$$b_E \in B_E, \quad b_E = (e^0_E, c^0_E)\ (e^1_E, c^1_E))$$

$$\begin{cases} e^0_E & : \text{ current effector state} \\ c^0_E & : \ i_E = 0, \ \delta_E = \delta^i \\ e^1_E & : \ e^1_x = e^0_x - \Delta e_x \\ c^1_E & : \ i_E = 1, \ \delta^i = -1 * \delta_E \end{cases} \tag{11}$$

where

i_E is the reaction B_E execution step number,

δ_E is the reaction's local search-direction marker, and

Δe_x is the position increment in the west (i.e. $X-$) direction.

Reaction B_E is activated, if an obstacle is detected during the motion to the east. In this case the probe moves slightly to the west and changes the sign of the search-direction marker. The definition of reaction B_W is similar to the above definition of B_E. The only difference is that instead of $e^1_E \ : \ e^1_x = e^0_x - \Delta e_x$, $\ e^1_W \ : \ e^1_x = e^0_x + \Delta e_x$ is used. Obviously, subscript E has to be changed throughout to W.

Reaction B_{ERR} is defined as follows:

$$b_{ERR} \in B_{ERR}, \quad b_{ERR} = (e^0_{ERR}, c^0_{ERR})\ (e^1_{ERR}, c^1_{ERR})$$

$$\begin{cases} e^0_{ERR} & : \text{ current effector state} \\ c^0_{ERR} & : \ i_{ERR} = 0 \\ e^1_{ERR} & : \ e^1_z = e^0_z + \Delta e_z \\ c^1_{ERR} & : \ i_{ERR} = 1, \ i = i_T \end{cases} \tag{12}$$

where

i_{ERR} is the reaction B_E execution step number,

Δe_z is the position increment in the upward (i.e. $Z+$) direction.

The main advantage of the above specification of robot actions triggered by sensors is the easiness of transformation of this specification into a robot control program. The following listings of procedures implementing reactions coded in pseudo-C show how the above definitions of B_0, B_N translate into programs. Reactions B_E, B_W and B_{ERR}, because of lack of space, have been omitted. To make the transformation easy to follow, the names of variables have been retained as in the definitions of reactions, and hence pseudo-C not C.

```
.........
int            δᵗ ;    /* search direction marker */
double         eᵗ ;    /* touch probe pose */
unsigned char  vᵗ ;    /* virtual sensor reading */
.........
δᵗ = 1 ;
get_virtual_sensor_reading ( &vᵗ ) ;
/* eᵗ = entry – the touch probe is at the entrance to the maze */
while ( eᵗ ∉ E*ᵗT  /* goal not attained */ )
  {
   switch ( vᵗ )
        case V0  :  eᵗy += Δey ;                                   /* continue B0 —  */
                    execute_motion_to ( eᵗ, &vᵗ ) ; break;        /* pursue task goal */
        case V1  :  reaction_BN( eᵗ, &δᵗ, &vᵗ ) ; break ;         /* avoid northern wall */
        case V2  :  reaction_BW( eᵗ, &δᵗ, &vᵗ ) ; break ;         /* avoid western wall */
        case V3  :  reaction_BE( eᵗ, &δᵗ, &vᵗ ) ; break ;         /* avoid eastern wall */
        default  :  reaction_BERR( eᵗ, &vᵗ ) ; exit ;             /* error detected */
     } ; /* end: switch */
  } ; /* end: while */
exit ;
.........
void execute_motion_to ( double eˡ,  unsigned char *v )
  { /* begin: execute_motion_to */
   do
     move_by_an_increment ( Δe ) ;
     get_virtual_sensor_reading ( v ) ;
   until ( *v ∉ V0 or e = eˡ ) ;
   return ;
  }; /* end: execute_motion_to */


void reaction_BN( double e,  int *δN,  unsigned char *v )
 { /* begin: reaction_BN */
   ey  -=  Δe'y ;
   execute_motion_to ( e, v ) ;
   if ( *v ∉ V0 )
     return ;
   ex  +=  *δN * Δex ;
   execute_motion_to ( e, v ) ;
   return ;
 }; /* end: reaction_BN */
```

3 Conclusions

The advantage of the presented method is the easiness of transforming the reaction definitions into control programs and the ability of incremental system design. First a simple version of the system can be obtained by selecting only a few subspaces of the virtual sensor reading space and assigning only a few reactions to them. The remaining portion of the space can be assigned a single reaction (e.g. B_{ERR}). While the system is being developed new subspaces can be extracted from this portion and new reactions can be assigned to them. The previously implemented reactions remain unaltered. Indeed, if a reaction is inadequate it can be modified or its subspace further divided and so a single reaction will be divided into several but more specific ones.

A simple maze running task was used as a proof-of-the-concept benchmark (fig.1). This task was implemented on the research oriented robot controller [9, 15] which can be tailored exactly to the needs of the task at hand. The system consisted of a 5 d.o.f. robot, 4-bit touch sensor, and an IBM/486/33MHz computer running a multiprocess software of the controller. The software was coded in a concurrent version of C.

References

[1] J. Pinkava, "Towards a Theory of Sensory Robotics", Robotica, Vol.8, pp.245-256, 1989.

[2] S. R. Ruocco, "Robot Sensors and Transducers", Open University Press, 1987.

[3] E. Grant, "Uncertainty in Robot Sensing", Sensor-Based Robots: Algorithms and Architectures, Ed. C. S. G. Lee, NATO ASI Series F, vol.66, pp.41-57, 1991.

[4] R. C. Luo, M-H. Lin, R. S. Scherp, "Dynamic Multi-Sensor Data Fusion System for Intelligent Robots", IEEE J. Robotics and Automation, Vol.4, no.5, pp.386-396, August 1988.

[5] R. C.Luo, M. G. Kay, "Multisensor Integration and Fusion in Intelligent Systems", IEEE Trans. Systems, Man, and Cybernetics, Vol.19, no.5, pp.901-931, September/October 1989.

[6] Y. F. Zheng. "Integration of Multiple Sensors into a Robotic system and its Performance Evaluation", IEEE Trans. Robotics and Automation. Vol.5, no.5, pp.658-669, October 1989.

[7] U. Rembold, P. Levi, "An Integrated Sensor System for Robots", Sensor-Based Robots: Algorithms and Architectures, Ed. C. S. G. Lee, NATO ASI Series F, vol.66, pp.3-24, 1991.

[8] Y. Wang, S. Butner, "RIPS: A Platform for Experimental Real-Time Sensory-Based Robot Control, IEEE Trans. Systems, Man, and Cybernetics", Vol.19, no.4, pp.853-860, July/August 1989.

[9] C. Zieliński, "Controller Structure for Robots with Sensors, Mechatronics, Pergamon Press, Vol.3, no.5, pp.671-686, 1993.

[10] D. M. Lyons, M. A. Arbib, "A Formal Model of Computation for Sensory-Based Robotics". IEEE Trans. Robotics and Automation, Vol.5, no.3, pp.280-293, June 1989.

[11] D. M. Lyons, M. A. Arbib. "A Task-Level Model of Computation for Sensory-Based Distributed Control of Complex Robot Systems", Symposium on Robot Control, Barcelona, Spain, November 1985.

[12] R. A. Brooks, "Intelligence Without Representation", Artificial Intelligence, No.47, pp.139-159, 1991.

[13] R. A. Brooks, "A Robust Layered Control System For a Mobile Robot". IEEE J. Robotics and Automation, Vol. RA-2, no.1, pp.14-23, March 1986.

[14] J. H. Connel, "A Behavior-Based Arm Controller", IEEE Trans. Robotics and Automation, Vol.5, no.6, pp.784-791, December 1989.

[15] C. Zieliński, "Flexible Controller for Robots Equipped with Sensors", Proc. 9-th CISM-IFToMM Symp. on Theory and Practice of Robots and Manipulators, Ro.Man.Sy'92, 1-4 September 1992, Udine, Italy, Lecture Notes in Control and Information Sciences 187, Springer-Verlag, 1993, pp.205-214.

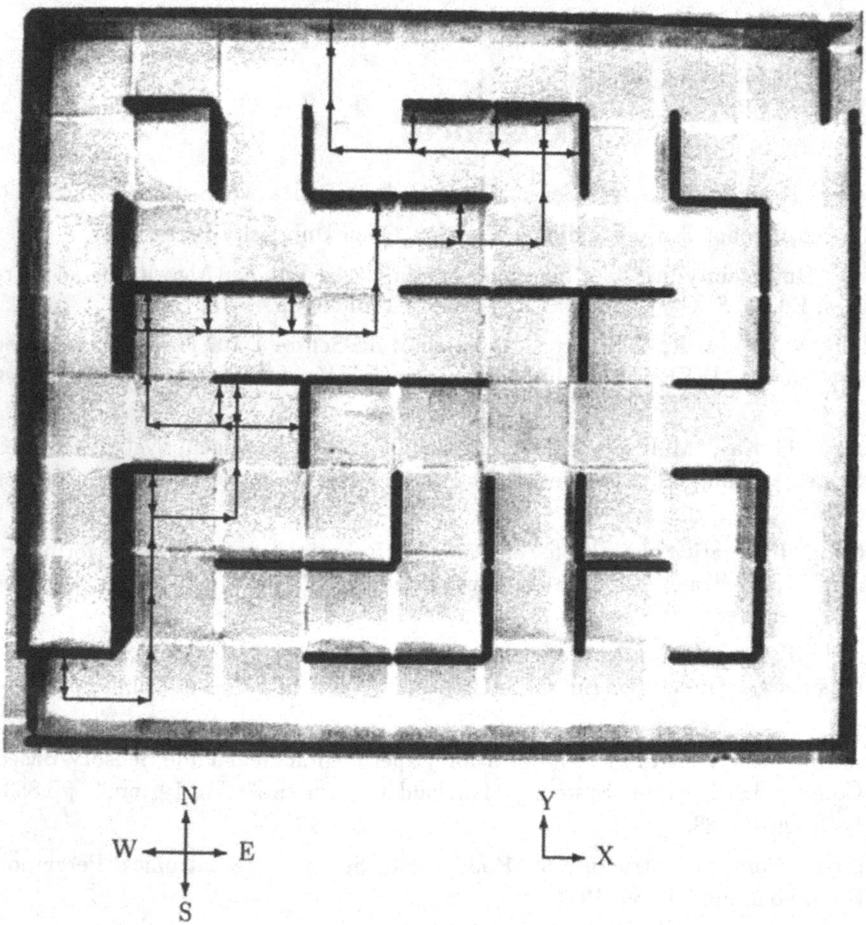

Figure 1: The probe trajectory in the exemplary maze (arrow heads indicate the end of each step)

SENSING THE POSITION OF A REMOTELY OPERATED UNDERWATER VEHICLE

E. Kreuzer and F.C. Pinto
Technical University of Hamburg-Harburg, Hamburg, Germany

Introduction

The importance of remotely operated underwater vehicles for construction, inspection, and maintenance of off-shore structures has been continuously increasing in the recent years. Especially oil prospecting in water depths inaccessible for human divers (>300m) needs unmanned vehicles. For a great variety of tasks a robust positioning control of the vehicle is required. This control is complicated not only due to nonlinearities involved in the system including the umbilical and parameter uncertainties [1,7] but also because of the lack of precise position information. The positioning problem can be split into two domains:

- the region far from the structure, where the navigation from the basis ship to the structure takes place and
- the region close to the structure, where the position of the vehicle relative to the structure must be kept for a task to be performed.

Each of them has its own characteristics and demands different properties from the sensor systems.

Navigation from the Basis Ship

The navigation to the structure does not demand a high precision of positioning but the distances involved are of the magnitude of hundreds of meters. What is actually needed to be determined is the relative position between the vehicle and the structure. This can, however, also be obtained if the absolute positions of both are known. The operator is still responsible for correcting eventual errors during the navigation until the desired point is reached.

Automatic navigation can be possible with the development of adequate sensors. This is one of the greatest problems in developing an autonomous vehicle.

One can classify the sensor systems for this area in : *Optic Systems, Acoustic Systems* and *Inertial Systems*. Optic and acoustic systems are both based on transmission and reflection phenomena of (light or sound) waves. The distances are then obtained through the measurement of the elapsed time between the emitted and reflected waves or through the analysis of the *image* produced by the reflected waves [2,3]. The precision and the range of these measurements are influenced, among other factors, by the velocity and the absorption of the waves in water, the distance between emitters or receptors and the precision of the positioning of these emitters or receptors [4,5]. Spurious reflections and shadows caused by the presence of other equipment in the area or even by parts of the structure are the main problems with the image analysis. Nevertheless these systems are very helpful for the driver.

Inertial systems have one major advantage: besides the initialisation, where the initial position must be given, they do not depend on any information from the outside. They are based only on the inertial forces that act on the ROV. The position and orientation of the vehicle can be described by a position vector $\boldsymbol{r}$ and a transformation matrix $\boldsymbol{S}$ which relates a body-fixed frame B and an inertial reference frame I, Figure 1. The matrix $\boldsymbol{S}$ can be determined in a variety of ways. In this case the use of cardan angles appears to be the best possibility as these angles are closely related to the body-fixed system B. Such systems normally use gyroscopes to measure the rates of change of the orientation of the vehicle $\dot{\alpha}, \dot{\beta}, \dot{\gamma}$. An integration is necessary to keep track of the actual orientation [10,11]. The linear accelerations $\ddot{x}_B, \ddot{y}_B, \ddot{z}_B$ are also measured, transformed into the inertial coordinate system, and then integrated twice to obtain the absolute position x, y, z. Gyroscopes are sensitive instruments that require expensive maintenance. Another possibility to determine the rates of change would be with specially located accelerometers or with accelerometer-like devices to sense the angular velocities.

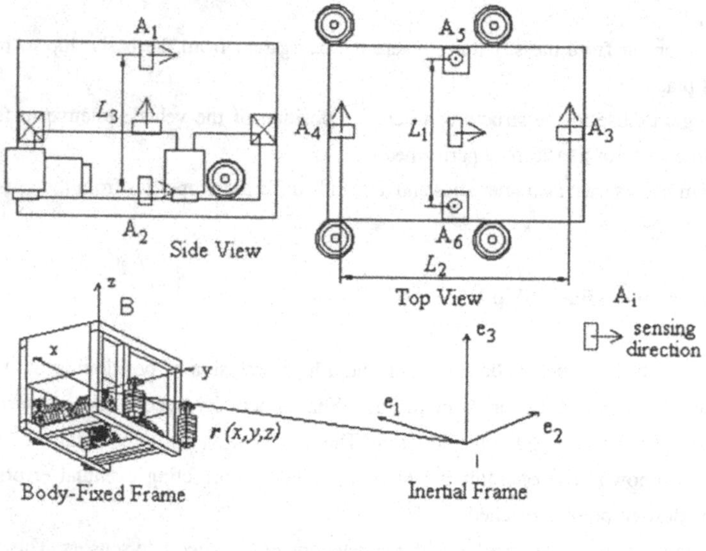

Figure 1 - Location of the accelerometers and reference frames

With accelerometers $A_1 \ldots A_6$ arranged as shown in Figure 1, the linear accelerations $\ddot{x}_B, \ddot{y}_B, \ddot{z}_B$ and the angular accelerations $\ddot{\alpha}, \ddot{\beta}, \ddot{\gamma}$ of the cardan angles in the body fixed frame B can be determined [11] as functions of the measured accelerations $a_1 \ldots a_6$ and the distances L_1, L_2, L_3 between the accelerometers.

$$\ddot{x}_B = \frac{1}{2} \cdot a_1 + \frac{1}{2} \cdot a_2, \qquad \ddot{\alpha} = \frac{1}{L_1} \cdot a_5 - \frac{1}{L_1} \cdot a_6,$$

$$\ddot{y}_B = \frac{1}{2} \cdot a_3 + \frac{1}{2} \cdot a_4, \qquad \ddot{\beta} = \frac{1}{L_3} \cdot a_1 - \frac{1}{L_3} \cdot a_2,$$

$$\ddot{z}_B = \frac{1}{2} \cdot a_5 + \frac{1}{2} \cdot a_6, \qquad \ddot{\gamma} = \frac{1}{L_2} \cdot a_3 - \frac{1}{L_2} \cdot a_4.$$

The angular accelerations can now be twice integrated over a time interval to obtain the change in orientation and the change ***dS*** in the transformation matrix ***S***. With the updated matrix ***S*** one can now project the linear accelerations in I and integrate them twice to obtain the absolute position of the ROV. In order to control the vehicle the integration scheme including measurements and all coordinate transformations must be fast enough to keep the control system informed about the position and to allow small integration time steps. Simulations have shown, Figure 2, that even a simple Euler integrator can perform this task without significant rounding errors. This has the advantage of being very simple and very fast. The disadvantage from such an integrator, the small step size needed, is less important in this case. The necessary position updates, for the control system of the ROV, already make use from the small step size. Discontinuities in the acceleration of the ROV due to subtle changes in the motion can also be better interpolated during the integration with small step size.

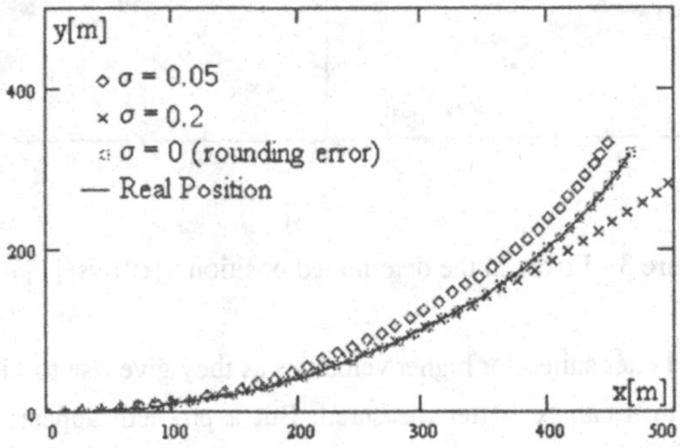

Figure 2 - Simulation of a inertial navigation system $\{\sigma[m/s^2]\}$

In practice, however, the accelerometers do have their own errors which must be modelled in the simulation. These errors can be described as an off-set error, which may even drift, and a superimposed gaussian white noise with variance σ. The off-set error is serious because even a small error will be quadratically amplified due to the twice time integration. To overcome this problem one can measure the angular velocities instead of the angular accelerations. In doing so the number of integrations necessary to determine the update of the matrix ***S*** decreases and so also the amplification of the off-set. This measurement can be made with accelerometer-like devices which are sensitive to the rates of change of orientation. Such sensors are not so precise as gyroscopes but have cost advantages and are thus still practicable for the short-time navigation from basis-ship to the structure. As in the case of the gyroscopes, the error increases with time due to the numerical integration, which may require a new initialisation of the system.

Since the accelerations, or angular velocities, can only be measured with respect to a body fixed coordinate frame, it is important to precisely calculate the orientation of the ROV in order to allow the transformation into the inertial reference frame. The performance of the system, Figure 3, decays rapidly when the errors in the orientation reach about 4 degrees. This error can establish an operation time limit after that the system must be reinitialised.

The reinitialisation could be achieved through satellites if the ROV came to the surface within this time limit, or through reaching the structure on a determined point. In this case the sensor system for the region close to the structure can be used for a new initialisation of the inertial navigation system.

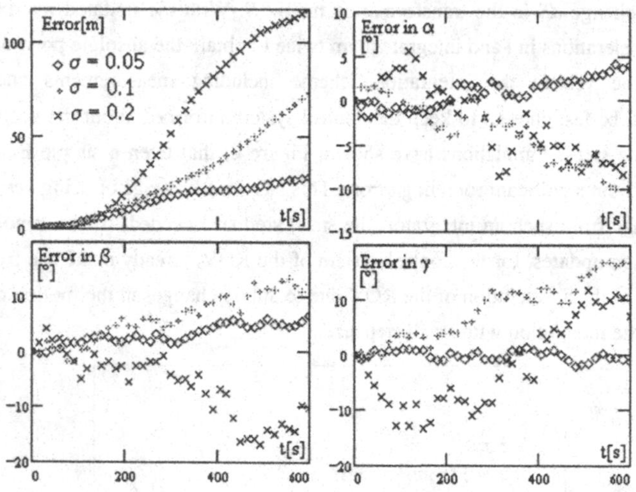

Figure 3 - Errors in the determined position {σ[m/s^2]}

The inertial system is better suited for higher velocities as they give rise to higher accelerations and angular velocities, which in turn can be better measured. But a problem appears due to the necessary subtraction of the gravity which requires a very precise knowledge of the respective value for the actual location.

Positioning Close to the Structure

A high precision in positioning the ROV relative to the structure is necessary for executing inspection or maintenance tasks. However, the involved distances are in the range of meters. Inspection and maintenance are the most time-consuming part during the operation. Therefore, either automatically or at least semi-automatically positioning is of great practical importance.

Here the same type of sensor systems can be used, in addition to sensors that mechanically connect the ROV to the structure. The light based sensors do not need great power in this region but the effect of shadows can be very severe [6]. The sonar systems must in turn cope with much more spurious reflections [5]. Another possibility would be the use of thin wires under tension (*taut-wires*) between the ROV and the structure. With the measurement of suitable angles and the length of the wires the position of the vehicle can be determined [9]. The great disadvantage is the lack of precision in the location of the ends of the wire. For calculation purposes it is assumed that the wire is straight, which does never happen due to the water flowing around it. This measurement can only be improved by increasing the tension of the wires. This, however, leads to greater forces acting on the vehicle which is also undesirable.

The use of a mechanical manipulator as a sensor is an interesting alternative. If the manipulator grasps the structure at a point, which is already known, through simple kinematic considerations one can obtain the position and orientation relative to the structure. The disadvantage is the limited range due to the restricted workspace of the manipulator. With the use of two or more manipulators this problem can be minimised by sequentially grasping different points of the structure. If at least two manipulators simultaneously grasp the structure both can be independently used to determine the position and so to diminish the error. Figure 4 shows a projection of the workspace of a manipulator and the error for the calculation of the position.

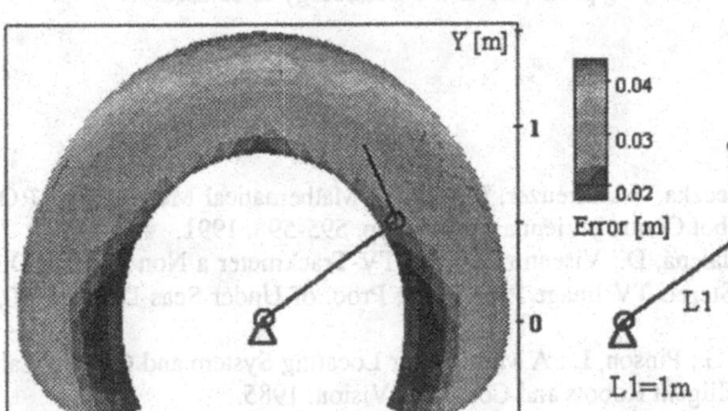

Figure 4 - Manipulator workspace and position measurement errors

Experimental Model

In order to verify the results an experimental model (600mm x 450mm x 400mm) has been developed. The model is designed to serve as a test-bed for different control, sensor, and propulsion systems. The tests are to be performed in the still water basin of the Ocean Engineering Section II of the Technical University of Hamburg-Harburg with size 2m x 2m x 2m. The hydrodynamic coefficients of the model were measured in a wind tunnel in order to allow a better prediction of its behaviour by means of computer simulations.

Conclusions

There are many types of sensors that can be used to determine the position and orientation of an underwater vehicle. These systems range from sonar and inertial navigation devices, to triangulation with laser beams. Requirements for the sensor systems vary depending on the distance between the ROV and the structure. Close to the structure the use of a mechanical connection, either by means of the aforementioned taut-wires or the manipulator, can perform well. If the navigation is to be performed automatically, an inertial navigation system appears to be better suited, specially for long ranges.

In fact the problem of vehicle positioning is very complex and none of the available technologies is clearly superior to the others. Therefore, the simultaneous presence of different redundant sensor systems is recommended as a help for the operator, and also to make the control of the vehicle more reliable. Simulations have shown that an inertial navigation system can be built without gyroscopes and hence is suited for short term navigation from the basis ship to the structure. A manipulator used as sensor can then measure the position close to the structure and eventually reinitialise the inertial system for the return. The precision of the positioning depends not only on the quality of the individual components but also on the understanding of the underlying principles of the technology to be used.

Literature

[1] Bevilacqua, L.; Kleczka, W.; Kreuzer, E.: On the Mathematical Modelling of ROV`s; Proc. of the Symposium on Robot Control, Vienna, Austria, pp. 595-598, 1991.

[2] Prendin, W.; Maddalena, D.; Visentin, R.: The TV-Trackmeter a Non Contact Distance Measuring Device Based on Stereo TV Image Processing; Proc. of Under Seas Defence '90, San Diego, CA, 1990.

[3] Wolfe, W.; White, G.; Pinson, L.: A Multisensor Locating System and Camera Calibration Problem; SPIE Vol.579, Intelligent Robots and Computer Vision, 1985.

[4] Blizard, M.A.: Ocean Optics: Introduction and Overview; 8th SPIE Conf. Ocean Optics, Orlando, FL, pp. 2-17, 1986.

[5] Jaffe, J.S.: The Domains of Underwater Visibility; 8th SPIE Conf. Ocean Optics, Orlando, FL, pp. 283-287, 1986.

[6] Winther, S.: SPOTSCAN 2D - System Description and Off-Shore Trials; Seatex A/S, Trondheim, 1990.

[7] Lewis, D.J.; Lipscomb, J.M.; Thompson, P.G.: The Simulation of Remotely Operated Underwater Vehicles; ROV '84, Marine Technology Society, 1984.

[8] Catipovic, J.: Performance Limitations in Underwater Acoustic Telemetry; IEEE Journal of Oceanic Engineering, Vol.15, No.3, pp. 205-216, 1990.

[9] Hsu, L.; Costa, R.; Cunha, J.P.V.: Medição de Posição Pela Estratégia "Taut-Wire" Para o Controle de Posição de Um VOR; Relatório interno COPPE/UFRJ, Rio de Janeiro, 1990.

[10] Magnus, K.: Kreisel, Theorie und Anwendungen; Springer Verlag, Berlin, 1971.

[11] Wrigley, W.; Hollister, W.M.; Denhard, W.G.: Gyroscopic Theory, Design and Instrumentation; M.I.T. Press, Massachusetts, 1969.

MICRO-ENERGY FLOW BEAM-SHAPED MICROACTUATORS AND SENSORS

K. Itao

Chuo University, Tokyo, Japan

and

H. Hosaka and H. Ukita

Nippon Telegraph and Telephone Corporation, Tokyo, Japan

Abstract

From the viewpoint of a micro energy flow, the microsystems design method is investigated for the microactuators and sensors. A simple microsystem is developed as our model, which is excited photothermally by the light from high power laser diode. The exciting point to vibrate a cantilever at maximum amplitude is obtained near the clamped position, which means the maximum efficiency of energy transmission. The energy dissipation characteristics of the cantilever is analytically evaluated. The quantities of dissipantion caused by airflow force, squeeze force internal friction and support loss are calculated and compared with experimental results.

As a result, micro energy input position and the relationship between microbeam size and dissipation ratio are evaluated, which makes progress in a design method of the microsystems excited by a wireless energy supply.

1. Introduction

At present, the energy to drive the precision mechanisms is usually provided with the battery stored in the mechanisms or external power supply through lead wires. Microsystems, however, are not easy to be excited by the above mentioned method, because the weight of the wire or battery is not able to be neglected. If the wireless energy supply systems are introduced, the microsensors, micro robots and micro rotary systems should be improved dramatically at their performance.

Towards this goal, semiconductor laser beam is adopted here as a promising means for the transmission of both energy and information, and also a cantilever is adopted as a simple structure of the microsystems. The examples of microsystems excited by a light without wire are shown in Fig.1 as a microsensor in (a), a microactuator in (b) and a micro mobile mechanism in (c).

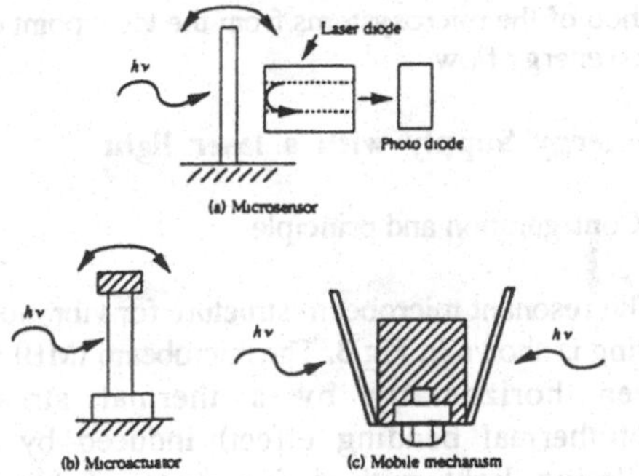

Fig1. Microsystems excited photothermally

The system (a) is composed of a cantilever resonant microbeam, a laser diode and a photodiode which are fabricated on the surface of a gallium arsenide substrate [1]. The microbeam vibration excited by a high power laser is detected by the photodiode as a light output variation due to the optical length difference between microbeam and another laser diode. The system (b) is the bending motion mechanism using thermal expansion on one side of the actuator [2]. Thermal energy is given to the actuating part and it causes rapid extension and generates impulsive force. The system (c) is the mobile mechanism with two cantilevers whose resonant frequencies are a little different. When the

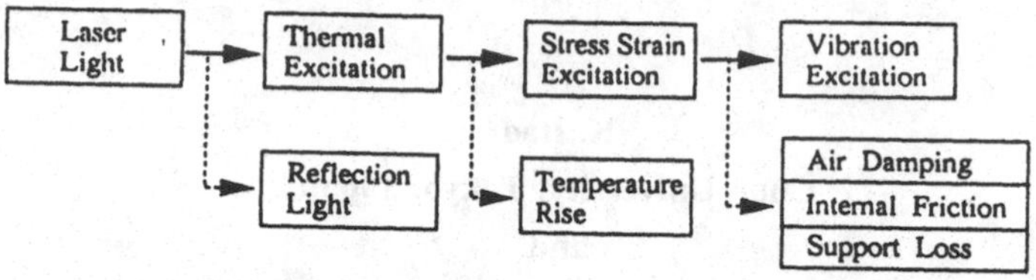

Fig2. Micro Energy Flow between a Semiconductor Laser and a Cantilever

cantilever of one side is excited photothermally, the mechanism runs ahead turning to the left and vise versa.

The energy flow of the microsystems illustrated in Fig.1 is expressed as shown in Fig.2. When the laser light strikes the cantilever, thermal energy brings about stress-strain on the surface of the cantilever and causes vibration.

We study the phenomena of the energy flow from the light energy to the kinematic energy of the cantilever through thermal energy. The energy transmission condition is experimented in order to determine how to supply the light energy to the cantilever at high efficiency. Then, the energy dissipation characteristics of the vibrational cantilever is analytically evaluated.

Finally, we approaches to the optimum design method of the microsystems from the view point of micro energy flow.

2. Energy Supply with a laser light

2.1 Configuration and principle

The resonant microbeam structure for vibration sensing is shown in Fig.3. The microbeam (MB) is driven horizontally by a thermal stress (photothermal bending effect) induced by a oscillating light with a frequency of beam resonance. The vibration is sensed by a laser diode (LD1) and a photodiode (PD) from the variation of external cavity length (phase difference) as shown in Fig.4. In the figure, the light (wavelength λ) reflected from the LD facet and that from the MB interfere in the laser diode, resulting a peak in the light output every distance of $\lambda/2$. The MB wall and laser diode facets form a composite cavity. Initially, the external cavity length h is set to 3μm. When the microbeam vibrates around that point with amplitude Δh, the light output varies as shown in Fig.4.

2.2 Microbeam vibration excited by laser light

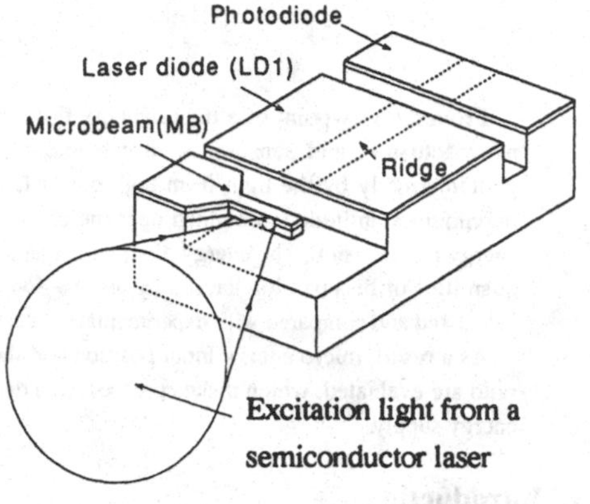

Fig3. Structure of a resonator sensor driven photothermally

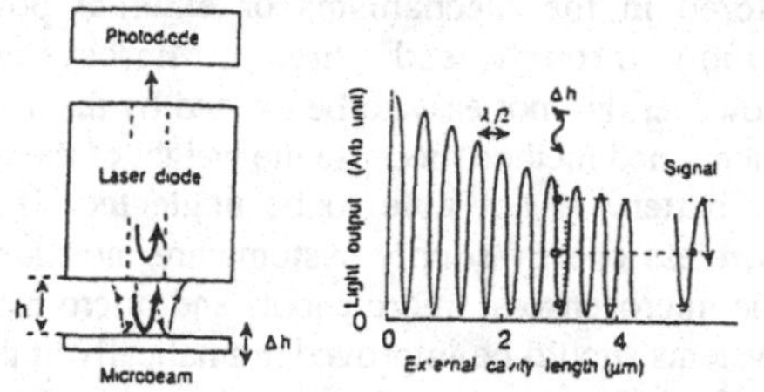

Fig4. Sensing principle of microbeam vibration

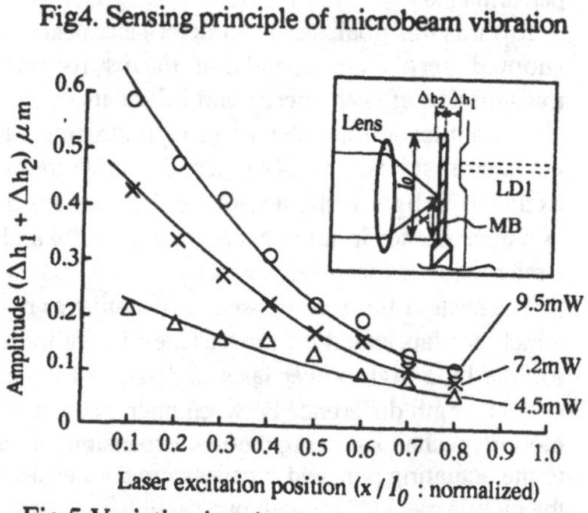

Fig.5 Variation in microbeam vibration amplitude as a function of laser excitation position

To measure the MB vibration properties, we used a laser diode of an optical pickup (not shown in Fig.3) to excite the MB. When the sinusoidal light intensity frequency coincided with the microbeam mechanical resonant frequency, the amplitude of the PD signal exhibited a maximum. The signal-to-noise ratio of the amplitude signal was more than 45 dB.

The excitation light is focused to a spot size of abut 1μm on different microbeam positions. As the incident light power rises, producing greater thermal expansion (stress) in the microbeam, the vibration amplitude increases. We can measure the absolute amplitude from the fact that the peak signal amplitude corresponds to $\lambda/4$ (0.195μm). Figure 5 shows the vibration amplitude versus laser excitation position with laser power as a parameter. The closer the light strikes to the microbeam support the higher the excitation efficiency becomes.

Figure 6 shows the resonant frequency spectra for a 110μm long microbeam with the amplitude signal photograph in the figure. The resonance of the microbeam is 200.6kHz. The mechanical Q value in air is about 250. Figure 7 shows resonant frequency as a function of cantilever microbeam length. The circles are experimental measurements, and they are in good agreement with the theoretical results. To increase the sensor sensitivity, the resonant frequency should be raised by shortening the cantilever length.

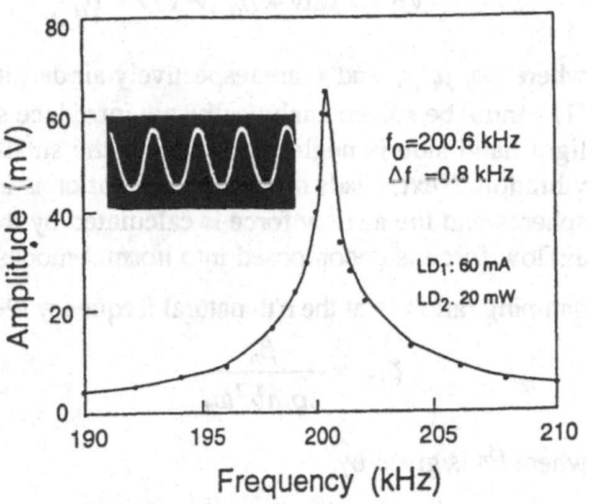

Fig6. Frequency response curve for a microbeam 110μm long, 5μm wide, and 3μm thick

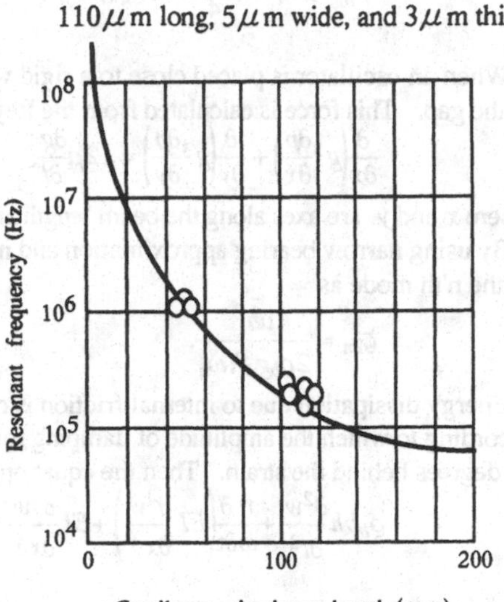

Fig.7 Resonance frequency as a function of Cantilever microbeam length for GaAs

3. Energy Dissipation Mechanisms of a Cantilever

3.1 Analytical and solution

Here we evaluate the characteristics of a beam oscillator (Fig.8) with constant width b and thickness h. We study two cases in which there is a rigid wall close to the actuator (as for an electrostatic actuator), and in which there is not (as for a magnetostatic or piezo actuator). This model takes into account damping forces generated by airflow, squeeze effect, internal friction, and support loss [3].

First we study airflow force in free space. Airflow around the oscillator is described by the Navier-Stokes equation.

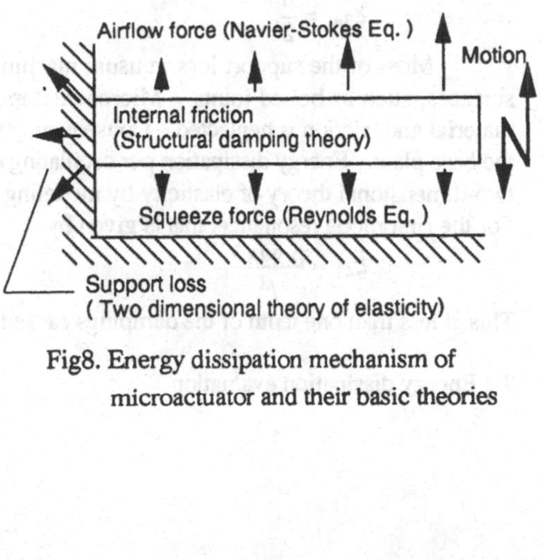

Fig8. Energy dissipation mechanism of microactuator and their basic theories

$$-\nabla p = -\mu\Delta v + \varrho_a (v \bullet \nabla)v + \varrho_a \frac{\partial v}{\partial t} \quad , \tag{1}$$

where ϱa, μ, p, and v are respectively air density, viscosity, pressure, and velocity. Because Eq. (1) cannot be solved analytically, we introduce some approximations. First, the second term of the light hand side is neglected by using the smallness of the Reynolds number in the microsystem vibration. Next, beads model approximation is introduced where the beam is replaced by a string of spheres and the airflow force is calculated by the sum of forces working to each sphere. Also the airflow force is decomposed into normal modes by using the linearity of basic equation. Then the damping ratio ζ_{1n} at the n'th natural frequency ω_n is obtained as

$$\zeta_{1n} = \frac{\beta_n}{2\varrho_b h b^2 \omega_n}, \tag{2}$$

where β_n is given by

$$\beta_n = 3\pi\mu b + \frac{3}{4}\pi b^2 \sqrt{2\varrho_a \mu \omega_n} \ , \tag{3}$$

When an oscillator is placed close to a rigid wall, a new airflow force -squeeze force- appears due to the gap. This force is calculated from the Reynolds equation:

$$\frac{\partial}{\partial x}\left(g^3 \frac{\partial p}{\partial x}\right) + \frac{\partial}{\partial y}\left(g^3 \frac{\partial p}{\partial y}\right) = 12\mu \frac{\partial g}{\partial t}, \tag{4}$$

where x and y are axes along the beam length and width and g is gap length.

By using narrow bearing approximation and modal decomposition, we can write the damping ratio of the n'th mode as

$$\zeta_{2n} = \frac{\mu b^2}{2\varrho_b g_\delta^3 h \omega_n}. \tag{5}$$

Energy dissipation due to internal friction is calculated by using the structural damping theory[4], according to which the amplitude of damping force is proportional to internal strain and its phase lags 90 degrees behind the strain. Then the equation of motion is given by

$$\varrho_b b h \frac{\partial^2 w}{\partial t^2} + \frac{\eta}{\omega} \frac{\partial}{\partial t}\left(EI \frac{\partial^4 w}{\partial x^4}\right) + EI \frac{\partial^4 w}{\partial x^4} = f e^{i\omega t} \ , \tag{6}$$

where η is a structural damping coefficient determined by material properties. By using modal decomposition, we can write the damping ratio ζ_{3n} at $\omega = \omega_n$ as

$$\zeta_{3n} = \frac{\eta}{2}. \tag{7}$$

Most of the support loss in usual machinery is generated by the friction between connected surfaces, such as bolted joints. Microactuators, however,are usually made from a single piece of material and friction is neglected. Consequently the main cause of damping is the elastic vibration of the base plate. Energy dissipation per oscillating cycle of a cantilever can be calculated according to a two-dimensional theory of elasticity by modeling the base plate as an infinitely large elastic body [5]. For the first mode resonance, this is given by

$$\zeta_{41} = \frac{0.23h^3}{l^3}. \tag{8}$$

This is less than one tenth of the dampings caused by other factors and can therefore be neglected.

3.2 Energy dissipation evaluation

The ratio of energy dissipation per vibrational cycle, ΔW, to the total elastic energy W is given by $\Delta W/W = 4\pi\zeta$. And the relationship between beam length and energy dissipation ratio for silicon cantilevers is shown in Fig. 9 for the first and the second mode resonances. Beam shape is proportionally miniaturized or enlarged according to the rule shown in the figure. The structural damping coefficient η is the one given in Ref.[6]. The result show that the squeeze force ζ_{2n} is greater than the airflow force ζ_{1n}, which is greater than the internal friction ζ_{3n}. This means that for electrostatic actuators, where there is a rigid wall nearby, the energy dissipation is determined by squeeze force and this is inversely proportional to the beam length. When there is not a wall; for example, when the oscillator is accelerated by a laser or a piezo actuator, energy dissipation is determined by the airflow force and this is inversely proportional to beam length (or to its square root). When the oscillator is used under vacuum conditions, the damping is determined by the internal friction and is independent of the beam length. we also see that the squeeze and airflow damping decrease as the mode number n increases.

Calculated damping ratios are plotted in Fig.10 against measured values [6] when there was no rigid wall. The vibration amplitude in the experiment was smaller than 80 nm and the Reynolds number was less than 0.1. All the circles are close to the line of 45 degrees, which means the calculated and the measured values are similar: the greatest difference between them is about 40%.

Measured [7] and calculated damping ratios of a Permalloy cantilever are shown in Fig.11. Permalloy is a soft permeable metal typically used in magnetostatic microactuators. Since these actuators do not need a stator or a wall, the squeeze damping ζ_{2n} is neglected. The material density and Young's modulus used in calculating the curves in these figures are those of balk Permalloy, ϱ_b=8600 kg/m^3,and E=186 GPa. Because the structural damping coefficient η has not yet been reported, we used a value at which the calculated damping best fits the experimental results:

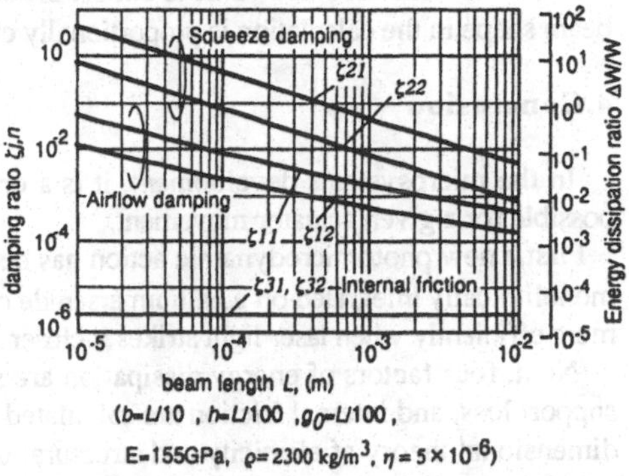

Fig. 9. Damping and energy disspation ratios of silicon cantilever.

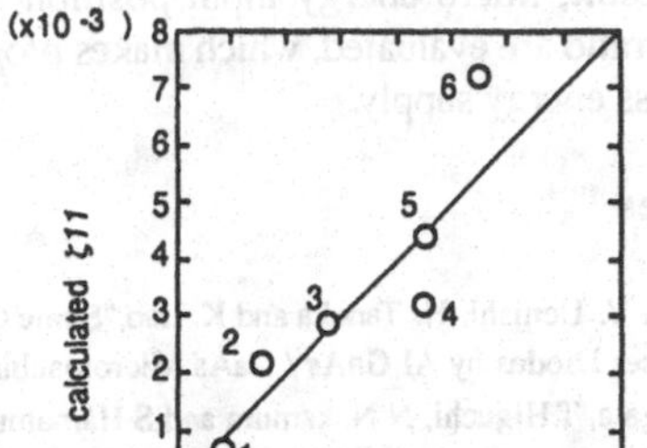

No.	L (mm)	b (mm)	h (mm)	Rv
1	2.8	0.4	34	430
2	1.5	0.5	9	590
3	2.0	0.5	10	360
4	4.0	0.5	15	150
5	3.0	0.5	10	170
6	4.0	0.2	11	16

Fig. 10. Comparison between calculated and measured airflow damping in a silicon cantilever.

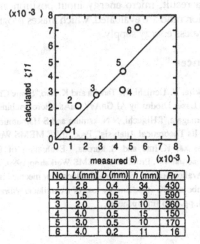

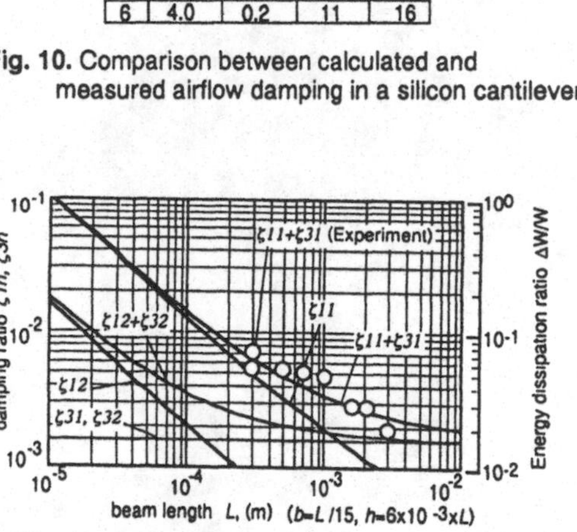

Fig. 11. Damping and energy dissipation ratios of Permalloy cantilever.

$\eta = 3 \times 10^{-3}$. This assumed value is almost the same as those of duralumin or phosphor bronze. The beam shape in the calculation is proportionally changed as shown in the figure.

4.Conclusion

In the microsystems development, it is a crucial problem to reduce driving power as much as possible for a given actuator movement.

First,a new photomicrodynamic action has been presented through a composite cavity laser diode monolithically integrated on a gallium arsenide chip. The microbeam of 110 μm length was vibrated more efficiently when laser light strikes it closer to the microbeam support point.

Next, four factors of energy dissipation are studied theoretically. Airflow force, squeeze force, support loss, and internal friction are calculated by Navier-Stokes equation. Reynolds equation, two dimensional theory of elasticity and structural damping theory, respectively, and their closed form solutions are obtained using appropriate approximations. With these solutions, relationship between structure size and dissipation ratio is illustrated. Good agreement between theoretical and experimental damping is obtained.

As a result, micro energy input position and the relationship between microbeam size and dissipation ratio are evaluated, which makes progress in a design method of the microsystems excited by a wireless energy supply.

References

[1] H. Ukita, Y. Uenishi, H. Tanaka and K. Itao,"Some Characteristics of a Micro-Mechanical Resonator Integrated with Laser Diodes by Al GaAs / GaAs Micromachining", J. JSPE 59, 9,1993,pp.1560-1565.

[2] Y. Yamagata, T.Higuchi, N.Nakamura and S.Hamamura, "A Micro Mobile Mechanism Using Thermal Expansion And Its Theoretical Analysis", Proc. IEEE MEMS Workshop,1994, pp.142-147.

[3] H. Hosaka, K. Itao and S. kuroda, " Evaluation of Energy Dissipation Mechanisms in Vibrational Microactuators", Proc. IEEE MEMS Workshop,1994, pp.193-198.

[4] E.A.Avallone edited,"Standard Handbook for mechanical Engineers", MacGraw-Hill, New York, 1987, p . 5-69.

[5] Y. Jimbo and K. Itao,"Energy loss of a cantilever vibrator,", Journal of the Horological Institute of Japan, 47, 1968, pp.1-15 (in Japanese).

Part X
Applications & Performance Evaluation

APPLICATION OF POLYMER GEL TO MICRO-ACTUATOR

T. Hayashi
Toin University of Yokohama, Yokohama, Japan
and
K. Yoshida
Tokyo Institute of Technology, Yokohama, Japan

[Abstract]

The gel of super absorbent Polymers as Polyacrylic Sodium Resin absorbs the water by 2,000 ~ 3,000 times of the material itself in volume and the water can be transferred around the cathode by electric current, then the gel is deformed as thick around the cathode and thin around the anode. For using the deformation to the actuators the following studies have been done. (1)Fundamental equations that express the progress of the deformation were induced. (2)Three kinds of application to a)on surface traveling worm, b)in tube traveling worm and c)rotary actuator"gel motor" will be introduced. (3)About the water flow inside and around the gel was investigated for discussing the possibility of using the flow to the power sources of microacutuators.

1. Introduction

The author had been shown at Sand Dunes Laboratory of Tottori University the small grain of plastic which was trying to use for afforestation and the grain absorbed water more the 2,000 times of its own volume and changed to the gel and it stored water and few years after that the author was known in the paper by Dr.T.Tanaka[1] about interesting phenomena of the plastic gel that sent out water by the electric current. The author began to study to use the plastic gel for the actuator. The sodium polyacrylate grain with about 0.5 mm diameter absorbed water in thin solution of electrolyte Na_2SO_4 and it changed to ball of gel with 6 mm diameter after tenth hours. Inserting a platinum electrode into the grain and giving the electric current from an electrode to another electrode that is set close to the gel, the diameter of the gel decreases and giving the current to the opposite direction it increases and the speed of the deformation is about 6% per 20 sec at increasing and 6% per 60 sec at decreasing. The authors could obtain the fundamental equation between the time difference of the diameter and the amount of electric current. Based on the equation fundamental experiments on artificial worm and rotary actuator were done. Also the possibility of use the water current in the gel for the energy source of liquid actuator was discussed.

2. Fundamental equations[2]

For applying the gel to an actuator, it needs to know the quantitative relation between the amount of deformation of the gel and the effective factors to it. Then, assuming that the gel has a ball shape and that the speed of volume changing dV(t)/dt is expressed by the sum of the current term which is proportional to the electric current I and the free absorption term which is proportional to the deference of the values of the surface areas at present A(t) and it's basic value A_0, the following equation was obtained.

$$\frac{dV(t)}{dt}=K_n I\frac{v}{e}+K_i[A_0-A(t)]. \tag{1}$$

There, V(t): volume of the gel ball$(=(4\pi/3)r^3(t))$, r(t): radius of the ball, A(t): surface area of the ball$(=4\pi r^2(t))$, A_0: basic value of the surface area, v: volume of a water molecular$(=1.13\times10^{-33}$ $m^3)$, e: electric charge unit$(1.60\times10^{-19}$ C), K_n: number of water molecules transferred per an electric charge of Na^+, K_i: volume of water going into a gel per unit surface area per sec.

Expressing Eq.(1) with r(t),

$$\frac{d[4\pi r^3(t)/3]}{dt}=K_i[\frac{K_n}{K_i}\frac{v}{e}I+4\pi r_0^2]-4\pi r^2(t)$$

or

$$r^2(t)\frac{dr}{dt}=K_i[\frac{k_n}{4\pi K_i}\frac{v}{e}I+r_0^2]-r^2(t). \tag{2}$$

Solving Eq.(2) on t, and giving the boundary condition that $r=r_\infty$(saturated value) when $t=\infty$, the following equations are obtained.

$$t=\frac{1}{K_i}[r_0-r(t)+\frac{r_\infty}{\overline{2}}\ln\frac{(r_\infty-r_0)(r_\infty+r(t))^*}{(r_\infty-r(t))(r_\infty+r_0)^*}] \tag{3}$$

and

$$K_n=\frac{\overline{4}\pi K_i}{I}\frac{e}{v}(r_\infty-r_0)(r_\infty+r_0)^* \tag{4}$$

From the experiment, the values of r_∞, r_0 and r(t) at a given time t can be obtained easily, then Eq.(3) is fixed and the deformation at given time can be predicted. Equation(4) will be used to discuss on the water flow. Figures 1 to 4 show some examples for comparing the experimental and theoretical results. Figures 1 and 2 show the cases of swelling and shrinking on balls respectively and Fig.3 and 4 show the case on cylinder and on square rod respectively. However, the equation must be a little modified as asterisked terms in Eq.(3) and (4) must put to 1 and overlined number 2 and 4 must put to 1 and 2 respectively. The figures tell us the usefulness of the equations not only on the ball but also on the cylinder.

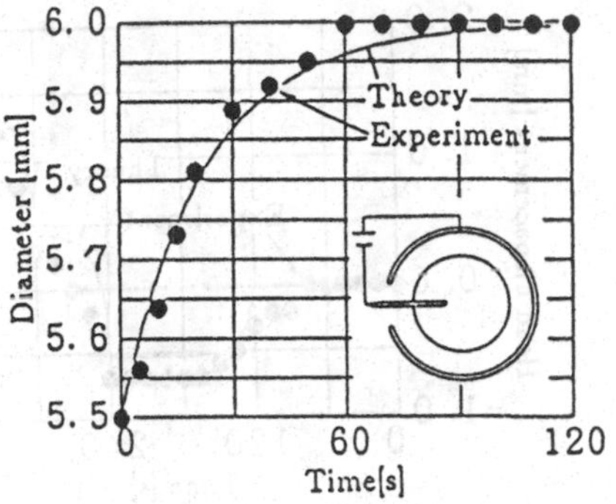

Fig.1 Swelling of gel ball

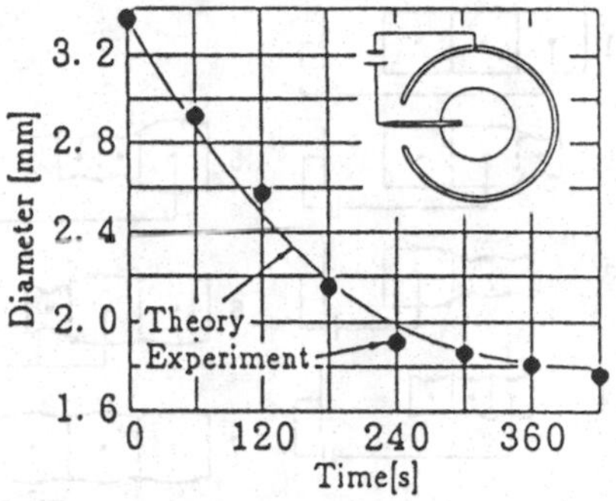

Fig.2 Shrinking of gel ball

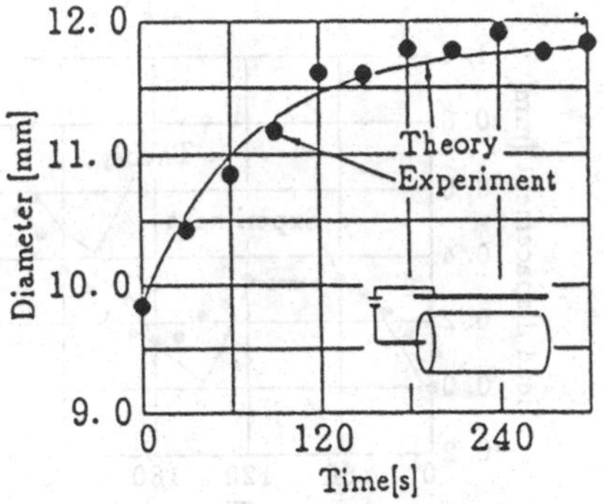

Fig.3 Swelling of gel cylinder

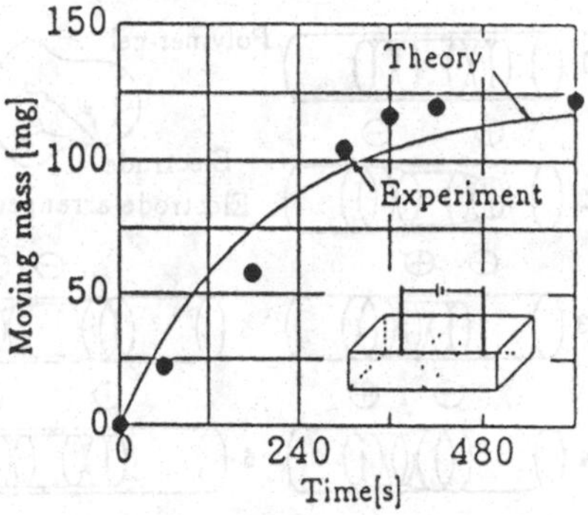

Fig.4 Swelling of gel prism

3. Application to actuator

Using the polymer gel which shape is controlled by electric current between the electrodes in the main body. That is, as the body itself pertly swell around (-) electrode and shrink around (+) electrode, the body can be moved by controlling the current with the partial difference of frictional forces between the body and outer boundary. Of course the deformation itself can be used for driving the other parts.

3.1 On-surface traveling worm : The artificial worm with 6.6×2.2×2.2 mm prismatic body with three electrodes and it's traveling sequence are shown in Fig.5. The experimental and theoretical results of traveling performance are shown in Fig.6. It takes six steps in one cycle and each steps take 60 sec. The traveling distance was 1.5 mm for the first cycle which distance corresponds 20 % of full length and 2.2 mm for the end of the second cycles. It was seen that the contact of the gel with electrodes was prevented by the bubbles occurred by current after some minuets.

3.2 In tube traveling worm : Figure 7 shows artificial worm with 5.6 mm diameter cylindrical body with three electrodes for traveling in tube. The inner diameter of the tube is 5.9 mm. When the middle part of the gel swells, the rear end part should swell and contact with the wall of the tube and support the gel. The expanded diameter hence should be larger than the inner diameter of the tube. Either end should

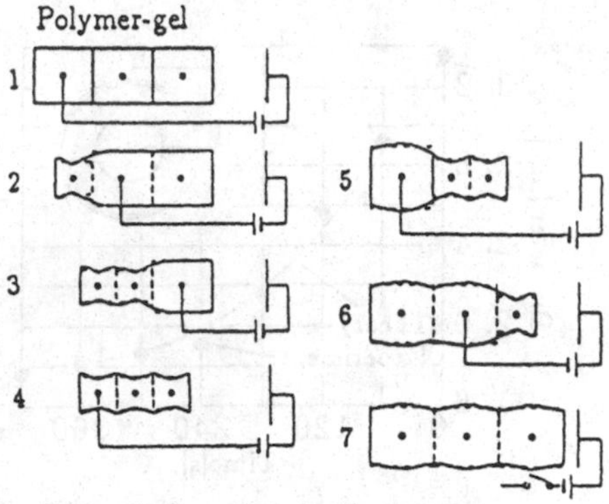

Fig.5 Surface Worm
and its sequence

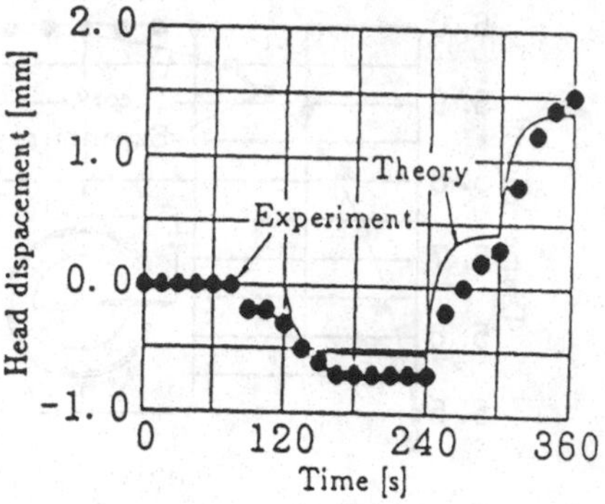

Fig.6 Performance
of Surface Worm

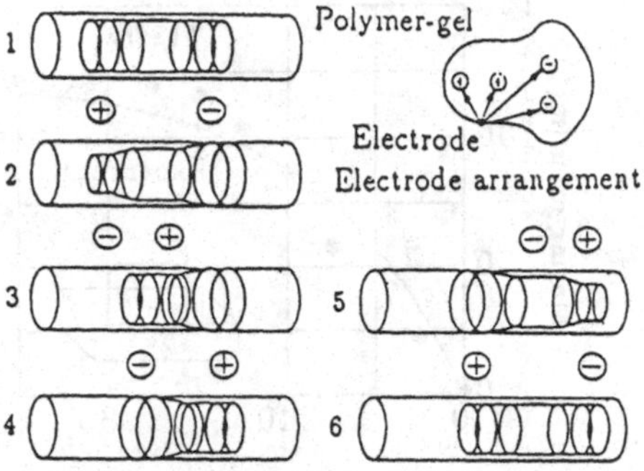

Fig.7 In-tube Worm
and its sequence

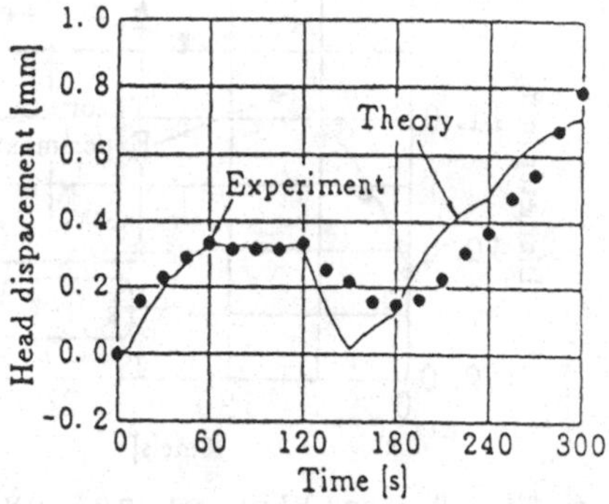

Fig.8 Performance
of In-tube Worm

contact to the inner wall during traveling sequence. The initial diameter of the gel was obtained based on the Eq.(4). Figure 8 shows the theoretical traveling distance compared with the experimental values. There happened the problem that when the gel diameter increased too bigger than the tube diameter the gel breaks. About the possibility of miniaturization of the In-tube travelling worm, Fig.9 shows studied result that the normalized displacement(traveled distance/initial length) increases with decreasing the size. That is, the performance is expected to be improved with minuter of the size.

3.3 Rotary actuator "Gel Motor" : Figure 10(a) and (b) show the principle and the schematic drawing of the Gel Motor respectively. There, a crank mechanism and four spherical gel actuators with diameter of 2 mm are used and the electric current was switched manually. The efficiency of the mechanism is given as the following Eq.(5).

$$\eta = 1 - \frac{\mu r + \mu_c r_c + T_0/W}{e} \qquad (5)$$

Where, (μ, r), (μ_c, r_c) are (frictional coefficient, radius) of

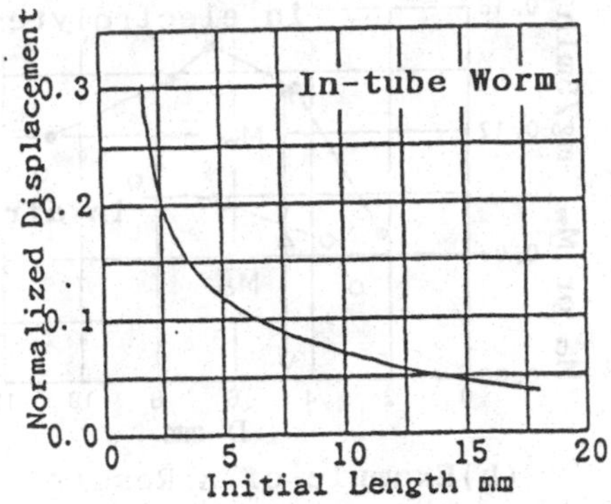

Fig.9 Performance change by miniaturization

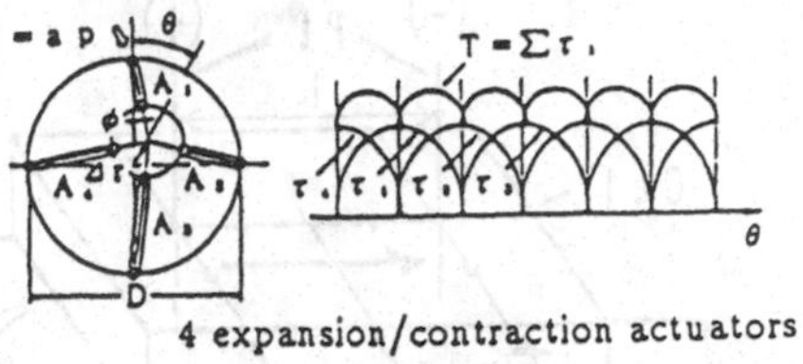

(a)Principle

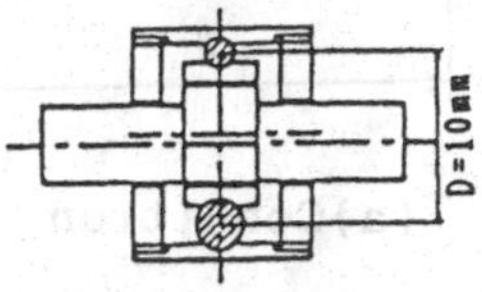

(b)Structure

Fig.10 Gel Motor

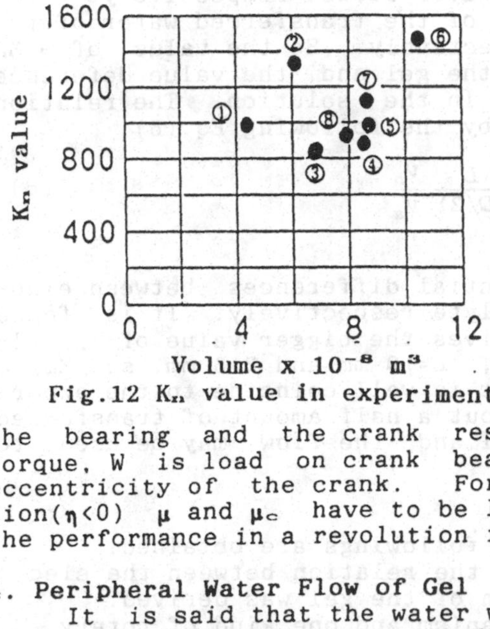

Fig.12 K_n value in experiment

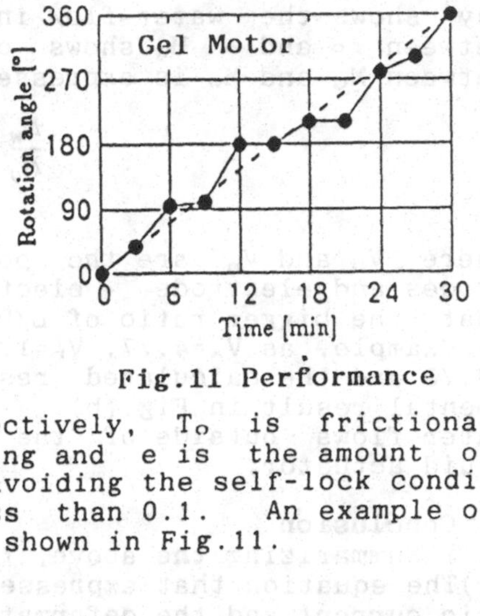

Fig.11 Performance

the bearing and the crank respectively, T_O is frictional torque, W is load on crank bearing and e is the amount of eccentricity of the crank. For avoiding the self-lock condition($\eta<0$) μ and μ_c have to be less than 0.1. An example of the performance in a revolution is shown in Fig.11.

4. Peripheral Water Flow of Gel

It is said that the water which is used for swelling and shrinking of the gel is carried by sodium ion Na^+. Following to the theory, the coefficient K_n is the number of water molecular that is carried by a Na^+ ion. Figure 12 shows the experimentally obtained values of K_n which distribute 800 ~ 1200. These values are equivalent to 2.4 ~ 3.6×10^{-26} m^3 per a Na^+ ion providing that the volume of water molecular is 3.0×10^{-29} m^3. As the value of unit charge of Na^+ is 1.60×10^{-19} C, the amount of the water carried by one Coulomb becomes 1.5 ~ 2.3×10^{-7} m^3/C. Thinking the experiment that was done at ambient pressure of 2.68 kPa and the electric current of 7 mA(mC/s), the power of the water flow W_w become 2.81 ~ 4.31×10^{-3} mW. On the other hand, about the electric source power W_e on the experiment was 2 ~ 60 mW then the efficiency of the gel actuator should be 0.1 ~ 1.0×10^{-3}.

Figure 13(a) and (b) show the condition and the result of

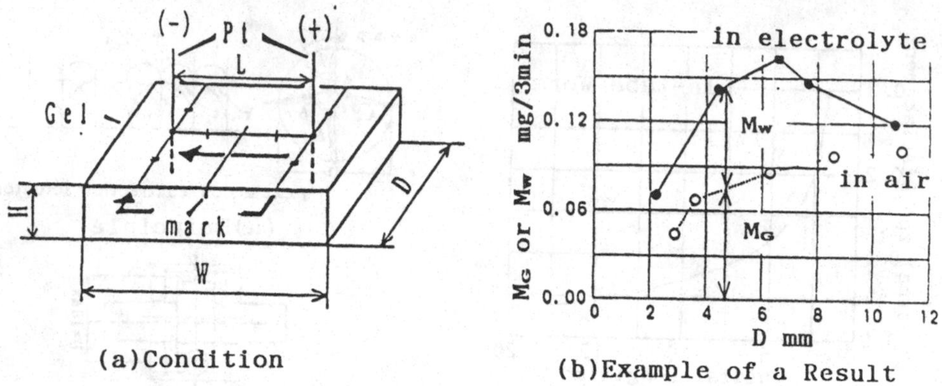

Fig.13 M_G and M_W value in experiment

the experiment of peripheral water flows respectively. The signs • and ∘ show the amounts of the transferred water of the gel in solution and in air respectively. So the value of ∘ M_G may shows the water flow in the gel and the value deference between ∘ and • M_W shows one in the solution. The relation between M_G and M_W is expressed by the following Eq.(6).

$$\frac{M_W}{M_G} = \frac{L}{(D/2)}\,\frac{V_b}{V_a} \tag{6}$$

Where V_b and V_a are the potential differences between electrodes and electrode ~ electrolyte respectively. It is found that the bigger ratio of L/D gives the bigger value of M_W. In an example, as V_a=4.77, V_b=1.04, L=10 mm and D=6 mm, so M_W/M_G =0.73 and the calculated result is well coincide to the experimental result in Fig.(b). About a half amount of transferred water flows outside of the gel and the flow may be used to fluid actuator.

5. Conclusion

Summarizing the above, the followings are obtained.

(1)The equation that expresses the relation between the electric current and the deformation of the gel was derived.

(2)Two kinds of traveling mechanism and one kind of rotary motor were offered and confirmed by experiment for it to work.

(3)The efficiency of the gel actuator was estimated as only less than 0.1% at present. Also, the power of the water flow at the outside of the gel was estimated as the half of all.

It is largely expected for the performance of gel actuator to improved much by improving the materials and by miniaturizing.

[References]

(1)T.Tanaka;Physics of Super Absorbent Polymer Gel, J.Chem-.Phys, Vol.59, No.9(1973),

(2)T.Hayashi,et al;Proceedings,IFToMM-jc International Symposium on TMM, Nagoya(1992)

HISTORY-DEPENDENT ASSEMBLY PLANNING

R. Caracciolo, E. Ceresole, F. Fanton and A. Rossi
University of Padua, Padua, Italy

Abstract
An history dependent planner for mechanical assembly is presented. Also non-mating operations are considered. The main feature of the proposed approach is the dynamic generation of the assembly plan. Only the best sequence is found. Selecting criteria are based on mechanical and technological constraints. Three modules are developed to interface the planner with a CAD environment, to determine all possible subassemblies allowing non linear planning and to manage the dynamic expansion of the assembly tree. A planning result is reported.

1. Introduction

The realization of an assembly planner entails a profound knowledge of the assembly processor features and of the rules which control the mating and auxiliary operations.
The design of a general system, which is able to deal with different kind of assembly, is difficult to formulate. In fact many differences exist among the different classes of assembly, like the assembly of mechanical or electronic components. The restriction to a single class of assembly allows the use of the features and the typical solutions of this assembly class, making the problem easier.
In this approach the planning problem is more structured than in a general approach and a big number of rules and constraints may be introduced. The present work concerns the assembly class of mechanical components. This assumption enables one to introduce a set of constraints and to consider a set of technological features which supplies a knowledge level higher than geometrical data.
Many efforts have been spent by a large number of authors to realize a system which is able to produce all possible assembly sequences for a given set of components [1][2]. The compact representation of the assembly sequences is one of the main problems faced by several authors.
The main works oriented to define and use a compact structure are proposed by Homem de Mello and Sanderson. They propose [3] an AND-OR graph based structure in which the number of nodes grows linearly with the number of the components. In another work [4] this structure is compared with the different ones based on sets of state sequences and sets of subassembly trees.
History independence is the main constraint introduced in this structure to limit the number of nodes: an *history-independent* planner simply represents the assembly status by the set of the assembled components; it is useful when dealing with geometric constraints and mating operations only.
The development of a mechanical oriented planner compels one to keep in mind a set of auxiliary operations different from the mating operations (as tool changes, subassembly reorienting and fixturing). Therefore the planner must have an history-dependent structure. Only a history-dependent planner can determine the best sequence distinguishing different ways to achieve the same assembly. A planner is *history-dependent* if each status of the partial assembly depends on the partial sequence and not only on the already assembled components. The plan representation is strictly connected to history-dependence problem [5]. In fact with a graph representation, extensively utilised in the history-independent approach, it is hard to manage the assembly history: an external data structure is needed.
On the contrary sequences differing in the mating operation order and in the auxiliary operations can be easily described by a tree representation. Generally the auxiliary operations have a remarkable weight in terms of executing time. In the mechanical domain the execution times of the mating and auxiliary operations are comparable; therefore auxiliary operations cannot be neglected.
Furthermore the introduction of these operations greatly increases the number of possible sequences and the number of nodes of each sequence, so the search of all possible sequence is not reasonable.
The main idea of this work is the dynamic generation of the sequence tree; the main tool is the use of an hypothetical reasoning technique implemented into an expert system shell; the result is an history-dependent planner able to determine the best assembly sequence without generating all possible

sequences. Technological and mechanical constraints, like the stability of the connections, allow the inference engine to reduce the number of created and expanded nodes.
Many authors face the problem of the assembly planning following a disassembly based approach to determine geometric constraints and to build the sequence data structure [6]. Backward approach is preferable for determining the geometric constraints. If only mating operations are used, the sequence determination can be executed during the backward process. According to this approach a disassembly sequence is determined beginning from the state in which all components are assembled: the corresponding assembly sequence is determined by backward reading the disassembly sequence.
The introduction of different kinds of operations and the evaluation of several type of constraints, like stability [7], suggests the *forward* approach, since the constraint check performed in a previous state can be used without recalculation.
The solution of non linear assembly plans is achieved in the present work. In the mechanical environment, where the use of subassemblies is useful and sometimes necessary, the limit of linearity is not acceptable. A proper module for the subassemblies search and a non linear planner are proposed.
A planning result obtained by a planner prototype based on an expert system shell is presented.

2. Planner structure

The planner is composed by three integrated modules subsequently acting. They are called:

- Constraints Search Module
- Subassembly Search Module
- Planning Module.

2.1 The Constraints Search Module (CSM)

The first module can be defined as a data pre-processor. Its task is to prepare all data, coming from different input sources, into a relational data structure. These data will be used in the following steps.
The CSM applies a set of rules oriented to determine:

- mating operations
- assembly directions
- precedence relations.

To describe the action of the CSM a set of terms which will be used later on is defined.
Each basic element of the assembly is a *Component*. The i-th Component of the assembly is generally indicated with C(i). A list of *Global Attributes* (GA), with technological and functional meaning, can be associated with each component.
Each component is bounded by a set of surfaces representing the B-Rep macro-faces. The j-th surface of the i-th Component is indicated with **su(C(i),j)**. A list of Local Attributes and a Direction are associated with each surface: *Local Attributes* generally describe both the technological properties and the processing type of the surfaces; the *Direction* is referred to the local frame and represents a significant direction of the considered surface. The meaning of this vector depends on the particular type of surface (for instance the normal of a plane surfaces, the axis of a cylindrical surface etc.).
A *Contact* is the coincidence of a surface of a Component with a surface of another Component. These surfaces are called *Contact Surfaces*.
A Contact between the m-th surface of the i-th Component and the n-th surface of the j-th Component is indicated with:

ct(su(C(i), m), su(C(j), n))

If there exists one or more Contacts between two Components then an *Adjacency* between the considered Components is established. An Adjacency is represented by two Components and by the set of the involved Contact Surfaces.
An Adjacency is then indicated with:

A{ ct(su(C(i), k), su(C(j), m)) , ct(su(C(i), n), su(C(j), p)) }

The CAD system provides the CSM with all the Components disposed in their final configuration and referred to the global reference frame. The description of the assembly is completed by the list of all

adjacencies. The basic information of the CSM is supplied by a module that executes a component feature identification and an identification of the Adjacencies between the Components of the assembly. Form feature identification is based on a teaching by examples approach [8].
In the first step the CSM determines the operation that joints each couples of adjacent Component. It works analysing the Contacts for each Adjacency.
The coexistence in a Contact of two surfaces having particular Local Attributes (LA) and belonging to Components having particular Global Attributes (GA) allows one to deduce a possible mating operation. For instance, a Contact between two cylindrical and threaded surfaces allows one to determine a screwing operation.
The search sequence of the mating operations can be represented as follows:

```
For ALL A into ASSEMBLY
  {
  If there EXIST into A:
      {
      - (a component whit a GA = A_FUNCTION)
           then ADD OPERATION of A_FUNCTION into OP_LIST(A)
      - (a ct such that EXIST into ct SET_OF_LA1)
           then ADD OPERATION of SET_OF_LA1 into OP_LIST(A)
      - (a ct such that EXIST into ct SET_OF_LA2)
           then ADD OPERATION of SET_OF_LA2 into OP_LIST(A)
      ....
      SELECT into OP_LIST(A) the OPERATION whit the largest PRIORITY
      SET MAIN_OP(A) by SELECTED OPERATION
      }
  else OPERATION NOT FOUND
  }
```

Therefore the CSM finds a mating operation for each contact. A *Priority* value is associated with every mating operation. The experience enables one to impose a Priority for each operation. If a operation is more restrictive than another one, then it has a greater Priority. The mating operation associated with the Adjacency is selected by considering the Priority values.
The CSM analyses all couples of adjacent Components and determines a mating operation. This operation will not always be the operation to be applied to the component during the assembly process. The real operation is deduced on-line by analysing the mating operations determined by the CSM.
The CSM infers all possible mating directions for each operation previously determined. Generally several possible mating directions are associated with each mating operation. Mating directions are determined by means of geometrical considerations on the Directions of each Contact. Also Precedence Relations are determined by the CSM [9]. The *Precedence Relations* summarise the events in which the existence of an assembled component binds the assembly operation of another component. If the events are verified for all possible mating directions then the Precedence Relation is an hard constraint:

```
If C(i) ASSEMBLED C(j) NOT ASSEMBLABLE
```

Instead, if the constraint is true only for one or more mating directions then the CSM applies a rule of inhibition of the forbidden directions.
The search of the Precedence Relations can be represented as follows:

```
For ALL C(i) into ASSEMBLY
  {
  For ALL DIRECTION(j) of C(i)
        if(C(i) COLLIDE with a COMPONENT C(j))
                then DIRECTION(j) of C(i) is BOUND by C(j)
  If ALL DIRECTION(j) of C(i) is BOUND
        then C(i) is BOUND by C(j)
  }
```

In the Precedence Relations search the CSM interacts with the solid modeller. In this way it is possible to verify the collisions between the components during the assembly operations.

2.2 Subassemblies Search Module (SSM)

The use of the subassemblies enables one to manage non linear plans. Most industrial assemblies are achieved by means of subassembly insertions.

The subassemblies can be carried out either into a single robotized cell or into further work stations. The shown planner considers only a single cell.

The Components making up a subassembly are selected verifying the following constraints:

- dynamic stability
- precedence relations
- possibility of insertion into the partial assembly
- contact coherence.

Dynamic stability is the main constraint used for determining subassemblies. A subassembly is Dynamically Stable if and only if all its Components does not change the relative positions under the effect of arbitrarily oriented forces. A Dynamically Stable subassembly can be arbitrarily reoriented without any change in the relative positions between its Components so that it can be used as a single Component. The i-th subassembly composed by n Components is indicated with $\mathbf{S_{i,n}(C(j), \ldots)}$.

A subassembly is Dynamically Stable if the relation $\mathbf{ds[S_{i,n}(C(j), \ldots)]}$ is verified [7].

The following rules make use of **ds**:

```
if NOT self stable adjacencies into Si,n(...) then ds[Si,n(...)] = FALSE
if NOT ds[Si,n(...)]
    {
    if ADD C(i) to Si,n(...) such that A{ct(C(i), m), ...} self stable
              verify ds[Sj,n+1(...)]
    else      ds[Sj,n+1(...)] = FALSE
    }
if ds[Si,n(...)] AND ds[Sj,m(...)]
    {
    if EXIST C(i) such that C(i) ∈ Si,n(...) AND C(i) ∈ Sj,m(...)
              ds[Si,n(...) ∪ Sj,m(...)] = TRUE
    }
```

These rules allow one to reduce the time of search of the Dynamically Stable subassemblies.

In this way all Dynamically Stable subassemblies are determined. The Planning Module works with Dynamically Stable subassemblies only: Dynamically Stable subassemblies can be used as they were single Components.

2.3 Planning Module (PM)

History dependence is achieved by means of the dynamic expansion of the assembly tree: this approach performs the best sequence search without calculating all possible sequences. The constraints introduced by the auxiliary operations allow the definition of a set of rules for the plan generation, while costs associated to each operation direct the selection criterion.

The assembly plan representation is an "hybrid" one: a mixing between a set of state sequences (Borjault tree [10]) and a subassembly tree is used. The assembly state is made up of three parts:

- assembled component history
- tool change history
- fixturing history.

The assembly state is associated to a tree node; every arc connecting two nodes represents either the mating operation or the auxiliary operation which performs the state transition.

The tree is dynamically created, using a Best-First Search method, which allows the best sequence determination without generating all possible assembly sequences. To do that a cost (proportional to the time spent to reach a state) is associated to every node. It is also considered a future cost, proportional to the number of not assembled components, which gives a measure of the distance of the state from the solution node. The total cost, obtained as the sum of the global and the future costs, is an index that characterizes each node. At every step of the planning process the node having the minimum total cost is chosen to be expanded. Nodes are created using a set of rules managing the state transition. Every state evolves in a set of nodes, each one associated with an assemblable Component: so it is important to correctly define the rules of assemblability of a component. Two kind of constraints are considered in the creation of a new node:

- soft constraints
- hard constraints.

Soft constraints take into account the execution time associated with the operation which generates the new state; they can be seen as the operation "cost". Such cost could also be chosen by heuristic considerations, provided by the user to support (or not) the use of particular kind of operations. These considerations become part of the knowledge base of the expert system. It is important to underline that soft constraints do not prevent the execution of any operation, but make it less probable.

On the contrary each hard constraint is able to prevent the creation of a new node when its activating conditions are true. In particular hard constraints are linked with the satisfaction of the Precedence Relations, the existence of Dynamically Stable subassemblies and the *Static Stability* of the components just assembled (those components which are in a stable equilibrium under the effect of the gravity once added to a partial assembly). Static Stability check must be done on-line, during the planning process; this requires an higher execution time due to extra calculation. On the other hand, the static stability condition reduces the size of the tree, cutting down the number of sons generated from each state: many sons are not created, being the new states associated with them statically not-stable. So planning execution time remains limited, but the planner runs closer to reality.

3. Planning example

An example of a simple mechanical assembly is presented in fig 1. This example has been chosen because it is simple and at the same time emphasizes some features of the proposed planner. The assembly process requires auxiliary operations like reorientations and tool changes.

Foreseen directions are shown in Table 1 and detected mating operations are in Table 2. These data are determined by the CSM. The SSM determine the subassemblies:

$$su_{1,4}(C(4), C(7), C(5), C(6)) \quad ; \quad su_{2,3}(C(1), C(2), C(3))$$

Table 3 describes the sequence determined by the planner. The operations CHANGING, FIXTURING and REORIENTING are auxiliary operations. The sequence emphasizes the number of auxiliary operations required to realize the assembly.

Mating Operations			TABLE 2
Ci	**Cj**	**Operation**	**Direction**
C	A	screwing	6
C	B	inserting	6
E	G	screwing	3
E	D	inserting	3
F	G	screwing	3
F	D	inserting	3
H	B	screwing	6
H	G	inserting	6
I	B	screwing	6
I	G	inserting	6
A	B	inserting	3
G	B	leaning	6
D	G	leaning	6

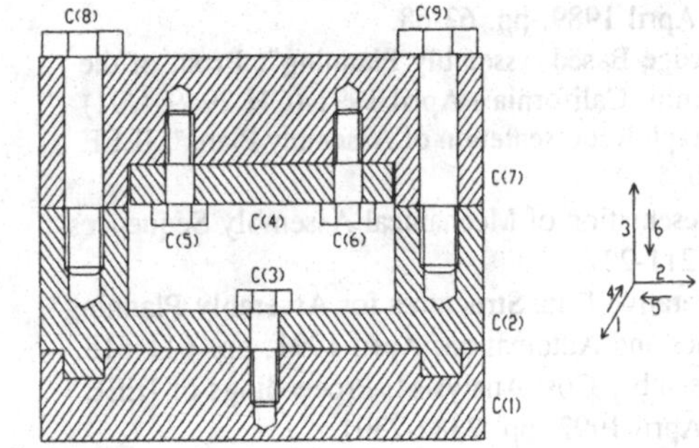

Fig 1 - Example of mechanical assembly

Directions						TABLE 1	
N°	X	Y	Z	N°	X	Y	Z
1	1.000	0.000	0.000	4	-1.000	0.000	0.000
2	0.000	1.000	0.000	5	0.000	-1.000	0.000
3	0.000	0.000	1.000	6	0.000	0.000	-1.000

Main Sequence			TABLE 3
Component	**Task Comp.**	**Operation**	**Dir.**
Fixture 1	-	fixturing	-
Tool 1	-	changing	-
C(1)	Fixture 1	inserting	6
Tool 2	-	changing	-
C(2)	C(1)	leaning	6
Screwdriver	-	changing	-
C(3)	C(1)	screwing	6
Tool 1	-	changing	-
$Su_{1,4}$	-	reorienting	-
$Su_{1,4}$	C(2)	leaning	6
Screwdriver	-	changing	-
C(5)	C(2)	screwing	6
C(6)	C(2)	screwing	6

$Su_{1,4}$ Sequence			TABLE 4
Component	**Task Comp.**	**Operation**	**Dir.**
Fixture 2	-	fixturing	-
Tool 1	-	changing	-
C(7)	Fixture 2	inserting	3
C(4)	C(7)	leaning	3
Screwdriver		changing	-
C(8)	C(7)	screwing	3
C(9)	C(7)	screwing	3

4. Conclusions

An assembly planner oriented to the mechanical domain has been presented in this work. The planner considers a set of auxiliary operations, useful to obtain efficient and closer to reality assembly sequences. The consideration of the auxiliary operations forces one to realize an history-dependent planner. Since the generation of all possible sequences is not proposable in this case, a Best-First Search method for the dynamic generation of the sequence tree is used, enabling one to search for the best sequence without generating all possible sequences. The main constraints used to limit the tree expansion are the Static Stability of the Component and the Precedence Relations.
A module for the determination of all possible subassemblies is proposed. The main constraint which is used to define the subassemblies is the Dynamic Stability. A planning example is presented.

Acknowledgement
This research has been partially sponsored by a grant of the Italian National Coucil of Research C.N.R. as part of the "Progetto Finalizzato Robotica" project.

References
[1] Jan D. Wolter, "On the Automatic Generation of Assembly Plans" ; Proceeding of IEEE International Conference on Robotics and Automation, April 1989, pp. 62-68.
[2] Y.F. Huang - C.S.G. Lee "A framework of Knowledge-Based Assembly Planning", Proc. of the 1991 IEEE Intern. Conf. on RA, pp. 599-604, Sacramento, California - April 1991 (CH 2969-4/91)
[3] L S Homem De Mello - A.C Sanderson, "And/Or Graph Representation of Assembly Plans", IEEE Transaction on RA, vol. 6, no. 2, pp. 188-198, April 1990
[4] L. S. Homem de Mello and A.C. Sanderson, "Representation of Mechanical Assembly Sequences", IEEE Transaction on RA, Vol 7, N° 2, April 1991, pp 211-227.
[5] Jan D. Wolter, "A Combinatorial Analysis of Enumerative Data Structures for Assembly Planning", Proceeding of IEEE International Conference on Robotics and Automation, April 1991, pp. 611-618.
[6] Sukhan Lee,"Backward Assembly Planning with Assembly Cost Analysis", Proceedings of IEEE International Conference on Robotics and Automation, April 1992, pp. 2382-2391.
[7] R. Caracciolo, E. Ceresole, "Forward assembly planning based on stability", Report n.1/93 of the Dept. of Innovation and Management, Padova Italy, 1993.
[8] R.Caracciolo, E.Ceresole, T.De Martino, F.Giannini, "From CAD Models to Assembly Planning", Proc. of the International Workshop on Graphics and Robotics, Schloss Dagstuhl, Germany, April 1993.
[9] Y.F.Huang, C.S.G. Lee, "Precedence Knowledge in Feature Mating Operation Assembly Planning" Proceeding of IEEE International Conference on Robotics and Automation, August 1989, pp. 216-221.
[10] A. Bourjault, "Contribution à un Approche Methodologique de l'Assembly Automatisé: Elaboration Automatique des Sequences Operatoires", Thesis to obtain grade of Docteur in Phisical Science at University of Franche-Comte, November 1984.

ROBOTIC DEBURRING TASK DESIGN BASED ON THE NATURE OF BURR

M. Mizukawa, E. Mitsuya, S. Ohara, S. Iwaki, S. Matsuo, T. Shakunaga and T. Okada
NTT Human Interface Laboratories, Tokyo, Japan

Abstract

In this paper, we propose the concept of robotic task design based on the nature of burr that it is formed within the constraint plane called parting palne. The deburring task is designed to use this constraint as a task guide. This system design consists of the following steps: (1) Robotic Task Design, (2) Teaching (Programming) and (3) Task Execution . These have equal importance instead of over emphasizing the task execution system. The designed robotic task system is implemented on a robotic task execution controller called NOAC. The main concept in the structure of NOAC is to separate function (1) which is commonly needed by every task in the task class and function (2) which is a task-oriented function. This system is shown to be simple and expandable to other tasks such as tool manipulation.

1. Introduction

Many trials have been made to design robots for deburring task[1-6]. However, not so many robot systems are in actual use because of the difficulties caused by the variety of tasks. We focus on the nature of burrs in metal castings in order to develop a methodology for designing a robotic deburring system.

2. Characteristics of Deburring Task

As the nature of the casted products, they inevitably have fluctuations in dimensions and shape so that they never coincide with geometric models even if it is possible to construct such models. Trajectories which the robot traces, consisting of complex three-dimensional lines fixed on an object (workpiece), are specified by a series of teaching points. Their dimensions tend to vary because castings are usually designed with large tolerance. This makes it difficult to use a fixed robot program even if the objects are made in the same lot. Consequently, it is necessary to teach an enormous number of points every time an object is placed in front of the robot. This is the dominant factor which prevents robots from being widely used for deburring.

Burrs tend to be formed in "parting planes" which coincide with two mating die faces. In most cases, the die surfaces for mating is planer because of the advantage of fabrication costs and keeping die accuracy. Therefore, it is reasonable to use this parting plane as the constraint plane of robotic motion.

3. Task Design

The structure of deburring task is shown in Fig. 1. The robot motion is constrained to the parting planes. This constraint reduces the degree of freedom, so it is easier to specify the motion of the robot. The task procedure consists of the following subtasks.

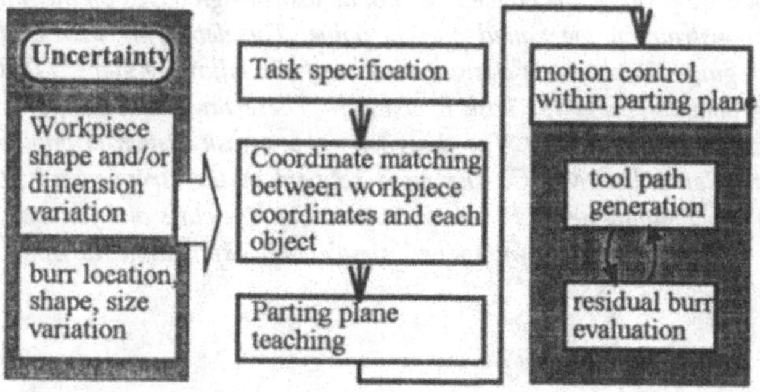

Fig. 1 Structure of deburring task

(1) teach the parting plane

To construct a rough geometric model of the workpiece which is to be used as a reference of the workpiece and to teach the location of the parting plane in the local coordinates defined on the workpiece. This enables us to significantly reduce the teaching time and to prepare the environment.

(2)coordinate matching between rough geometric model and actual workpiece

To use a vision system to roughly measure the position and orientation of the workpiece which has size fluctuation and positioning error.

(3) compliance control within the parting plane with force/torque and laser range sensors

A deburring tool attached to a manipulator, which has a force/ torque sensor and a laser range sensor, tracks the parting line defined as the intersection of the parting plane and the work surface with compliant motion constrained to the parting plane.

To use the parting plane as a reference in the workspace, task design process, and task description, meanes that the deburring motion regularity can be achieved in a way matched to the nature of the burr and/or deburring task. Also the 2D constraint reduces the degree of freedom, thus making it easier to specify the motion of the robot. So, in this paper, the constraint plane has the role of *guide plane*.

4. Teaching System

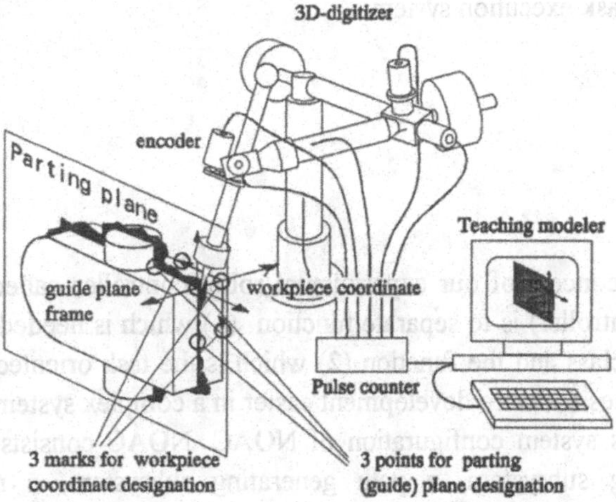

Fig.2 Schematic view of teaching the deburring task using constraint plane

The constraint plane such as parting plane is represented in the form of a "frame" which has an origin and three axes like ordinary coordinates. The origin of this frame indicates the position of the plane and the z-axis indicates the normal vector of the plane as shown in Fig. 2. This formulation has following advantages.

(1) Compact data set representation.
(2) Explicit position and posture expression.
(3) Coordinate provision for constrained motion execution.

The teaching procedure is carried out in the following sequence.

(1) generation of rough geometric model of the workpiece
(1-1) Define a workpiece local coordinates
(1-2) Measure several marks on the workpiece, which define the workpiece position and posture and in its local coordinates.
(2) parting plane definition in the workpiece local coordinates
Measure several points on the workpiece, which define the parting plane and then a

matrix expression of the frame is calculated. If it is possible to pre-define a CAD model, this information can be obtained without these measurements.

(3) coordinate matching between rough geometric model and actual workpiece in the robot coordinates

Place the workpiece in the robot working area and measure the marks with the robot and/or fixed vision system to determine the position and orientation of the workpiece in the robot coordinate system. Thus, the guide plane is turned to be used as the constraint plane for the robot task execution system.

5. Execution System

5.1 System Controller

The structural concept of our experimental robot controller called NOAC (NTT Open Architecture Controller) is to separate function (1) which is needed in common by every task in the task class and the function (2) which is the task oriented function. This separating process makes software development easier in a complex system.

Figure 3 shows system configuration of NOAC. NOAC consists of four major subsystems: a sensing subsystem, a path generating subsystem, a motion control subsystem and a teaching & monitoring subsystem. To maintain modularity and high processing power, the sensing, path generating and motion control subsystems use MC68040 processors and each subsystem is connected to a VME bus. A real time operating system (VxWorks) is used with these processors. A SUN Sparcstation is used in

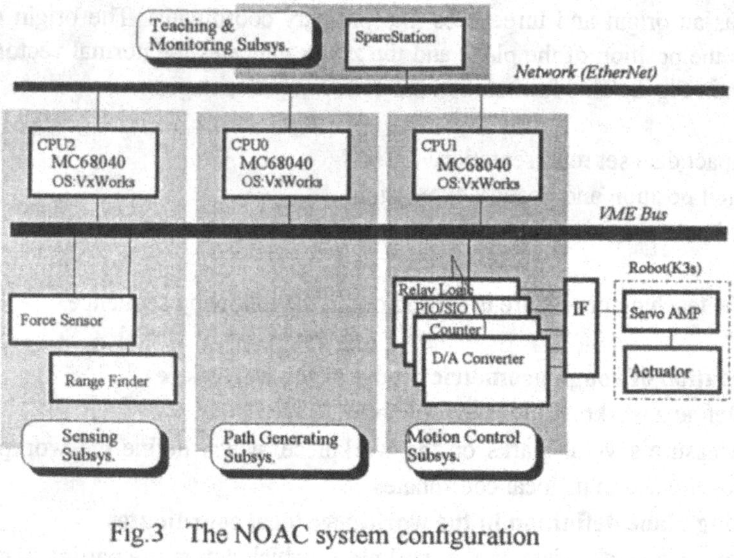

Fig.3 The NOAC system configuration

the teaching & monitoring subsystem which is connected to other subsystems through Ethernet.

5.2 Task Execution

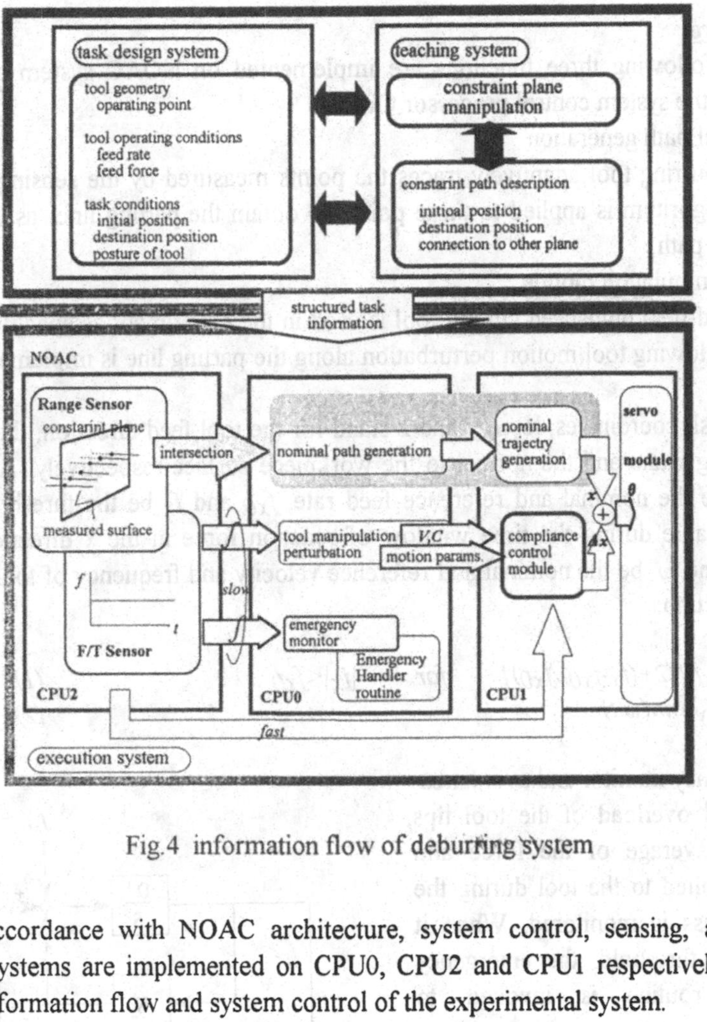

Fig.4 information flow of deburring system

In accordance with NOAC architecture, system control, sensing, and motion control subsystems are implemented on CPU0, CPU2 and CPU1 respectively. Figure 4 shows the information flow and system control of the experimental system.

sensing

In this system, it should be pointed out that the most important information from a laser range sensor is not the shape of burr but the position of the workpiece surface because burrs are removed and their shape is changed during the task execution and the objective of deburring is to finish the surface within a specified tolerance. A laser range-finder attached on the robot wrist is used and the 3-dimensional workpiece surface is obtained as the intersection of the laser-beam plane and the workpiece surface. To recognize the surface of the workpiece, the measured data is fitted by a straight line with the least-square method. This set of fitting lines is used to calculate the reference path by calculating the intersections with the parting plane previously discussed.

system control

The following three functions are implemented on NOAC system and parallel processed in the system control processor CPU0.

(1) Nominal path generation

The deburring tool adaptively traces the points measured by the sensing subsystem. Fairing algorithm is applied to these points to obtain the parting line as the nominal reference path.

(2) Tool manipulation motion

The hand-held highspeed cutting tool is used in the experiment as specified in Table 1. The following tool motion perturbation along the parting line is implemented in this case.

In the task coordinates, let x, y and z stand for the tool feed direction, the normal to the parting plane and the normal to the workpiece surface respectively. Also let V_{x0} and V_x be the nominal and reference feed rate, f_{x0} and f_x be the threshold and the average value during the time window of reaction force in the x direction, and let V_{y0} ,V_y and ω be the nominal and reference velocity and frequency of tool motion in the y direction.

$$V_x=V_{x0}\{(2+(f_x-f_{x0})/f_{x0})\} \quad \text{for} \quad |f_x|<f_{x0} \tag{1}$$

$$V_y= V_{y0}\sin(\omega t) \tag{2}$$

(3) Emergency monitor and avoidance

To avoid overload of the tool tips, the time average of the force and torque applied to the tool during the task process is monitored. When it exceeds a threshold, the emergency handling routine is invoked to manage the situation.

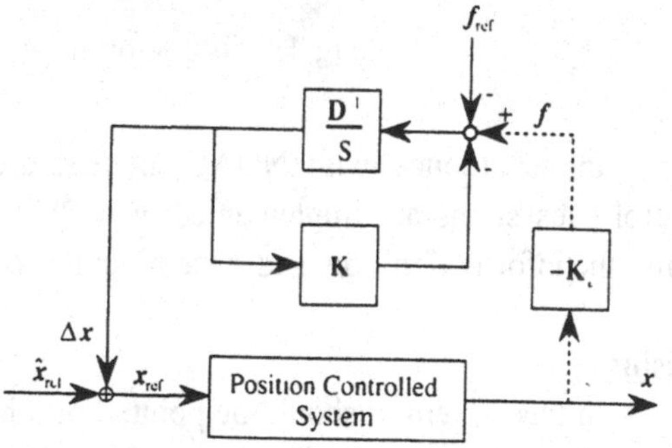

Fig.5 Block diagram of compliance control

motion control

Figure 5 shows the compliance control scheme in CPU1. A force control loop is set as the outer loop of a position controlled robot system. By choosing appropriate damping **D**, stiffness **K** and force reference f_{ref} in the force control loop, the reference x_{ref} for the position controlled system is generated by compensating the nominal trajectory $\hat{x}_{ref}$ calculated in CPU0 by using the range-finder readings and parting plane. The cycle time of CPU1 is 5 ms.

Fig. 6 shows the experimental setup of the proposed robotic deburring system and Table 1 is an outline of the specifications.

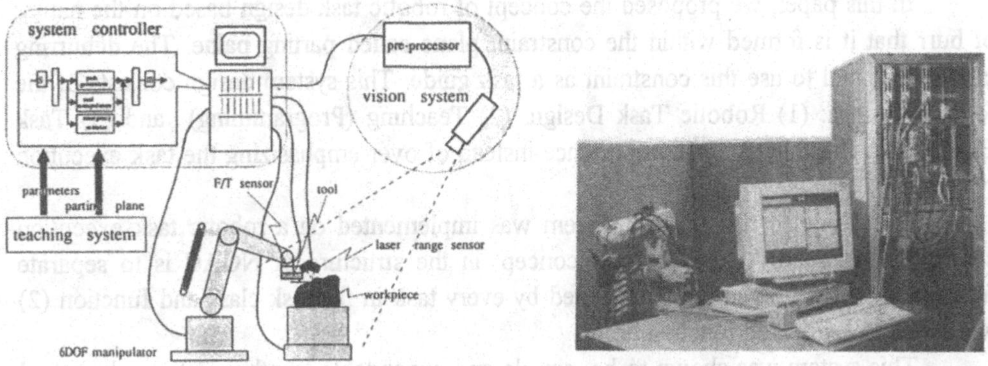

Fig.6 Experimental system setup

5. Discussion

Table 1: system outline

	Vendor/Type	specifications
Robot	Yaskawa K3S	loadcapacity: 30N
F/T sensor	Nitta UFS3012A-15	load capacity: 60 N(x,y),120 N(z), 5 Nm(Mx,My,Mz) accuracy: 0.05 % sampling rate: 5 ms
Laser Range sensor	Toyota RF-1M	measure distance: 66-94 mm accuracy: 0.1 mm max scan rate: 30 Hz
tool	Minitor M26A (a hand held cutter)	torque: 0.06 Nm speed: 1420-14200 rpm
cutting tip	high speed cutter	diameter: 6 mm

In the previous sections, we proposed a task structure design for deburring task with a constraint plane as a task guide. the task class in which a manipulator operates the tool along a specified line on a workpiece surface. Such a constraint plane can be used as a guide plane in teaching tasks other than deburring such that the task class in which a manipulator operates the tool along a specified line on a workpiece surface as polishing in which physical constraints on the workpiece and a manipulator are induced to specify the tooling paths. Also, the path generation algorithm can be applied to tasks in this category, since the tracking path generation and tool operation manipulation around the tracking path are explicitly separated and replacing the tool manipulation operation is only required if other kinds of tasks in the class are to be carried out.

6. Concluding Remarks

In this paper, we proposed the concept of robotic task design based on the nature of burr that it is formed within the constraint plane called parting palne. The deburring task is designed to use this constraint as a task guide. This system design consists of the following steps: (1) Robotic Task Design, (2) Teaching (Programming) and (3) Task Execution . These have equal importance instead of over emphasizing the task execution system.

The designed robotic task system was implemented on a robotic task execution controller called NOAC . The main concept in the structure of NOAC is to separate function (1) which is commonly needed by every task in the task class and function (2) which is a task-oriented function.

This system was shown to be simple and expandable to other tasks such as tool manipulation.

Acknowledgment

The authors wish to express their gratitude to Dr. Tadashi Naruse, Mr. Ken'ichiro Shimokura, Mr. Yoshito Nanjo and Mr.Yukihiro Kubota for development of the robot controller, also to them and Dr. Masashi Okudaira for their support and useful discussions.

References

[1] H.Asada, H.-H.Yang, "Skill Acquisition from Human Experts Through Pattern Processing of Teaching Data", *Proc. IEEE Int. Conf. on Robotics and Automation*, pp1302-1307, 1989

[2] H.Ito,Y.Saito, et.al., "A Development of Torque Equilibration Type Deburring Robot System", *Proc. 20th Int. Symp. on Industrial Robots(ISIR)*, pp901-908, 1989

[3] K.Kashiwagi, et.al,"Force Control Robot for Grinding", *Proc. IEEE Int. Workshop on Intelligent Robots and Systems 90(IROS'90)*, pp1001-1006, 1990

[4] M.Ichinohe, K.Ohara, et.al," Development of Deburring Robot for Casst Iron with Vision and Force Sensing", *Proc. 24th Int. Symp. on Industrial Robots (ISIR)*, pp49-54, 1993

[5] Y.Saito, T.Haneyoshi, et.al, "A Periodic Damping Effect with Multi-Force Control to Robotic Deburring and Polishing", *Proc. 24th Int. Symp. on Industrial Robots (ISIR)*, pp55-62, 1993

[6].T.Yoshimi, M.Jinno, et.al.,"Application Research on a Grinding Robot System -Grinding of Rolling Stock Parts-", *Proc. 24th Int. Symp. on Industrial Robots (ISIR)*, pp63-72, 1993

NUMERICAL MODELLING OF A FLOATING ROBOT

R. Kościelny
Technical University of Gdansk, Gdansk, Poland

SUMMARY

The paper presents a spatial model of a floating robot at operation in maritime conditions. Influence of water surface, viscoelasticity of the arm and a specific arm suspending system has been regarded. Dry friction at the load-deck contact phase has been taken into account as well.

1. INTRODUCTION

A class of problems in applied mechanics comprises setting up a dynamic model of a mechanical device regarded as a subsystem implanted into an existing complex mechanical system.

In many cases, dynamic analysis of the device can be treated locally, nevertheless in respect of bilateral influences at the subsystem boundary. In particular, the influences are exciting forces and large displacement of the base.

The subsystem is nested in the system by nonlinear constraints. Thus dynamics of the subsystem influences the whole system depending on properties both the system and boundary constraints.

These problems should be taken into considerations in dynamic analysis of robot operation in open sea conditions. A floating robot mounted on a deck can be taken as an essential part of the subsystem. Depending on definition of the problem the subsystem boundary can be set at the robot base or on the ship "wet" surface.

An associated ship and the manipulated object form the other parts of the subsystem.

Waving water surface induces stochastically the motion of the robot base and consequently the vibration of its arm suspending system and steel structure.

In return, the robot response influences the pontoon motion and produces itself water waving.

In the paper the modification of the RFEM (rigid finite element method) is applied to discretize of a flexible crane jib. The presented method applied to analysis of mechanical systems with changing configuration is based on the formalism used in dynamic analysis of manipulators with rigid links (Craig, 1988). This method differs basically from the classical finite element approach (Kane, 1987, Du and Hitchings, 1992)

2. DESCRIPTION OF MECHANICAL MODEL

The model considered in the paper is shown in Fig. 1. It consists of the robot, the grabbing device, the ship on which the robot is settled, the object, the associated ship and the water.

In oblique irregular 3D waves the floating structure is excited in six motions. The elestic structure moves bodily and also distorts. The bodily motions are those of surge, sway, heave, roll, pitch and yaw and they are performed as if the hull does not distort.

The two hulls motions in the beginning of reloading operation is described in deterministic, quasi-static terms. For the purpose, such a simplified description of behaviour of a stiff hull heading in an irregular seaway remains effective.

The extended description of the behavior of the flexible ship structure introduces the additional motions of natural frequency, resonance frequency, ship-wave matching as well as statistical estimates of the response magnitudes. In addition, the structure distors in an infinite number of ways. In the paper, the influcncc of thc small dcflcction of thc hull structurc has bccn ncglcctcd. That means the floating bodies are regarded as stiff and under quasi-static external forcing.

While the operation is done, the ship is kept in proper position parallelly with the pontoon. To complete the operation, the robot is to grab the object from the pontoon, hoist it, slew as to take it over the ship deck, lower it and finally put it onto the deck.

Consequently, a spatial model of the subsystem (robot, grabing device, object, ship) has been employed. In order to allow the large displacement analysis the RFEM has been used to discretize the jib. Assuming the jib consists of n+1 RFEs (rigid finite elements) and n SDEs (spring-damping elements), the local coordinate systems has been introduced as in fig.2. The transformation formula is given below:

(1) $O_{r_1} = O_{{}_1A}\; {1}_{{}_2A}\; {2}_{{}_3A}\; A_1 A_2 \ldots A\; i_{r_i} = B\; i_{r_1},$

where :

$$O_{r_i} = O_{{}_1A(x_1{}^1, x_2{}^1, x_3{}^1 . \varphi_1{}^1, \varphi_2{}^1, \varphi_3{}^1)}$$

$$(O_{{}_1A} = O_{{}_1A(t)}$$

when base motion of pontoon is known)

$${1}_{{}_2A} = {1}_{{}_2A(\varphi)}$$

$${2}_{{}_3A} = {2}_{{}_3A(\psi)}$$

Let us assume that only the bending flexibility of arm is taking into account. In this case the transformation matrices A_i can be written as

(2) $A_i = A_{i1}(\varphi_{i1})\, A_{i2}(\varphi_{i2})$ where:

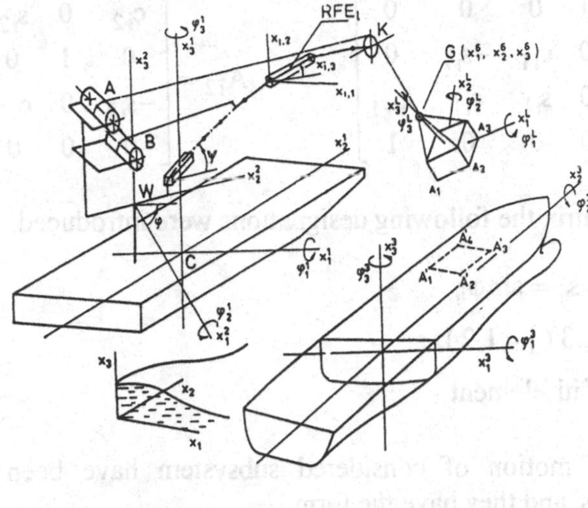

Fig. 1. Physical model of system.

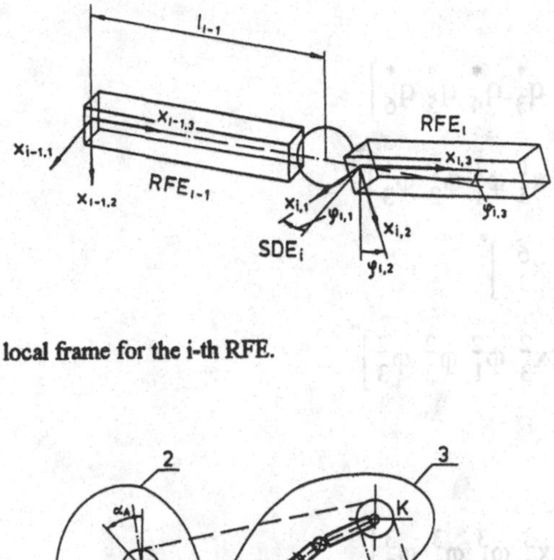

Fig. 2. The local frame for the i-th RFE.

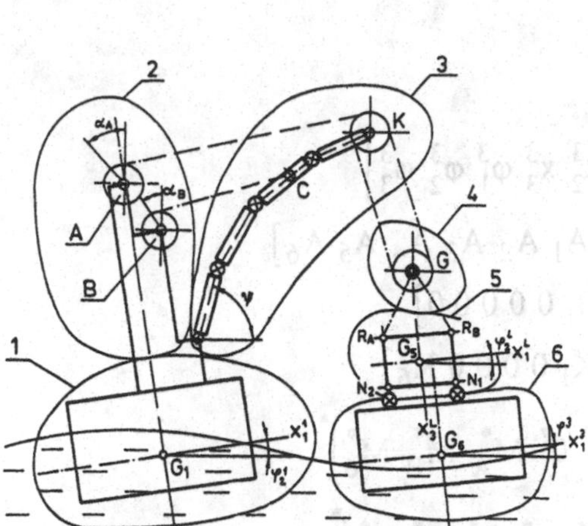

Fig. 3. Subsystems in the two trawlers model.

$$A_{i1} = \begin{bmatrix} 1 & 0 & 0 & 0 \\ 0 & c_{i1} & -s_{i1} & 0 \\ 0 & s_{i1} & c_{i1} & l_{i-1} \\ 0 & 0 & 0 & 1 \end{bmatrix} \qquad A_{i2} = \begin{bmatrix} c_{i2} & 0 & s_{i2} & 0 \\ 0 & 1 & 0 & 0 \\ -s_{i2} & 0 & c_{i2} & 0 \\ 0 & 0 & 0 & 1 \end{bmatrix}.$$

In the above matrix the following designations were introduced:

$c_{ij} = \cos\varphi_{ij}, \; s_{ij} = \sin\varphi_{ij}$

φ_{ij} – as in Fig. 3 (j = 1,2),

l_i = length of ith element.

The equations of motion of considered subsystem have been derived from Lagrange equations, and they have the form

(3) $M\ddot{q} + B\dot{q} + Kq = F - H$

where

$$q = \left[q_1^* \; q_2^* \; q_3^* \; q_4^* \; q_5^* \; q_6^*\right]$$

$$q_1 = \left[x_1^1 \; x_2^1 \; x_3^1 \; \varphi_1^1 \; \varphi_2^1 \; \varphi_3^1\right]^*$$

$$q_2 = \left[x_1^6 \; x_2^6 \; x_3^6\right]^*$$

$$q_3 = \left[x_1^2 \; x_2^2 \; x_3^2 \; \varphi_1^2 \; \varphi_2^2 \; \varphi_3^2\right]^*$$

(4) $q_4 = \varphi$

$q_5 = \psi$

$$q_6 = \left[x_1^3 \; x_2^3 \; x_3^3 \; \varphi_1^3 \; \varphi_2^3 \; \varphi_3^3\right]^*$$

$M = \mathrm{diag}\left[A_1 \; A_2 \; A_3 \; A_4 \; A_5 \; A_6\right]$

$B = \mathrm{diag}\left[B_1 \; 0\;0\;0\;0\;0\right]$

$K = \mathrm{diag}\left[K_1 \; 0\;0\;0\;0 \; K_6\right]$

$$F = \left[M_1^* \; 0^* \; 0^* \; M_A^* \; M_B^* \; M_6^*\right]^*$$

$$F = \left[H_1^* \; H_2^* \; H_3^* \; H_4^* \; H_5^* \; H_6^*\right]^*$$

In cases of large devices, the arm can be suspended on a rope system. The stiffnes of the suspending system can be expressed in form :

$$(5)\ H_1 = G_1 + D_1 + C_A \Delta_A \frac{\partial \Delta_A}{\partial q_1} + C_B \Delta_B \frac{\partial \Delta_B}{\partial q_1}$$

$$(6)\ H_2 = G_2 + C_A \Delta_A \frac{\partial \Delta_A}{\partial q_2} + \sum_{j=1}^{4} \delta_{Lj} c_{Lj} \frac{\partial \Delta_j}{\partial q_2}$$

$$(7)\ H_3 = G_3 + \sum_{j=1}^{4} \delta_{sj} c_{sj} \frac{\partial \Delta_{sj}}{\partial q_3} + \sum_{j=1}^{4} \delta_{Lj} c_{Lj} \frac{\partial \Delta_j}{\partial q_3}$$

$$(8)\ H_4 = \frac{dc_B}{dq_4} \frac{1}{2} c_A \Delta_A^2 + c_A \Delta_A \frac{\partial \Delta_A}{\partial q_4}$$

$$(9)\ H_5 = \frac{dc_B}{dq_5} \frac{1}{2} c_B \Delta_B^2 + c_B \Delta_B \frac{\partial \Delta_B}{\partial q_5}$$

$$(10)\ H_6 = \sum_{j=1}^{4} \delta_{sj} c_{sj} \frac{\partial \Delta_{sj}}{\partial q_6}$$

The rope elongations Δ_A, Δ_B, Δ_j and deflections Δ_{sj}, (j = 1,2,3,4) can be described as the functions of generalized coordinates. The coeficients dLj, dsj depend on q1 ... q6 and they are the Heaviside functions:

$$(11) \begin{cases} \Delta_A = \Delta_A(q_4\, q_1\, q_2) \\ \Delta_B = \Delta_B(q_5\, q_1) \\ \Delta_j = \Delta_j(q_2\, q_3) \\ \delta_{Lj} = \begin{cases} 1 & \text{if } r_{Aj} - r_{Aj0} > 0 \\ 0 & \text{in other case} \end{cases} \end{cases}$$

where $r_{Aj} = r_{Aj}(q_2, q_3) = \left|A_j G\right|$, or

$r_{Aj0} = \left|A_j G\right|$ if the rope slackes.

Another group of nonlinear constraints occurs in the phase of setting the object onto the deck. These constraint equations form the set of nonlinear algebraic equations

$$(12)\ N(q, \dot{q}, C_S, \mu) = 0$$

The number of equations (12) depend on the number of contact points and varies in time from 0 to 4.

The coeffiticnt of local deck stiffness C_S and the coeffiticnt of dry friction m (static or kinematic) are used in expressions for tangential components of reaction forces. The choice between static or kinematic friction depends an relative velocity in the contact area.

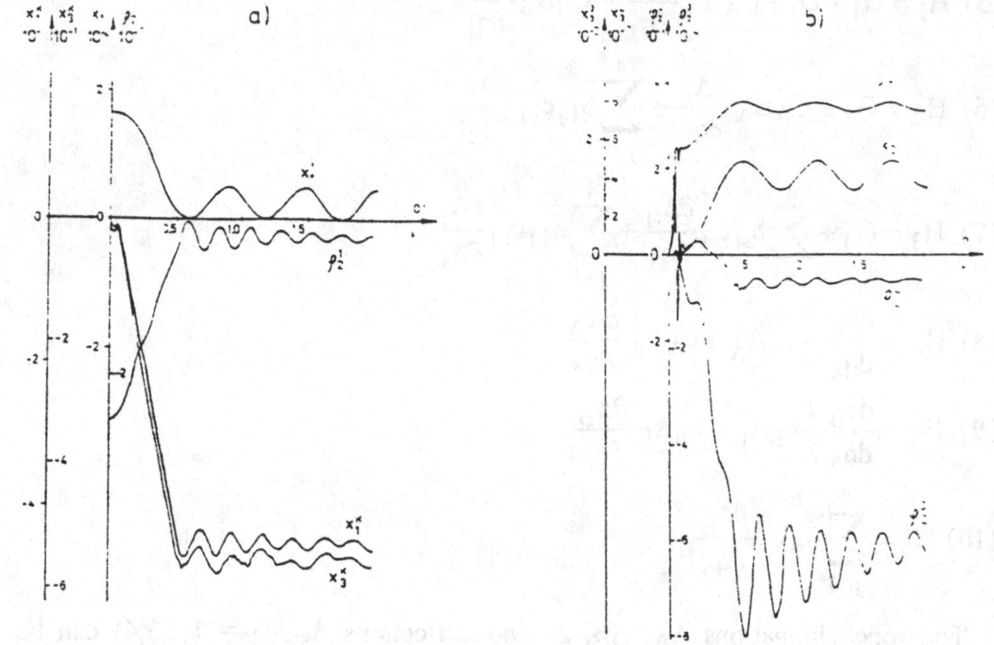

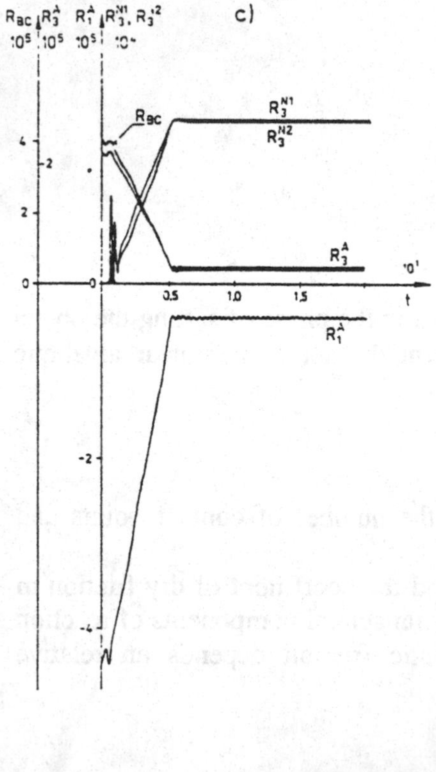

Fig.4. Results of calculations

a) displacments x_1^1, φ_2^1 of the mother-ship and x_1^k, x_3^k of the arm end,

b) displacements x_3^3, φ_2^3 of the accompaning ship and x_3^L, φ_3^L of the object.

c) forces : R_{BC} in the arm suspending system, R_1^A, R_3^A in the arm bearing and R_3^{N1}, R_3^{N2} the load-deck reaktions.

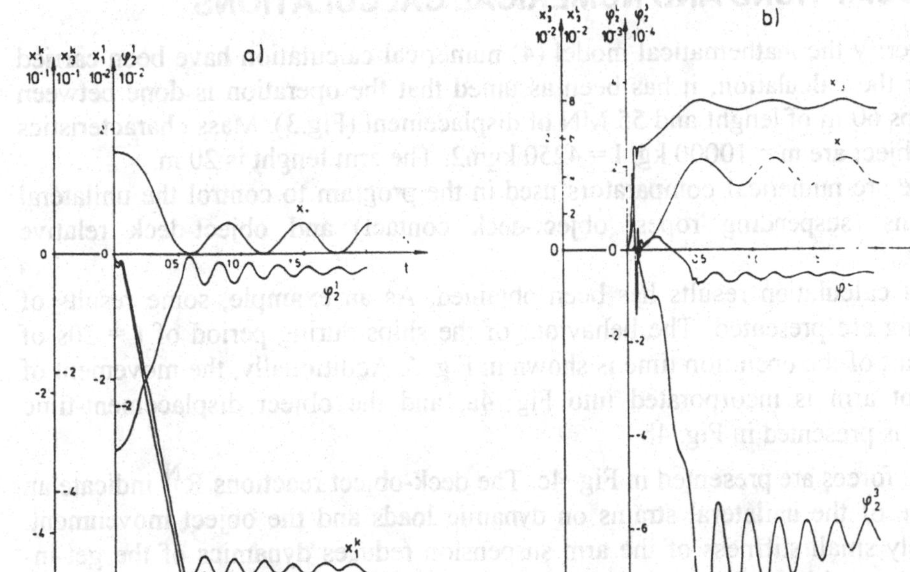

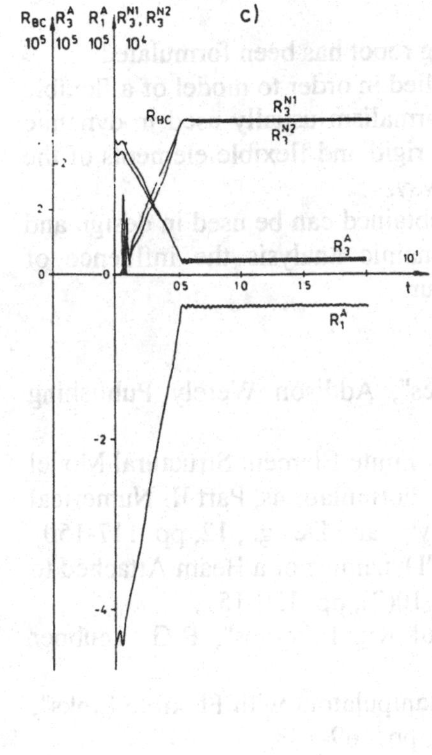

Fig.4. Results of calculations
a) displacments x_1^1, φ_2^1 of the mother-ship and x_1^k, x_3^k of the arm end,
b) displacements x_3^3, φ_2^3 of the accompaning ship and x_3^L, φ_3^L of the object,
c) forces : R_{BC} in the arm suspending system, R_1^A, R_3^A in the arm bearing and R_3^{N1}, R_3^{N2} the load-deck reaktions.

3. ASSUMPTIONS AND NUMERICAL CALCULATIONS

To verify the mathematical model (4) numerical calculation have been carried out. For the calculation, it has been assumed that the operation is done between two ships 60 m of lenght and 55 MN of displacement (Fig.3). Mass characteristics of the object are m = 10000 kg, I = 4250 kgm2. The arm lenght is 20 m.

There are numerical comparators used in the program to control the unilateral constrains (suspending ropes, object-deck contact) and object-deck relative velocity.

A set of calculation results has been obtained. As an example, some results of modelling are presented. The behaviour of the ships during period of t = 20s of initial part of the operation time is shown in Fig. 4. Additionally, the movement of the robot arm is incorporated into Fig. 4a, and the object displacement-time diagram is presented in Fig. 4b.

Reactive forces are presented in Fig. 4c. The deck-object reactions R^N indicate an influance of the unilateral strains on dynamic loads and the object movenment. Relatively small stiffness of the arm suspension reduces dynamics of the get-in-touch phase, which reflects the suspension load diagram R_{BC} and displacement of the end of the arm x^k.

4. CONCLUSIONS

In this paper the numerical model of a floating robot has been formulated.

The rigid finite element method has been applied in order to model of a flexible robot's arm. This method allows to apply the formalism usually used in dynamic analysis of manipulators with rigid links. So the rigid and flexible elements of the considered system could be treated in the same way.

The mathematical model of a floating robot obtained can be used in design and research, when it is important to take in dynamic analysis the influence of maritime conditions on ships and robots behaviour.

REFERENCES

Craig J.J., 1988 : "Introduction to Roborics", Addison Werely Publishing Company, Massachusetts,

Du H., Hitchings D., Davies G.A.O., 1992: "A Finite Element Structural Model of a Beam with an Arbitrary Moving Base, Part I: Formulations, Part II: Numerical Examples and Solutions", Finite Element in Analysis and Design, 12, pp. 117-150,

Kane T.R., Ryan R.R., Banerjee A.K., 1987 : "Dynamics of a Beam Attached to a Moving Base", J. Guidance, Control Dynamics, 10(2), pp. 139-151,

Wittenburg J.,1977 : "Dynamic of System of Rigid Bodies", B.G. Teubner Stuttgard,

Wojciech S , 1990 · "Dynamic Analysis of Manipulators with Flexible Links", Archive of Mechanical Engineering, XXXVII (1), pp. 169-188,

Wojciech S., Adamiec-Wójcik I., 1993 : "Nonlinear Vibrations of Spatial Viscoelastic Beams", Acta Mechanica 98, 15-25.

Part XI
Biomechanical Aspects of Robots and Manipulators

DEVELOPMENT OF ACTIVELY CONTROLLED ELECTRO-HYDRAULIC ABOVE-KNEE PROSTHESES

M.S. Ju, Y.F. Yang and T.C. Hsueh
National Cheng Kung University, Tainan, Taiwan

Summary

In previous works, a four-bar linkage above-knee (AK) prosthesis with an electro-hydraulic controller was built. A mechanical model of an amputee who wears the actively controlled prosthesis was developed and simulations of the man-machine system were performed. To obtain a reliable design a test rig is built for evaluating the knee controller performance. In this work gait testing of an amputee is performed and the human test results indicate that the amputee swing phase gait can be improved by the active knee controller based on either PD or fuzzy controls.

Introduction

The knee mechanisms of traditional above-knee (AK) prostheses can only control the motion of the artificial leg in a passive way, i.e., no feedback from the leg is provided to adjust the knee moment. This is different from our natural knee in that the knee moment is regulated by our motor control system according to the soma sensor feedback. Due to the loss of active knee control the AK amputee uses the compensation actions like circumduction and vaulting to maintain a minimum mechanical energy consumption in each gait cycle. With the passive control of a traditional AK prosthesis, the amputee can only walk with a limited range of cadence. Active control based on knee angular displacement and velocity feedback is necessitated for changing cadence.

Dyck suggests a design with four digital valves to adjust the knee damping force [1]. The electromyography (EMG) signals of the residual hip muscles are employed. Zarrugh and Radcliffe develop a mathematical model for the analysis of a four-bar knee which has a pneumatic damper [2]. Ishai and Bar build a micro-computer controlled artificial knee prosthesis [3]. They suggest the finite state approach and the controlled prosthesis can adapt for different walking speeds. Kognezawa and Kato design a knee mechanism which has two hydraulic cylinders, an accumulator and a control valve [4]. The prosthesis can provide multiple functions such as level walking, down-stair descending and up-stair climbing for an AK amputee. In these works only trial and error procedures are adopted to find out the best control strategy. Wang and Ju employed the adaptive control theory to develop an electro-hydraulic knee controller [5]. The control system can compensate for the variations of hip moment, hip trajectory and toe-off condition. However, their control design is complicated. Recently, Tsai and Ju developed a prototype electro-hydraulic above-knee prosthesis and a test rig is also built to evaluate the knee control system performance [6]. Their results indicated that by using a proportional plus derivative (PD) control better swing phase gait for amputee can be achieved. In existing two-dimensional simulations of amputee gait, the inputs for the model, i.e., the hip and knee moment trajectories are either calculated from the inverse dynamics of normal[7] or through the optimal control[8]. Ju et al [13] also developed a two-dimensional model for the prototype AK prosthesis developed in [6]. By utilizing the fuzzy control and the optimization technique the hip moment is generated by a fuzzy controller.

This work is an extension of previous works [6, 13] and the goals are threefold: first, the prototype electro-hydraulic AK prosthesis is fitted to an amputee and hardwares for gait testing of the subject are built. Second, gait test of the subject is performed and third, the controller designs reported in [6, 13] are realized for controlling the swing phase gait.

METHODS

In this section, a review of our previous works will be given followed by the experiment setup and procedures for human test of the prototype electro-hydraulic AK prosthesis.

Modeling of AK amputee-prosthesis system [13]

Figure 1 shows the configuration of the EH AK prosthesis system and the block diagram of the gait control system.

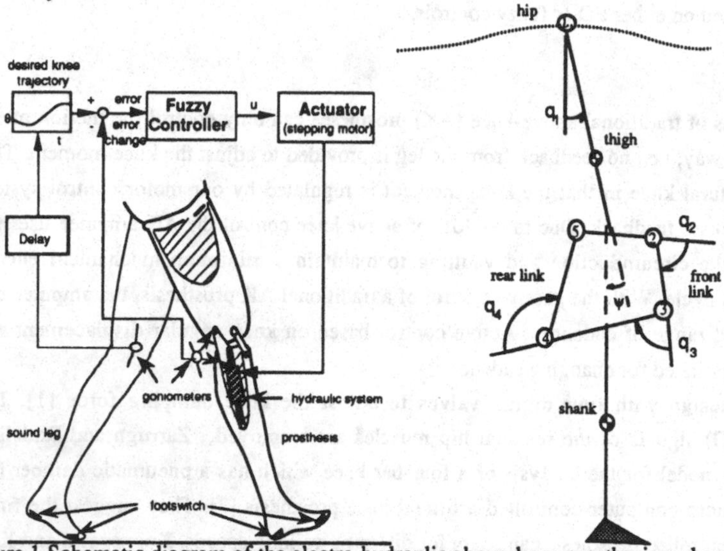

Figure 1 Schematic diagram of the electro-hydraulic above-knee prosthesis and mechanical model of the amputee-prosthesis man-machine system

A commercial four-bar knee is adopted for developing the prototype electro-hydraulic (EH) prosthesis [6]. A hydraulic cylinder installed between the rear linkage and the artificial leg of the prosthesis provides a damping force. Orifice of the puppet valve is adjusted by a micro-computer controlled stepping motor. Two foot switches are used to recognize the status of the gait, i.e., heel strike and toe-off conditions. The electronic goniometers measure the knee angles of the prosthetic knee. The Kane's Dynamic Equation [9] was employed to derive the equations of motion of the man-machine system as:

$$\mathbf{M}(\mathbf{q})\ddot{\mathbf{q}} + \mathbf{N}(\mathbf{q},\dot{\mathbf{q}}) = \boldsymbol{\Gamma}(t) + \mathbf{H}^T\boldsymbol{\lambda} \qquad (1)$$

where the generalized coordinate vector **q** contains the angular displacement of each link. **M** is the inertial matrix, N is a vector of Coriolis, centrifugal and gravitational forces and Γ is the generalized force vector. **H** is the Jacobian matrix associated with the holonomic constraints and **λ** is the Lagrange multiplier vector. The derivation of Eq. (1) are carried out using MACSYMA [10]. A detailed schematic diagram of the hydraulic circuit is shown in Fig. 2.

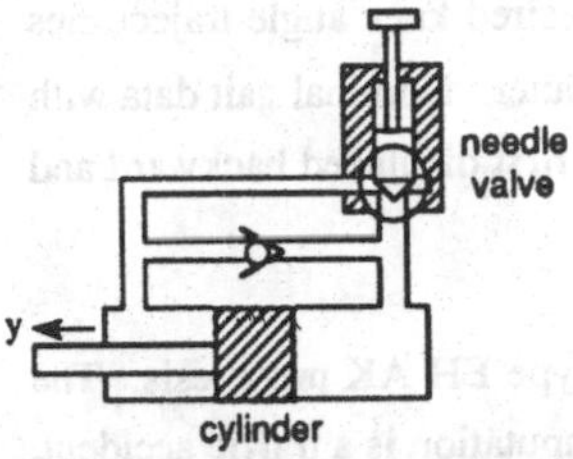

Figure 2 Detail of the hydraulic circuit

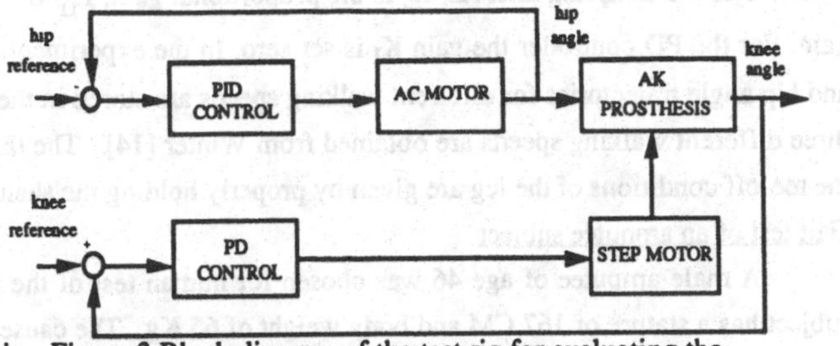

Figure 3 Block diagram of the test rig for evaluating the electro-hydraulic knee controller

The damping moment provided by the hydraulic system is found to be a nonlinear function of the puppet valve height and the relative speed between the piston rod and the cylinder. The damping torque can be written as:

$$\tau_4 = \left[\frac{\rho A^3}{2 C_d^2} \left(\frac{y}{a(x)}\right)^2 - C_k \dot{y} \right] c \cos(\phi - \theta) \qquad (2)$$

Detailed derivation of Eqs.(1)-(2) can be found in [6, 13].

Simulation of AK amputee-prosthesis system

By incorporating the damping moment term into the equations of motion the amputee-prosthesis system can be simulated on a digital computer. First, we define the state of the system as

$$x = \begin{pmatrix} q \\ \dot{q} \end{pmatrix} \qquad (3)$$

then Eq. (1) can be rewritten as

$$\dot{x} = \begin{pmatrix} \dot{q} \\ -M^{-1}N + M^{-1}\Gamma + M^{-1}H^T\lambda \end{pmatrix} \qquad (4)$$

where

$$\lambda = (H M^{-1} H^T)^{-1} [H M^{-1} (N - \Gamma) - \dot{H} \dot{q}] \qquad (5)$$

By using the Gear's method [11] Eqs. (4) and (5) can be integrated numerically. The input Γ is generated by the knee controller and the hip controller. The hip controller is a fuzzy controller which uses three triangular membership functions. By using the optimization technique suggested in [12] the optimal gains for the fuzzy controller and the initial knee velocity are obtained.

Test rig evaluation of knee controller [6]

For evaluating the performance of different control laws a test rig which can simulate the hip rotation of normal subjects and amputees was built (Fig. 3). The rig consists of a heavy platform of 300 kg weight and 1 m height and a computer-controlled alternating current (AC) servomotor system with a rated power of 5 hp. The EH prosthesis was fitted to an artificial thigh mounted on the rotating shaft. The AC servomotor system was controlled by a microcomputer to generate a hip angle trajectory close to that taken from normal gait or the amputee gait. The PD control law is employed to control the EH knee and the PID control is adopted for control of the AC servomotor. The transfer function of the PID controller is written as

$$D(z) = K_P + K_I \frac{T}{2} \left[\frac{z+1}{z-1}\right] + K_D \left[\frac{z-1}{Tz}\right], \qquad (6)$$

where T is the sampling interval, K_P is the proportional gain, K_D the derivative gain and K_I the integral gain. For the PD controller the gain K_I is set zero. In the experiment, the desired knee angle trajectories and hip angle trajectories for different walking speeds are stored in the computer. Nominal gait data with three different walking speeds are obtained from Winter [14]. The thigh is first displaced backward and the toe-off conditions of the leg are given by properly holding the shank.

Gait test of an amputee subject

A male amputee of age 46 was chosen for human test of the prototype EH AK prosthesis. The subject has a stature of 167 CM and body weight of 65 Kg. The cause of amputation is a traffic accident. Two custom-made foot switches are adhered to the heel and the toe of the artificial leg. Since the subject is not very active so only slow walking speed condition is tested at the beginning. The average of ten ensembles of knee angle trajectories of a normal subject is utilized as the reference input for the knee control system. The cadence is about 57 steps/min. The procedures for subject test are following:

(1) A vacuum type socket is fitted to the residual thigh of the subject. The superior edge of the socket is trimmed so it is comfort for the subject to wear.

(2) Only swing phase control is considered at this stage of experiment. With the aid of the foot switches the knee controller is triggered to action at heel strike.

(3) To maintain a constant walking speed the subject is asked to listen to the beat generated by a metronome and keep his cadence as close as possible to the desired speed.

RESULTS

Fig. 4 shows the controlled stick diagrams for fast walking speed (134 steps/min), natural walking speed (107 steps/min) and slow walking speed (89 steps/min). In all these simulations all the knee RMS errors are less than 3.2 degrees. The optimal initial knee velocities are 40% of the normal values for fast walking, 60% for natural walking and 70% for slow walking. The knee commands for fast and slow walking speeds are obtained from that of the normal by a scaling of time.

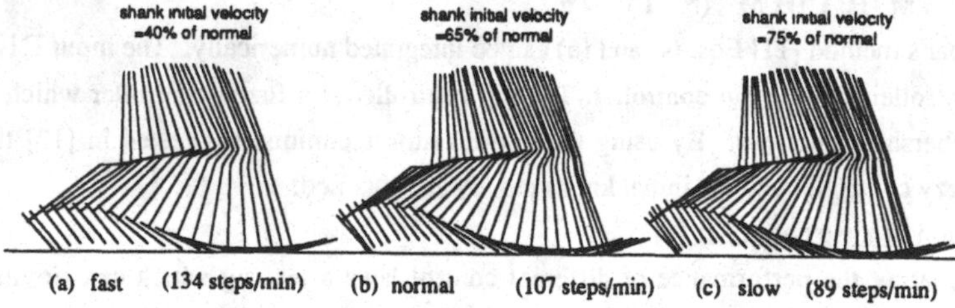

Figure 4 stick diagrams of the controlled gait at different walking speeds

Fig. 5 shows the controlled knee angle trajectories for different controllers at different walking speeds. Fig. 5(a) shows the controlled knee angle trajectory with a walking speed of 87 steps/min. This speed is the natural cadence for normal level walking. The RMS knee tracking error is 6.0 degrees for PD control, 11.9 for passive control with orifice half open and 14.2 degrees when the orifice is fully open. Among the three designs the PD control yields the smallest RMS tracking error. Fig. 5(b) shows a typical result for slow walking (75 steps/min). Note that the knee tracking errors of the various controls have a trend similar to that of the natural walking. The RMS error for PD controls are much smaller than the passive designs.

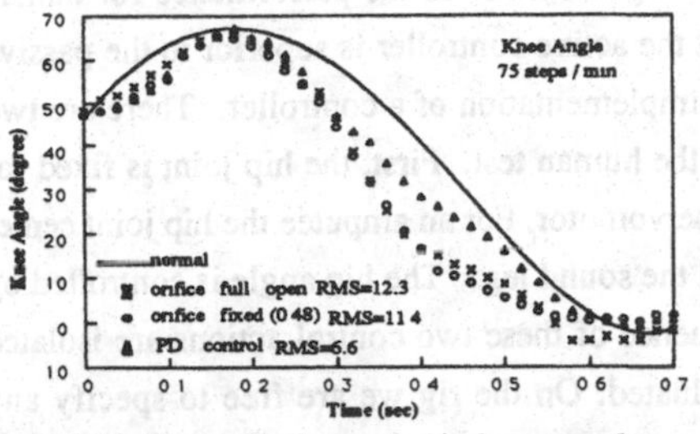

Figure 5(a) Controlled prosthetic knee angles at slow walking speed (experiment)

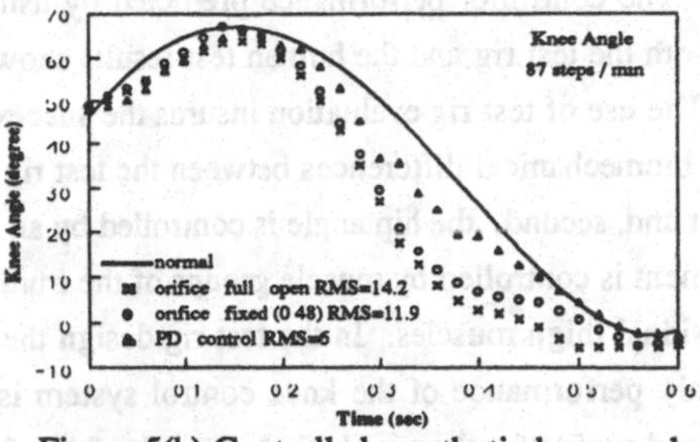

Figure 5(b) Controlled prosthetic knee angles at natural walking speed (experiment)

Fig. 6 compares the controlled prosthetic knee angles for the passive, P, PD and fuzzy controls. The walking speed is about 57 steps/min. The results have a similar trend to that of the test rig results. The RMS knee tracking error is 8.7 degrees for the passive control when the valve is fully open. For the PD control the RMS error is 3.3 degrees and for the fuzzy control the RMS error is 5.2 degrees. The results show that active controls are superior to passive controls.

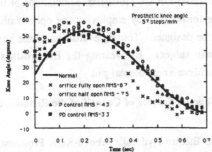

Figure 6(a) Human subject test results for passive control and PD control

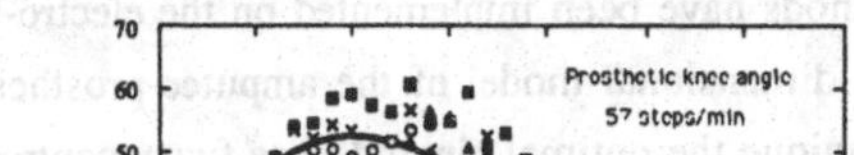

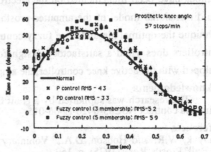

Figure 6(b) Human subject test results for fuzzy control and PD control

DISCUSSION

The simulated prosthetic knee angles are in good agreement with the normal gait. Although the equations of motion for the amputee-prosthesis system is nonlinear the fuzzy controllers do regulate the knee angle to track the normal knee angle. The optimization technique as suggested in [12] yields the optimal parameters for the hip controller, the knee controllers and the toe-off conditions. The optimal initial knee velocity does depend on the walking speed. For slow walking speed the initial knee speed is higher than that of the fast walking. From biomechanics point of view, higher initial knee velocity can maintain the swung leg in the air for a longer period of time. However, for an amputee this level of initial knee velocity may not be attainable. It depends on many factors, e.g., the length of stump, the power of the residual hip muscles and the degree of rehabilitation that an amputee achieved. Model simulation of the amputee-prosthesis system reveals that initial knee angular velocity is one of the dominant factors for gait normality. The use of optimization technique yields a fuzzy controller for generating the appropriate hip moment.

The controller performance predicted by using the rig is similar to the performance for human test. Both the test rig and the human test results show that the active controller is superior to the passive one. The use of test rig evaluation insures the successful implementation of a controller. There are two major biomechanical differences between the test rig and the human test. First, the hip joint is fixed for rig test and, second, the hip angle is controlled by an AC servomotor. For an amputee the hip joint center movement is controlled by muscle groups of the trunk and the sound leg. The hip angle is controlled by the residual thigh muscles. In the test rig design the influence of these two control actions are isolated and only performance of the knee control system is evaluated. On the rig we are free to specify any command profile for the prosthetic knee and safety of the subject can be guaranteed.

For this particular test, the PD controller performs better than the fuzzy controllers. Although further tuning of the fuzzy controller may improve its performance. In the human test the commands for the knee are taken from the averaged knee angle of a normal. The controlled knee angles are closer to the command than the test rig results. This may be that in the gait testing the human subject try to control not only the hip angle but also its vertical movement. The control actions improve performance of the knee controller.

In summary, a four-bar linkage above knee prosthesis has been developed. Active control methods have been implemented on the electro-hydraulic system of the prosthesis. Simulations of the two-dimensional model of the amputee-prosthesis are performed and by employing the optimization technique the optimal hip and knee fuzzy controllers can be designed. Test rig evaluation of the knee controllers does yield a satisfactory design for the human subject. By wearing the EH prosthesis equipped with the active knee controller, the amputee can achieve a near normal gait.

Acknowledgments

The research was supported by a grant from the National Science of Council of the Republic of China via contract NSC83-0420-E006-015.

References

1. Dyck, W.R. and Hobson, D.A., Voluntary Controlled Electro-Hydraulic Above-Knee Prosthesis, Bull. Prosth. Res., (1975), p. 169.
2. Zarrugh, M.Y and Radcliffe, C.W., Simulation of Swing Phase Dynamics in Above-Knee Prosthesis, J. of Biomechanics, Vol. 9 (1976), p. 283.
3. Bar, A. and Ishai, G., Evaluation of AK Prosthesis Comparing Conventional with Adaptive Knee Control Devices, Journal Biomed. Eng., Vol. 6 (1984), p. 27.
4. Koganezawa, K. and Kato, I., Control Aspects of Artificial Leg, in IFAC Control Aspect of Biomedical Engineering, edited by Maciej Nalecz, (1987), p. 71.
5. Wang, T. K., Ju, M.S. and Tsuei, Y.G., Adaptive Control of Above Knee Electro-Hydraulic Prostheses, Tran. ASME, Journal of Biomechanical Engineering, Vol. 114, n 3 (1992), p. 421.
6. Tsai, P. J., Ju, M.S. and Tsuei, Y.G. , Development of Electrohydraulic Controlled Above-Knee Prostheses, JSME International Journal, Series C, v 36, n 3 (1993), p. 347.
7. Tsai, C.S. and Mansour, J.M., Swing Phase Simulation and Design of Above Knee Prostheses, Tran. ASME, J. of Biomech. Engrg., Vol. 108 (1986), p.65.
8. Chao, E.Y.S. and Rim, K., Application of Optimization Principles in determining the Applied Moment in Human Leg Joint During Gait, J. of Biomechanics, Vol. 6 (1973), p. 497-510.
9. Kane, T. R., Spacecraft Dynamics, McGraw-Hill Book Company, (1983).
10. VAX UNIX MACSYMA Reference Manual, Version 11, Symbolics, Inc., (1985).
11. Ju, M.-S. and Mansour, J.M., Simulation of the Double Limb Support Phase of Human Gait, Tran. ASME, J. of Biomech. Engrg., Vol. 110 (1988), p.223.
12. Ju, M. S., Lin, J. H. and Chou, Y. L., Dynamical Optimal Design of Above-Knee Constant Friction Prostheses, ASME Advance in Bioengineering, (1989), p. 55.
13. Ju, M.S., Yi, S. H., Tsuei, Y.G. & Chou, Y.L., Fuzzy Control of Electro-Hydraulic Above-Knee Prostheses, submitted to JSME International Journal.
14. Winter, D.A., Biomechanics of Human Movement, John Wiley & Sons, N.Y. 1979.

INVESTIGATION OF TERRAIN ADAPTIVE CONTROL FOE A WALKING MACHINE MOTION

A.V. Bogutsky

Moscow State University, Moscow, Russia

Introduction.

The most appropriate application of walking machines is locomotion over difficult terrain. It demands an appropriate control system. Some designers try to build control systems providing locomotion of the vehicle in automatic mode using some wave gaits [2,3]. This approach presumes that information on the terrain is delivered by some visual or force sensors.

Another approach is based on the assumption, that suitable points for leg placement (footholds) are selected by the driver, while computer builds the robot motion and provides its execution [1,4]. In the work [4] it was assumed, that the vehicle moves using some gait from the "Follow-the-Leader Gaits" family. But it seems reasonable not to fix the gaits of a walking machine as it would limit its capability on a difficult terrain.

Following this approach, only the kinematic ability of the robot and the maintenance of the static stability will be regarded as criteria in route planning for a quasi-static motion. Under such conditions, the leg transfer sequence may be arbitrary. This method of motion is known as "free gait". This approach permits to find an optimal sequence of support and transfer phases for all legs. Moreover, it assures six-legged robot locomotion even if one or two legs are damaged.

We assume that it is not feasible today to develop an information system that would give the complete information about terrain. Therefore it is useful to construct a semi-automatic system. Our previous work [1] presented a step in this direction. It presented a control algorithm and its experimental verification. The work has shown that a problem of development of the operator-computer interface arise. This interface should provide a dialog between the operator and the control system, the main aim of which would be the selection of suitable set of footholds.

For this reason we develop a system of computer simulation of walking robot motion over preset footholds to study this control mode. We consider it as a part of control system, that represents the current situation. It gives us a possibility to study and to select an allowable motion over a given set of footholds if there are several possibilities. Finally, it is a tool, that enables us to study the impact of some parameters of a walking machine on its dexterity. This system can also be used as a training facility to learn the driver to estimate the terrain and to select footholds.

The control mode under consideration demands a large amount of calculations. Hence it does not seem feasible to provide motion planning in real time. So, it would be useful to have some hints to decide, when it is necessary to use this regime, and when it is not. This system could assist the driver to distinguish the situations, where it is necessary to use "free gait" approach and where it is not worth using it.

1. Description of a walking robot motion.

To describe the motion of a six-legged walking vehicle in the case when the robot uses footholds from the given set only, we use the notion of the support vector $Q=(q_1,q_2,q_3,q_4,q_5,q_6)$. Its component q_i equals k if the foot of the i-th leg is placed into the k-th foothold; q_i equals 0 if the i-th leg is raised $(i=1,...,6)$.

The goal of the motion can be formulated as follows: the state described by a support vector from the set $\bar{Q}$ of all possible final support vectors must be reached.

Organization of walking machine motion implies a construction of body and leg trajectories. If the legs during the transfer phase move with respect to the body along standard paths, the robot motion is completely specified by the prescribed body trajectory and by selection of footholds for supporting legs. Footholds for supporting legs may be given by a sequence of support vectors $\{Q_i\}_{i=0}^{n}$.

Let us consider that body inclination is changed in discrete instants only. So during motion planning we will suppose the orientation of the body to be constant over some time interval.

Because of the complexity of considered problem we solve it in the following way. First we explore the possible way of motion using a simplified model of the vehicle. Then computer verifies and specifies in details the preliminary solution using the full model. The first stage is the main and the most difficult. Therefore, below we consider only this stage. The main features of simplified model are: 1) legs are massless; 2) the leg working zones are convex.

2. Motion planning.

Let the legs be located in the footholds according to the support vector Q. Taking into account the finite leg working zones, the center of the body can be located at some point of a set E(Q). We will call this set of points maneuvre field. Using Q, we now define a support polygon as a set of points of a horizontal plane bounded by a closed broken line with nodes in the footholds specified by Q. Let us denote a set of the points such that the projection of any point onto a horizontal plane belongs to the support polygon together with its ε-vicinity, as support cylinder C(Q).

During the walking, when only points from the given set are used, one support vector is replaced by another at the t_i instants. During the motion planning, a problem of verification is unavoidable: whether the walking machine can perform the transition from the Q_i state to the Q_j state or not. If it is possible, such support vectors will be called stably connected.

Under condition of simplified walker model, the criterion of stable connection can be expressed in the following form [5]:

$$E(Q_i) \cap E(Q_j) \cap C(Q_{ij}) \neq \varnothing, \qquad (1)$$

where: Q_{ij} - is a vector whose components satisfy the condition $q_l^{ij}=0$, if $q_l^i \neq q_l^j$; $q_l^{ij}=q_l^i$, if $q_l^i=q_l^j$ $(l=1,...,6)$. Let us denote Q_{ij} as $Q_i \cap Q_j$.

For convex leg working zones a broken line will be an admissible body center path if all adjacent support vectors in the sequence $\{Q_i\}_{i=0}^{n}$ are stably connected.

Since the initial support vector Q_0 and some set of footholds have been given, we can construct a connected graph $G(Q_0)$ of support vectors starting from node Q_0. This graph is characterized by the following properties: 1) any two nodes connected by an edge are stably connected; 2) if Q_i is a node of $G(Q_0)$, then $G(Q_0)$ contains all Q_j stably connected with Q_i. Therefore if even one element from $\bar{Q}$ is a node of $G(Q_0)$, the selected set of footholds will provide a desirable motion of the vehicle and the task of motion planning can be easily completed. The case when $G(Q_0)$ does not include any vector from $\bar{Q}$ is of greatest interest. Then the initial set of footholds must be corrected.

3. Extension of footholds set.

Similarly to the graph $G(Q_0)$, we can construct a $G(\bar{Q})$ graph starting from $\bar{Q}$ set. In the case under consideration, $G(Q_0)$ and $G(\bar{Q})$ have no common nodes, otherwise there would be a support vector from $\bar{Q}$ among the nodes of the $G(Q_0)$ graph.

We shall perform a correction of the set of footholds by adding new ones to those available. Addition of footholds is to be done so that any vector Q_1 from $G(Q_0)$ and any vector Q_2 from $G(\bar{Q})$ be connected by a path in the $G(Q_0)$ graph completed after new footholds have been added. Simultaneously Q_1 and Q_2 are connected in the completed graph $G(\bar{Q})$. In so doing, it is reasonable to consider such pairs of support vectors (Q_1,Q_2) that Q_1 and Q_2 have the least number of different components for the particular $G(Q_0)$ and $G(\bar{Q})$.

The presented method of correction of a set of footholds is based on a criterion that permits the selection of a new foothold K numbered with k making the following transitions possible:

$$\begin{gathered} Q_1=(\ldots \overset{j}{0} \ldots \overset{i}{m_1} \ldots) \longrightarrow Q_1^*=(\ldots \overset{j}{k} \ldots \overset{i}{m_1} \ldots) \longrightarrow Q_1^{**}=(\ldots \overset{j}{k} \ldots \overset{i}{0} \ldots), \\ Q_2=(\ldots 0 \ldots m_2 \ldots) \longrightarrow Q_2^*=(\ldots k \ldots m_2 \ldots) \end{gathered} \tag{2}$$

Here $m_1 \neq m_2$, $m_1 \neq 0$, $q_j^1 = q_j^2 = 0$, $q_i^1 = m_1$, $q_j^{*2} = q_j^{*1} = q_j^{**1} = k$, $q_i^{*1} = m_1$, $q_i^{**1} = 0$, $q_i^2 = q_i^{*2} = m_2$, where $q_l^1 = q_l^{*1} = q_l^{**1}$ and $q_l^2 = q_l^{*2}$, if $l \neq i$ and $l \neq j$.

It can be seen easily that to achieve a desirable goal, it is sufficient to select a k point providing the feasibility of chains (2).

Assume that $U_j(Q)$ is the set of all the points into which the end of the j-th leg can be placed in the case when the robot is in the state Q (meanwhile the center of the body may be located in any point of the $E(Q) \cap C(Q)$ set). Now let us formulate the criterion to be satisfied by the K point:

The K point ensures the feasibility of the pair of chains (2) if and only if the following conditions are satisfied:

$$\begin{gathered} K \in U_j(Q_1) \cap U_j(Q_2), \\ E(Q_1^*) \cap C(Q_1^{**}) \neq \varnothing. \end{gathered} \tag{3}$$

■ **Proof.** Let us prove sufficiency. Due to the first condition of (3) the transitions $Q_1 \rightarrow Q_1^*$ and $Q_2 \rightarrow Q_2^*$ are possible. It can be seen easily that if robot puts one of its leg on the ground, its maneuvre field becomes smaller. In the case of Q_1^{*} and Q_1^{**} it means the following: $E(Q_1^{*}) \cap E(Q_1^{**}) = E(Q_1^{*})$. Obviously $Q_1^{*} \cap Q_1^{**} = Q_1^{**}$. Therefore, the criteria of stable connection of Q_1^{*} and Q_1^{*} can be expressed in the form: $E(Q_1^{*}) \cap C(Q_1^{**}) \neq \emptyset$. So, (3) means that Q_1^{*} and Q_1^{**} are stably connected.

Let us prove necessity. As transitions $Q_1 \rightarrow Q_1^{*}$ and $Q_2 \rightarrow Q_2^{*}$ are possible then $K \in U_J(Q_1)$ and $K \in U_J(Q_2)$, i.e. $K \in U_J(Q_1) \cap U_J(Q_2)$. Transition $Q_1^{*} \rightarrow Q_1^{**}$ is possible if and only if $E(Q_1^{*}) \cap E(Q_1^{**}) \cap C(Q_1^{*} \cap Q_1^{**}) \neq \emptyset$. Because $E(Q_1^{*}) \cap E(Q_1^{**}) = E(Q_1^{*})$ and $Q_1^{*} \cap Q_1^{**} = Q_1^{**}$ relation $E(Q_1^{*}) \cap C(Q_1^{**}) \neq \emptyset$ is true. ■

The verification of the condition (3) can be reduced to the solution of a system of inequalities. Unfortunately it is impossible to find a complete set of K points using this criterion. But it is useful to supply to the operator the information on where it is necessary to look for a new foothold. For this reason computer constructs some region, all points of which satisfy necessary (but not sufficient) conditions for some point obey conditions (3). This region is an estimation for the set of points satisfying (3).

4. Results.

So, we have a formalized procedure for correction of initial set of footholds. Namely, when graphs $G(Q_0)$ and $G(\bar{Q})$ are constructed, computer selects the pairs $(Q_1 \in G(Q_0),\ Q_2 \in G(\bar{Q}))$ of support vectors having the least number of different components. Then it proposes to the operator regions where to select new footholds which would provide sequential realization of transitions of type (2).

The key point of the considered approach is that information on the terrain is represented only as a set of footholds. So, we simulate a terrain this way i.e., as a set of columns standing on some reference plane, the height of a column being the vertical position of foothold. The system also provides some additional information such as body position and its velocity, footholds coordinates.

Figure below presents an example. There the distance between footholds 1 and 6 is only a little bit less than the length of vehicle body. It is supposed that it is impossible to find any additional footholds between them. The walker should move from the state $Q_0=(1,2,0,0,3,4)$ to some state, where the first and the second legs would occupy the 14th and the 15th footholds respectively. Footholds 16, 17 and 18 have been added on demand of the control system. This example demonstrates the situation, when using the "free gait" is the single possibility.

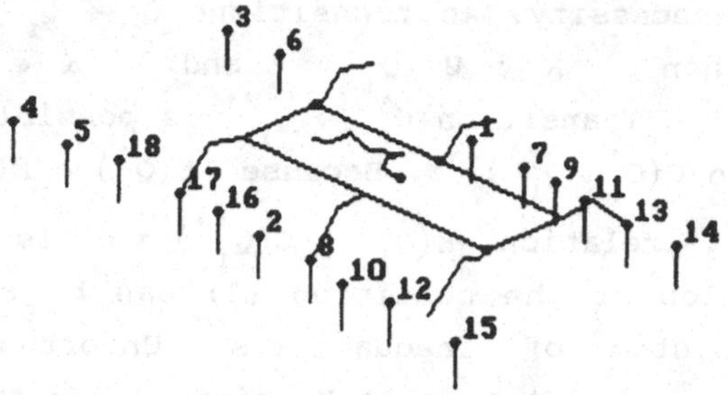

References.

1. Bogutsky A.V. 1993, Control of a walking machine on extra-complex terrain. The Second European Control Conference, V.3, pp. 1261-1264. Groningen, the Netherlands.

2. Devjanin E.A., Zhitomirskij S.W., Zhiharev D.N., Lensky A.V., Shtilman L.G., Gurfinkel V.S., Gorinevsky D.M., Gurfinkel E.V., Grishin A.A., Shneider A.Yu. 1987, Control of Adaptive Walking Robot. The 10th World Congress on Automatic Control. IFAC, Munich. V.4, pp.218-225.

3. Hirose S. 1984, A Study of Design and Control of a Quadruped Walking Vehicle. //The Int. Journal of Robotics Research. The MIT Press. V.3, N.2. pp. 113-134.

4. Ozguner F., Tsai S.J., McGee R.B. 1984, An Approach to the Use of Terrain-Preview Information in Rough-Terrain Locomotion by a Hexapod Walking Machine. //The Int. Journal of Robotics Research. The MIT Press. V.3, N.2. pp. 134-146.

5. Procopchuk Yu.A. 1986, Organization of Walking Machine Motion on Given Footholds. Ph.D. thesis. Moscow State University. (In Russian).

DESIGN OF THE PROGRAM REGIME FOR BIPED WALKING ANTHROPOMORPHIC APPARATUS

V.M. Budanov and E.K. Lavrovsky
Moscow State University, Moscow, Russia

Extensive literature exists on the dynamics and control of two-legged walking apparatuses, and their mechanical design. Many works contain theoretical studies [1, 2], some experimental works are also known [3-9]. Several prototype biped walkers and algorithms of their control have been designed. Some of them have been tested experimentally. Unlike multi-legged walking apparatuses that can move within the static stability, a biped with uncontrollable feet can maintain dynamic stability only. This makes the control of bipeds difficult. Two apparatuses were designed at the Institute of Mechanics of the Moscow University: one having telescopic legs [8-10] and another of anthropomorphic kinematic scheme [10].

In the present paper we consider an algorithm of locomotion control for one of them, namely, with anthropomorphic legs. The designed algorithm is based on the tracking of the commanded trajectory. They are planned in advance using a mathematical model of the apparatus. The mathematical model includes nonlinear equations of double- and single-support motion of mechanism. But the model doesn't account for a possible impact as the planned trajectories do not presume any impacts. Another feature of our algorithm is the programming of nominal motion in terms of cartesian mass center coordinates. The nominal trajectory is programmed in terms of the motion of the mass center which is close to a rectilinear uniform motion.

Mechanical design. The considered apparatus [10] is represented in the Fig.1. Its total mass is about 8 kgs. The apparatus is actuated by four DC drives with reduction gears. Two DC drives turn the thighs relative to the torso. Two other drives affect the knee angles. There are planar rotational joints in the pelvis, knees and ankles. Thus the apparatus can perform planar motion in sagittal plane only. The feet are attached to the shins by a passive (uncontrolled) rotary joint. They extend in the frontal plane, thereby the stability of the motion in the sagittal plane is ensured.

The kinematic scheme of the apparatus is represented in the Fig.2. The position of the apparatus is defined by five generalized coordinates - angles with the vertical (generalized angles) $-\psi$, α_1, α_2, β_1, β_2. The lengths of thighs and shins are about 0.25 m, the height of the torso is about 0.2 m. We calculated the coefficients in the equations of the dynamical model using values of inertia moments, inertia radii and coordinates of mass center for each link. To control the apparatus walking it is necessary to know the generalized angles and their derivatives. These signals are read from four potentiometers and tachometers in all controlled joints as well as from two potentiometers in ankles. Each shin is additionally equipped with a two-component force sensor that enables to estimate ground reactions applied to the feet.

Control system. We control the biped walking using a PC AT-286/287 having an ADC board for measurement of sensor signals. Power amplifiers for DC drives are controlled through the parallel output port. The control algorithm is based on the servoing of program trajectories formed for each controlled interlink angle of the apparatus. We calculate the nominal trajectory in terms of ψ, α_1, α_2, β_1, β_2 and then convert them into the interlink angles $\delta_1=\psi-\alpha_1$, $\delta_2=\psi-\alpha_2$, $\delta_3=\alpha_1-\beta_1$, $\delta_4=\alpha_2-\beta_2$. The servoing of the programmed trajectories is software-implemented. For a comfortable work with laboratory-scale model we employ an interactive software package. It assures the testing of each drive and of the whole apparatus, real-time control of the walking as well as the recording of various experimental data for later analysis.

Concept of a nominal program regime. We design the mathematical model to form a nominal trajectory of apparatus motion. This model includes the complete non-linear equations of the planar motion of the apparatus. We take into account the DC drive masses as well as the inertia of their rotors (but not of the gears). Electric characteristics of motors are also accounted for in the model.

The found periodic solutions consist of alternating phases of single and double support. In the beginning and at the end of the double-support phase acceleration and deceleration take place; these intervals are treated in a special manner. The single- to double-support phase transition is planned to occur without any impact.

To describe both phases it is convenient to use the mass center coordinates x_c, y_c together with cartesian coordinates of the endpoint of the leg being transferred. Here and later we denote by "x" and "y" the horizontal and the vertical cartesian coordinates in an absolute frame. The origin of the frame is in the support point of one of the legs. Specifying these coordinates as functions of time, we prescribe the motion.

To consider single-support phase we introduce the variables

$$\mu_1 = x_1 - x_c, \quad \mu_2 = x_2 - x_c, \quad \mu_3 = y_c - y_1, \quad \mu_4 = y_c - y_2$$

Substituting the x's and y's by their evident expressions containing the abovementioned angles ψ, α_1, α_2, β_1, β_2 and differentiating these relations, we get

$$(\dot{\mu}_1, \dot{\mu}_2, \dot{\mu}_3, \dot{\mu}_4)^T = A(\dot{\psi}, \dot{\alpha}_1, \dot{\alpha}_2, \dot{\beta}_1, \dot{\beta}_2)^T \qquad (1)$$

Here A is some matrix (4X5) that depends on mass geometry and the generalized angles; the values with subscript "1" relate to the supporting leg and the values with subscript "2" to the leg being transferred; "T" means transposition. If the current configuration of the apparatus is known, equations (1) present four relations containing five unknown angular velocities. To find the missing relation we consider the equation of momentum for the whole dynamic system relatively to the point of support

$$dK/dt = - Mg\,\mu_1 \qquad (2)$$

Here g is the gravity acceleration, M is total mass of the apparatus. The right-hand side of (2) is a function of time. Momentum K on the left can be presented as a function of angles ψ, α_1, α_2, β_1, β_2 and angular velocities in the following form

$$K = c_1\dot{\psi} + c_2\dot{\alpha}_1 + c_3\dot{\alpha}_2 + c_4\dot{\beta}_1 + c_5\dot{\beta}_2 \qquad (3)$$

Here coefficients "c" depend on the generalized angles and mass geometry of the apparatus. We can consider the equations (1)-(3) as the system for the determination of six unknown time functions $\dot{\psi}$, $\dot{\alpha}_1$, $\dot{\alpha}_2$, $\dot{\beta}_1$, $\dot{\beta}_2$ and K over the single-support phase. Its solution corresponds to the motion defined in terms of x and y.

Nominal trajectory for the single-support phase is built as

follows. Having selected a double-support configuration, corresponding to the beginning of the single-support phase and set an initial value of $\dot{\psi}$, using (1), (3) we evaluate the initial momentum K . After that integration of (1), (3) yields the single-support trajectory in state-variables α_1, α_2, β_1, β_2.

Similarly to above, for the double-support phase let us define a program trajectory in terms of cartesian coordinates x_c, y_c and angle ψ. Let us denote:

$$\mu_1 = \psi,\ \mu_2 = x_2 - x_c,\ \mu_3 = y_c - y_2,\ \mu_4 = (a \sin\alpha_1 + b \sin\beta_1) - (a \sin\alpha_2 + b \sin\beta_2),\ \mu_5 = (a \cos\alpha_1 + b \cos\beta_1) - (a \cos\alpha_2 + b \cos\beta_2) \qquad (4)$$

where a and b are lengths of thigh and shin of leg. Here subscript "2" corresponds to the forward leg, therefore $x_2 = y_2 = 0$. It can be seen easily, that functions μ_4 and μ_5 are in fact constants, as they describe the horizontal and vertical displacements of one foot relative to another. Differentiating both sides of all the equations (4), we get a system of linear algebraic equations (5x5), similar to (1). Further, double-support trajectory in state-variables α_1, α_2, β_1, β_2 can be obtained similarly to the single-support phase.

Testing of algorithm. The proposed algorithm was initially implemented for the following program regime. Single-support phase was defined as

$$x_c = -x_c^0(\sigma_c t/T - 1),\ y_c = a_* \sin^2\omega t + y_c^0$$
$$x_2 = \sigma - 0.5\ \ell \cos\omega t,\ y_2 = b_* \sin^2\omega t \qquad (5)$$

where t is current time, T - the single-support phase duration, $\omega = \pi/T$; σ , σ_c, ℓ, a_*, b_*, x_c^0 , y_c^0 - parameters, while ℓ is the stride length of leg being transferred, x_c^0 and y_c^0 are initial values of coordinates at the beginning of single-support phase; non-zero values of σ and σ_c lead to trajectory asymmetry; a_* is altitude of center mass vertical motion; b_* is the maximal height of foot transfer path. It should be pointed out that (5) assure a zero-impact foot placement of the transferred leg onto the surface.

The double-support phase was defined by means of three functions ψ , x_c, y_c having the form of a fourth-order polynomials in time

$$a_1 + a_2(T-t) + a_3(T-t)^2 + a_4(T-t)^3 + a_5(T-t)^4$$

Here T is the duration of the double support. It is selected as to make the horizontal motion of mass center x_c as close to uniform as possible. Coefficients a_1 - a_5 are imposed by four boundary conditions as state coordinates and velocities must coincide at t=0 and t=T with those selected for the single-support phase. One of the coefficients may be selected arbitrarily to make the motion more attractive. It should be noted that in the case when the acceleration or deceleration is performed during the double-support phase, the boundary conditions and the number of arbitrary coefficients "a" do not change, but velocities at the beginning or at the end of phase are equal to zero.

Proposed algorithm has been simulated beforehand and best nominal trajectories were recorded for the subsequent implementation on the laboratory-scale model. These trajectories include the stage of acceleration , the sequence of alternating single and double-support phases and the deceleration stage. At the same time we considered the complete dynamic model, resolved it to find six or eight (depending on the motion phase) unknown variables of the support reactions and torques in the joints. Thus it becomes possible to check whether there is a contact between the supporting legs and the surface (positive vertical reactions in the supporting legs), to find required drive control voltages , and to calculate extreme values of interlink angles (as the apparatus design restricts them). We also pay attention to the attractive character of the walk. The free parameters of model are chosen as to satisfy these various conditions.

We carried out the experiment on the laboratory-scale model using the simulated trajectory. Initially it was shortened i.e. the trajectory consisted of an acceleration stage (0.7 s long), a single-support phase (step size 0.2 m, 0.4 s long) and a deceleration stage (0.7 s long). The apparatus motion confirms the validity of dynamic model and control algorithm on the whole. However to realize the simulated periodic walking requires some improvements . So it is necessary to specify a more detailed dynamic model. It might be necessary to introduce new types of feedback to eliminate the influence of random perturbations arising during the walking.

References

1.Beletsky V.V. Biped walking. Model problem of dynamics and control. M., "Nauka", 1984 (in Russian).

2.Formal'sky A.M. Locomotion of the anthropomorphic apparatus. M:"Nauka", 1982 (in Russian).

3.Kato T.,Takanishi A.,Naito G., Kato I. The realization of the quasi dynamic walking by the biped walking machine. Proc. of Int. Symp. on Theory and Practice of Robots and Manipulators ROMANSY, 1981.

4.Mita T., Yamaguchi T., Kashiwase T., Kawase T. Realization of high speed biped using modern control theory. Int.J. of Control 1984, V.40, No 1.

5.Miuro H., Shimoyama I. Dynamic walk of biped. Int.J. of Robotics Research, 1984, V.3, No 2.

6.Yamada M., Furusho J., Sano A. Dynamic control of walking robot with kickaction. Proc. of ICAR, Tokyo, 1985.

7.Furusho J., Mashubushi M. Control of dynamical biped locomotion system for study walking. Trans. ASME: J.Dyn.Syst.Meas. and Contr. 1986, V.108, No 2.

8.Grishin A.A., Zhitomirsky S.V., Lensky A.V., Formal'sky A.M. Walking of biped vehicle with two telescopic legs. Izv. AN USSR , Tech. kibern., No 2,1991 (in Russian).

9.Grishin A.A., Formal'sky A.M., Lensky A.V., Zhitomirsky S.V. Control of a biped vehicle with telescopic legs. Proc. of the Second Eur.Contr.Conf., Groningen, 1993, V.3, p.1265- 1268.

10.Devjanin E.A. et al. Control of adaptive walking robot. Preprints 10-th World Congress IFAC. Munich, FRG, V.4, p. 218-225.

The study was partially supported the Fundamental Research Fund of Russian Academy of Sciences.

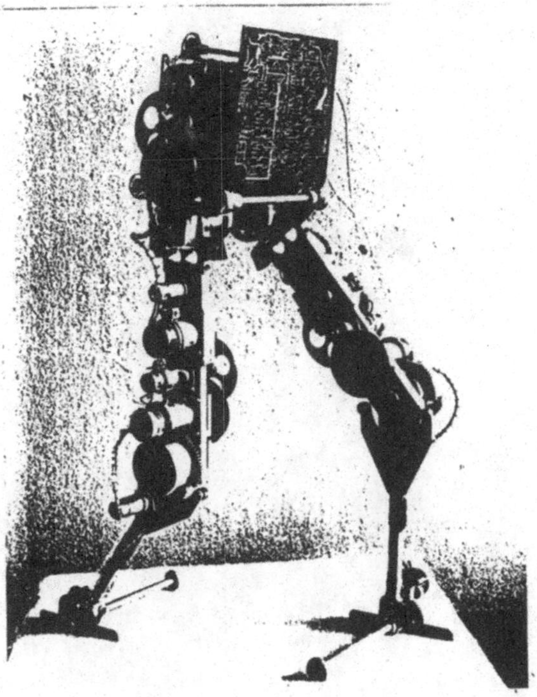

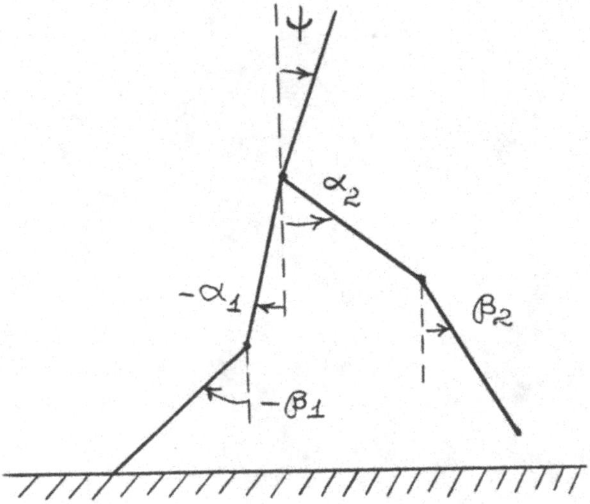
ψ
α2
-α1
β2
-β1

IMPULSIVE CONTROL FOR ANTHROPOMORPHIC BIPED

A.M. Formalsky
Moscow State University, Moscow, Russia

Introduction

Many authors study the dynamics of two-legged walk theoretically. They are, for instance, Beletsky, Berbjuk, Bolotin, Farnsworth, Frank, Gubina, Hemami, Larin, Morecki, Moreinis, Novozhilov, Schiehlen, Seireg, Townsend, Vukobratovich, Zackiorski.

Most researchers investigate the dynamics of two-legged walk using the inverse or semi-inverse approach. They prescribe the motion of the biped, fully or partly, and then, using the motion equations, find the forces applied to the biped, evaluate the consumed energy.

Several control algorithms for dynamic walk of two- and one-legged mechanisms have been designed and tested experimentally (Kato et al, Katoh and Mori, Miura and Shimoyama, Raibert, Furusho and Mashubushi, Furusho and Sano, Zheng and Sias, McGeer, Kajita et al).

In the present theoretical investigation direct method of biped locomotion design is used.

Biped and its motion equations

Model description.

We study the mathematical model of biped walk in sagittal plane. The investigated model of biped contains the torso and two identical legs. Each leg consists of the thigh and the shin (Fig. 1). All these five links are massive and absolutely rigid. We neglect the friction in the hip and knee joints considering them as ideal. Note, that in human joints the friction is very small.

A seven-link model with massless feet added is considered a well.

We describe the position of the considered biped mechanism in the plane *XY* by seven generalized coordinates: x, y, ψ, α_1, α_2, β_1, β_2 (Fig. 1).

Let q_1 and q_2 be the torques of the forces acting between the trunk and thighs, u_1 and u_2 be the torques of the forces in the knee joints *B* and *D*, Π_1 and Π_2 be the torques in the ankle joints *A* and *E*, $\mathbf{R}_1(R_{1x}, R_{1y})$ and $\mathbf{R}_2(R_{2x}, R_{2y})$ be the forces applied to the leg tips *A* and *E* (Fig. 2). Similarly to the human walk, the walk modeled here consists of alternating phases of single and double support. In the double-support phase both legs are on the bearing surface (on the ground), the forces $\mathbf{R}_1$ and $\mathbf{R}_2$ are the support reactions, and they are non-zero. During the single-support motion, one of reactions $\mathbf{R}_1$, $\mathbf{R}_2$ is zero.

Equations of planar motion of five-link biped.

The equations of the planar motion of the described biped mechanism can be obtained by the second Lagrange method. Omitting the intermediate calculations, we write the equations in the matrix form

$$B(z)\ddot{z} + gF\,\sin z_i + D(z)\,\dot{z}_i^2 = C(z)Q \tag{1}$$

where

$$z = \|x,\ y,\ \psi,\ \alpha_1,\ \alpha_2,\ \beta_1,\ \beta_2\|^*$$

$$\sin z_i = \|0,\ 1,\ \sin\psi,\ \sin\alpha_1,\ \sin\alpha_2,\ \sin\beta_1,\ \sin\beta_2\|^*$$

$$\dot{z}_i^2 = \|0,\ 0,\ \dot{\psi}^2,\ \dot{\alpha}_1^2,\ \dot{\alpha}_2^2,\ \dot{\beta}_1^2,\ \dot{\beta}_2^2\|^*$$

$$Q = \|u_1,\ u_2,\ q_1,\ q_2,\ \Pi_1,\ \Pi_2,\ R_{1x}, R_{1y},\ R_{2x},\ R_{2y}\|^*$$

Here the asterisk denotes transpose. The symmetrical, positive definite matrix of kinetic energy *B(z)* is of the size (7×7), matrices *F*, *D(z)* and *C(z)* are of the sizes (7×7), (7×7) and (7×10) respectively and *g* is the gravity acceleration.

To describe the motion during the single or double support phase this system must be complemented by constraint equations, defining the conditions for fixation of one or both feet on the support.

Single-support motion.

The scheme of the mechanism standing on one leg is shown in the Figure 3. The tip *A* is motionless on the bearing surface. We can assume that it is connected to the surface by an ideal joint.

But during the locomotion the bearing surface doesn't keep the supporting leg. Therefore the ground reaction in the point *A* must be directed upwards. In the single-support phase, the biped has five degrees of freedom and five general coordinates: ψ, α_1, α_2, β_1, β_2 (Fig. 3).

Now we write the equations of single-support motion excluding the reaction force in supporting leg. We can write these equations in matrix form too

$$H(\zeta)\ddot{\zeta} + gL \sin\zeta_1 + M(\zeta)\, \dot{\zeta}_1^2 = N(\zeta)W \qquad (2)$$

where

$$\zeta = \zeta_1 = \psi, \alpha_1, \alpha_2, \beta_1, \beta_2^{*}$$

$$\sin\zeta_1 = \sin\psi, \sin\alpha_1, \sin\alpha_2, \sin\beta_1, \sin\beta_2^{*}$$

$$\dot{\zeta}_1^2 = \dot{\psi}^2, \dot{\alpha}_1^2, \dot{\alpha}_2^2, \dot{\beta}_1^2, \dot{\beta}_2^{2\,*}$$

$$W = u_1, u_2, q_1, q_2, \Pi^{*}$$

Here we denote by Π the torque in the ankle joint *A* (Fig. 4). No force is applied to the tip *E* of the leg being transferred. The symmetrical, positive definite matrix of the kinetic energy *H(ζ)* and matrices *L*, *M(ζ)*, *N(ζ)* are of the sizes (5×5). Having solved the equations (2) we can check whether the supporting leg loses contact with the surface or slips.

Statement of the problem

Reasons.

When formulating this problem the following considerations were taken into account. Some authors (Bernstein, Bogdanov and Gurfinkel, Gurfinkel and Fomin, Mochon and McMahon, McGeer, Vitenzon) suppose that in many human motions muscle activity alternates with some period of relaxation. Apparently, the motions of that kind are less energy-consuming. During human walk the considerable efforts are acting in the double-support motion on the whole (Vitenzon). The motion of the leg being transferred, for instance, is very close to that of a free pendulum (Gurfinkel and Fomin). Therefore we believe that in the single-support motion of our model there is no active torques in joints, but they are acting in double-support phase. Of course, during the single-support phase the ground reaction in supporting leg is non-zero. But no active forces are acting, and we call this single-support motion ballistic.

Thus we consider a ballistic, in other words, free single-support motion, when active forces are zeros, that is $W(t) = 0$. The equations of ballistic single-support motion can be obtained from (2)

$$H(\zeta)\ddot{\zeta} + gL \sin\zeta_1 + M(\zeta)\, \dot{\zeta}_1^2 = 0 \tag{3}$$

Single-support phase.

At the beginning of the step ($t = 0$) let the biped be in the configuration shown in Fig. 5 left (position i). It is described by some vector

$$\zeta(0) = \psi(0),\ \alpha_1(0),\ \alpha_2(0),\ \beta_1(0),\ \beta_2(0)\ ^{*} \tag{4}$$

Assume that in this configuration the front and hind legs are both on the surface. For $t > 0$ the front leg is supporting and hind leg is being transferred.

We want to transfer the biped into the final configuration shown in Fig. 5 right (position f) in a fixed time $t = T$. It is described by vector

$$\zeta(T) = \psi(T),\ \alpha_1(T),\ \alpha_2(T),\ \beta_1(T),\ \beta_2(T)\ ^{*} \tag{5}$$

We assume that the final configuration coincides with the initial one but the legs are swapped

$$\psi(T) = \psi(0),$$

$$\alpha_1(T) = \alpha_2(0),\ \alpha_2(T) = \alpha_1(0),$$

$$\beta_1(T) = \beta_2(0),\ \beta_2(T) = \beta_1(0)$$

If on the single-support phase there is no active forces, the biped can be transferred from the given initial configuration to the final one by selection of initial values for angular velocities of links only. Thus it is required to find the suitable vector $\dot{\zeta}(0)$ of angular velocities. In other words, the problem of ballistic single-support motion design is formulated mathematically as a boundary-value problem for the equation (3). It is required to find the solution $\zeta(t)$ of this equation under boundary conditions (4) and (5).

The problem as stated above leaves no means to ensure that the leg tips move above the support, the legs bend "knee forward", the torso don't fall down.

We formulated above the first part of the problem statement. After the single-support phase comes the double-support one and the second part is the following.

Double-support phase.

During the human double-support motion, there is a transition of the support from one leg to another. The double support time in human locomotion is less than 20 per cent of the whole step period. In our study the double support time is assumed infinitely small, that is, the double-support phase is regarded as instantaneous.

Assume that T is the instant when the double-support phase occurs. At that instant a constraint is imposed on the transferred leg, that is, its tip E becomes stationary with respect to the surface. This results in an impulsive reaction (impact), and the velocities of the biped links get instantaneously changed. In addition to this constraint let us apply impulsive efforts in the joints, *i.e.*

$$Q(t) = I\delta(t-T) \tag{7}$$

where δ-function is non-zero for $t = T$ only and

$$I = I_{u_1}, I_{u_2}, I_{q_1}, I_{q_2}, I_{\Pi_1}, I_{\Pi_2}, I_{R_{1x}}, I_{R_{1y}}, I_{R_{2x}}, I_{R_{2y}}$$

is vector of intensities of impulsive actions (weights of δ-functions). We hope that under impulsive control (7) the desired link velocities can be achieved just before the beginning of the next single-support phase.

The relations between the velocity jumps and intensities of impulsive efforts can be obtained easily from equation (1)

$$B\big(z(T)\big)[\dot{z}] = C\big(z(T)\big)I \tag{8}$$

Here $[\dot{z}] = \dot{z}(T+0) - \dot{z}(T)$ - vector of velocity jumps.

Let the boundary-value problem formulated above be solved. Then the velocities vectors $\dot{z}(0)$ and $\dot{z}(T)$ at the start and the end of the single-support phase are known.

We assume that the bearing surface is a plane. Let the gait of the biped be regular, that is its single-support motions coincide on all the steps. Then the equality

$$\dot{z}(T+0) = \dot{z}(0)$$

holds. Hence the vector of velocity jumps $[\dot{z}]$ can be computed. From the relation (8) we can find vector I required to produce desired link velocities at the beginning of the next single-support phase.

Results

We investigate the boundary-value problem formulated above both analytically and numerically. It has some symmetry properties which are important to reveal the general pattern of the gait. In the case of linearized motion equations the solution is found analytically. It is unique for any boundary configurations (4) and (5). For the complete nonlinear boundary-value problem a numerical solution can be found using an iterative procedure. It should be noted that several solutions are possible for the same boundary configurations (4), (5) and the time T.

Having solved the boundary-value problem for the model with anthropomorphic parameter values, we use the display to estimate the pattern of the synthesized locomotion. At first, we look at the solutions for linearized equations of the five-link model (without feet), then at the solution for complete nonlinear equations of five-link model and finally at the solution for nonlinear equations describing the seven-link model (with massless feet).

We can choose the configurations (4), (5) and time T so that the leg tips move above the support, the mechanism moves with knees forward, the torso doesn't fall and slightly oscillates near the vertical position once per step. It is necessary to point out that the reactions are directed upwards both in the single and in the double support. It is important to note that these "human" features of the gait are not inferred by the problem statement in itself. Hence these are the intrinsic features of ballistic motion. The designed ballistic gaits are in some sense human-like. This resemblance gives additional hints to suppose that human walk contains some intervals of ballistic movement as well.

For the model with feet we estimated the work of control forces performed during one step. It amounts near 70 J. This value is less than the energy consumption estimated by others authors for different (non-impulsive) controls.

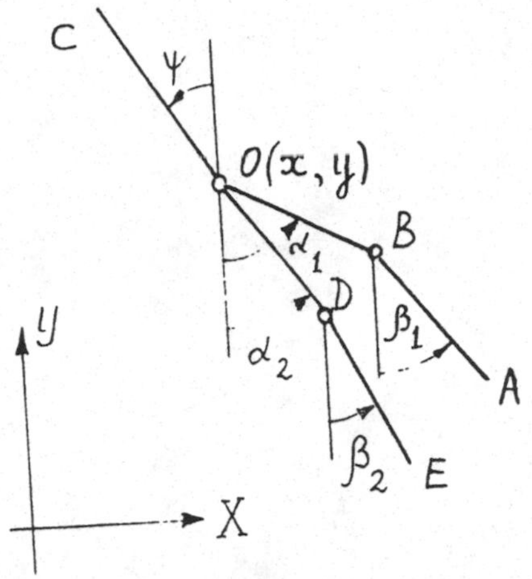

Fig.1

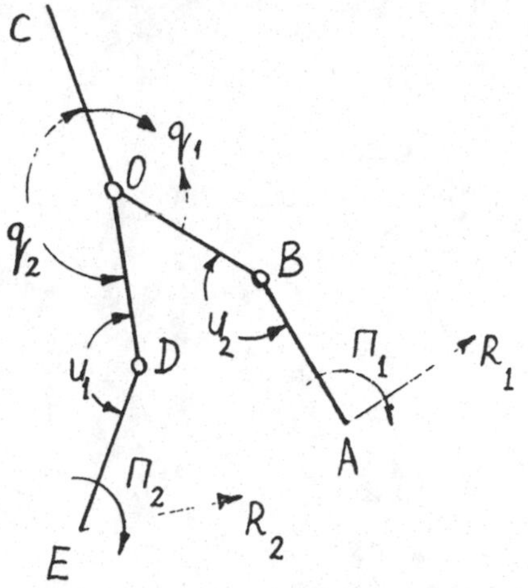

Fig.2

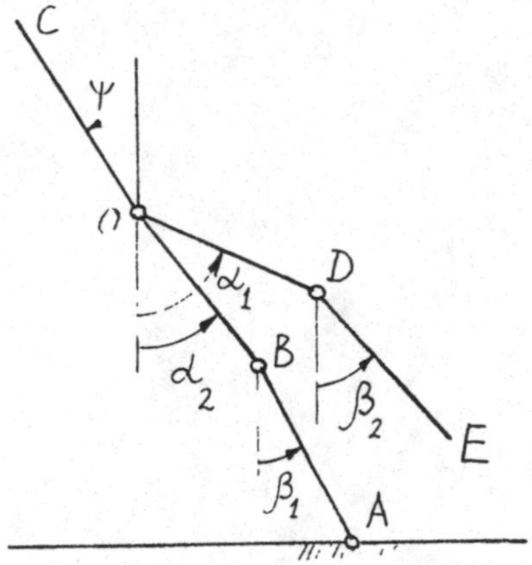

Fig.3

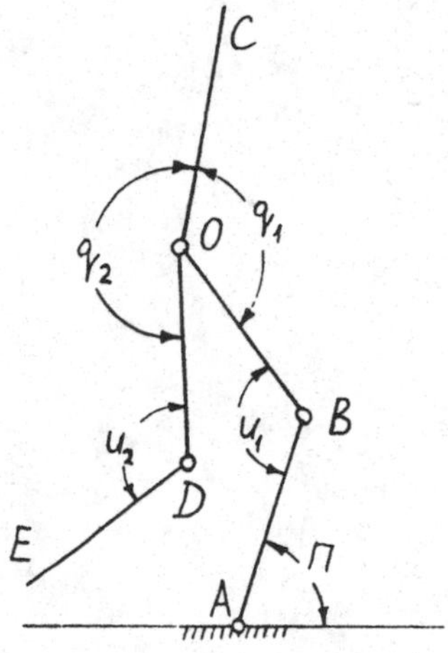

Fig.4

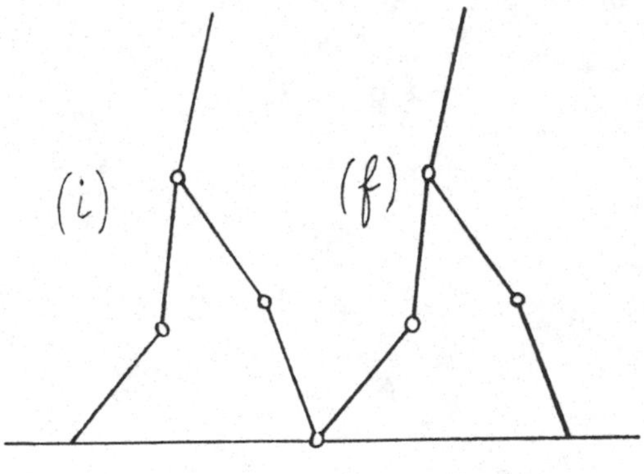

Fig.5

ESTIMATING POWER OF WALKING: METHOD AND INSTRUMENTATION

K. Jaworek and A. Ferenc

Warsaw Institute of Technology, Warsaw, Poland

ABSTRACT

The paper presents an unconventional method of assessing the average power developed in the three main leg joints, and the hip joint separately, of a normal middle-aged and young man, during walking.
Up till now measurement and calculation of parameters of human gait has required very expensive devices.
The method enables the exclusion of very expensive measuring equipment, produced only in highly developed countries, from the power assessment process.
Using this method as a foundation, a new type of wireless electrogoniometer for dynamic multi-step human bipedal locomotion under field conditions was build. Currently investigations generalising the proposed method are under way.

1. INTRODUCTION

During walking power is developed by the human leg muscle actuators of the hip, knee and ankle joint. The paper presents the method of assessing the average power developed in the three main leg joints together and in the hip joint separately during foot support phase. The methods were verified experimentally using a very expensive human motion recording and analysis system named ELITE-3D installed in SAFLO·

*/ This project is supported by the KBN (the State Committee for Scientific Research) in Warsaw - project nb. PB0674/P4/92/03

laboratory in Centro di Bioingegneria in Milan [1]. The proposed assessment methods are relatively simple and do not require expensive equipment manned by a lot of operators. It can even be operated by the person whose motion apparatus is assessed [2].

2. METHOD

2.1. ASSESSMENT OF THE AVERAGE POWER DEVELOPED IN THE HIP, KNEE AND ANKLE JOINT.

Up till now to assess the average power developed by the muscles actuators driving the hip, knee and ankle joint during the foot support phase formula (1) has been used [1,5,6].

$$P_{2-D} = \sum_{i=1}^{3} \sqrt{\frac{1}{T_-} \int_{t_1}^{t_2} (M_i \, \omega_i)^2 \, dt} \quad , \text{ watts} \qquad (1)$$

where:

M_i - muscle moment developed in the i-th joint,

ω_i - angular velocity of the i-th joint axis,

$T_- = t_2 - t_1$ - duration of the foot support phase,

i - subscript indicating the hip, knee or ankle joint.

Using Winter's monograph [3] the average power developed by the muscles actuators driving the hip, knee and ankle joint during the foot support phase for normal and middle-aged individuals of both sexes was determined. A similar procedure was followed to assess the gait of normal elderly individuals of both sexes [4].

Next using Jaworek's indices method [1,2,5,6] of human motion during walking the power indicator P_{2-D} for a constant gait velocity can be described by:

$$P_{2-D} = 2.1 \; m \, \bar{v} \, f \quad , \; m^2 kg \, s^{-3} \qquad (2)$$

where:

m - mass of the body of examined individuals,

$\bar{v}$ - is the absolute value of the vector describing the average velocity of motion of the examined individuals;

$\bar{v} = |\vec{v}|$ - where vector $\vec{v}$ is bound to the hip joint axis,

f - is the gait frequency calculated from the formula f=n/t where n- number of double steps, t-duration of walking [1]

unit 2.1 is m/s.

2.2. ASSESSMENT OF THE AVERAGE POWER DEVELOPED IN THE HIP JOINT.

From formula (1), by substituting i=H for a hip joint we obtain the following formula:

$$P_H = \sqrt{\frac{1}{T_-} \int_{t_1}^{t_2} (M_H \omega_H)^2 \, dt} \quad . \qquad (3)$$

Formula (3) is not very useful for differentiating the hip joints of individuals with masses or velocities differing a lot.
Formula (3) was modified in the following way:

$$\epsilon_H = \frac{P_H}{m \bar{v}} , \quad \frac{J}{kg\ m} \qquad (4)$$

where:
m - mass of the body of the examined individual,
$\bar{v}$ - is the absolute value of the vector describing the average velocity of motion of the examined individuals;
$\bar{v} = |\vec{v}|$ - where vector $\vec{v}$ is bound to the hip joint axis.
It was assumed that the indicator ϵ_H will be called the motion indicator of the order 2, for the motion of an individual in the sagittal plane [1,2,5,6]. Formula (4) can be transformed into the following form:

$$\epsilon_H = 0.5 \frac{P_H}{\bar{E}_K} \bar{v} , \qquad (5)$$

If we assume that:

$$k_H = 0.5 \frac{P_H}{\bar{E}_K} , \quad s^{-1} \qquad (6)$$

then the formula (5) assumes the following form:

$$\epsilon_H = k_H \bar{v} , \qquad (7)$$

where:
k_H - is called motion coefficient of a hip joint.
Basing on the power indicator P_{2-D} it is possible to obtain the hip joint motion indicator in the form:

$$\eta_H = \frac{P_H}{P_{2-D}} \ \% . \qquad (8)$$

If we associate equation (8) with formulas (2) and (6) we obtain:

$$\eta_H = 0.476 \frac{P_H}{m f \bar{v}} \ \% , \qquad (9)$$

where: $[0.476] = s\ m^{-1}$.

2.3. APPLICATION OF LOWER ORDER PHASE DIAGRAMS IN THE ASSESSMENT OF THE VALUE OF POWER INDICATOR P_H.

Using Winter's monograph [3] the hip power indicator P_H divided by the examined individual's mass was determined.
Table 1 presents the numerical values of the indicator P_H/m, W/kg, for the hip joint.

TABLE I. Values of the normalized power indicator for the hip joint (data quoted by Winter [3]).

$\bar{v}$ m/s	1.097 ± 0.05	1.33 ± 0.05
Δt s	0.0265± 0.001	0.0348± 0.001
P_H/m W/kg	0.30 ± 0.05	0.695 ± 0.055

Utilizing the data from the table of instantaneous power $p_H(t)$ and moment $M_H(t)$, and the formula for the instantaneous power developed in the hip joint:

$$p_H(t) = M_H(t)\ \omega_H(t)\ , \qquad (10)$$

the angular velocity of the hip joint axis $\omega_H(t) = \dot{\phi}_H(t)$ was determined.
Utilizing the sampling period Δt from the table 1 and the formula prposed in [6] the angular acceleration of the hip joint axis $\varepsilon_H(t)=\ddot{\phi}_H(t)$ was determined too.
Next the phase diagrams of lower order [6] were made for the free and fast gait of young and middle-aged individuals.
Figures 1 (a) and (b) present the phase diagrams $\ddot{\phi}_H(\dot{\phi}_H)$ of lower order for young and middle-aged individuals. The shape of the diagrams and the position of the small loop in the phase plane results from the Winter's method of determining the angular position of segments.

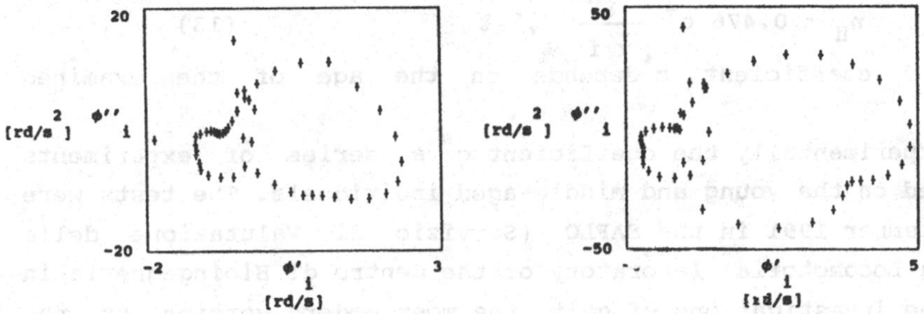

Fig.1. The hip phase diagram $\ddot{\phi}_H(\dot{\phi}_H)$ of the gait of young and middle-aged individuals.
(a) - normal motion,
(b) - fast motion.

Table II contains data pertaining to the analyzed gaits and phase diagrams.

TABLE II. Values of kinematic and geometric parameters for young and middle-aged individuals (according to Winter and Jaworek [3,6]).

$\bar{v}$ m/s	1.097 ± 0.05	1.33 ± 0.05
f Hz	0.89 ± 0.05	0.975 ± 0.05
T s	0.842 ± 0.02	0.733 ± 0.02
S^H rd^2/s^3	7.35 ± 0.04	21.68 ± 0.03

The calculations and the inspection of small loops in the hip joint phase diagrams $\ddot{\phi}_H(\dot{\phi}_H)$ enable us to notice that:

- the larger the gait velocity and frequency the larger is the area S^H of the small loop in the phase diagram,
- the larger the gait velocity and frequency the larger is the normalized power P_H/m for the hip joint.

The above two observations and the values contained in tables I and II enabled the formulation of the following relationship:

$$P_H = c^* \, m \, S^H \quad , W \qquad (11)$$

where: the coefficient $c^* = 0.03 \ m^2 \ rd^{-2}$.

Formula (11) and formula (6) enable us to write down the hip joint motion coefficient in the following form:

$$k_H = c^* \frac{S^H}{\bar{v}^2} \quad , s^{-1} \qquad (12)$$

The formula (8) assumes the following form:

$$\eta_H = 0.476\, c^* \frac{s^H}{\bar{v}\, f} \quad , \quad \% \qquad (13)$$

The value of coefficient c^* depends on the age of the examined individual.

To verify experimentally the coefficient c^* a series of experiments was performed on the young and middle-aged individuals. The tests were done in September 1991 in the SAFLO (Servizio di Valutazione della Funzionalita Locomotoria) laboratory of the Centro di Bioingegneria in Milan. In the investigations of gait, the most modern version of the ELITE-3D system adopted to clinical conditions was used [1]]. The sampling frequency of the motion phases was 50Hz.

Table III contains the kinematical and dynamic gait parameters of a normal man aged 45, with a mass of 76.6 kg.

TABLE III. Table of kinematic and dynamic parameters of the examined individual computed using exact formulas.

$\bar{v}$ m/s	f Hz	T s	s^H rd^2/s^3	P_H/m W/kg	P_{2-D}/m W/kg	k_H s^{-1}	η_H %
1.56	1.02	0.579	37.801	1.092	3.17	0.449	34.45

Table III enabled the evaluation of the coefficient c^* (see formula (11)), which for the examined individual is equal to:

$$c = 0.0289\ m^2 rd^{-2}.$$

Kinematical and dynamic data contained in the table III and the examined individual's hip joint phase diagram $\ddot{\phi}_H(\dot{\phi}_H)$ (see fig.2) were computed using the formulas (1), (3), (6) and (8). In the calculations it was assumed that the coefficient c^* has the value $c=0.0289\ m^2 rd^{-2}$.

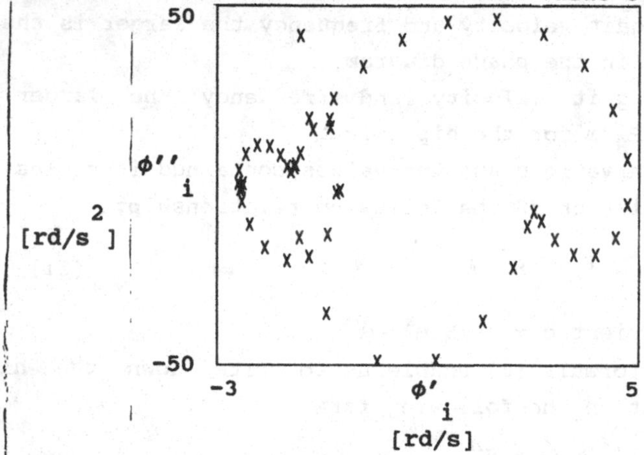

Fig.2. Examined individual's hip joint phase diagram; angle ϕ_H is measured between trunk of the body and thigh of the examined individual.

3. INSTRUMENTATION - A CORDLESS ELECTROGONIOMETER

3.1. A MECHANISM NAMED THE MECHANISM "POLITECNICO DI TORINO"

A small metal box approx. 40x40x8mm contains a winding mechanism.

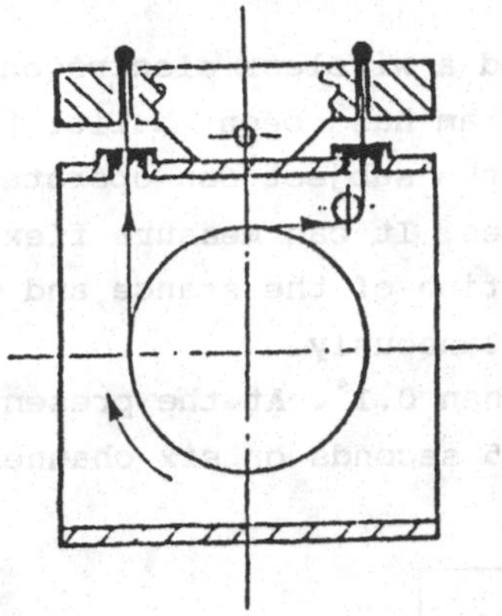

Fig.3. The "Politecnico di Torino" mechanism.

On its rotating part a plastic pulley is mounted and this pulley is able to accommodate two thin nylon wires. One of these wires leaves the pulley and directly passes through a very small hole drilled on the upper side of the box. The other follows a slight longer route, bending around a pin a leaving the box at another hole drilled on the upper side. The wires outside the box are permanently secured to the bar. The bar is positioned exactly parallel to the line joining the upper side of the box, and which is fixed to the pin of a potentiometer.

Since the two wires are simultaneously wound on the same pulley and in the same sense, in a plane motion the bar remains always parallel to the box, thus constituting a variable length parallelogram. The fact that two long arms of the parallelogram are made in wire allows also free displacement of the plane of rotation of the mechanism, and also angular movements in torsion, thus giving all the necessary degrees of freedom.

In order to reduce hysteresis problems, and to increase the life of the wires, the holes on the upper part of the box were provided with aluminum oxide thread guides of the type used in the textile industry [7,8].

3.2 THE ACQUISITION SYSTEM.

The proposed method of estimating the average power developed by man during motion may significantly simplify the measuring equipment that has been used up till now in the investigations of human bipedal locomotion.

For using the proposed method a wireless electrogoniometer of a new kind with elastic parallelogram has been built [2]. The device is designed in such a way that the subject can operate it himself after capturing human gait sequences. It can measure flexion angeles for a knee and hip joints and duration of the stance and the swing phases of the foot on both limbs simultaneously.

Angular resolution is less than 0.1°. At the present configuration the measurement can last up to 15 seconds on six channels at sampling rate of 100 Hz (Fig.4).

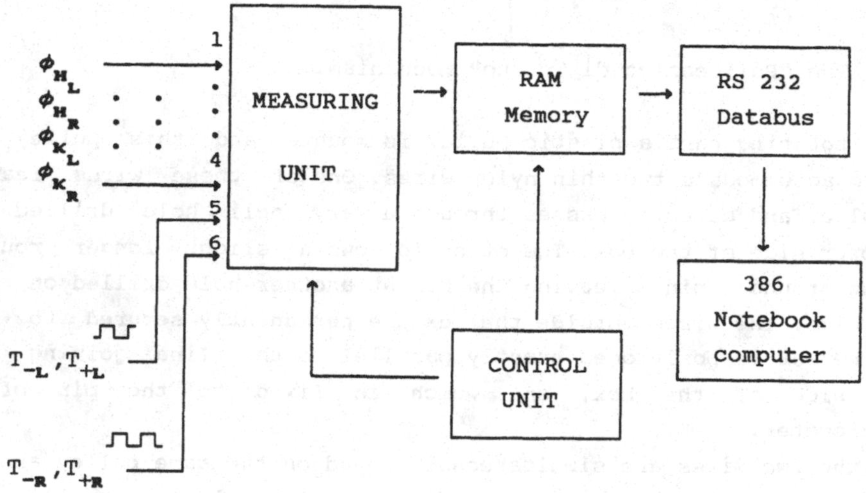

Fig.4.The block diagram of a wireless electrogoniometer.

The portable acquisition box enables the data to be transferred to a computer for analysis and storage via an RS232 Databus.

4.RESULTS

A series of experiments were performed on the young and middle-aged individuals of both sexes. The tests were done from February to May 1993 at Warsaw University of Technology. In the investigations of gait, a wireless electrogoniometer adopted to clinical conditions was used.

Table IV contains the kinematical and dynamic gait parameters of a normal woman aged 24, with a mass of 62 kg. The dynamic values contained in the table IV were computed using the formulas (2), (11), (12) and (13).

In the calculation of the power P_H it was assumed that the coefficient c^* has the value $c^*=0.03\ m^2 rd^{-2}$.

TABLE IV. Table of kinematic and dynamic parameters of the examined a hip joint individual using approximate formulas [$\bar{v}$=1,32m/s (±0.05), f≅0.9695Hz (±0.01), m_B=64kg (±0.5)].

Number of step	T_- s	S^H rd^2/s^3	P_H W	k_H s^{-1}	η_H %
1	0.721	16.44	30.57±0.43	0.283±0.03	9.17±0.1
2	0.721	11.08	20.59±0.35	0.191±0.03	6.18±0.1
3	0.721	26.74	49.73±0.59	0.460±0.03	14.92±0.1

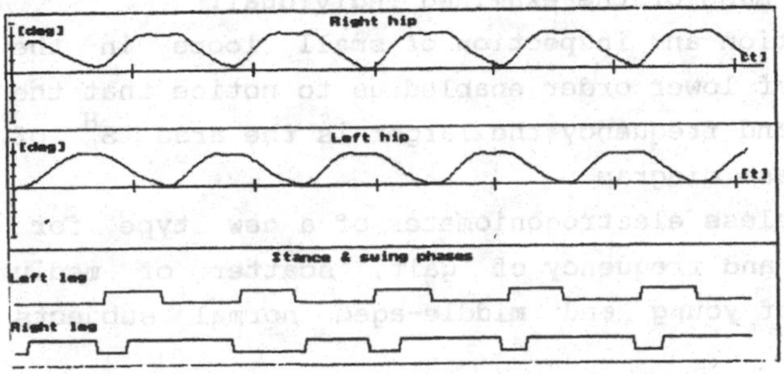

Fig.5. Recording flexion of the hip joint during normal walking of a woman with a relatively constant speed and frequency of walking.

Figure 5 shows the recording for a hip flexion. The recording examplifies the difference in variability between the steps.

All data were plotted versus time. Plots were also made of joint angular velocities versus time. It allowed to plot a phase diagram of first order $\ddot{\phi}_H(\dot{\phi}_H)$ for a hip joint.

A typical phase diagram of the hip joint during normal walking of a woman with a relatively constant speed and frequency of gait is shown on the figure 6.

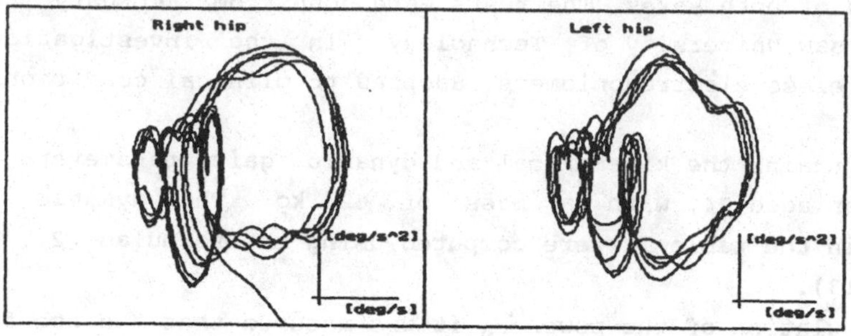

Fig.6. A typical phase diagram of the hip joint obtained for several steps during normal walking of a woman.

5.CONCLUSIONS AND DISCUSSION

1. The hip joint phase diagram of lower order enabled the determination of the average power developed in this joint during the foot support phase as a function of kinematic and geometric parameters as well as the mass of the examined individual.
2. The calculation and inspection of small loops in the hip joint phase diagram of lower order enabled us to notice that the larger the gait velocity and frequency the larger is the area S^H of the small loop in the phase diagram.
3. Using a wireless electrogoniometer of a new type for a relative constant speed and frequency of gait, scatter of medium power P_H between steps of young and middle-aged normal subjects have been observed.
4. The proposed method of assessing the average power developed by man at the hip joint during motion may significantly simplify the measuring equipment that has been used up till in the investigations of human bipedal locomotion.

REFERENCES

[1] Jaworek K.: Methods of evaluation of the power developed in a man's leg during normal walking. Proc. of 9th CISM-IFToMM Symp. on Theory and Practice of Robots and Manipulators, Ro.Man.Sy-9, CISM, Udine, Sept 1-4, 1992, ed. by Springer-Verlag, London, 1993, pp.322-332.

[2] Jaworek K., Ferenc A.: A simplified assessment's method of human gait. Proceedings of the 2nd Polish-Italian Seminar, April 20-21, 1993, edited by G.Belingardi & F.A.Raffa, Politecnico di Torino, Torino, Italy, 1993, pp.89-96.

[3] Winter D. A.: The Biomechanics and Motor Control of Human Gait. Ed. by Waterloo University, Canada, 1987.

[4] Jaworek K., Olney S.: Gait Assessment and Human Motion Indicators of the Elderly Persons. Proc. of the 8th World Congress on the TMM, Prague, August 26-31, 1991, vol.V, pp.1455-1457.

[5] Jaworek K.: Indices of human motion during walking and running. Abstract Book of XIII th International Congress on Biomechanics, Perth, Western Australia, December 9-13, 1991, pp.351-352.

[6] Jaworek K.: Indices method of assessing human gait and run. D.Sc. Dissertation. Ed. by International Center of Biocybernetics, Warsaw, Poland, 1993 (in polish).

[7] Gola M. M., Guglietta A.: An improved experimental device for simplified gait analysis. Int.Conf.Exper.Mech., Beijing, 1985.

[8] Gola M. M., Guglietta A.: An improved experimental device for simplified gait analysis. Mat. Ogólnopolskiej Konf. Biomechaniki, Gdańsk, Poland, July 22-24, 1987, pp.631-639.

CAD/CAM AND ROBOTICS: A BIOENGINEERING APPLICATION IN BONE PROSTHESIS REALIZATION

X. Wen, A. Rovetta and K. Lucchesi Cavalca
Polytechnical University of Milan, Milan, Italy

ABSTRACT. This paper deals with the application of CAD/CAM to the realization of an artificial bone for a prosthesis. The first prototype is in construction at the Politecnico di Milano, Department of Mechanics, Laboratory of Robotics, using a C.A.T.I.A. system and an industrial IBM robot with special mills.

1. Introduction.

In a surgical intervention, the pieces of bone necessary to the patients prosthesis are always made by handiwork. Moreover, the material is recovered from another part of their own bodies.

The biomaterials offer the possibility of avoiding the heavy and so long handiwork, arising the velocity of the bone reconstruction, which also allows a better operation safety. The development of the computer science technology and robotics can support the fulfilment of reconstructing a piece of bone which can replace a missing or defective one in the patient body.

The human body is one of the most complex known structures. In several cases, the robot flexibility and versatility permit to face the problem of forming a very complex shaped object. By using the results already verified industrially, a series of applications were developed for surgery.

The measured data can be translated, modified and adapted to the C.A.T.I.A. software format. On this way, a piece of human bone can be formed and the output file is sent to a robot for construction.

Figure 1 shows a scheme of the steps to perform a piece of bone and transfer the data to the robot for the construction.

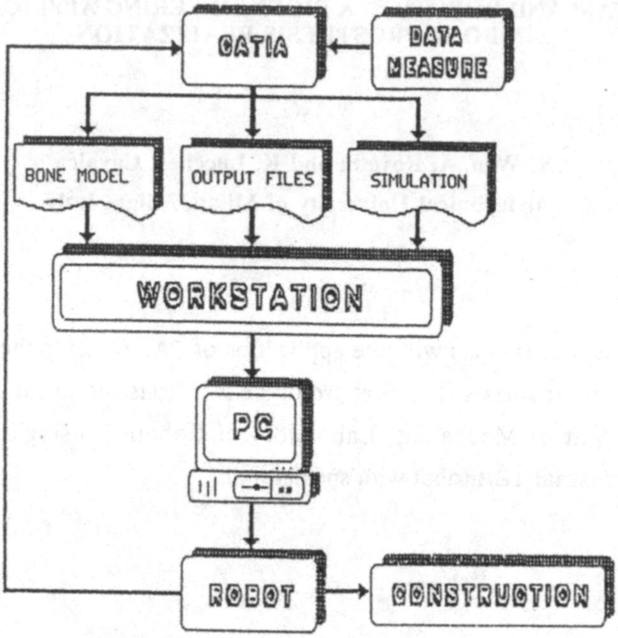

Figure 1. Diagram of the procedure.

2. The Bone Model.

Normally, bones have very complicated shapes and to form them it is necessary to apply a high-level surface module. The C.A.T.I.A. software is used to prepare an human piece of bone, putting this output file and the simulated robot model together and sending the whole system to the robot for construction.

2.1. The Bone Surface Performance.

A piece of bone which will replace of a defective or absent bone is measured. The measured data are transferred in the C.A.T.I.A. software by the POINT function using the coordinates system. Afterwards, applying the SPLINE function, a series of curves are settled on these points (figure 2), defining the piece of bone surface, which allows to create a volume or a solid element.

Figure 3 shows a piece of animal bone surface performed by the C.A.T.I.A. software.

2.2. The Robot Trajectory.

When the bone surface is defined, the robot trajectories on it can be established by dividing the bone surface with a series of planes, like shown in figure 4.

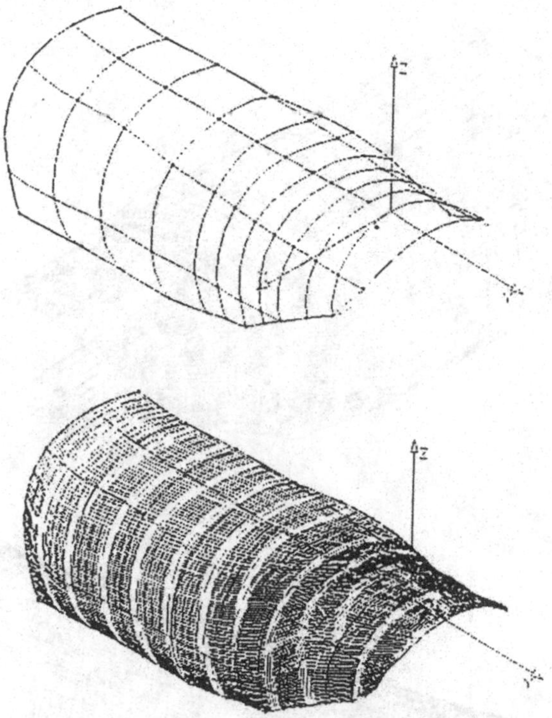

Figure 2. Spline curves settled on the generated points.

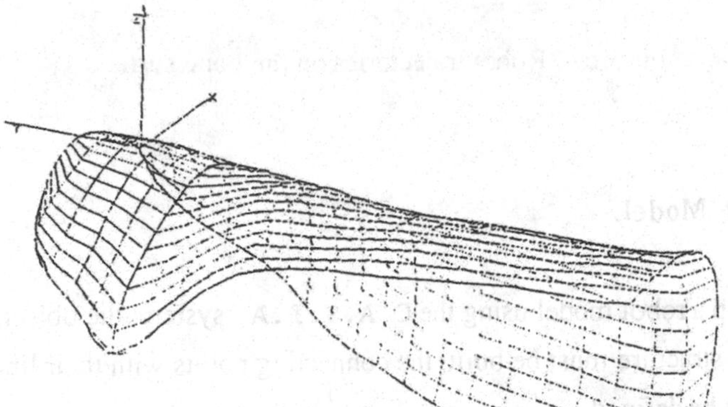

Figure 3. Bone surface.

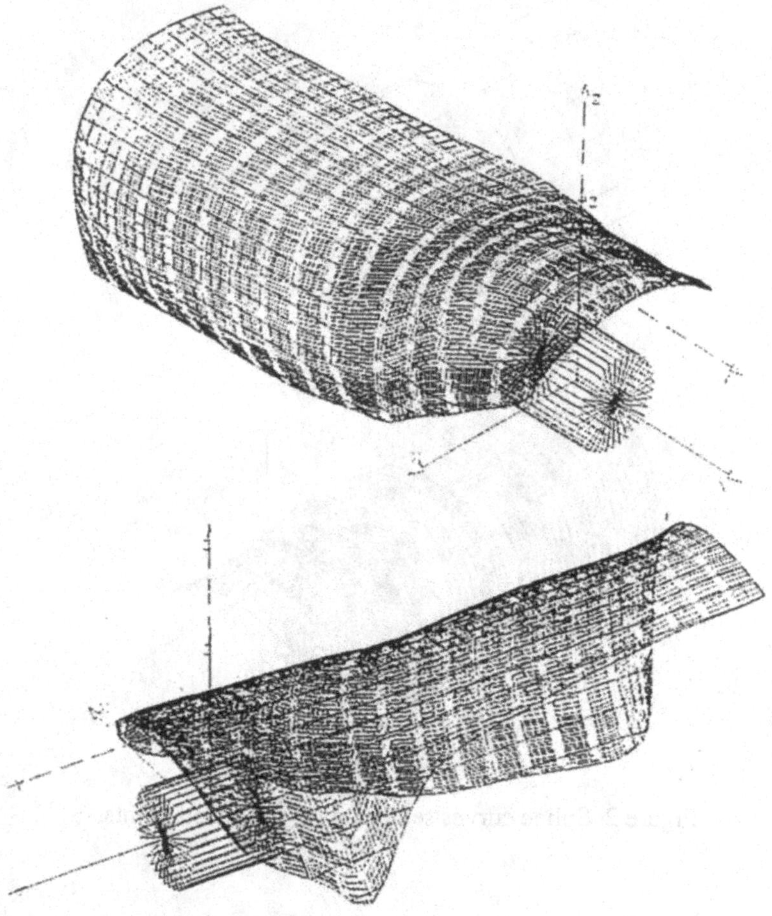

Figure 4. Robot trajectories on the bone surface.

3. The Robot Model.

To set up a robot model using the C.A.T.I.A. system, the object, the links and the mechanical structure must be built; the connecting points with their limits, space and references must be defined.

3.1. The Robot Mechanical Structure and its Work Space.

The connecting points and their motion limits, which can be either a prismatic translation or a rotation of each mobile element, can be defined by the C.A.T.I.A. software.

The kinematic parameters can be determined after the complex model has been built. The coordinates systems, inertial or auxiliaries, are necessary to the data output.

Figure 5 illustrates a ROBOT SCARA 7575 model with four degrees of freedom.

3.2. Static Forces in the Connecting Points.

At first, the external load is calculated by the C.A.T.I.A. software. The application point of the load (force or moment) , is defined and the static forces and moments in the robot structure are obtained in each joint. If the application point and the load vary, the static characteristics of the robot are determined and the static equilibrium conditions are evaluated.

3.3. The Motion in the Work Space.

The robot must be verified to make sure that the necessary movements are possible; otherwise, some changements must be done.

The study of the robot motion is possible either in the joint space or in the general cartesian space. In the point space, every link motion can be studied.

At the same time, the work space is analysed, considering the whole motion of all joints.

In the cartesian space, the trajectory of the end effector is analysed, for both position and orientation.

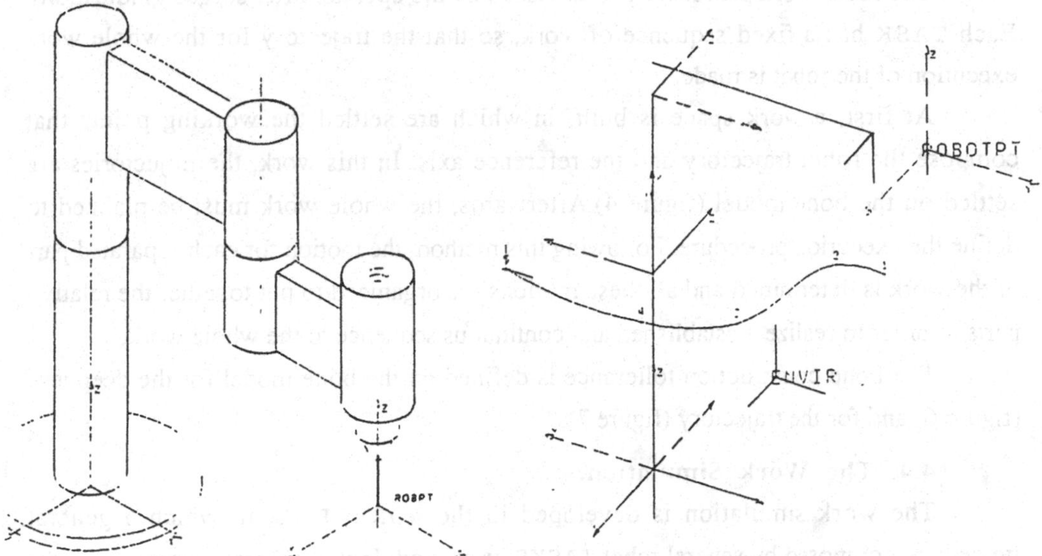

Figure 5. Model of the ROBOT SCARA 7575.

4.Simulation.

4.1. The Work Analysis.

The work analysis decides the sequence of the robot movements. The complete work is divided in several parts with respect to their level into the whole execution.

For example, a no-loaded displacement is considered a simple motion without motion quality requirements, while a loaded displacement is a higher level motion. The displacement following a certain trajectory can be a motion which requires a high quality of execution. Therefore, this classification can be established taking into account the respective work characteristics, because the realization of bones requires particular characters due to the biomaterials features.

4.2. The Work Planning.

It is possible to define several kinds of motion of the robot, considering either the work level or the economy and time of execution for cycles.

To the simple work without motion quality requirements, it can be imposed a point to point movement, neglecting the robot extremity trajectory and using high velocity and acceleration. For a complicated work, the motion following a determined trajectory is imposed and the velocity and acceleration can be modified to obtain an optimal result for the work of produce a bone in biomaterial.

4.3. The Work Fulfilment.

The robot work procedure is established by the operator after surgeon indications. Each TASK has a fixed sequence of work, so that the trajectory for the whole work execution of the robot is made.

At first, a work space is built, in which are settled the working points that compose the robot trajectory and the reference axis. In this work, the trajectories are settled on the bone model (figure 4).Afterwards, the whole work must be planned to define the execution procedure. Following this method, the motion for each separated part of the work is determined and all these motions are organized to put together the relative parts in order to realize a established and continuous sequence to the whole work.

The bone construction tollerance is defined on the bone model for the deepness (figure 6) and for the trajectory (figure 7).

4.4. The Work Simulation.

The work simulation is developed in the WORKCELL, in which a general procedure, composed by several robot TASKS, is created. In this general procedure, all the possibilities of the established work are shown to verify if the robot satisfies the work requirements. If the quality level is not enough, it is possible to modify either the TASKS or their own distribution, so that a better procedure can be obtained.

Then, it is possible to verify the quality of the manipulation of the robot end effector. The analysis between the general procedure and the robot work space takes into account the necessity of avoiding a collision between the robot and the object or any other obstacle present in the work space. The analysis also observes the human safety conditions.

During the simulation, the time for each completed cycle is studied, besides the robot velocity and acceleration. If it is necessary, they can be modified to obtain a correct time for cycle in each fase of the work. Figure 8 shows the work simulation for a robot SCARA 7575 on the modelled piece of bone.

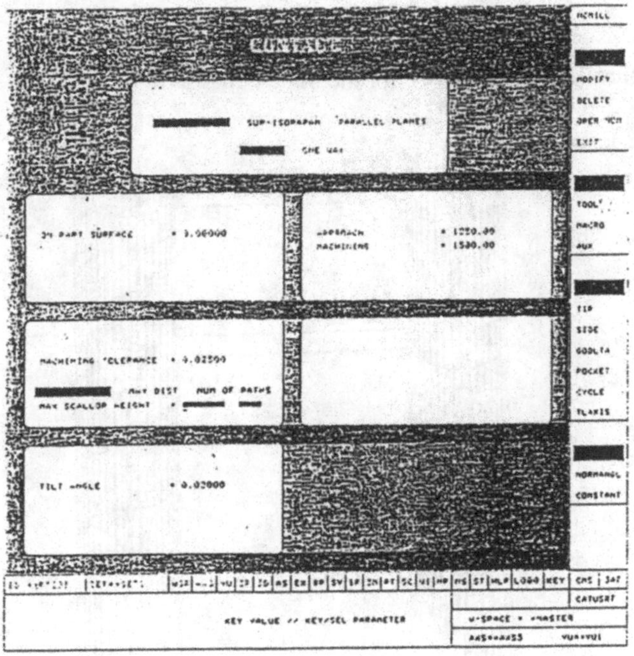

Figure 6. The deepness tollerance for the bone construction.

5.Transferring Data into the Robot.

The C.A.T.I.A. software permits to get output data from the models and the simulations. These output data must be transferred from the VM system to the DOS system and, then, to the robot.

In this paper, a robot SCARA 7575 is used to execute the work on the modelled bone surface. The tool machine is a pneumatic milling machine.

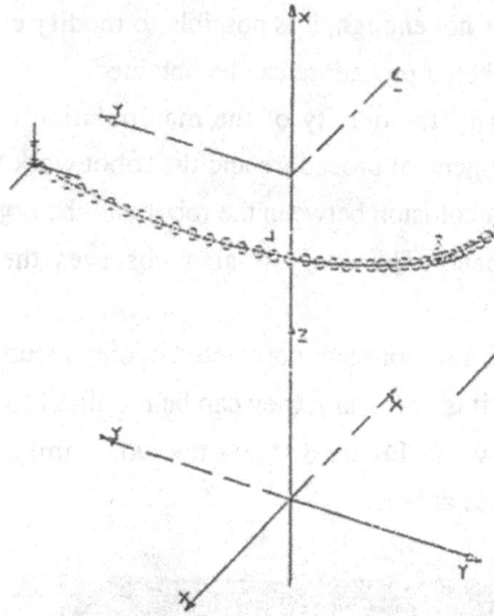

Figure 7. The trajectory for the bone construction.

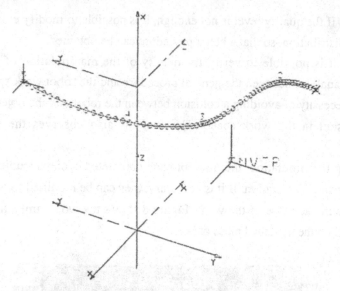

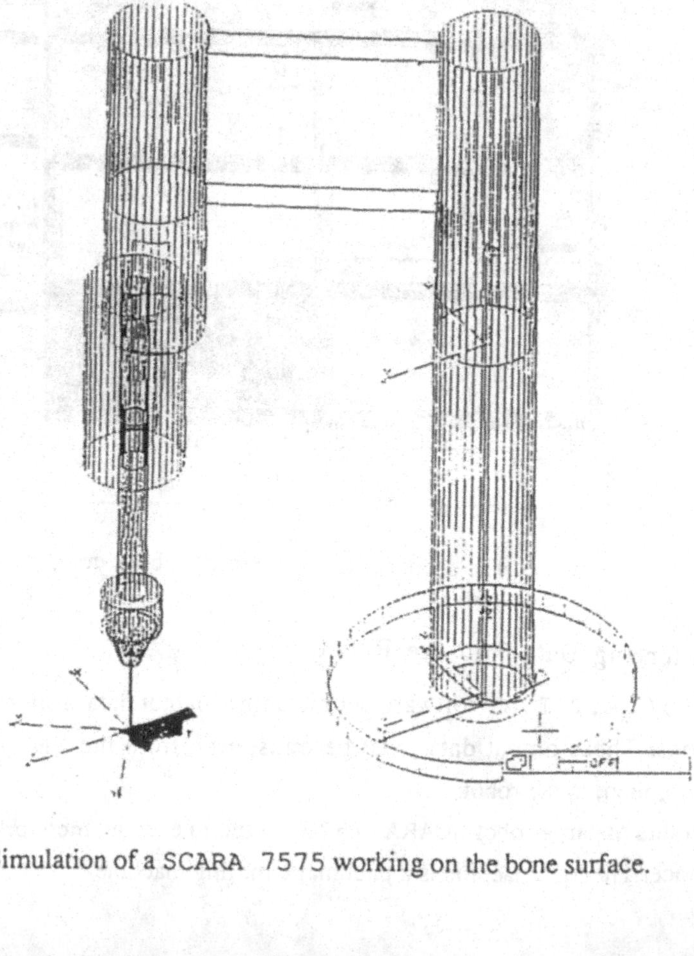

Figure 8. Simulation of a SCARA 7575 working on the bone surface.

The transformed output data are introduced in a PC computer (DOS system) and, successively, transfered to a robot SCARA 7575 .

At this point, the robot is ready to start the work execution automatically. The first realization of a prototype is on development.

6.Conclusion.

The present paper describes the use of CAD / CAM technology for application in bioengineering, which can be useful to the next step: to introduce directly the measurements in the C.A.T.I.A. software and to define a post-processor to translate the output data from C.A.T.I.A. to the robot, which means to realize an automatic procedure to work on pieces of bone in surgery. A series of experimental tests have been developed in the Department of Mechanics, Laboratory of Robotics, with a robot SCARA 7575.

Acknowledgements.

The authors thank the staff of Department of Mechanics for the co-operation on the realization of the prototype, Paolo Bernardelli and Remo Sala for developments of software and hardware.

References.

/1/ Wen Xia "Application of CAD/CAM in Biomechanics " , 8th World Congress on the Theory of Machines and Mechanisms, Prague Czechoslovakia, August 1991.

/2/ Rovetta A., Wen Xia " Realization of a Prosthesis of Lower Limb on Development of Kinematics " , 2nd International Conference on Robotics, Dubrovnik Yugoslavia, June 1989.

/3/ Muller W. " The Knee, Form, Function, and Ligament Reconstruction " , Springer Verlag, Berlin Heidelberg, 1983.

/4/ Cigada A., Pedeferri P., Bedini E. " Materiali Metallici Utilizzati in Ortopedia-Parte II" Materiali - Il Filo Mettalico, n.34 vol.4, 1987.

/5/ Pezzi D. " Caratterizzazione Tribologico-Corrosionistica di Componenti Protesici, Rivestiti Superficialmente, attraverso lo sviluppo di un Simulatore dell'Articolazione Coxo-Femorale " , Graduation Thesis, Politecnico di Milano - Ingegneria delle Tecnologie Industriali, Milan Italy, 1991.

/6/ " Manual C.A.T.I.A. " , Dassault Systemes, France, 1989.

Programme and Organizing Committee

Chairmen

Prof. A. Morecki
Warsaw. University of Technology
Al. Niepodległości 222, 00-663 Warsaw, Poland
Prof. G. Bianchi
CISM, Piazza Garibaldi 18
33100 Udine, Italy

Members

Prof. A. P. Bessonov
Russian Academy of Sciences
Griboedova 4, Moscow-Centre, 101000, Russia
Prof. J. C. Guinot
Universite P. et Marie Curie
Laboratoire de Mécanique et Robotic, IMTA
4, Place Jussieu - Tour 66, 75230 Paris Cedex 05, France
Prof. B. Heimann
Institute of Mechanics Universitat Hannover, Appelstr. 11
30167 Hannover,Germany
Prof. J.R. Hewit
Dept. of Applied Physic and Electronic & Manufacturing Engineering
University of Dundee, Dundee
DD14HN Scotland
Prof. O. Khatib
Stanford University
Dept. of Computer Science,
Stanford CA 94305, USA
Prof. A.I. Korendyasev
Russian Academy of Sciences
Griboedova 4, Moscow-Centre,
101000 Russia
Prof. W.O.Schiehlen
Universitat Stuttgart-Inst. B für Mechanik
Pfaffenwaldring 9, D-70550 Stuttgart, Germany
Prof. K. Tanie
Robotics Dept.-Mech.Eng.Lab.
Agency of Industrial Science and Technology,
Namiki 1-2 Tsukuba 305 Japan
Prof. M. Vukobratovič
Institute Mihailo Pupin
Volgina 15 - P.O.Box 906,
11000 Beograd, Yugoslavia
Prof. K.J. Waldron
Dept. of Mechanical Engineering
The Ohio State University Columbus
Ohio 43210, USA

Scientific Secretary:
Dr h. K. Jaworek
Warsaw University of Technology
Al. Niepodległości 222
00-663 Warsaw, Poland

Secretary:
Dr A. Bertozzi
CISM, Piazza Garibaldi 18,
33100 Udine, Italy

Local Organizing Committee:

Prof. E.Wittbrodt, dr R.Kościelny, G.Szczęsna
A.Kozikowski, A.Ruszkowski, M.Grochowski, W.Portee, D.Krawczenko, J.Libudzki, A.Kuźmicki, P.Jagodziński
Faculty of Mechanical Engineering,
Gdańsk University of Technology, Gdańsk, Poland

Participants

Austria

Wohlhart K
Graz University of Technology
Institute of Mechanics
Graz

Belorussia

Pashkevich A.
Minsk, Radioengineering Institute
6, P. Brovka str., Minsk 220600

Bulgaria

Chacarov D
Institute of Mechanics
Bulgarian Academy of Science
Block 4, Acad.G. Bonchev St.
1113 Sofia

Canada

Jankowski K.
Flexible Manufacturing Centre
Faculty of Engineering,
McMaster University
Hamilton, Ontario, Canada L8S 4L7

Kujath M.
Department of Mechanical Engineering
Technical University of Nova Scotia Halifax
Nova Scotia

Finland

Leinonen T.
University of Oulu
Department of Mechanical Engineering
FIN-90570 Oulu

Louhisalmi Y.
University of Oulu
Laboratory of Machine Design
Linnanmaa
FIN-90570 Oulu

Nevala K.
VTT Automation
Machine Automation
P.O. Box 13023, FIN 90571 Oulu

France

Ariscault M.
Laboratoire de Mécanique des Solides
U.R.A. 861 du C.N.R.S.
40 av. du Recteur Pineau
86022 Poitiers Cedex

Barraco A.
Laboratoire de Mecanique des Structures
Ecole Nationale Superieure d`Arts et Metiers
75013 Paris

Danescu G.
Laboratoire de Mécanique Appliquée
Faculté des Sciences - Route de Gray
F 25030 Besançon

Fayet M.
Institut National des Sciences Appliquees de Lyon
Laboratoire Automatique Industrielle
151 Bd de l`Hopital.
Bat. 303 20 av. Albert Einstein
69621 Villeurbane/Cedex

Gaborieau J.N.
Laboratoire de Mécanique des Solides
U.R.A. 861 du C.N.R.S.
40 av. du Recteur Pineau
86022 Poitiers Cedex

Gaudry O.
L.I.R.M.M.
161 rue Ada
34392 Montpellier Cedex 5

Guinot J.C.
Université P. et M. Curie
Laboratoire de Mecanique et Robotique IMTA
4 Place Jussie - Tour 66
75230 Paris Cedex 05

Hervé J.
Ecole Central de Paris
92295 Chatenay Malabry Cedex

Lellemand J.P.
Laboratoire de Mécanique des Solides
U.R.A. 861 du C.N.R.S.
40 av. du Recteur Pineau
86022 Poitiers Cedex

Morel G.
Université de Versailles St. Quantin
Laboratoire de Robotique de Paris
URA 1778 du CNRS
10 - 12 Av. de l`Europe
78140 Velizy

Ombede G.
Intitut National des Sciences Appliquees de Lyon
Laboratoire Automatique Industrielle
Bat 30320 av. Albert Einstein
69621 Villeurbane/Cedex

Pfister F.
Intitut National des Sciences Appliquées de Lyon
Laboratoire Automatique Industrielle
Bat. 30320 av. Albert Einstein
69621 Villeurbane/Cedex

Regnier S.
Centre Universitaire de Technologie
Laboratoire de Robotique de Paris
10 - 12 Av de l'Europe
78140 Velizy

Germany

Haug J.
University of Stuttgart
Institute B of Mechanics
Pfaffenwaldring 9 D-70550 Stuttgart

Heimann B.
University of Hannover
Institute of Mechanics
Appelstr. 11
30167 Hannover

Müller P.
University of Wuppertal
Institute of Robotics and Safety Control Engineering
Gausstr 20
42097 Wuppertal

Pinto C.F.
Technical University Hamburg - Harburg
Ocean Engineering Section II
Eisendorfer Strasse 42
21071 Hamburg

Söffker D.
University of Wuppertal
Safety Control Engineering
5600 Wuppertal

Hungary

Filemon E.
Technical University of Budapest
Budapest

Italy

Bertozzi A.
CISM, Piazza Garibaldi 18
33100 Udine

Bianchi G.
CISM. Piazza Garibaldi 18
33100 Udine

Ceccarelli M.
University of Cassino
Dept. of Industrial Engineering
Via Zamosch 43
03043 Cassino

Gola M.
Dip. di Meccanica
Politecnico di Torino
Corso Duca degli Abruzzi. 24
10129 Torino

Rossi A.
Dept. of Innovation in Mechanics and Management
University of Padua
Via Venezia 1
35131 Padova

Togno A.
Politecnico di Milano
Dipartimento di Meccanica
Via Bonardi 9
20133 Milano

Japan

Hayashi T.
Toin University of Yokohama
Dept. of Control and System Engineering
1814 Kuroganecho. Midoriku
Yokohama 225

Ikeda K.
Agency of Industrial Science & Technology
Mechanical Engineering Laboratory
Tokyo

Itao K.
Chuo University
Faculty of Science & Engineering
1-13-27 Kausuga. Bunkyo-ku
Tokyo 112

Kato A.
Aichi Institute of Technology
Dept. of Electronics
Yakusa. Toyota 470 -03

Katoh R.
Kyushu Institute of Technology
Department of Control Engineering
Sansui. Tobata. Kitayushu 804

Kondo N
Aichi Institute of Technology
Dept. of Electronics
Yakusa. Toyota 470 -03

Luo Zhi-Wei
Bio-Mimetic Control Research Center (RIKEN)
8-31. Rokuban 3-Chome. Atsuta-ku
Nagoya 456

Matsushima K.
Toin University of Yokohama
Faculty of Engineering
1614 Kurogane-Cho Midori-Ku
Yokohama 225

Ohara S.
Autonomous Robot Systems Lab.
NTT Human Interface Laboratories
3-9-11. Midori-cho. Musashino-shi
Tokyo 180

Sankai Y.
University of Tsukuba
Institute of Engineering Mechanics
Tsukuba 305

Tanie K
Robotics Dept -Mech. Eng. Lab.
Agency of Industrial Science and Technology
Namiki 1-2
Tsukuba 305

Uchiyama M.
Dept. of Aeronautics & Space Engineering
Faculty of Engineering. Tohoku University
Aramaki-aza Aoba. Aoba-ku
Sendai 980-77

Yano H.
Research Institute
National Rehabilitation
Center for the Disabled
Tokio University

Yano T
Mechanism Division. Robotics Dept
Mechanical Engineering Laboratory
Namiki 1-2. Tsukuba Ibaraki 305

Poland

Ferenc A.
Warsaw University of Technology
Institute of Aeronautics and Applied Mechanics
24. Nowowiejska str.
00-665 Warsaw

Frączek J
Warsaw University of Technology
Institute of Aeronautics and Applied Mechanics
24. Nowowiejska str
00-665 Warsaw

Garus J.
Navy Academy
Gdynia - Oksywie

Gosiewski A
Warsaw University of Technology
Plac Politechniki 1
00-661 Warsaw

Jaworek K.
Warsaw University of Technology
Institute of Aeronautics and Applied Mechanics
24. Nowowiejska str.
00-665 Warsaw

Kitowski Z
Navy Academy
Gdynia - Oksywie

Kościelny R
Technical University of Gdansk
11/12. Narutowicza str
80-952 Gdansk

Mianowski K
Warsaw University of Technology
Institute of Aeronautics and Applied Mechanics
24. Nowowiejska str.
00-665 Warsaw

Miller K
Warsaw University of Technology
Institute of Aeronautics and Applied Mechanics
24. Nowowiejska str
00-665 Warsaw

Morecki A.
Warsaw University of Technology
Institute of Aeronautics and Applied Mechanics
24. Nowowiejska str
00-665 Warsaw

Nazarczuk K
Warsaw University of Technology
Institute of Aeronautics and Applied Mechanics
24. Nowowiejska str
00-665 Warsaw

Olędzki A.
Warsaw University of Technology
Institute of Aeronautics and Applied Mechanics
24. Nowowiejska str
00-665 Warsaw

Rzymkowski C.
Warsaw University of Technology
Institute of Aeronautics and Applied Mechanics
24. Nowowiejska str.
00-665 Warsaw

Szymkat M.
University of Minning & Metallurgy
Robotics & Machine Dynamics Group
Av. Mickiewicza 30
30-059 Cracow

Wittbrodt E.
Technical University of Gdańsk
11/12. Narutowicza str.
80-952 Gdańsk

Zielińska T.
Warsaw University of Technology
Institute of Aeronautics and Applied Mechanics
24. Nowowiejska str.
00-665 Warsaw

Zieliński C.
Warsaw University of Technology
24. Nowowiejska str.
00-665 Warsaw

Russia

Bogutsky A.
Institute of Mechanics of Moscow State University
Dept 301
Bld. 1. Mitchurinsky Prosp.
117192 Moscow

Budanov V.
Institute of Mechanics of Moscow State University
Michurinsky Prosp. 1
119899 Moscow

Formalsky A.
Institute of Mechanics of Moscow State University
Mitchurinsky Prosp. 1
119899 Moscow

Taiwan

Ju Ming-Shaung
Control Research Laboratory
Dept. of Mechanical Engineering
National Cheng Kung University
Tainan 701. R.O.C.

U. K.

Gilbert J.
The University of Hull
Dept. of Electronic Engineering
Hull HU6 7RX

Grindley J.
The University of Hull
Dept. of Electronic Engineering
Hull HU6 7RX

USA

Khatib O.
Stanford University
Dept. of Computer Science
Stanford CA 94305

Yugoslavia

Vukobratovič M.
Institute "Mihailo Pupin"
Volgina 15 P.O. Box 906
11000 Belgrade

The proceedings of the previous symposia may be obtained from:

Symposium	Proceedings distributed by
RoMansy 1 *(5-8 September 1973, Udine)*	Springer Verlag, Vienna; CISM, Udine
RoManSy 2 *(14-17 September 1976, Jadwisin)* and RoManSy 3 *(12-15 September 1978, Udine)*	Elsevier, Amsterdam; PWN-Polish Scientific Publishers, Warsaw
RoManSy 4 *(8-12 September 1981, Zaborow)*	PWN, Warsaw
RoManSy 5 *(26-29 June 1984, Udine)* and RoManSy 6 *(9-12 September 1986, Cracow)*	Editions Hermes, Paris; Kogan-Page, London; MIT Press, Cambridge (MA)
RoManSy 7 *(12-15 September 1988, Udine)*	Editions Hermes, Paris
RoManSy 8 *(2-6 July 1990, Cracow)*	CISM, Udine; Warsaw Technical University
RoManSy 9 *(1-4 September 1992, Udine)*	Springer Verlag, Berlin

Printed in the United States
By Bookmasters

Printed in the United States
By Bookmasters